D0838547

PETROLEUM RESERVOIR ROCK *and* FLUID PROPERTIES

PETROLEUM RESERVOIR ROCK *and* FLUID PROPERTIES

Abhijit Y. Dandekar

A CRC title, part of the Taylor & Francis imprint, a member of the Taylor & Francis Group, the academic division of T&F Informa plc.

Published in 2006 by
CRC Press
Taylor & Francis Group
6000 Broken Sound Parkway NW, Suite 300
Boca Raton, FL 33487-2742

Printed in the United States of America on acid-free paper
10 9 8 7 6

International Standard Book Number-10: 0-8493-3043-2 (Hardcover)
International Standard Book Number-13: 978-0-8493-3043-8 (Hardcover)
Library of Congress Card Number 2005035149

Library of Congress Cataloging-in-Publication Data

Dandekar, Abhijit Y.
Petroleum reservoir rock and fluid properties / Abhijit Y. Dandekar.
p. cm.
Includes bibliographical references and indexes.
ISBN-13: 978-0-8493-3043-8 (acid-free paper)
ISBN-10: 0-8493-3043-2 (acid-free paper)
1. Petroleum reserves--Fluid dynamics. 2. Petroleum reserves--Mechanical properties. 3. Petroleum engineering. I. Title.

TN870.57.D36 2006
553.2'82--dc22 2005035149

Visit the Taylor & Francis Web site at
http://www.taylorandfrancis.com

and the CRC Press Web site at
http://www.crcpress.com

This book is dedicated to my grandfather
Purushottam V. Dandekar

Preface

During my teaching of the undergraduate course in petroleum reservoir rock and fluid properties, at the beginning of the semester the first question students ask me is, "What book would you recommend?" This book is specifically designed to answer that question and provides students with a strong foundation in reservoir rock and fluid properties. Such a foundation will be linked to almost every course in the petroleum engineering curriculum and is, in fact, the backbone of almost all the activities in the petroleum industry.

Even though this work is primarily aimed at undergraduate students, graduate students in petroleum engineering, especially those with a nonpetroleum engineering undergraduate degree, will also find it beneficial for preparing a strong foundation for their future studies. Additionally, personnel engaged in laboratory studies in support of special core analysis and PVT and reservoir fluid studies, as well as reservoir simulation engineers, reservoir engineers, and chemical engineers diversifying into the petroleum engineering field, will find this book a valuable resource. In fact, the book is structured in such a manner that anyone with an engineering background who wants to learn about petroleum reservoir rock and fluid properties will be able to do so with relative ease.

This book is comprised of 17 chapters, arranged in a logical sequence to ease understanding of the entire text. Chapter 1 serves as an overall introduction and begins with a brief discussion of the formation and characteristics of petroleum reservoirs. It concludes by discussing the significance of petroleum reservoir rock and fluid properties, thus setting the stage for the remaining chapters. Chapters 2–9 are dedicated to various reservoir rock properties, while Chapters 10–17 are dedicated to various reservoir fluid properties. Wherever necessary, all the chapters include schematic illustrations, various types of plots based on experimental data, and practical examples to enhance the understanding of key concepts. Finally, for effective learning, most chapters end with a wide variety of practice or exercise problems based on the topics covered. A brief synopsis of Chapters 2 to 17 follows.

Chapter 2 provides a preamble to petroleum reservoir rock properties. This chapter begins with a discussion of coring methods used for recovering physical rock or core samples from hydrocarbon formations, which, in fact, constitute a crucial step toward deriving various reservoir rock properties from laboratory tests. Important issues related to various coring methods, types of core samples, and how core samples are handled are examined. The chapter ends with a brief discussion of conventional core analysis and special core analysis, respectively.

Chapters 3 and 4 address the two most fundamental reservoir rock properties: porosity and absolute permeability. Both chapters begin with a discussion of the significance and definitions of these properties. For porosity, various types of porosities, such as absolute porosity and effective porosity, are defined and mathematical expressions are provided. In Chapter 4, Darcy's law is introduced and the mathematical expression for permeability is derived. The application of Darcy's law to inclined flow and radial flow is also examined. Chapters 3 and 4 end with discussions on the averaging of porosity and permeability, laboratory measurement of porosity and permeability, and factors affecting these two properties.

Chapter 5 focuses on the mechanical and electrical properties of reservoir rocks. It begins with an introduction to and a discussion of the significance of these properties. First, definitions of stress and strain and their relationship to one another, and various rock mechanics parameters such as Poisson's ratio and Young's modulus are presented. Laboratory measurement of compressive strength by the conventional triaxial compression test is also discussed. The discussion of mechanical properties ends with the definition of reservoir rock compressibility; empirical correlations for the estimation of formation compressibility are also presented. The section on electrical properties begins with a discussion of fundamental concepts such as formation factor, cementation factor, and the Archie equation. Next, the resistivity index is defined, and its relationship with water saturation and the saturation exponent is presented in the form of a generalized equation. Chapter 5 ends with an examination of the effect of factors such as wettability and conductive solids on the electrical properties of reservoir rocks.

Chapter 6 is devoted to fluid saturation. In order to set the foundation for this chapter, a discussion on the distribution of fluid saturation in a petroleum reservoir is first provided. Laboratory methods for the measurement of fluid saturation in rock or core plug samples are presented. A methodology for the assessment of the validity of fluid saturation data measured on the plug-end trims of core plug samples is also provided. Next, three special types of fluid saturation are defined: critical gas saturation, residual oil saturation, and irreducible water saturation. The chapter ends with a detailed discussion of various factors that affect the determination of fluid saturation.

The subjects of Chapters 7 and 8 are interfacial tension and wettability and capillary pressure, respectively. Wettability and capillary pressure properties are the consequence of the presence of more than one fluid phase in a reservoir rock, which is also influenced by interfacial forces existing between various fluid–fluid pairs, the significance of which is presented at the beginning of both chapters. Chapter 7 makes a distinction between interfacial tension and surface tension and also distinguishes between various types of wettabilities such as water-wet, oil-wet, mixed-wet, and fractional-wet. Various laboratory methods for the measurement of interfacial tension, surface tension, and wettability are provided next. The effect of pressure and temperature on interfacial tension and wettability is also examined. Chapter 7 ends with a discussion of the relationship between wettability and irreducible water saturation and residual oil saturation.

Chapter 8 first presents the basic capillary pressure equation and then discusses its dependence on rock and fluid properties of pore radius, contact angle, and interfacial tension. Next, the salient features of capillary pressure curves and hysterisis are given. Various laboratory measurement methods, beginning with Leverett's capillary pressure experiments, are presented. With consideration of the dependence of capillary pressure on rock and fluid properties, methodology for the conversion of laboratory-measured capillary pressures to reservoir condition capillary pressures is given. The averaging of capillary pressures through the *J* function is also discussed. Chapter 8 ends with a discussion of the effect of wettability on capillary pressure and the practical application of capillary pressure data in various reservoir engineering calculations.

Chapter 9 is devoted to relative permeability; both two-phase and three-phase relative permeabilities are discussed. First, the distinction between absolute permeability and relative permeability is drawn, following which the salient features of gas–oil and oil–water relative permeability curves, end-point saturations, and end-point permeabilities are presented. The discussion on laboratory measurement of two-phase relative permeability begins with topics such as the type of core samples (native or restored) used, displacement fluids and test conditions, establishment of initial water saturation, and determination of base permeability. The two commonly used laboratory techniques for the measurement of relative permeability, steady state and unsteady state, are discussed in significant detail. For both methods, the discussion focuses on practical (experimental) aspects as well as theoretical aspects; the former provides detailed experimental methodology whereas the latter focuses on the derivation (Buckley–Leverett to Welge to Johnson–Bossler–Naumann) and the application of mathematical expressions used to calculate relative permeabilites. A very detailed discussion of various factors, such as wettability, fluid saturations, and capillary numbers, that affect relative permeability is also provided. The fitting of two-phase relative permeability data for determination of Corey exponents and the practical application of relative permeability data are presented at the end of the chapter. The remainder of Chapter 9 discusses three-phase relative permeability, including fundamental concepts, the representation of three-phase data, and an empirical model used to determine three-phase data from two-phase relative permeabilities.

Chapter 10 basically sets the stage for the material that follows by introducing some key aspects related to the overall characteristics of petroleum reservoir fluids. This chapter provides a broad classification of petroleum reservoir fluids in terms of their basic chemical and physical properties.

Chapter 11 is devoted to the fundamental aspects of phase behavior, beginning with the phase behavior of pure substances followed by synthetic or model hydrocarbon mixtures. Important concepts such as phase rule; critical properties; behavior in critical, subcritical, and supercritical regions; saturation pressures; and retrograde behavior in a two-phase region are covered in significant detail. Chapter 12 smoothly transitions into the phase behavior of the five petroleum reservoir fluids: black oils, volatile oils, gas condensates, wet gases, and dry gases. The coverage of all the important concepts from Chapter 11 makes the understanding of the phase behavior of petroleum reservoir fluids relatively simple.

Chapter 13 deals with the most important aspects of PVT and phase behavior of petroleum reservoir fluids, such as sampling. The chapter begins with practical considerations of reservoir fluid sampling, which is followed by a presentation of commonly used methods employed for sampling. The various issues related to reservoir fluid sample representativity or nonrepresentativity are also discussed.

Chapter 14 delves into the actual beginning of PVT and phase behavior analysis of petroleum reservoir fluids. Methods employed to determine the overall single-phase composition and two-phase composition of reservoir fluids are provided. The latter part of the chapter deals with issues related to the single carbon number fractions and the plus fractions of a reservoir fluid and sheds light on the wide variation observed in their properties when different reservoir fluids are compared.

Chapter 15 is the culmination of the reservoir fluid properties section of the book. The chapter begins with a discussion of various important elements of PVT analysis and reservoir fluid properties and presents a discussion of the fundamental aspects related to the properties of gases (ideal and real) and liquids in general. The chapter then gradually moves into the definitions of all the important reservoir fluid properties. Before discussion of various laboratory tests that are conducted to obtain reservoir fluid properties, a description is provided of PVT equipment in which various tests are conducted. Adjustments of black oil laboratory data to determine fluid properties for reservoir engineering calculations are also discussed. Finally, empirical correlations and prediction methods for obtaining fluid properties are examined.

Chapter 16 deals with hydrocarbon vapor liquid equilibria. This is yet another important part of this field, considering the fact that all petroleum reservoir fluids undergo pressure and temperature changes during production and transportation, resulting in a process called PT flash, which in turn results in the formation of hydrocarbon vapor and liquid phases. The properties of the resulting phases are important in recovery methods and surface processing. Equations-of-state (EOS) models commonly used in the petroleum industry to perform these vapor-liquid equilibria calculations are discussed. The chapter begins with the fundamental concepts of the ideal solution principle, equilibrium ratios, calculation of saturation pressures, and equilibrium-phase compositions and densities. The use of K-value charts for phase equilibria calculations is also examined. The application of EOS models for calculating reservoir fluid properties is demonstrated through solved examples.

Chapter 17 focuses on properties of formation or oil field waters. The chapter primarily discusses empirical correlations used for obtaining the formation water properties of reservoir engineering significance, such as formation volume factor, compressibility, viscosity, density, solubility of hydrocarbons in water phase, and water solubility in hydrocarbon gases and liquids.

Finally, I would like to present a note about the system of units used in this book. This book does not use any particular system of units such as SI, MKS, CGS, or FPS. The choice of a particular system of units is made on the basis of the variable or quantity under consideration. For example, reporting the weight of a core plug, which is generally a small section of the whole core, in units of grams would be much more logical than using kilograms or pounds. However, wherever possible, field units with which a petroleum engineer is more conversant and which are generally accepted are used. I strongly believe that an engineer should be well conversant with a variety of systems of units and should feel comfortable dealing with them (e.g., knowledge of conversion factors and reduction of all variables in an equation to a consistent set of units).

Acknowledgments

I have enjoyed the pleasure of teaching the material covered in this text to petroleum engineering students at the University of Alaska Fairbanks, and my research work at the Technical University of Denmark, and my Ph.D thesis work at Heriot-Watt University. I am sincerely grateful to all three institutions for providing me with a wonderful opportunity to learn, which has been instrumental in bringing this book to reality.

Since much of the material covered in this book is based on publications of the Society of Petroleum Engineers (SPE), tribute is due to the petroleum engineers, scientists, and authors who have made numerous and significant contributions to the petroleum literature. Particularly noteworthy are the works of Professors James Amyx, Daniel Bass Jr., William D. McCain Jr., and Robert Whiting.

I would like to express my appreciation to a number of people from academia: Professor Ali Danesh at Heriot-Watt University; Professors Erling H. Stenby and Simon I. Andersen at the Technical University of Denmark; and Professors Sukumar Bandopadhyay, Godwin A. Chukwu, Santanu Khataniar, David O. Ogbe, Shirish Patil, and Tao Zhu at the University of Alaska Fairbanks. Special thanks are due to Professor Gang Chen at the University of Alaska Fairbanks for providing the data on triaxial compression tests and for useful discussions on rock mechanics.

I would like to thank the staff of CRC Press. I greatly appreciate the assistance of T. Michael Slaughter, Yulanda Croasdale, Theresa Delforn, Elise Oranges, and all others involved in this project. I also appreciate the work of Mr. Pandian and his staff at Macmillan India Ltd. for handling the copy editing, typesetting, and production of this book.

Finally, I would like to thank all my family members for their support and encouragement during the course of writing this book. Most deeply, my special thanks go to my wife, Mrudula Dandekar, son Shamal Dandekar, and daughter Rama Dandekar, who walked with me through the tortuous paths of writing a book. I am particularly grateful for their patience and, most importantly, their tolerance for putting up with my long hours of work, without which the journey would have never ended.

Abhijit Y. Dandekar
Fairbanks, Alaska

Author

Abhijit Dandekar is associate professor of petroleum engineering at the University of Alaska Fairbanks (UAF), where he has taught since January 2001. Before joining UAF, he was an assistant research professor at the Technical University of Denmark. In the summer of 2002, he also worked as visiting faculty at the University of Petroleum, Beijing, P.R.C. He holds a B.Tech. degree in chemical engineering from Nagpur University, India, and a Ph.D. degree in petroleum engineering from Heriot-Watt University, Edinburgh, United Kingdom. Dandekar is a member of SPE and the author or co-author of more than 30 technical papers in the petroleum literature and numerous research reports in areas as diverse as special core analysis, PVT and phase behavior, gas-to-liquids, gas hydrates, viscous oils, wettability alteration, and CO_2 sequestration.

Table of Contents

1 Introduction

1.1 THE FORMATION OF PETROLEUM RESERVOIRS

Like coal, petroleum reservoir fluids (oils and gases) are, fossil fuels that formed millions of years ago and are found, trapped in rocks in the earth, in both offshore and onshore locations. According to the popular organic theory of formation of petroleum, these petroleum reservoir fluids were formed millions of years ago when animal and plant matter settled into the seas together with sand, silt and rocks, subsequently resulting in a build-up of several layers along the coastline and on the sea bottom. Some of these layers were buried deep in the earth due to geological shifts. Variables, such as geological time scales, pressure (due to burial and depth), and temperature (due to geothermal gradient related to depth) resulted in the conversion of the organic material into petroleum reservoir fluids and the mud, sand and silt into rock. The rock containing the organic material that converted into petroleum reservoir fluids is referred to as a *source rock.*

Due to the organic origin of both petroleum reservoir fluids and coal, obvious similarities and links exist between the former and latter, and a number of obvious differences. Chemically, petroleum reservoir fluids are principally hydrocarbons—compounds of carbon and hydrogen—whereas with coal, much of the corresponding hydrogen has been eliminated. Although all petroleum reservoir fluids are constituted primarily of carbon and hydrogen, a widely different molecular constitution or chemical composition makes every petroleum reservoir fluid unique in nature..

Physically, petroleum sources are fluids where coal is a solid; this has important consequences. As a fluid, petroleum can migrate and the rocks from that it is produced are usually not the same as the ones (source rocks) from that it was formed. These petroleum reservoir fluids that formed deep within the earth, began moving upwards (due to lower gravity) through tiny, connected pore spaces in the rocks. In the absence of impermeable barriers, some of these fluids seeped to the surface of the earth; however, most petroleum hydrocarbons were trapped by nonporous rocks or other barriers that did not allow any further migration. It is these underground traps of oil and gas that are called *petroleum reservoirs.* These reservoirs are made up of porous and permeable rocks that can hold significant amounts of oil and gas within the pore spaces of those rocks; much like water is soaked up in a sponge. The schematic of an idealized petroleum reservoir is shown in Figure 1.1.

1.2 TYPICAL CHARACTERISTICS OF PETROLEUM RESERVOIRS

Petroleum reservoirs are created through three sequential steps: (1) deposition, (2) conversion/migration, and (3) entrapment. A typical *trap* is an anticline, where

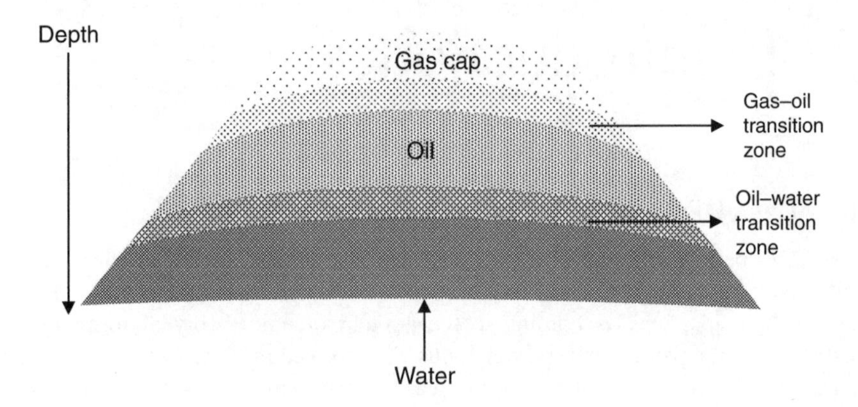

FIGURE 1.1 Schematic of an idealized petroleum reservoir showing gas, oil, and water distribution. Note that due to capillary forces, which resist complete gravity segregation, water is also found (in small amounts) in all zones of the reservoir, including the gas cap.

rocks have been buckled into the form of a dome. If the anticline has a seal of impermeable rock, hydrocarbons remain in this trap until they are drilled for and brought to the surface. However, petroleum reservoir fluids contained in the pore space or *interstices* of the reservoir rock must negotiate tortuous passageways through the rock to travel from the reservoir into the well bore and on to the surface.

These oil and gas reservoirs can be found at depths as shallow as 40 ft (West Africa–Gabon) and as deep as 21,000 ft (North Sea–UKCS). In 1859, the first commercial oil well drilled by Col. Edwin L. Drake in Titusville, Pennsylvania was merely to a depth of 69.5 ft.

Most petroleum reservoirs contain at least two fluid phases, either gas and water or oil and water; however, some contain all three phases; gas, oil, and water. In principle, if placed in an open container, gravity segregation should separate all fluid phases into distinct layers as per their densities with gas on top, followed by oil and water. However, other parameters, such as rock/fluid properties and solubility, restrict complete gravitational segregation (see Figure 1.1).

The reservoir rocks, that hold petroleum, are mostly sandstone and limestone (also called *carbonates* including dolomites). Less than 1% of the world's oil has been found in fractured igneous or metamorphic rocks, that typically lack the pore or void space needed to be successful reservoir rocks.

Three different geothermal processes impart particular characteristics or properties to petroleum reservoir fluids present in these reservoirs. The processes that transform the organic matter (known as *kerogen*) into hydrocarbons are

- digenesis (100 to 200°F) resulting in biochemical methane;
- catagenesis (200 to 300°F) resulting in oil or wet gas; and
- metagenesis (300 to 400°F) resulting in dry gas.

1.3 THE SIGNIFICANCE OF PETROLEUM RESERVOIR ROCK AND FLUID PROPERTIES

It is virtually impossible to obtain visual information of rock structures that contain petroleum reservoir fluids due to, for example, the overlying sea and other rock layers. Therefore, when searching for oil and gas, random holes cannot be drilled to explore a formation, and hence a more precise survey of the area is carried out. The quest for petroleum reservoir fluids normally begins with geologists and geophysicists and a variety of scientists and engineers who study sedimentary rocks and the hydrocarbon fluids contained in them.

A petroleum geologist studies the physical and chemical characteristics of the rocks. A geophysicist in the petroleum industry, primarily carries out the processing and interpretation of seismic data and generation of subsurface maps on the basis of seismic data. These interpretations enhance the understanding of subsurface geology from a hydrocarbon resource potential perspective.

Based on the evaluation of the geologist and the geophysicist, a decision is made regarding an exploratory well drilling program following that various methods and tools are used to locate and evaluate the commercial significance of the rocks, since the mere knowledge of the presence of oil and gas is not sufficient.

These various methods and tools are part of a process called *formation evaluation*. Generally during this formation evaluation process, reservoir rock and fluid samples are recovered and studied from engineering and commercial evaluation standpoints. Detailed information regarding the type and physical properties of the reservoir rocks, the petroleum reservoir fluids present in them, and interaction of the former and the latter is essential in understanding and evaluating the potential performance or productivity of a given petroleum reservoir.

Reservoir rock samples recovered during the formation evaluation process are the only actual physical samples of the reservoir and therefore are a direct source of valuable data such as the nature and physical characteristics of reservoir rocks that enable them to store fluids and to allow fluids to flow through them. Therefore, core analysis is an essential, basic tool for obtaining direct and valuable data concerning drilled rock formations.

Similar to reservoir rock samples, petroleum reservoir fluid samples that are recovered during the formation evaluation process are used for studying their nature and physical and chemical characteristics. This is especially important because these petroleum reservoir fluids are mixtures of diverse organic chemical species, that exhibit a multiphase behavior over wide ranges of pressure and temperature. Moreover, it is also important to identify the state in which these hydrocarbon accumulations exist, that is, gaseous state, liquid state, solid state, or in various combination of gas, liquid and solid. The conditions under which these phases exist (due to chemical composition, temperature, and pressure conditions) are a matter of considerable practical importance. Therefore, the ability to measure and predict the thermodynamic behavior of petroleum reservoir fluids is vital to the development of new oil and gas fields, the design and selection of transmission and processing facilities, and the evaluation of production techniques.

In addition to the independent study of reservoir rock and fluid samples, equally important are the properties that are based on rock-fluid interactions. The prominent examples of these include wettability (affinity of a rock to a particular type of fluid; in the presence of another fluid), which depends on properties of both the reservoir rock and the various fluid(s) with which it is saturated. Hence, knowledge of the reservoir rock wettability is essential in determining the location where gas, oil, and water would exist in the reservoir rock pore space.

As mentioned earlier, petroleum reservoir fluids (and the accompanying water in most cases) reside in the tiny pore spaces of the reservoir rocks. If all three phases, of gas, oil, and water were to be present in an open container, they would readily separate or segregate in to three distinct layers due to their density difference. However, that would not be the case in a petroleum reservoir containing all three phases, that is, the phases will not be completely segregated as per their densities because of the presence of an additional force called *capillarity* or *capillary pressure* that will simply resist complete gravity segregation. This so-called capillary pressure is basically a result of the tiny pore spaces that store the petroleum reservoir fluids. Due to this very significant rock–fluid property (capillarity) or a balance of gravity and capillarity, the location of contacts between various fluid phases, such as gas–oil contact or oil–water contact, and the respective transition zones in a petroleum reservoir will vary; this is again a matter of considerable practical importance from exploration and production standpoints.

Based on the foregoing, the significance of reservoir rock and fluid properties in the exploration and production of petroleum reservoirs is clearly evident. Therefore, detailed knowledge of reservoir rock and fluid properties is the backbone of almost all exploration and production-related activities such as reservoir engineering, reservoir simulation, well testing, production engineering, production methods, and so on. In other words, petroleum reservoirs can be effectively described and efficiently managed only when suitable data are available at all levels such as field, well, core, and pore levels. The level and quality of the data also dictate the degree to which reserves can be correctly estimated. The success of defining an optimum field development plan and reservoir management strategy for any field is crucially dependent on our knowledge and understanding of the reservoir rock and fluid properties.

This book addresses the various aspects related to reservoir rock properties in Chapters 2 through 9; those related to reservoir fluid properties are covered in Chapters 10 through 17.

2 Preamble to Petroleum Reservoir Rock Properties

2.1 INTRODUCTION

As discussed in Chapter 1, if geologists find something particularly interesting about a formation, they will normally ask for a reservoir rock core sample from that particular formation for a more thorough study with regard to potential commercial productiveness of the formation. The amount of reservoir rock core taken is usually dependent on the basis of a technical argument between data collection, technical difficulty and economic considerations. However, it should also be realized that the opportunity to recover reservoir rock core samples arises only once in the lifetime of a well.

Geologists and reservoir engineers require reservoir rock samples for reservoir description and definition, reservoir characterization, and to enhance the geological and petrophysical models. More specifically, the recovery of a physical sample of reservoir rock core is essential to evaluate the two most significant characteristics: the capacity and ability of the reservoir rock to store and conduct petroleum reservoir fluids through the matrix. In addition to these characteristics, data on the formation's lithology and production potential (primary, secondary, and tertiary) are just a few of the valuable types of information obtained through a successful coring program.

While some estimates of reservoir rock properties can be made from indirect methods such as electrical and radioactive log surveys, accurate determination of various important properties that are discussed in this book, can only be obtained from physical rock samples. In fact the data obtained from core analysis are actually used for calibration of the indirect methods such as well logs.

Reservoir rock samples are obtained by a process called *coring*; that is the removal of continuous formation samples from a well bore. To the extent possible, core samples are recovered undamaged, preserving the physical and mechanical integrity of the rock. Formation material may be solid rock, friable rock, conglome-rates, unconsolidated sands, shales, or clays. The reservoir rock core sample is generally obtained by drilling into the formation with a hollow-section drill pipe and drill bit. A facility is also available to retain the drilled rock as a cylindrical sample with the dimension of the internal cross-sectional area of the cutting bit and the length of the hollow section.

In some cases, reservoir rock material is also recovered in the form of cuttings (chips of rock) on which some basic properties are measured. However, this book limits the discussion to only geometrically well-defined cylindrical core samples on which various rock properties are measured. With conventional equipment, core samples upto 10 m in length and up to 15 cm in diameter can be obtained. The recovery of a reservoir rock sample according to this procedure is somewhat analogous to using a giant apple corer.

2.2 CORING METHODS

Essentially three different types of coring methods are used to recover formation samples from petroleum reservoirs. Out of these three methods, two are conventional types, such as the *rotary method* and *sidewall coring*. The third method called *high-pressure coring* is a much more advanced technique of recovering formation samples. These three methods are briefly discussed in the following sections.

2.2.1 Rotary Method

In this method, cores are obtained by a coring bit (which has a hole in the center) in combination with a core barrel and a core catcher. The provision of a hole in the center of the coring bit allows the drilling around a central rock cylinder. The rock cylinder is collected in the core barrel through the coring bit. The retrieved core is stored in the core barrel. The bottom of the core is tightly held by the core catcher. As soon as tension is applied to the drillstring the core breaks away from the undrilled formation underneath. The retrieved core is eventually lifted to the surface. The core retrieved using the rotary method is called as the whole core.

2.2.2 Sidewall Coring

Taking a full core from a formation by the rotary method is an expensive operation; hence, the other inexpensive coring method called as sidewall coring is used. This type of coring method obtains smaller samples, ranging from ¾ in. in diameter and 2 in. long to about 1 in. in diameter and upto 6 in. long. The method employs hollow cylindrical core barrels (also called as bullets), which can be shot in sequence, from the gun into the formation. The coring gun containing the bullets is lowered to the bottom of the well and the bullets are fired individually as the gun is pulled up the hole. The coring tools used in sidewall coring typically holds upto 30 bullets. Advantages of this technique include low cost and the recovery of core samples from the formation after it has been drilled. Disadvantages are possible non recovery because of lost or misfired bullets and a slight uncertainty about the sample depth. The samples of formation obtained by this method are called as sidewall cores.

2.2.3 High-Pressure Coring

The two conventional methods discussed earlier suffer from some inherent problems: Formation samples recovered are subject to loss of fluids due to pressure reduction as these are brought to the surface. However, the *high-pressure coring method* attempts to circumvent this problem. The pressure barrel collects the reservoir fluids in their natural container, that is, the reservoir rock, by maintaining the core specimen at bottom hole or reservoir conditions, until the core fluids can be immobilized by freezing. Additionally, pressure coring offers a method for obtaining *in situ* reservoir fluid (gas, oil, and water) saturations.

The technology used in cutting a pressure core is essentially the same as cutting a conventional core. The use of a pressure-retaining core barrel is certainly not new. Sewell[1] of Carter Oil Company reported the first design and application of such a core barrel in 1939. Various other investigators[2–5] have also reported the use of pressure core barrel. Coring rates and core recovery are comparable to conventional coring since the pressure core barrel retains the basic structure of conventional equipment. An additional requirement in high-pressure coring is the necessity of freezing the core in order to immobilize the hydrocarbon fluids within the core. Once these fluids are immobilized, the core can be removed from the barrel after depressurization and subsequently transported (in a frozen state) for laboratory analysis, without the loss of valuable *in situ* fluid saturation information, as discussed in Chapter 6 on fluid saturations.

2.3 IMPORTANT ISSUES RELATED TO CORING METHODS

Despite the fact that core samples recovered are representative of the physical properties of the given formation, the petroleum reservoir fluid contents of that particular core sample are not those of the native rock. Basically, two different factors play an important role in effecting the changes that take place in the recovered reservoir rock sample. First, the core sample on its trip to the surface experiences a reduction in pressure as well as temperature, thereby allowing the fluids contained within the formation to expand and be expelled from the core. Secondly, drilling fluids used in recovering the core samples also interact with the fluids contained within the pore spaces of the core sample (and also the formation), which may cause the displacement of native core fluids by the drilling fluid. Therefore, as a net effect, the recovered core sample may not contain the representative petroleum reservoir fluids.

The problem of loss of native reservoir fluids due to pressure and temperature changes is, however, greatly alleviated in the pressure coring system where formation fluids are kept intact within the core sample. The invasion of drilling fluid/mud filtrate to some extent can be mitigated by selecting appropriate drilling muds or by using special techniques to encapsulate the core. These two issues are discussed in detail in Chapter 6.

2.4 TYPES OF CORES

Generally, petroleum reservoir rock properties can be measured either on whole core samples or small core plugs that are drilled from the whole core samples. A brief discussion regarding whole core and core plug samples is provided in the following two sections.

2.4.1 WHOLE CORE

A whole core sample is basically a complete section of a conventionally drilled core from a given formation. The importance of whole core analysis lies in the fact that small-scale heterogeneity (e.g., for variations in rock properties as a function of position) may not be appropriately represented in measurements on small core plug samples. The advantage of whole core analysis is that it measures properties on a larger scale, somewhat closer to that of the reservoir. Currently, many commercial laboratories are equipped to conduct various rock property measurements on whole core samples.

The determination of rock properties using whole core samples is, however, a much more demanding task considering the sample dimensions, larger size equipment, and additional time are necessary and hence the control of experimental conditions, such as stabilizations, flowrates, pressure, temperature, and so on, can be rather tricky. Moreover, cleaning of whole cores can also be difficult and time consuming, and laboratory analysis is generally significantly more expensive than conventional core plug analysis. In summary, whole cores or full diameter cores are tested only when there is a reason to believe that smaller samples (core plugs) do not reflect average properties.

2.4.2 Core Plug

A core plug sample refers to a much smaller portion of the whole core sample. A core plug sample is obtained by cutting cylindrical plugs of typically 1 or 1.5 in. in diameter and of lengths upto 3 in., from a whole core. All necessary rock properties are typically measured on a number of such core plug samples. Generally, core plugs are cut from whole core samples in two different orientations: perpendicular or parallel to the axis of the whole core. These core plugs, when drilled from a whole core from a vertical well bore, are called *horizontal* and *vertical plugs,* respectively. The determination of rock properties using core plugs has some distinct advantages such as relatively short amount of test duration and ease of maintaining experimental conditions. A diagrammatic representation of core plugs cut from a whole core sample is shown in Figure 2.1. The measurement of rock properties on core plugs is probably the most common practice in the petroleum industry.

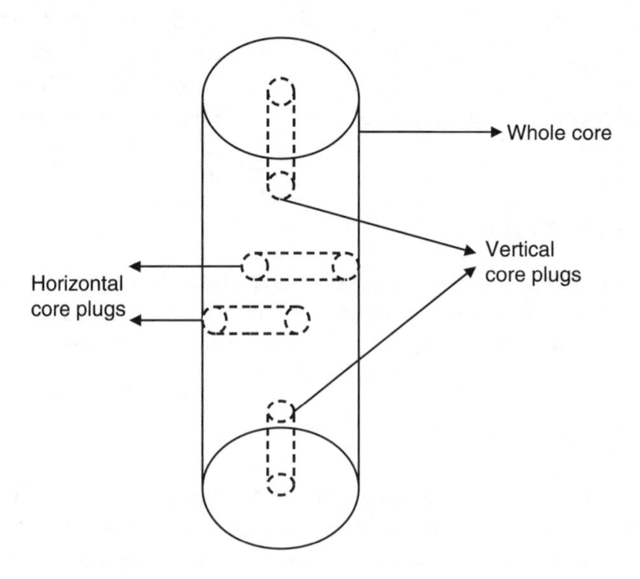

FIGURE 2.1 Core plugs drilled from a whole core sample.

2.5 ALLOCATION OF CORE DATA FOR MEASUREMENT OF RESERVOIR ROCK PROPERTIES

Core sample measurements are intended to achieve several goals and are distinguished by their order of urgency and whether it is a question of an exploratory well or a development well. The data derived from core analysis are typically utilized by geologists, petrophysicists, completion engineers, and reservoir engineers. This particular data allocation is best described by Figure 2.2.

2.6 HANDLING OF RESERVOIR ROCK CORE SAMPLES

A preliminary discussion regarding core handling is given here. However, other specific implications of core handling are discussed in the individual chapters where various rock properties are presented. The reservoir rock core sample is only as good as the various rock properties from which it can be measured. There is no guarantee that good quality core samples always yield reliable rock properties representative of the formation. However, if core samples are properly handled in the laboratory, the probability of obtaining accurate and reliable rock properties that are representative of the formation is much greater. Therefore, proper handling of the core is as critical as its acquisition and, core quality and its handling are key because they directly impact the measured properties.

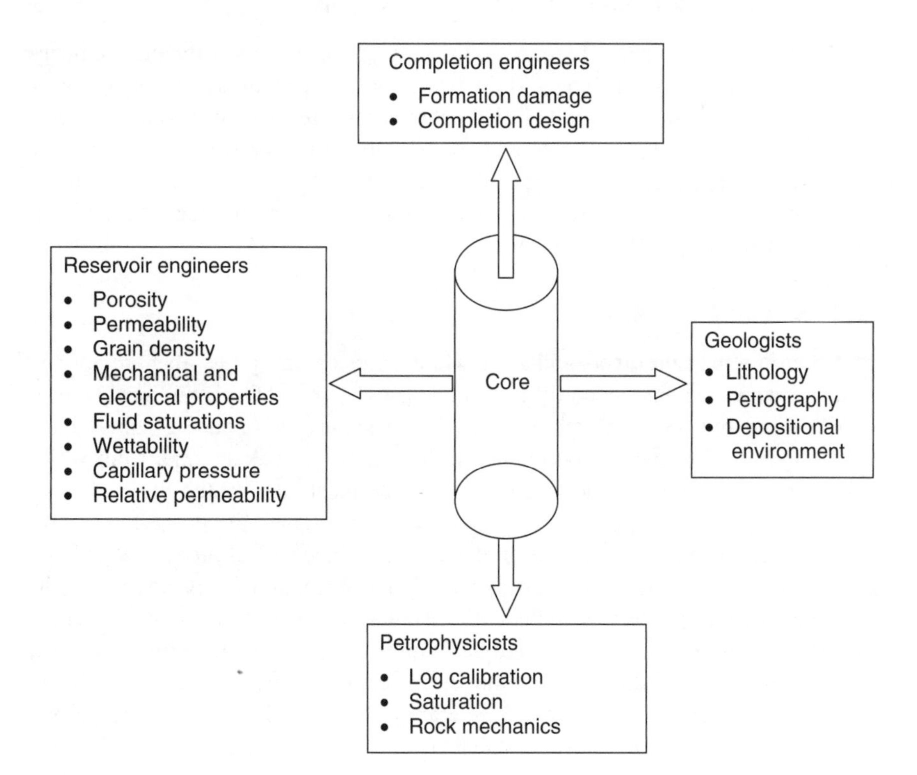

FIGURE 2.2 Allocation of core or core data.

As far as handling core samples from well site to laboratory is concerned (and also in the actual laboratory analysis), it is desirable to preserve the "native" state of the core samples in order to maintain important properties of the core sample such as "wettability". If original wettability of the system is not maintained; it may significantly impact measured rock properties. Therefore, it is a common practice to store the core plug samples immersed in formation fluids (normally crude oil).

2.7 TYPES OF CORE TESTS

Since core samples represent the ground truth in the evaluation of petroleum reservoirs, a thorough laboratory analysis specifically termed *core analysis* is normally carried out. Through such laboratory analysis, a variety of tests are carried out on either the whole core samples or core plug samples. The information obtained from core analysis aids in formation evaluation, reservoir development, and reservoir engineering. The type of data measured as part of this analysis is porosity, permeability, fluid saturations, capillary pressure, relative permeability, and so on. Core analysis is generally categorized into two groups: *routine or conventional core analysis* and *special core analysis or SCAL*. The following paragraphs briefly discuss theses categories of analysis.

2.7.1 ROUTINE OR CONVENTIONAL CORE ANALYSIS

Routine core analysis generally refers to the measurement of porosity, grain density, horizontal permeability (absolute), fluid saturations, and a lithologic description of the core. These measurements are carried out either on the whole core sample or core plug samples at ambient temperature (also sometimes at reservoir temperature) and at either atmospheric confining pressure, formation confining pressure (preferred), or both. Routine core analyses also often include a core gamma log and measurements of vertical permeability (absolute).

2.7.2 SPECIAL CORE ANALYSIS

Any laboratory measurements, either on whole cores or core plugs, that are not part of routine core analysis generally fall under the category of special core analysis (SCAL). Probably the most prominent SCAL tests are two-phase or three-phase fluid flow or displacement experiments in the formation rock sample, from which reservoir engineering properties such as relative permeability, wettability, and capillary pressure are determined. In addition to reservoir engineering properties, SCAL tests also include the measurement of electrical and mechanical properties and petrographic studies. Electrical properties include the formation factor and resistivity index; mechanical properties include the evaluation of various rock mechanics parameters. Petrographic and mineralogical studies basically include imaging of the formation rock samples through scanning electron microscopy, thin-section analysis, x-ray diffraction, and computerized tomography (CT scanning). Sometimes, a preliminary fluid characterization such as density and viscosity measurement of formation hydrocarbon samples and water samples, surface and interfacial tension measurements, and the chemical analysis of water samples are also considered as part of SCAL.

REFERENCES

1. Sewell, B.W., The Carter pressure core barrel, *API Drilling and Production Practice*, 69–78, 1939.
2. Mullane, J.J., Pressure core analysis, *API Drilling and Production Practice*, 163–179, 1949.
3. Willmon, G.J., A study of displacement efficiency in the Redwater field, *J. of Pet. Techno.*, 449–456, 1969.
4. Murphy, R.P. and Owens, W.W., The use of special coring and logging procedures for defining reservoir residual oil saturations, Society of Petroleum Engineers, (SPE) paper number 3793, 1972.
5. Sandiford, B.B. and Eggebrecht, N.M., Determination of the residual oil saturation in two Aux vases sandstone reservoirs, Society of Petroleum Engineer's, (SPE) paper number 3795, 1972.

3 Porosity

3.1 SIGNIFICANCE AND DEFINITION

The petroleum reservoir rocks, for example shown in Figure 3.1, are a sandstone sample that appear to be solid but are often not so solid. Sandstone started out as individual sand particles of varying grain sizes that were buried and compressed as part of the depositional process, resulting in spaces remaining between the particles. Therefore, even though a reservoir rock looks solid to the naked eye, a microscopic examination reveals the existence of tiny openings in the rock. These spaces, or pores, or the tiny openings (typically up to 300 μm) in petroleum reservoir rocks are the ones in which petroleum reservoir fluids are present, much like a sponge holds water. A schematic representation of a pore space is shown in Figure 3.2. This particular storage capacity of reservoir rocks is called as *porosity*. The more porous a reservoir rock material is, the greater the amount of open space or voids it contains, hence greater the capacity to store petroleum reservoir fluids. From a reservoir engineering perspective, porosity is probably one of the most important reservoir rock properties.

The specific definition of porosity is the ratio of the pore volume (or void space) in a reservoir rock to the total volume (bulk volume) and is expressed as a percentage. The pore volume basically refers to the summation or combined volume of all the pore spaces in a given reservoir rock. This significant reservoir rock property is denoted by ϕ and is mathematically expressed by the following relationship

$$\phi = \frac{\text{pore volume}}{\text{total or bulk volume}} \tag{3.1}$$

3.2 TYPES OF POROSITIES

Consideration of the fact that the reservoir rocks were formed by the deposition of sediments in past geological times gives rise to three different types of pores or void spaces. Some void spaces that developed were interconnected with other void spaces forming a network, some were connected with other void spaces but with a deadend or a cul-de-sac; some pores became completely isolated or closed from other void spaces because of cementation. Almost every porous medium or reservoir rock has three basic types of pores: interconnected pores, deadend pores, and isolated or closed pores. A diagrammatic representation of such a pore space is shown in Figure 3.3. Based on these three different types of pores, the total or absolute porosity of a reservoir rock comprises of effective and ineffective porosity, which are defined in the following sections.

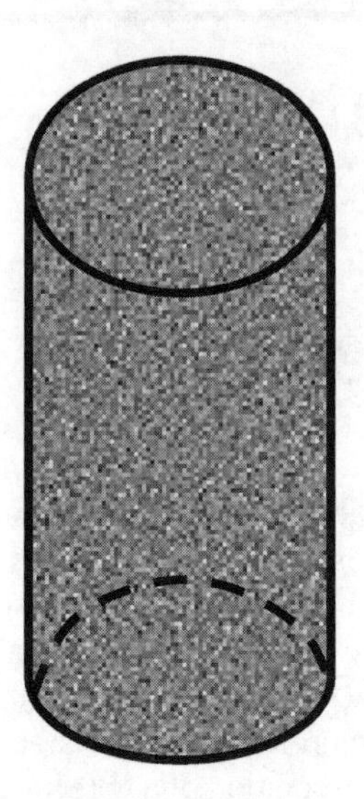

FIGURE 3.1 A sandstone core plug sample.

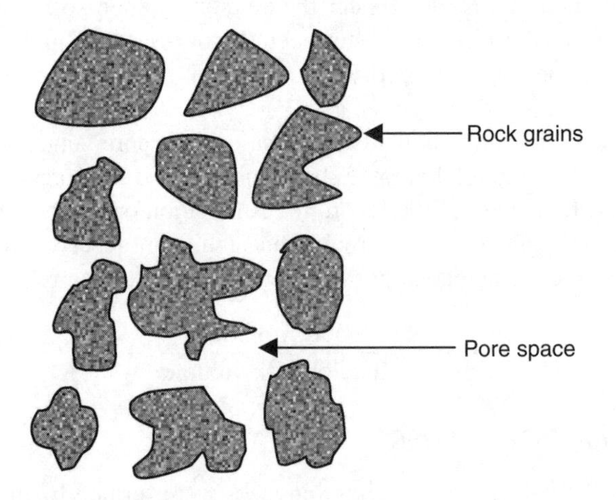

FIGURE 3.2 Conceptual representation of a pore space.

3.2.1 Total or Absolute Porosity

The total or absolute porosity is the ratio of the total void space in the reservoir rock to the total or bulk volume of the rock:

$$\phi = \frac{\text{total pore volume}}{\text{total or bulk volume}} \tag{3.2}$$

or

$$\phi = \frac{\text{vol. of interconnected pores + vol. of deadend or cul-de-sac pores + vol. of isolated pores}}{\text{total or bulk volume}} \tag{3.3}$$

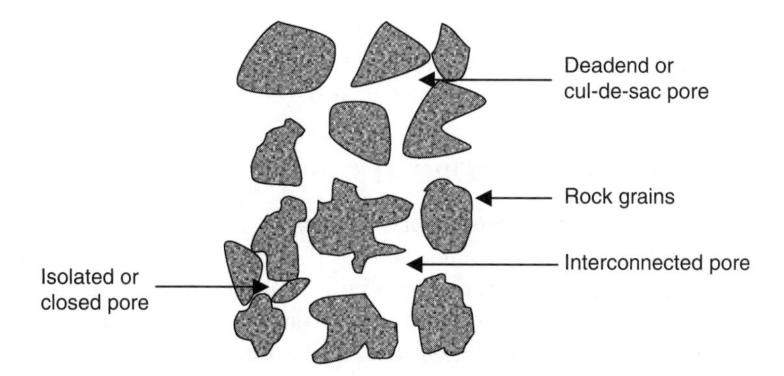

FIGURE 3.3 Conceptual representation of different types of pores in reservoir rock.

Therefore a reservoir rock may have a very high total porosity and no conductivity to fluids residing in the pores due to the lack of interconnectivity. As a result, petroleum reservoir fluids may remain trapped inside the isolated pore spaces and hence immobile or unrecoverable.

3.2.2 Effective Porosity

The effective porosity is defined as ratio of the volume of interconnected pores and the deadend or cul-de-sac pores to the total or bulk volume:

$$\phi = \frac{\text{vol. of interconnected pores + vol. of deadend or cul-de-sac pores}}{\text{total or bulk volume}} \tag{3.4}$$

From a reservoir engineering standpoint, effective porosity is the quantitative value desired and is used in all calculations because it represents the pore space that is occupied by mobile fluids. However, Fatt et al.[1] found that in two limestone reservoir rock samples that they studied, about 20% of the pore volume was in the deadend or cul-de-sac pores. It is important to recognize that even though deadend pores are not "flow through pores", that is, they cannot be flushed out, they can still produce petroleum reservoir fluids by pressure depletion or gas expansion.

3.2.3 Ineffective Porosity

Ineffective porosity is defined as the ratio of the volume of isolated or completely disconnected pores to the total or bulk volume:

$$\phi = \frac{\text{vol. of completely disconnected pores}}{\text{total or bulk volume}} \tag{3.5}$$

However, the closed or isolated or completely disconnected pores are ineffective in producing any petroleum reservoir fluids due to their isolation.

In summary, generally, for poorly or moderately well-cemented material, the total porosity is approximately equal to the effective porosity, however, for highly cemented material significant differences between the total and effective porosity can occur.

3.3 CLASSIFICATION OF POROSITY

Reservoir rock porosity can be generally classified by the mode of origin, as either original or induced. *Original or induced porosity* is also sometimes called *primary or secondary porosity*, respectively. Original porosity resembles a native porosity, that is., developed in the deposition of the material. Induced porosity is developed by some geological process following the deposition of the rock. A common example of induced porosity is the development of fractures or vugs commonly found in limestones.

A good example of the concept of original or induced porosity would be the manner in which loose or unconsolidated sand particles (of varying sizes) are packed in a container. For example, if the container is simply filled with the sand particles (without the use of any force or compaction) original porosity is obtained, while induced porosity results if the container is shaken or sand particles are compacted in the container. In this case, original porosity is higher than the induced porosity because pore spaces between sand particles or grains are due to a higher degree of packing and rearrangement of sand grains. Reservoir rocks that have original porosity are more uniform in their characteristics than those rocks in which a large part of the porosity is induced.

3.4 PARAMETERS THAT INFLUENCE POROSITY

Many factors affect the porosity of reservoir rocks including grain size, grain shape, sorting, clay content, compaction, and cementation. The effect of these factors is briefly discussed in the following text.

In order to understand the effect of grain size on porosity, let us first consider a system of well-rounded sediments that are packed in a cubical arrangement (the least-compact arrangement), as shown in Figure 3.4, which results in a porosity of 47.64%. Rhombohedral packing (the most-compact arrangement), as shown in Figure 3.5, yields a porosity of 25.96%. In other words, a room full of bowling balls and a room full of baseballs would have the same porosities if the respective balls were packed the same way. Therefore, the porosity of such a system is independent of grain size (sphere/ball diameter in this case). However, if smaller spheres are mixed among the spheres of either system, the pore space or the porosity is reduced. If smaller particles are mixed with larger grains, the effective porosity is considerably reduced.

The shapes of sediments affect the porosity of a rock. In reservoir rocks, grains can be numerous different shapes. Generally, sediments are not perfectly round, but occur in many shapes. In the case of irregular grains, they tend not to pack as neatly as rounded particles, which results in a higher proportion of void space or higher porosity.

The effect of sorting on porosity is that, well-sorted sediments (similar range of particle sizes) generally have higher porosity than poorly sorted sediments simply because if a wide range of particle sizes exists then the smaller particles fill in the

FIGURE 3.4 Cubic packing of spheres resulting in a least-compact arrangement with a porosity of 47.64%.

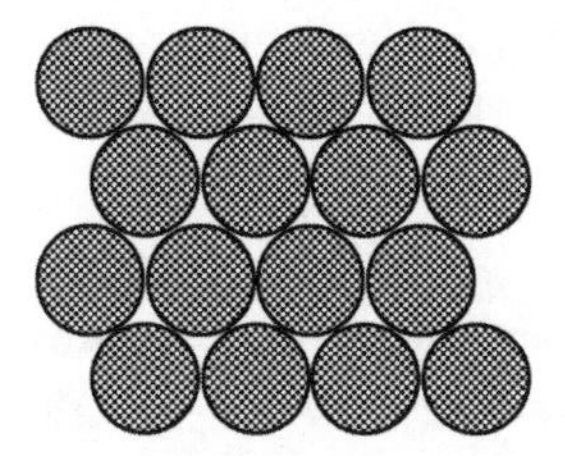

FIGURE 3.5 Rhombohedral packing of spheres resulting in a most-compact arrangement with a porosity of 25.96%.

voids between the larger particles thus resulting in lower porosities. For unconsolidated or loose sand, the grain size distribution is generally measured by sieve trays of graded mesh size by a technique called *sieve analysis*.

The presence of clay and organic content generally tends to increase the void space because clay particles tend to electrostatically repel one another resulting in higher porosities. On the other hand, compaction and cementation tend to actually decrease porosity. Compaction creates a certain mechanical arrangement of grains and a new pore system pattern because of sediment compaction, which is marked by a reduction in pore volume. Cementation refers to the deposition of minerals within the pore space eventually causing a decrease in the pore volume. Both these parameters are in fact related to primary and secondary porosities.

3.5 LABORATORY MEASUREMENT OF POROSITY

The porosity of reservoir rocks can be determined by essentially two different processes: routine core analysis (laboratory measurements on core plugs drilled from whole core samples) and well logging techniques. Between these two processes, routine core analysis is probably the most common method used in the determination of porosity of reservoir rocks. Rock samples used in porosity measurements are called core plugs. Sometimes subsamples called *end trims* sliced from core plugs are used in routine core analysis (porosity, absolute permeability, and saturation measurement). See Chapter 6 for a discussion on core plugs and end trims.

Well logging techniques are indirect in nature and usually porosity is measured *in situ*, that is, actual physical samples of the reservoir rock are not tested in laboratories like routine core analysis. Porosity determination using well logging techniques are not discussed here; Brock[2] gives a detailed discussion of this subject. In addition to the two methods of routine core analysis and well logging, other nonconventional techniques of porosity determination exist, such as x-ray computerized tomography (CT) scanning that is discussed in Section 3.6.

The routine core analysis method of porosity determination is discussed in the following sections.

3.5.1 Porosity Determination Using Routine Core Analysis

A given reservoir rock sample basically comprises three different volumes: bulk volume (BV), pore volume (PV), and grain volume (GV). These three volumes are related by the following simple relationship:

$$BV = PV + GV \tag{3.6}$$

Therefore, in the laboratory measurement of porosity, it is necessary to determine only two of the three volumes: BV, PV, or GV. The various methods described in the following text for determining BV, PV, or GV are for dry, cleaned reservoir rock samples.

3.5.1.1 Bulk Volume Measurement

The most common types of samples used in routine core analysis are cylindrical core plugs which allow the determination of BV from the dimensions of the sample (length and diameter). This is the easiest and simplest method of determining the BV of a reservoir rock sample. Although this method works well for perfectly cylindrical samples, inaccuracies in the computed BV are evident in the case of chipped samples or slight geometric irregularities, usually resulting in a nonrepresentative BV and incorrect porosities. Therefore, in order to avoid such uncertainties in the BV measurement, a procedure that utilizes the observation of the volume of fluid displaced by the sample is employed (Archimedes principle). The fluid displaced by a sample can be observed either volumetrically or gravimetrically. This procedure has obvious advantages because the BV of irregular shaped samples as well as geometrically well-defined or symmetric samples can be determined with the same accuracy and speed. However, it is very important to prevent the penetration of the fluid used in observing the displacement into the pore space of the rock specimen because this affects the BV measurement. This can be accomplished by either coating the sample with paraffin wax or by saturating the sample with the same fluid used for observing the displacement, or by using mercury, which owing to its wetting characteristics, does not tend to enter the pore spaces unless it is forced. Saturating the sample with the fluid that is used for observing the displacement has a clear advantage, that as part of the BV measurement, PV is also measured, which actually allows the determination of sample porosity.

3.5.1.2 Pore Volume Measurement

All methods used for determining pore volume are based on either extraction of a fluid from the rock sample or introduction of a fluid in the pore spaces of the rock sample. It is noteworthy that all methods measuring pore volume yield effective porosity; simply because the fluid either extracted from the pore spaces of the rock sample or introduced into the pore spaces of the rock sample will always be from interconnected and deadend or cul-de-sac pores. It should also be mentioned here that since pore spaces in reservoir rocks are quite small (of the order of 300 μm), the determination of pore volumes of such samples involves measuring the volume of literally thousands of pores.

In the extraction methods, the rock sample (in most cases saturated with native or original reservoir fluids) is subjected to an extraction procedure that uses suitable solvents to recover the fluids contained in the pore spaces. The volume of the extracted fluids is determined and that in itself represents the pore volume. This method is sometimes called the *summation of fluids* and is applied to as-received core plug samples in which the pore volume is considered to be equal to the sum of any gas, oil, and water volume occupying the rock sample. This particular technique is in fact part of the fluid saturation determination of as-received core plugs samples and is discussed in detail in Chapter 6.

When introducing fluids into the pore spaces of the rock sample, a number of methods are used for the determination of pore volume of reservoir rocks. These methods use three different types of fluids: helium, water or synthetic oil, and mercury. The porosity measured is, however, effective porosity because the saturating fluids only penetrate the interconnected and deadend pore spaces. Mercury it is not commonly used today due to health hazard concerns. The various methods that employ these saturating fluids are in the following text.

3.5.1.2.1 Helium porosimeter

The use of helium in the determination of porosity has certain obvious advantages over other gases: its small molecules rapidly penetrate the small pores, its inertness does not allow the adsorption on rock surfaces, and it can be considered an ideal gas for pressures and temperatures usually employed in the test. The helium porosimeter actually employs the principles of Boyle's law (PV = constant, where P is the pressure and V the volume) for the determination of porosity of rock samples.

A variety of desktop-type helium porosimeters are available on the market. Figure 3.6 shows, in principle, that the apparatus consists of two equal volume chambers or cells called the *reference chamber* and the *sample chamber*. The reference chamber has a volume V_1 at initial pressure P_1 and the sample chamber has an unknown volume V_2 and initial pressure P_2 (normally atmospheric). The system is then brought to equilibrium by opening the valve to the sample chamber allowing the determination of the unknown volume V_2 by noting the resultant equilibrium pressure P. The application of Boyle's law allows the equalization of pressures (for isothermal conditions) before and after the opening of the valve to the sample chamber, as per the following equation:

$$P_1V_1 + P_2V_2 = P(V_1+V_2) \tag{3.7}$$

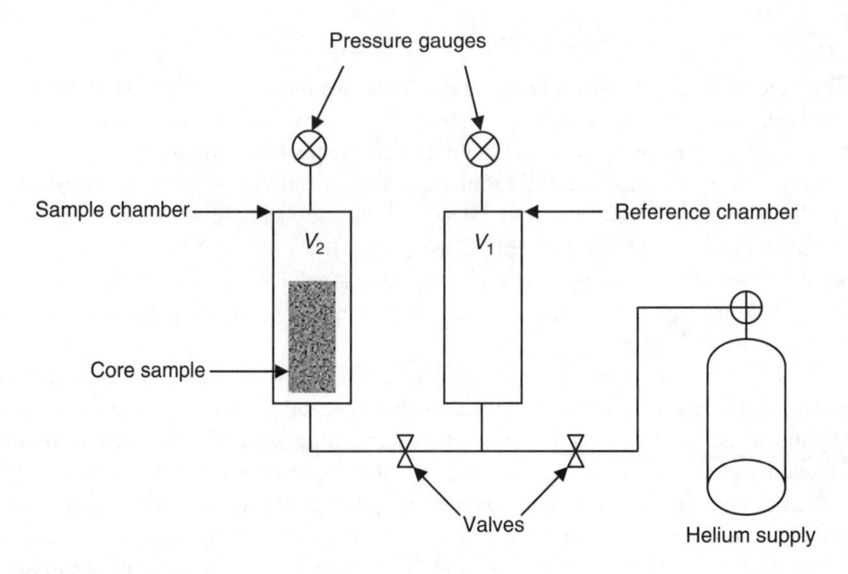

FIGURE 3.6 Schematic illustration of a helium porosimeter.

or by rearrangement

$$V_2 = \frac{V_1(P_1 - P)}{(P - P_2)} \tag{3.8}$$

The calculated unknown volume V_2 can in fact be expressed as

$$V_2 = V_1 - \text{BV} + \text{PV} \tag{3.9}$$

which allows the calculation of PV or porosity,

$$\phi = \frac{V_2 - V_1 + \text{BV}}{\text{BV}} \tag{3.10}$$

where BV is the bulk volume of the sample measured (i.e., from sample dimensions), V_1 is known and V_2 is determined from Equation 3.8. The sample porosity calculated in Equation 3.10 can be multiplied by 100 to report the value in percentage.

3.5.1.2.2 Vacuum saturation

The vacuum saturation method is in fact one of the primitive methods of obtaining the pore volume of a rock sample. The only advantage is the fact that pore volumes of multiple samples can be determined in one step. The method uses a large enough vacuum flask or a beaker, filled with a degassed liquid, normally water, in which dry rock samples are placed. Subsequently, as soon as the evacuation of the vacuum flask is initiated, air bubbles are seen in the saturating liquid as it displaces air from the pore spaces of the rock samples. The disappearance of the air bubbles gives an indication that the saturation is complete and at this point the evacuation is terminated, and porosity is calculated as follows:

$$\phi = \frac{(\mathrm{WW} - \mathrm{DW})}{\mathrm{BV}\,\rho_{\mathrm{w}}} \times 100\% \tag{3.11}$$

where WW is the wet weight of the sample, after vacuum saturation; DW is the dry weight of the sample, before vacuum saturation; and ρ_{w} is the saturating fluid (water) density. The time required for completion of the saturation is directly proportional to the sample size and pore sizes. This simple method may work well for small samples and those having reasonably large pore spaces.

3.5.1.2.3 Liquid saturation by other methods

The other methods of introduction of a liquid into the pore spaces of a rock sample include forced saturation by either water or a synthetic oil. The rock sample is held in a special device called a *core holder* and a given liquid is forced through the sample by use of a pump. This method, however, requires advanced apparatus called *a core flooding rig* or a *displacement apparatus* (see Figure 4.7 in Chapter 4), compared to the techniques discussed earlier.

A volume balance, based on the injection rate and total time between the injected liquid and the produced liquid from the rock sample can give an indication of the saturation of the sample. In some cases, the vacuum saturation technique discussed earlier may be used as a forerunner for this technique to speed up the saturation process. On completion of the saturation sample porosity is determined using Equation 3.11.

3.5.1.3 Grain Volume Measurement

All methods measuring grain volume usually yield total or absolute porosity, simply because the rock samples are normally crushed for grain volume measurements which actually destroys all pores; thus resulting in total porosity as grain volume is subtracted from the bulk volume. Although only the effective pore space has direct application in most reservoir engineering calculations, knowledge of the magnitude and distribution of the isolated pore spaces can lead to crucial information about the reservoirs.

Grain volume of rock samples is sometimes calculated from dry sample weight and knowledge of average density. For example,in the case of sandstone, average density of quartz (2.65 g/cm^3) can be used as the sand grain density to calculate the grain volume. However, formations of varying lithology and grain density limit applicability of this method.

The Boyle's law technique for pore volume measurement that was discussed earlier can in fact also be construed as a method that determines the grain volume. This is clear from Equations 3.8 and 3.9, that is, the volume, V_2 occupied by helium in the sample chamber is equal to the difference between the overall chamber volume V_1 because both reference as well as sample chambers have equal volume and the volume of solids in the sample chamber. This particular volume of solids in the sample chamber is nothing but the grain volume, which allows Equation 3.9 to eventually permit the determination of porosity,

$$V_2 = V_1 - \mathrm{GV} \tag{3.12}$$

The measurement of the grain volume of a cleaned and dried crushed core sample may also be based on the loss in weight of a saturated sample plunged in a liquid, which is similar to the bulk volume measurement using the principle of buoyancy.

3.6 NONCONVENTIONAL METHODS OF POROSITY MEASUREMENTS

Apart from the conventional methods described previously to determine the reservoir rock porosity, a number of non-conventional methods are increasingly being used. One such popular technique is based on x-ray computerized tomography or x-ray CT scanning which was invented in 1972 by a British engineer Sir Godfrey Hounsfield of EMI Laboratories in England. Computerized tomography is based on the x-ray principle: As x-rays from a rotating frame pass through the rock sample that is placed on a special mounting table that can be moved in and out, they are absorbed or attenuated (weakened) at differing levels creating a matrix or profile of x-ray beams of different strengths. This x-ray profile is registered on film creating an image. Every time the x-ray tube and the detector completes a full circle (360°) an image or a slice is acquired which is typically 1 to 10 mm in size. These particular images are processed and used in the calculations of various rock sample properties.

Although used primarily in the medical field, the use of CT scanning in routine core analysis and even special core analysis is overwhelming. A paper by Withjack et al.[3] provides an excellent overview of CT core analysis studies. The application of CT scanning for the determination of porosity is discussed briefly here. The basic principle behind use of the CT scanner for porosity measurement involves imaging the core plug sample when it is clean and dry and then when it is fully saturated with either oil or water. The image of the clean sample is then subtracted from that of the saturated sample to obtain porosity. Such a scanning process is carried out slice by slice and porosity is determined for each of the slices. The porosity determined for each of the slices also provides an indication of porosity variation, which is something that cannot be accomplished by conventional methods. However, it should be mentioned here that the accuracy of a CT-determined porosity depends greatly on factors such as complete saturation of the rock sample by either oil or water. Finally, porosities determined by the CT scanning technique can also be compared with the average sample porosity obtained by conventional methods described earlier.

Various images that are obtained by a CT scanner have a characteristic *CT number* expressed in Hounsfield units. This CT number is actually a normalized value of the calculated x-ray absorption coefficient of a pixel (picture element). Pure water typically has a CT number of 0; air has a CT number of −1000. Similar to pure water and air, CT numbers are also obtained for oil and rock samples saturated with either oil or water. Various CT numbers are employed in the following simple equations, which allow the determination of porosity. For example, the CT number for an oil-saturated rock sample is a combination of CT number for rock and CT number for oil expressed as

$$CT_{o,r} = (1-\phi)\,CT_r + \phi CT_o \tag{3.13}$$

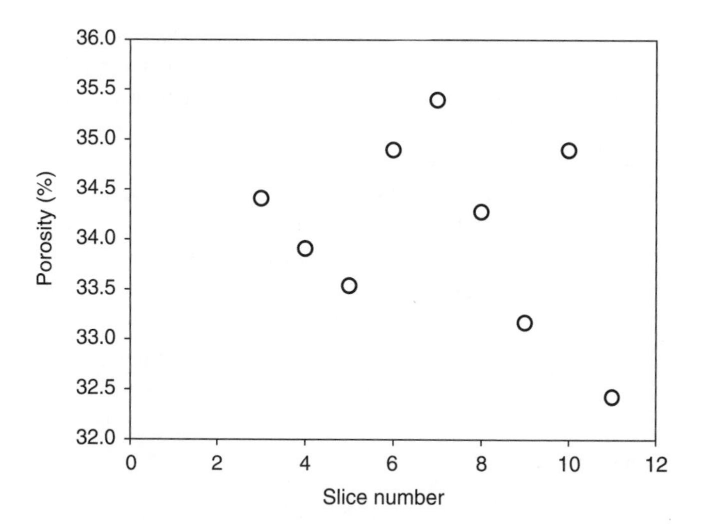

FIGURE 3.7 Porosities for a North Sea chalk core plug[4] measured using x-ray CT scanning. The average porosity determined using a helium porosimeter for the same sample is 34.41%.

For water-saturated rock,

$$CT_{w,r} = (1-\phi)\, CT_r + \phi CT_w \tag{3.14}$$

For air-saturated rock,

$$CT_{a,r} = (1-\phi)\, CT_r + \phi CT_a \tag{3.15}$$

where $CT_{o,r}$, $CT_{w,r}$, and $CT_{a,r}$ are CT numbers for oil–rock, water–rock, and air–rock systems, ϕ is the porosity, and CT_o, CT_w, CT_a, and CT_r are CT numbers for oil, water, air, and rock, respectively.

Therefore, based on the CT number data for each slice of the rock sample, porosity can be easily determined by rearranging Equations 3.13 or 3.14 and 3.15,

$$\phi = \frac{CT_{o,r} - CT_{a,r}}{CT_o - CT_a} = \frac{CT_{w,r} - CT_{a,r}}{CT_w - CT_a} \tag{3.16}$$

As an example, the CT determined porosity values for a 5 mm slice thickness, 3.8 cm diameter of North Sea chalk (limestone) core plug[4] is shown in Figure 3.7.

3.7 AVERAGING OF POROSITY

The porosity data that are measured as part of the routine core analysis is obtained from core plug samples that actually represent a very small fraction of the entire reservoir rock. Therefore properties that are measured as part of the routine core analysis must be averaged and scaled up from the core scale to the reservoir scale for use in reservoir engineering and reservoir simulation. This is accomplished by employing different types of averaging methods. If the reservoir rock shows large variations in porosity vertically but has fairly uniform porosity parallel to the bedding planes, the arithmetic average porosity or the thickness-weighted average porosity is used to describe the average reservoir porosity. However, due to the changes in sedimentation or depositional conditions, significant variations in porosity can be observed in different section of the reservoir. In such cases, the areal-weighted or the volume-weighted average porosity is employed to describe the average reservoir rock porosity. The mathematical equations used for averaging the porosity data have the following forms:

Arithmetic average:

$$\phi = \Sigma\phi_i/n \tag{3.17}$$

Thickness-weighted average:

$$\phi = \Sigma\phi_i h_i/\Sigma h_i \tag{3.18}$$

Areal-weighted average:

$$\phi = \Sigma\phi_i A_i/\Sigma A_i \tag{3.19}$$

Volumetric-weighted average:

$$\phi = \Sigma\phi_i A_i h_i/\Sigma A_i h_i \tag{3.20}$$

where n is the total number of core samples, h_i is the thickness of core sample i or reservoir area i, ϕ_i is the porosity of core sample i or reservoir area i, and A_i is the reservoir area i.

3.8 EXAMPLES OF TYPICAL POROSITIES

The porosity of reservoir rocks differs from one material to another. Specifically, porosity depends on various factors such as grain shape, sorting, cementation, and compaction. More precisely, the magnitude of porosity largely depends upon the geometrical arrangement of particles in the sediment. All these factors substantially vary from formation to formation which results in a wide range of rock porosities. Therefore, considering the complexity of the variables involved, porosity cannot be correlated to any specific properties, thereby allowing only a certain range of porosities for particular type of reservoir rock such as sandtsones or carbonates. Although, the porosities of petroleum reservoir rocks typically range from 5 to 40%, with a range of 10 to 20% being more common. Porosities as high as 48% have been reported for Ekofisk chalk formations in the North Sea area.[5]

As an example the porosity data for North Sea chalk is shown in Figure 3.8, in which porosity is plotted as a function of sample depth. Generally, reservoir rock porosity shows a trend of decreasing porosity with depth because overlying rock layers cause compaction. However, sedimentary rock porosities are a complex function of many factors, including but not limited to: rate of burial, depth of burial, the nature

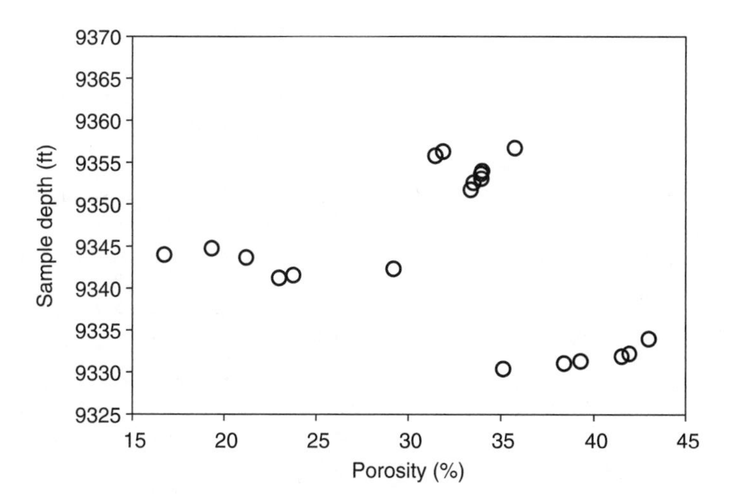

FIGURE 3.8 Porosity as a function of sample depth for North Sea chalk. (From Dandekar, A.Y., unpublished data, 1999.)

of the reservoir fluids, and the nature of overlying sediments. A commonly used relationship between porosity and depth is given by an equation developed by Athy.[6] However, the trend of decreasing porosity with depth is not evident in Figure 3.8 mainly because the data is in a rather narrow depth range. The work of Krumbein and Sloss,[7] even though for sandstones, indicates that porosity decreases from about 42% to only 32% in a depth range of 0 to 6000 ft. Similarly in the case of South Florida basin chalk[8] the porosity decreases (based on their fitted expression of porosity versus depth) where depth is almost insignificant in the range of 9330 to 9360 ft.

PROBLEMS

3.1 A petroleum reservoir has an areal extent of 20,000 ft^2 and a pay zone thickness of 100 ft. The reservoir rock has a uniform porosity of 35%. What is the pore volume of this reservoir?

3.2 Assuming unit formation thickness, determine the average porosity for the following system when $\phi_a = 0.20$, $\phi_b = 0.11$, $\phi_c = 0.29$, $L_1 = 0.35L$, and $h_a = h_b = 0.5h_c$.

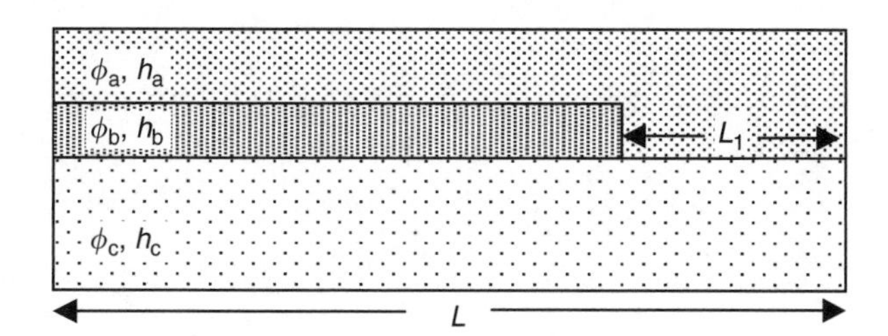

3.3 A 37.5485 g cleaned and dried core plug was flooded with a 0.75 g/cm^3 crude oil for several days to ensure complete saturation. On termination of the flood, the plug weighed 44.4178 g. What is the oil storage capacity of this plug?

3.4 Assuming a sandstone grain density of 2.65 g/cm^3, calculate the porosity of a 3 in. long sandstone core sample of 1.5 in. width and breadth, respectively, if the grains weigh 250.0 g?

3.5 Calculate the weight of 1 m^3 sandstone of 14% porosity, assuming a sand grain density of 2.65 g/cm^3?

3.6 Calculate the arithmetic average and thickness-weighted average porosity for the following core data:

Sample Number	Depth (ft)	Porosity (%)
1	3705.5	40.1
2	3706.5	35.1
3	3707.5	39.3
4	3708.5	36.5
5	3709.5	29.1

REFERENCES

1. Fatt, I., Maleki, M. and Upadhyay, R.N. Detection and estimation of dead-end pore volume in reservoir rock by conventional laboratory tests, *Soc. Pet. Eng. J.*, 206, 1966.
2. Brock, J. *Applied Open-Hole Log Analysis*, Vol. 2, Contributions in Petroleum Geology & Engineering, Gulf Publishing Company, Houston, 1986, chap. 9.
3. Withjack, E.M., Devier, C.O. and Michael, G. The role of X-ray computed tomo-graphy in core analysis, presented at the Society of Petroleum Engineers Western Regional/AAPG Pacific Section Joint Meeting, Long Beach, CA, SPE 83467, 2003.
4. Dandekar, A.Y. Unpublished data, 1999.
5. Suiak, R.M. and Danielsen, J. Reservoir aspects of Ekofisk subsidence, *J. Pet. Technol.*, 709, 1989.
6. Athy, L.F. Density, porosity, and compaction of sedimentary rocks. *AAPG Bull.*, 14, 1930.
7. Krumbein, W.C. and Sloss, L.L. *Stratigraphy and Sedimentation*, 1st ed., W.H. Freeman Publishing Company, New York, 1951, p. 218.
8. Schmoker, J.W. and Halley, R.B. Carbonate porosity versus depth: a predictable relation for south Florida, *AAPG Bull.*, 66, 2561, 1982.

4 Absolute Permeability

4.1 SIGNIFICANCE AND DEFINITION

Chapter 3 addressed porosity, or basically, the storage capacity of reservoir rock. However, merely having a large enough porosity of reservoir rock is not sufficient because the petroleum reservoir fluids contained in the pore spaces of reservoir rock have to flow, so that they can be produced or brought to the surface from the reservoir. This particular property of a reservoir rock, denoted by k, is called *permeability*, which is conceptually illustrated in Figure 4.1. Permeability of a petroleum reservoir rock is one of the most influential parameters in determining the production capabilities of a producing formation.

Unlike porosity, which is a static property of the porous medium, permeability is basically a flow property (dynamic) and therefore can be characterized only by conducting flow experiments in a reservoir rock. This chapter discusses *absolute permeability* or simply, *permeability* of the porous medium, that is, when a reservoir rock is 100% saturated with a given fluid. The permeability measure of a rock filled with a single fluid is different from the permeability measure of the same rock filled with two or more fluids, called *relative permeability*, which is discussed in Chapter 9. At the outset, it should be mentioned that absolute permeability is a property of the rock alone and not the fluid that flows through it, provided no chemical reaction takes place between the rock and the flowing fluid.

When defining permeability, various ways in addressing this important reservoir rock property should be used:

- The permeability of a rock is a measure of its specific flow capacity.
- Permeability is basically a property of the medium and is a measure of the capacity of the medium to transmit fluids.
- Permeability signifies the ability to flow or transmit fluids through a rock when a single fluid, or phase, is present in the rock.
- Permeability is a measure of the fluid conductivity of a particular porous medium.
- By analogy with electrical conductance, permeability represents the reciprocal of the resistance that the porous medium offers to fluid flow.
- Permeability is the proportionality constant between the fluid flow rate and an applied pressure or potential gradient.

4.2 MATHEMATICAL EXPRESSION OF PERMEABILITY: DARCY'S LAW

It was the pioneering work of Henry Darcy,[1] a French civil engineer, that led to the development of the mathematical expression, still used today by the petroleum

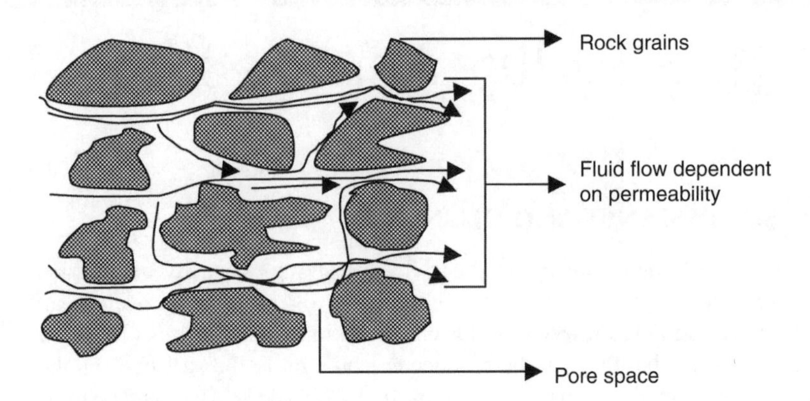

FIGURE 4.1 Conceptual illustration of permeability of a reservoir rock.

industry, that allows the calculation of permeability from flow experiments. This mathematical equation for calculating the permeability of a porous medium originated from Darcy's investigations on flow of water through sand filters for water purification for the city of Dijon in France.

Darcy's original experiment of the flow of water through sand is analogous to the flow of a fluid through a core plug and can be schematically represented as shown in Figure 4.2. The only difference between Darcy's original experiment and the schematic shown in Figure 4.2 is the orientation, that is, vertical in the former and horizontal in the latter case.

Darcy expressed the results of his flow experiments in the following mathematical form:

$$Q = KA\frac{(h_1 - h_2)}{L} \tag{4.1}$$

However, with reference to Figure 4.2, Q is the volumetric flow rate through the core plug (in m^3/sec or ft^3/sec), K the proportional constant also defined as hydraulic conductivity (in m/sec or ft/sec), A the cross-sectional area of the core plug (in m^2 or ft^2), L the length of the core plug (in m or ft), and h_1 and h_2 represent the hydraulic head at inlet and outlet, respectively (in m or ft).

Alternatively, Equation 4.1 can also be expressed in terms of the pressure gradient dP over a section dL as

$$Q = -KA\frac{\mathrm{d}P}{\mathrm{d}L} \tag{4.2}$$

where

$$\mathrm{d}P = \Delta h \rho g \tag{4.3}$$

dP is the difference between the upstream and downstream pressures (in N/m^2), Δh the difference between the upstream and downstream hydraulic gradients (in m), ρ the fluid density (in kg/m^3), and g is the acceleration due to gravity (9.81 m/sec^2).

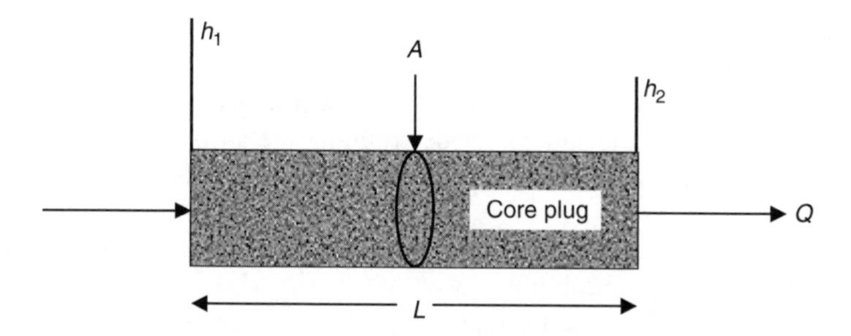

FIGURE 4.2 Darcy experiment expressed by a schematic representation of fluid flow through a core plug.

It should be noted that Darcy's investigations were restricted to the flow of water through sand packs that were 100% saturated by water. However, later investigators found out that Darcy's law could be extended or generalized to other fluids if K is expressed as a ratio of k/μ, where μ is the viscosity of a given fluid and k the *permeability* of the porous medium, allowing Equation 4.2 to be written as

$$Q = -\frac{k}{\mu} A \frac{dP}{dL} \tag{4.4}$$

Equation 4.4 can now be integrated between the limits of length from 0 to L and pressure from P_1 (upstream) to P_2 (downstream), for a fluid flow case, such as the one shown in Figure 4.2, under the following assumptions:

- The core plug is 100% saturated with the flowing fluid.
- The flowing fluid is incompressible.
- The flow is horizontal, steady state and under the laminar regime.
- The flow of fluid through the porous medium takes place under viscous regime (i.e., the rate of flow is sufficiently low so that it is directly proportional to the pressure differential or the hydraulic gradient).
- The flowing fluid does not react with the porous medium because it may alter the characteristics of the porous medium thereby changing its permeability as flow continues.

$$\frac{Q}{A}\int_0^L dL = -\frac{k}{\mu}\int_{P_1}^{P_2} dP \tag{4.5}$$

$$\frac{Q}{A}(L-0) = -\frac{k}{\mu}(P_2 - P_1) \tag{4.6}$$

$$Q = \frac{kA}{\mu L}(P_1 - P_2) \quad \text{or} \quad Q = \frac{kA\Delta P}{\mu L} \tag{4.7}$$

where Q is the flow rate in m^3/sec, k the absolute permeability in m^2 (which can be converted to mD or Darcy; see Section 4.3 for conversion factors), A the cross-sectional area in m^2, $P_1 - P_2 = \Delta P$ is the flowing pressure drop in N/m^2, μ the fluid viscosity in N sec/m^2, and L the length in m.

Equation 4.7 is commonly known as *Darcy's law* and is extensively used in petroleum engineering calculations for determining the absolute permeability of a reservoir rock. Equation 4.7 represents a combination of the following:

- The property of the porous medium or the reservoir rock is represented by k, the absolute permeability.
- The property of the fluid is represented by μ, its viscosity.
- The geometry of the porous medium is represented by the ratio of A/L.
- The fluid flow characteristics is represented by Q, ΔP, and μ.

Equation 4.7 shows that the absolute permeability k is entirely a property of a porous medium and is independent of the properties of the flowing fluid because ΔP obtained for the flow of a particular fluid is scaled according to the flow rate and the viscosity of the fluid. For instance, if flow rate Q is increased, pressure drop ΔP increases; k is not an independent function of either flow rate or ΔP. Similarly, an increase in length L for the same ΔP results in a decrease in flow rate Q so that k is again unchanged.

4.3 DIMENSIONAL ANALYSIS OF PERMEABILITY AND DEFINITION OF A DARCY

The dimensions of permeability can be easily obtained by substituting the appropriate dimensions, such as, M for mass, L for length, and T for time for each of the quantities in Equation 4.7.

$$Q = \frac{L^3}{T}$$

$$A = L^2$$

$$\Delta P \text{ or } P = \frac{\text{force}}{\text{area}} = \frac{ML}{T^2L^2} = \frac{M}{LT^2}$$

$$\mu = \frac{\text{force}}{\text{area}}\text{time} = \frac{MLT}{T^2L^2} = \frac{M}{LT}$$

$$L = L$$

$$\frac{L^3}{T} = kL^2\frac{M}{LT^2}\frac{LT}{M}\frac{1}{L}$$

or

$$k = L^2$$

Therefore, permeability has the units ft^2 in the English system, cm^2 in the CGS (centimeter gram system) system, or m^2 in either the MKS (meter kilogram system) or SI units.

However, because these units are too large a measure with the porous medium, petroleum industry adopted the unit "darcy" for permeability in honor of Henry Darcy for his pioneering work that led to the development of the mathematical expression for the calculation of absolute permeability, also known as Darcy's law.

A porous medium is said to have a permeability of one darcy when a single phase fluid having a viscosity of one centipoise (cP) completely saturates the porous medium and flows through it at a rate of 1 cm^3/sec under a viscous flow regime and a pressure gradient of 1 atm/sec through a cross-sectional area of 1 cm^2

From Equation 4.7,

$$1 \text{ darcy} = \frac{(\text{cm}^3/\text{sec})(\text{cP})}{(\text{cm}^2)(\text{atm/cm})}$$

where

$$1 \text{ cP} = 1.0 \times 10^{-7} \text{ N sec/cm}^2$$

$$1 \text{ atm} = 10.1325 \text{ N/cm}^2$$

$$1 \text{ D} = \frac{(\text{cm}^3/\text{sec})(1.0 \times 10^{-7} \text{ N sec/cm}^2)}{(\text{cm}^2)(10.1325 \text{ N cm}^2/\text{cm})}$$

$$= 9.869 \times 10^{-9} \text{ cm}^2$$

$$= 9.869 \times 10^{-13} \text{ m}^2$$

$$= 1.062 \times 10^{-11} \text{ ft}^2$$

However, one darcy is a relatively high permeability because most reservoir rocks have permeabilities less than one Darcy. In order to avoid the use of fractions in describing the reservoir rock permeability, the term millidarcy (mD) is used.

$$1 \text{ D} = 1000 \text{ mD or } 1 \text{ mD} = 0.001 \text{ D}$$

4.4 APPLICATION OF DARCY'S LAW TO INCLINED FLOW AND RADIAL FLOW

The equation for calculating the absolute permeability (Equation 4.7) is applicable to a horizontal flow. However, in the case of an inclined flow or a dipping flow, the vertical coordinate or the gradient should also be accounted for by calculating the absolute permeability. This type of flow system is shown in Figure 4.3, for which the darcy flow rate is given by:

$$Q = \frac{kA}{\mu}\left[\frac{(P_1 - P_2)}{L} + \rho g \sin \alpha\right] \tag{4.8}$$

where Q is the flow rate in m^3/sec, k the absolute permeability in m^2 (can be converted into mD or D; see Section 4.3 for conversion factors), A the cross-sectional area in m^2, μ the fluid viscosity in N sec/m^2, $P_1 - P_2$ the flowing pressure drop in N/m^2, L the length in m, ρ the fluid density in kg/m^3, g the acceleration due to gravity (9.81 m/sec^2, and α the angle of inclination or dip. (The contribution from the vertical coordinate or gradient can also be expressed in terms of the sine of the angle of inclination).

The flow of reservoir fluids from a cylindrical drainage zone into a well bore is characterized by the *radial flow system*, which is shown in Figure 4.4. The Darcy equation for the radial flow system can be written in the flowing differential form:

$$Q = \frac{k}{\mu} A \frac{\mathrm{d}P}{\mathrm{d}r} \tag{4.9}$$

Since we are dealing with a radial flow system, the term dL in Equation 4.4 is now replaced with dr. Similarly, the area A open to flow is $2\pi rh$. Using these quantities the Darcy equation can be integrated between the well bore and the external boundary of the system as follows:

$$Q\int_{r_\mathrm{w}}^{r_\mathrm{e}} \frac{\mathrm{d}r}{r} = \frac{k2\pi h}{\mu}\int_{P_\mathrm{wf}}^{P_\mathrm{e}} \mathrm{d}P \tag{4.10}$$

$$Q(\ln r_\mathrm{e} - \ln r_\mathrm{w}) = \frac{k2\pi h}{\mu}(P_\mathrm{e} - P_\mathrm{wf}) \tag{4.11}$$

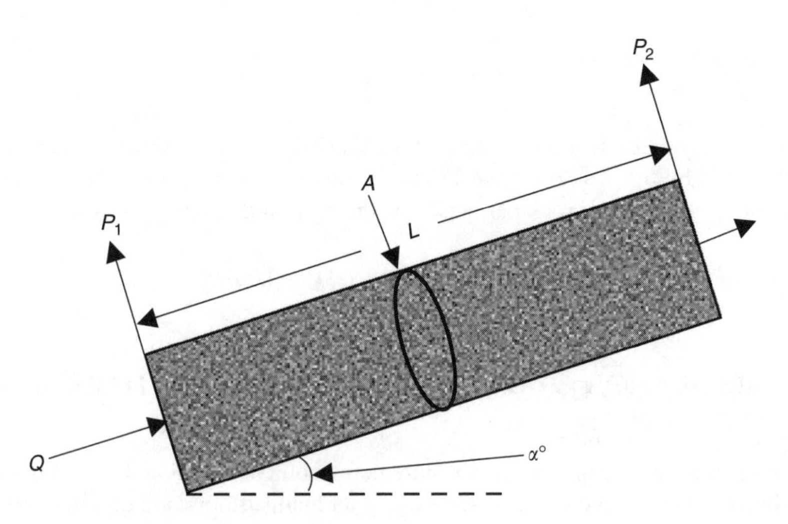

FIGURE 4.3 Inclined or dipping flow system.

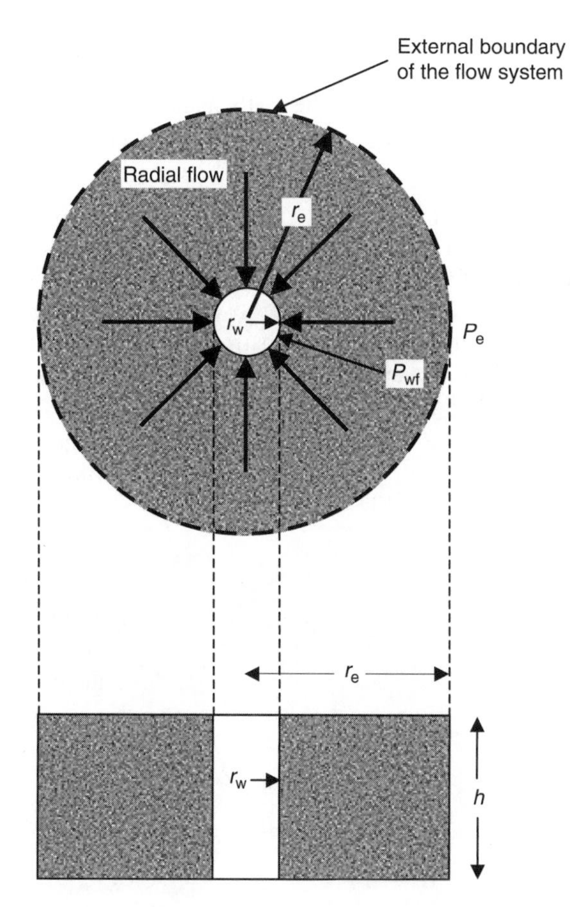

FIGURE 4.4 Radial flow system.

Solving for the flow rate,

$$Q = \frac{2\pi kh(P_e - P_{wf})}{\mu(\ln r_e - \ln r_w)} = \frac{2\pi kh(P_e - P_{wf})}{\mu \ln(r_e/r_w)} \tag{4.12}$$

where Q is the flow rate in m^3/sec, k the absolute permeability in m^2 (can be converted into mD or darcy; see Section 4.3 for conversion factors), h the thickness in m, P_e the pressure at drainage radius in N/m^2, P_{wf} the flowing pressure in N/m^2, μ the fluid viscosity in N sec/m^2, r_e the drainage radius in m, and r_w the well-bore radius in m.

4.5 AVERAGING OF PERMEABILITIES

The absolute permeability expression such as the one in Equation 4.7 is derived based on a fairly uniform or continuous value of permeability between the inflow and outflow faces. However, such uniformity and consistency is rarely seen in reservoir rocks. Most reservoir rocks have space variations of permeability. For example, reservoir rocks may contain distinct layers, blocks, or concentric rings of fixed permeability. In such cases the permeability values are averaged according to a particular type of flow: parallel or series. The mathematical expressions for averaging permeability for these cases are developed in the following text.

4.5.1 Parallel Flow

As shown in Figure 4.5, consider the case of fluid flow taking place in parallel through different layers of vertically stacked porous media. These individual layers of porous media, that have varying permeability and thickness, are separated from one another by infinitely thin impermeable barriers that preclude the possibility of cross flow or vertical flow. The average permeability for such a combination can be easily developed by applying Darcy's law to the individual layers. For layer 1,

$$Q_1 = \frac{k_1 W h_1 \Delta P}{\mu L} \tag{4.13}$$

For layer 2,

$$Q_2 = \frac{k_2 W h_2 \Delta P}{\mu L} \tag{4.14}$$

For layer 3,

$$Q_3 = \frac{k_3 W h_3 \Delta P}{\mu L} \tag{4.15}$$

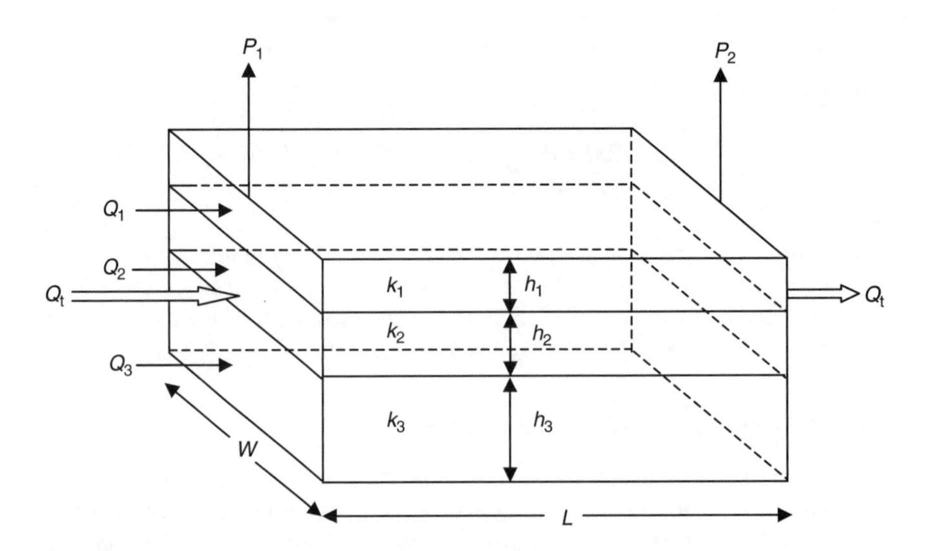

FIGURE 4.5 Fluid flow through a parallel combination.

However, since flow is taking place in parallel, the total volumetric flow rate can be equated to the summation of the individual flow rates through the three layers:

$$Q_t = Q_1 + Q_2 + Q_3 \tag{4.16}$$

Similarly, the total height is given by

$$h_t = h_1 + h_2 + h_3 \tag{4.17}$$

Based on Equations 4.16 and 4.17, Darcy's law can be written for the total flow rate for the entire systems using k_{avg} as the average absolute permeability:

$$Q_t = \frac{k_{avg} W h_t \Delta P}{\mu L} \tag{4.18}$$

$$\frac{k_{avg} W h_t \Delta P}{\mu L} = \frac{k_1 W h_1 \Delta P}{\mu L} + \frac{k_2 W h_2 \Delta P}{\mu L} + \frac{k_3 W h_3 \Delta P}{\mu L} \tag{4.19}$$

or

$$k_{avg} h_t = k_1 h_1 + k_2 h_2 + k_3 h_3 \tag{4.20}$$

that subsequently leads to the final generalized expression for calculating the average absolute permeability for a parallel system of n layers,

$$k_{avg} = \frac{\sum_{i=1}^{n} k_i h_i}{\sum_{i=1}^{n} h_i} \tag{4.21}$$

Equation 4.21 and the schematic used for deriving this equation demonstrates the practical significance from a petroleum reservoir point-of-view, that is, in horizontal flow situations, fluids travel through the reservoir strata to production wells and remain in the zone in which they originated. Or in other words, the case of fluid flow in parallel is relevant to conventional wells (vertical) where fluid flow takes place parallel to the bedding planes (horizontal).

4.5.2 Series Flow

The other type of flow encountered primarily in horizontal wells is vertical flow in which fluids must pass in series from one zone to the next. Figure 4.6 illustrates a series flow taking place through a stack of porous media of varying absolute permeabilities and lengths. The mathematical expression for calculating the average absolute permeability for a flow system shown in Figure 4.6 is developed in the

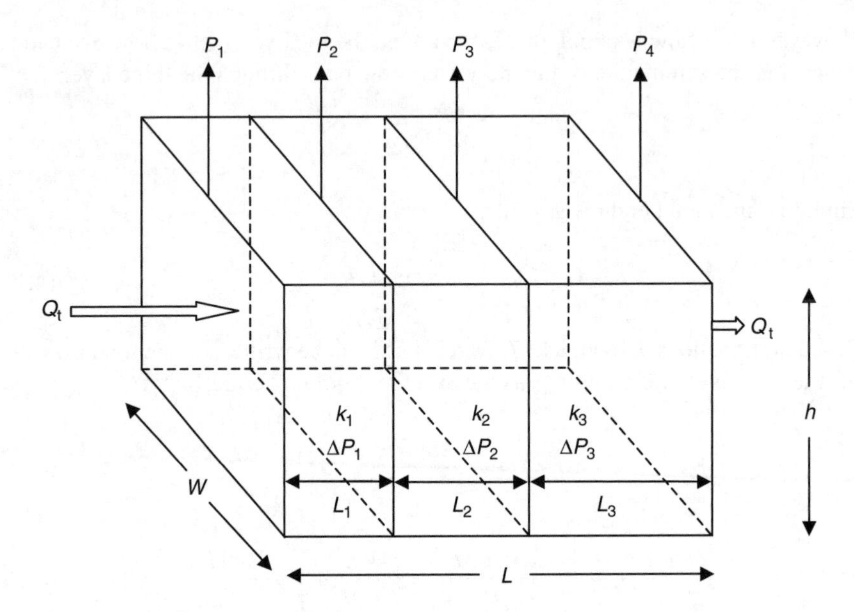

FIGURE 4.6 Fluid flow through a series combination.

following text. Again writing Darcy's law for each of the layers or blocks of porous medium stacked in series:

for layer 1,

$$Q_1 = \frac{k_1 Wh\Delta P_1}{\mu L_1} \tag{4.22}$$

for layer 2,

$$Q_2 = \frac{k_2 Wh\Delta P_2}{\mu L_2} \tag{4.23}$$

for layer 3,

$$Q_3 = \frac{k_3 Wh\Delta P_3}{\mu L_3} \tag{4.24}$$

It should be noted that for series flow, each of these layers or blocks has a different differential pressure and the summation of these is equal to the total or overall differential pressure of the entire flow system. Additionally, the total flow rate is also equal to the individual flow rates:

$$\Delta P_1 = P_1 - P_2,\ \Delta P_2 = P_2 - P_3,\ \Delta P_3 = P_3 - P_4 \tag{4.25}$$

or

$$\Delta P = P_1 - P_4 = \Delta P_1 + \Delta P_2 + \Delta P_3 \tag{4.26}$$

and

$$Q_t = Q_1 = Q_2 = Q_3 \tag{4.27}$$

Now, Darcy's law can be written for the total flow rate as,

$$Q_t = \frac{k_{avg} Wh \Delta P}{\mu L} \tag{4.28}$$

Subsequently, the differential pressures can be separated from Equations 4.22 to 4.24, and 4.28 and substituted into Equation 4.26,

$$\frac{Q_t \mu L}{k_{avg} Wh} = \frac{Q_1 \mu L_1}{k_1 Wh} + \frac{Q_2 \mu L_2}{k_2 Wh} + \frac{Q_3 \mu L_3}{k_3 Wh} \tag{4.29}$$

or

$$\frac{L}{k_{avg}} = \frac{L_1}{k_1} + \frac{L_2}{k_2} + \frac{L_3}{k_3} \tag{4.30}$$

$$k_{avg} = \frac{\sum_{i=1}^{n} L_i}{\sum_{i=1}^{n} \frac{L_i}{k_i}} \tag{4.31}$$

Equation 4.31 is used for calculating the average absolute permeability for serial flow system.

A similar mathematical treatment used here can also be used to derive equations for a radial flow system for parallel flow and serial flow. It should be noted that the radial flow system for a parallel flow resembles the horizontal flow of reservoir fluids from zones of varying permeability into the well bore of a conventional well (vertical). The radial flow system for a serial flow represents the horizontal flow of reservoir fluids in to the well bore of a vertical well when concentric rings of fixed permeability are present in the formation. However, the radial flow system for a serial flow is more analogous to a horizontal well because reservoir fluids must pass in series from one permeability zone to the next, and eventually into the well bore.

4.6 PERMEABILITY OF FRACTURES AND CHANNELS

The various aspects of absolute permeability looked at so far were confined to the matrix permeability. However, petroleum reservoir rocks such as sandstones and carbonates frequently contain solution channels and natural or artificial fractures. Therefore, the key to understanding the flow of fluids in petroleum reservoir rocks is to account for flow in all the three permeability elements: the matrix, channel, and fracture.

Matrix permeability refers to the flow in primary pore spaces in a reservoir rock. *Fracture* or *channel* permeability refers to the flow in cracks or breaks in the rock or in the so-called *secondary network*. However, these channels and fractures do not change the permeability of the matrix but do change the effective permeability of the overall flow network, mainly because the latter permeability is saturation dependent and pore space saturation distribution varies in matrix or fracture/channel network. The remainder of this section looks at the equations that can be used to describe the permeability of channels and fractures based on which their contribution to the overall or total conductivity of the system can de determined. The equations for channel permeability are based on Poiseuille's equation for fluid flow through capillary tubes

$$Q = \frac{\pi r^4 \Delta P}{8 \mu L} \tag{4.32}$$

or in terms of area, $A = \pi r^2$, Equation 4.32 becomes

$$Q = \frac{A r^2 \Delta P}{8 \mu L} \tag{4.33}$$

In fact, equating Equations 4.7 (Darcy equation) and 4.33, the permeability of a channel can be defined as

$$k_{\text{channel}} = \frac{r^2}{8} \tag{4.34}$$

So, if the channel (assumed to be a circular opening) radius r is in m, k_{channel} is in m^2.

Similar to the derivation of the equation for calculating the channel permeability, the fracture permeability equation is also developed by comparing flow equations for a simple geometry of slots of fine clearances[2] with that of a porous media.

For flow through slots of fine clearances (analogous to fractures),

$$\Delta P = \frac{12 \mu v L}{h^2} \tag{4.35}$$

and from the Darcy equation,

$$\Delta P = \frac{\mu v L}{k} \tag{4.36}$$

where v is the interstitial flow velocity.

Comparing Equations 4.35 and 4.36,

$$k_{\text{fracture}} = \frac{h^2}{12} \tag{4.37}$$

where h is the thickness of the slot or the fracture. The calculated permeability is m^2 if h is in m or cm^2 if h is in cm.

4.7 DARCY'S LAW IN FIELD UNITS

So far for the Darcy flow equations looked at (Equations 4.7, 4.8, and 4.12), consistent units were used for all the quantities that resulted in the volumetric flow rate, for example in m^3/sec. However, in the petroleum industry, reservoir fluid flow rates are almost always reported in barrels per day. Therefore, it is imperative to convert the volumetric flow rate of m^3/sec to barrels per day. Two methods can be used to keep a consistent set of units for all quantities and convert the flow rate or appropriately assign the pertinent field units to the individual quantities in the Darcy flow equation so that the final computed value of flow rate is obtained in barrels per day. Both methods are shown in the following text.

Conversion from m³/sec to barrels per day:

$$1\,\text{m}^3 = 6.2898 \text{ barrels}$$

$$1\,\text{sec} = 0.000011574 \text{ day}$$

$$1\,\text{m}^3/\text{sec} = 543438.72 \text{ barrels/day}$$

Note: use the last factor to convert the flow rate from m^3/sec to barrels/day.

Flow rate directly in barrels per day by using field units for other variables
We use the following conversion factors:

$$Q \text{ in barrels/day} = (1/543438.72)\ \text{m}^3/\text{sec}$$

$$= (100^3/543438.72)\ \text{cm}^3/\text{sec}$$

$$= 1.8401\ \text{cm}^3/\text{sec}$$

$$A \text{ in ft}^2 = 30.48^2\ \text{cm}^2$$

$$\Delta P \text{ in psi} = (1/14.696)\ \text{atm}$$

$$L \text{ in ft} = 30.48\ \text{cm}$$

However, do not convert the fluid viscosity μ because it is specified in cP both in the oilfield unit as well as in the original Darcy equation.

Using these conversion factors in the definition of a darcy:

$$Q \times 1.8401 = \frac{k(A \times 30.48^2)(\Delta P/14.696)}{\mu(L \times 30.48)} \tag{4.38}$$

or

$$Q = 1.1271\frac{kA\Delta P}{\mu L} \tag{4.39}$$

In Equation 4.39, if k is in darcies, A in ft^2, ΔP in psi, μ in cP, L in ft, and the calculated flow rate Q in barrels/day.

Using either of these two methods is obviously a matter of personal choice and convenience, however, always ensure that the consistent set of units are used in either of the methods so that a correct value of the flow rate (or permeability) is obtained.

4.8 LABORATORY MEASUREMENT OF ABSOLUTE PERMEABILITY

Laboratory measurement of absolute permeability usually involves the direct application of the Darcy equation, discussed in Section 4.2, based on the measurement of individual variables such as flow rate, pressure drop, sample dimensions, and fluid properties. Most laboratory measurements are carried out on formation samples of well-defined geometry, such as cylindrical core plugs discussed in Chapter 2. These core plugs are generally 1 or 1.5 in. in diameter with lengths varying from 2 to 4 in.

Prior to using core plug samples for permeability measurements, the residual fluids or *in situ* formation fluids are removed so that the sample is 100% saturated by air. Considering the fact that absolute permeability can only be measured by conducting a flow experiment in a porous media, gases or nonreactive liquids are commonly used as a fluid phase. Several commercial bench-top permeameters or minipermeameters that use gases or nonreactive liquids for permeability measurement, are available. However, the following sections discuss a step-by-step, practical procedure for the measurement of absolute permeability using nonreactive liquids and gases.

The apparatus used for conducting flow experiments on core plug samples is a core flooding rig or a displacement apparatus. The schematic of a typical displacement apparatus capable of using both liquids and gases for permeability measurement is shown in Figure 4.7.

4.8.1 MEASUREMENT OF ABSOLUTE PERMEABILITY USING LIQUIDS

The most common liquids used for the measurement of absolute permeability are formation waters (sometimes called *brine*) or degassed crude oil. Formation water or crude oil used is generally from the same formation for which the absolute permeability measurement is desired. In some cases, synthetic oil such as Isopar-L® (ExxonMobil Chemical Company, Houston, Texas) is also used. The typical steps involved in performing an absolute permeability measurement are as follows:

1. The dimensions (length and diameter) of the core plug are recorded.
2. The core plug sample is normally housed or snugly fit in a Viton® (DuPont Dow Elastomers L.L.C. Wilmington, Delaware) sleeve which in-turn is mounted in a Hassler core holder (see Figure 4.7).
3. An appropriate net overburden or confining pressure is applied radially to the core held in the Viton sleeve, via a hydraulic hand pump. This confining pressure is determined from the gross overburden (depth from where the core sample originated) and the reservoir pressure. The confining pressure also helps prevent the flow of liquid through the minute annular space between the core plug and the sleeve during the flow experiment.

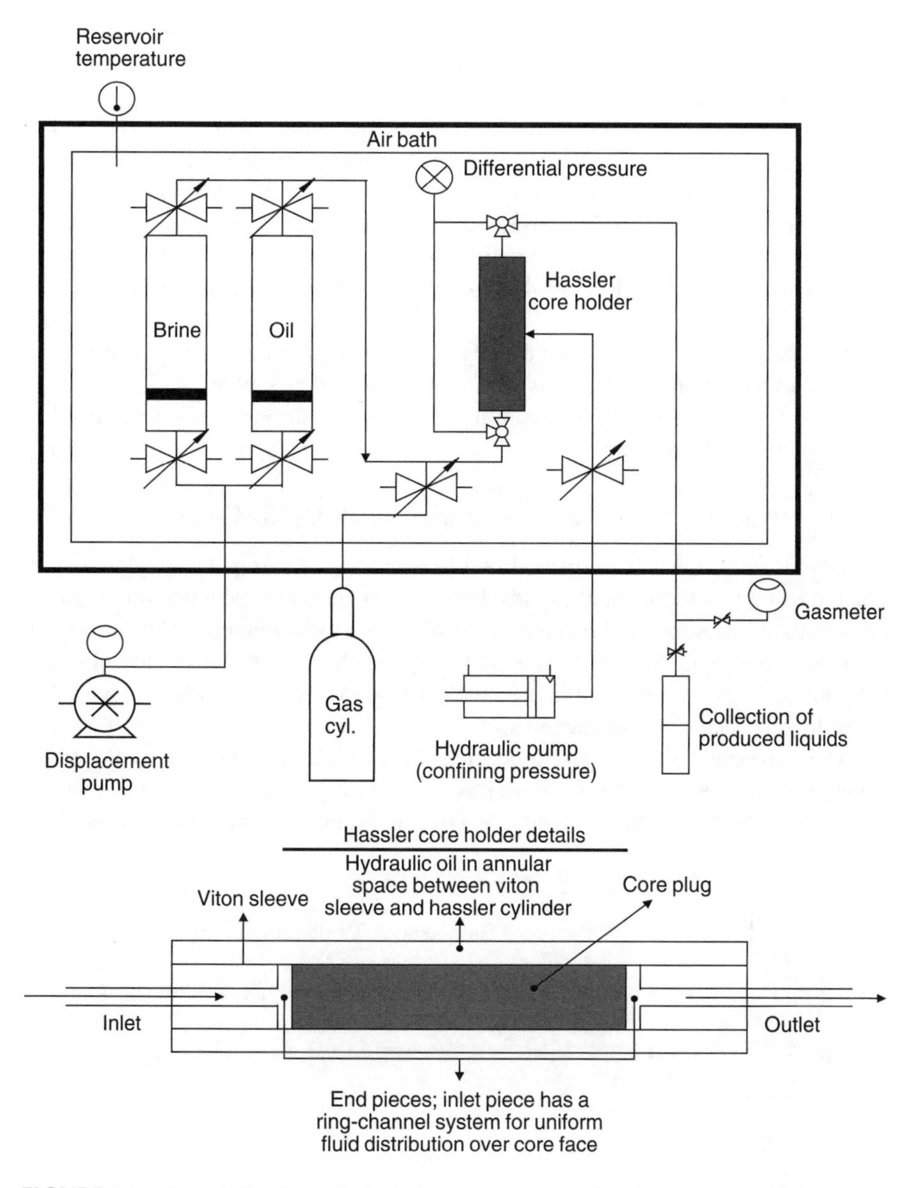

FIGURE 4.7 Schematic of a typical displacment apparatus for absolute permeability measurement using gases and liquids.

4. A constant reservoir temperature is maintained using the climatic air bath.
5. A displacement pump and floating piston sample cylinder (for storage of fluids) combination is used to initiate the flow of brine or degassed crude oil at either a constant rate or constant differential pressure. Usually, these

floods are carried out at constant flow rate rather than constant differential pressure where the inlet pressure is monitored and the outlet is normally at atmospheric pressure.

6. The pressure drop across the core plug is monitored using a computerized data logging system and a constant or steady pressure drop across the sample is recorded for calculations. As an example, the inlet pressure development for brine flood in a North Sea chalk sample[3] is shown in Figure 4.8.
7. The flow experiment is sometimes repeated by varying the liquid flow rates in order to determine the rate dependency, if any, on the absolute permeability.
8. The viscosity of the brine or the oil is measured at the flooding pressure and temperature conditions if unknown from other sources.
9. Finally, the absolute permeability of the core plug sample is determined using the Darcy equation.

4.8.2 Measurement of Absolute Permeability Using Gases

Quite frequently, absolute permeability measurements of core plug samples are carried out using gases instead of liquids. Dry gases such as nitrogen, helium, or air are commonly used as the fluid medium in permeability measurements. Choosing a gas is simply convenient and practical because a gas is clean, nonreactive, does not alter the pore network; in other words absolute permeability measurements are not influenced by any rock–fluid interactions.

The experimental setup and procedure for absolute permeability measurement using gases is similar to the one presented in the previous section. For example, the operation of the displacement apparatus shown in Figure 4.7 can be slightly modified

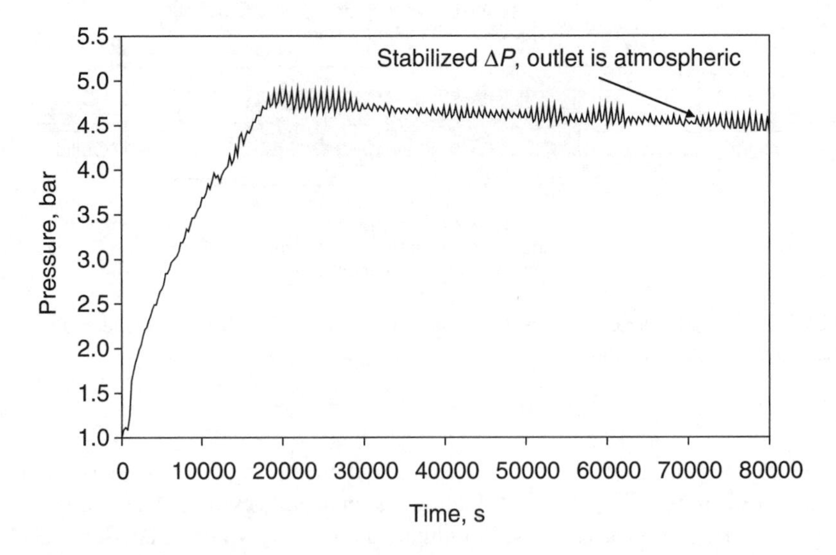

FIGURE 4.8 Differential pressure vs. time recorded during brine flood for absolute permeability measurement of a North Sea chalk sample.

by switching the fluid source from liquid to a gas by opening and closing the appropriate valves. Another minor variation is the use of constant differential pressure for performing the flow tests. A constant differential pressure can be easily maintained by setting a certain inlet pressure on the gauge of the gas cylinder and by keeping the downstream or outlet pressure atmospheric (sometimes under a back pressure), while monitoring gas the flow rate via a gas meter (at atmospheric conditions), as shown in Figure 4.7.

Even though the experimental procedures for permeability measurement using liquids and gases are basically similar, one major difference exists between liquids and gases in the approach used for the determination of absolute permeability: the compressible nature of gases. So far, issues related to the compressibility of the fluids have not been addressed because the original Darcy equation was developed under the assumption of an incompressible fluid flow.

When an incompressible fluid flow takes place through a porous medium of uniform cross section, the flux (Q/A) is constant at all sections along the flow path. However, when gases are used, the pressure drop along the flow path is accompanied by gas expansion that also results in an increase in the flux. Therefore, gas flux is not constant along the flow path. This of course necessitates the modification of Darcy equation for calculation of permeability.

First, the product of inlet and outlet flow rates (Q_1 and Q_2) and pressures (P_1 and P_2) are equated by using Boyle's law,

$$Q_1P_1 = Q_2P_2 \quad \text{(temperature is constant)} \tag{4.40}$$

This product can also be equated to the product of average gas flow rate (Q_{avg}) and average pressure (P_{avg}),

$$Q_1P_1 = Q_2P_2 = Q_{avg}P_{avg} \tag{4.41}$$

The Darcy equation can then be expressed in terms of the average gas flow rate to account for gas expansion in the sample,

$$Q_{avg} = \frac{kA(P_1 - P_2)}{\mu L} \tag{4.42}$$

However, the flow rate of gas is normally measured at the outlet of the core plug, Q_2. Therefore, Equations 4.41 and 4.42 can be rearranged as

$$\frac{Q_2P_2}{\left(\frac{P_1 + P_2}{2}\right)} = \frac{kA(P_1 - P_2)}{\mu L} \tag{4.43}$$

or

$$Q_2 = \frac{kA(P_1^2 - P_2^2)}{2\mu LP_2} \tag{4.44}$$

where Q_2 is the gas flow rate measured at the outlet of the sample in m^3/sec, k the absolute permeability in m^2 (can be converted to mD or darcy; see Section 4.3 for conversion factors), A the cross-sectional area in m^2, P_1 the inlet pressure in N/m^2, P_2 the outlet pressure in N/m^2, μ the gas viscosity in N sec/m^2, and L the length of the sample in m.

Equation 4.44 is used for the determination of absolute permeability of core plug samples using gases.

Another artifact associated with the use of gases for absolute permeability measurement is the higher permeability value obtained in comparison to the liquid flow for the same core sample. Kinkenberg[4] first reported this particular artifact in 1941 when he discovered that there were variations in the absolute permeability as determined using gases as the flowing fluid from those obtained when using nonreactive liquids.

Klinkenberg's observations were based on the measurement of absolute permeability for a certain core sample for which liquid (isooctane) permeability was reported as 2.55 mD; on the other hand the same core sample showed a trend of increasing permeability as a function of increasing reciprocal mean pressure $\{1/[(P_1+P_2)/2]\}$ when hydrogen, nitrogen, and carbon dioxide were used. These particular variations in permeability were ascribed to a phenomenon called *gas slippage* that occurs when the diameter of the capillary openings approach the mean free path of the gas.

The gas slippage phenomenon is sometimes also called the *Klinkenberg effect*. The Klinkenberg effect is a function of the gas with which permeability of a core sample is determined because the mean free path of the gas is a function of its molecular size and kinetic energy. This was clearly evident from Klinkenberg's experiments using three different gases of varying molecular sizes: hydrogen, nitrogen, and carbon dioxide. The Klinkenberg observations are schematically illustrated in Figure 4.9 and summarized in the following points:

- A straight line is obtained for all gases when gas permeabilities are plotted as a function of reciprocal mean pressures.
- The data obtained with the lowest molecular weight gas (hydrogen) result in a straight line with greater slope that indicates a higher slippage effect; whereas the highest molecular weight gas (carbon dioxide) data yield a straight line with the lowest slope indicative of a lesser slippage effect. The data for nitrogen (gas B in Figure 4.9) lie in between hydrogen and carbon dioxide.
- The straight lines for all gases, when extrapolated to an infinite mean pressure or zero reciprocal mean pressure, that is, $\{1/[(P_1+P_2)/2]\} = 0$, intersect the permeability axis at a common point. This common point is designated as a Klinkenberg corrected or equivalent liquid permeability because gases tend to behave like liquids at such high pressures.
- The previous point is also validated by the permeability value, directly measured by using a liquid (isooctane) found to be similar to the Klinkenberg corrected or equivalent liquid permeability.

Observations similar to that of Klinkenberg's experiments were also reported by Calhoun[5] for methane, ethane, and propane, where methane showed the greatest

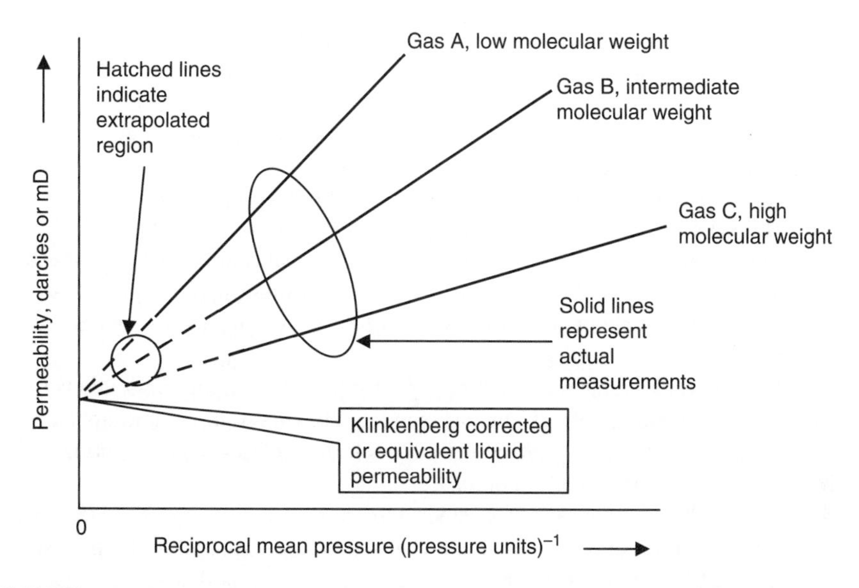

FIGURE 4.9 Schematic representation of Klinkenberg effect for absolute permeability measurement using gases.

slippage, the opposite was observed for propane. It is evident from this discussion that the procedure for measurement of absolute permeability using a gas involves several measurements at different mean pressures and extrapolation of the results to infinite mean pressure. The Klinkenberg effect can also be mathematically correlated by a straight-line fit of the relationship between the observed gas permeability data and the reciprocal mean pressure:

$$k_{\text{gas}} = k_{\text{liquid}} + m\left[\frac{1}{P_{\text{mean}}}\right] \tag{4.45}$$

where k_{gas} is the measured gas permeability, k_{liquid} the equivalent liquid permeability or the Klinkenberg corrected liquid permeability, m the slope of the straight-line fit, and

$$P_{\text{mean}} = \frac{P_1 + P_2}{2} = \text{mean pressure}$$

It should be noted that in Equation 4.45 the slope of the straight-line fit is a constant or a specific value valid for a given gas in a given medium, that is the straight-line fit cannot be generalized. However, the straight-line fit can be used for determining the equivalent liquid permeability when gas permeabilities are measured at other pressure conditions.

4.9 FACTORS AFFECTING ABSOLUTE PERMEABILITY

A number of factors affect the absolute permeability of a reservoir rock. The discussion of these factors clearly categorizes and subsequently reviews each factor which can be grouped as rock-related factors, fluid phase-related factors, thermodynamic factors, and mechanical factors.

- Rock-related factors are basic characteristics, structure, or indigenous properties of reservoir rocks, such as grain size and shape and clay cementing. These can in fact also be termed *natural factors*.
- The type of fluid medium (i.e., gas/brine/water) used for permeability measurement as well as the physical and chemical characteristics of these fluids are also major factors that affect the absolute permeability. These factors can be characterized as *artificial* or *laboratory* factors that affect permeability.
- The thermodynamic factors affecting absolute permeability basically consist of temperature effects, and as seen later, based on some literature data, these fall under the category of fluid–rock interaction-induced laboratory artifacts that affect permeability.
- The mechanical factors are related to the effect of mechanical stresses or confining pressures on absolute permeability, and also fall under the category of laboratory artifacts.

4.9.1 Rock-Related Factors

Before discussing the effect of rock-related factors on absolute permeability, revert to Figure 2.1 to consider horizontal and vertical permeabilities. As seen in this figure, core plug samples are normally drilled from the whole core in the horizontal direction, also parallel to the bedding planes. Therefore, permeability measured on such plug samples is called *horizontal permeability* or k_h. If plugs are drilled along the long axis of the whole core; these samples are perpendicular to the bedding planes and hence yield what is called *vertical permeability* or k_v.

Horizontal permeability, is significant from a conventional well (vertical) production point-of-view because fluids flow parallel to the bedding planes in a horizontal direction toward the well bore creating a natural pressure drop as fluids are produced. Vertical permeability is important when dealing with horizontal wells because fluids flow perpendicular to the bedding planes, or in series, toward the well bore, creating a natural pressure drop as reservoir fluids are produced.

Horizontal and vertical permeabilities are greatly impacted by the grain size and shape. In order to understand the effect of grain shape on horizontal permeability, a hypothetical porous medium that consists of uniformly arranged, identically shaped large grains, as shown in Figure 4.10, must be studied. This figure clearly shows that $k_h \approx k_v$, that is if the rock is primarily composed of large and uniformly rounded grains, its permeability is of the same order in both directions because flow paths are quite similar. However, if the grains in Figure 4.10 are now altered to uniformly arranged flat grains, as shown in Figure 4.11, then obviously the horizontal permeability is greater than the vertical permeability because the former is characterized by a relatively unrestricted flow path, whereas relatively restricted or tortuous path characterize the latter. Most reservoir rocks generally have much lower permeabilities in the vertical direction compared to the horizontal permeabilities because grains are small and irregularly shaped.

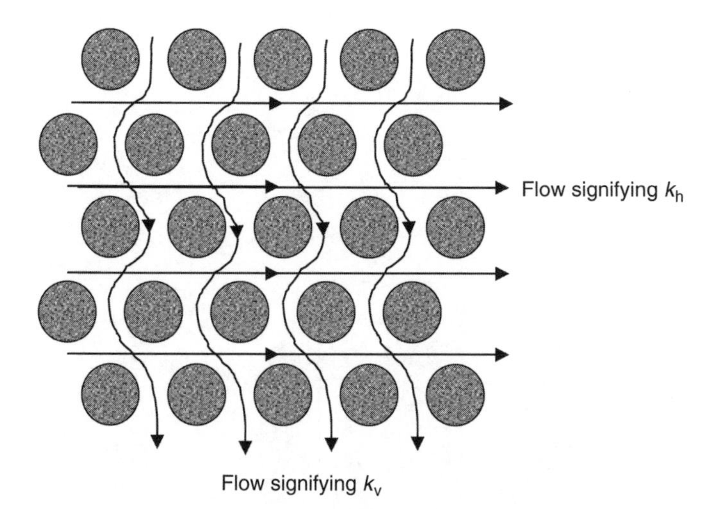

FIGURE 4.10 Schematic representation of a hypothetical porous medium consisting of uniformly arranged, indentically shaped large grains ($k_h \approx k_v$).

Clay cementing affects both the reservoir rock porosity as well as permeability because clay cementing basically coats or increases the grain size. This increase in the grain size obviously reduces the pore space and also alters the flow paths by constriction.

4.9.2 Fluid Phase-Related Factors

These are factors that consist of the physical or chemical characteristics of the fluid that affect the absolute permeability. One such factor already addressed is related to the use of gases in permeability measurement – the Klinkenberg effect. Other fluid-related factors are connected with the use of brine and degassed crude oil for permeability measurement.

Although water is generally considered as nonreactive in an ordinary sense, it can have significant impact on permeability, especially for those reservoir rocks that contain clays that swell after coming in contact with water. Clay swelling of course depends on the type of clay minerals present in the reservoir rock. The ion exchange between water and clay minerals is principally responsible for clay swell and enlargement. It is well known that kaolinite and illite are nonswelling clays; montmorillonite is a common swelling clay, according to Zhou et al.[6] Permeability reduces as a result of clay swelling. This particular reduction in the absolute permeability is also sometimes termed *formation damage*. Although it is not really a mechanical damage, it is rather a pore network alteration. Although when reactive liquids such as brine swell a clay and alter the internal geometry of the porous medium, they do not vitiate Darcy's law but create a a new porous medium with permeability characterized by the new internal geometry.

Another scenario of permeability reduction is related to mixing incompatible waters in pore spaces of a reservoir rock. For example, incompatibility of formation

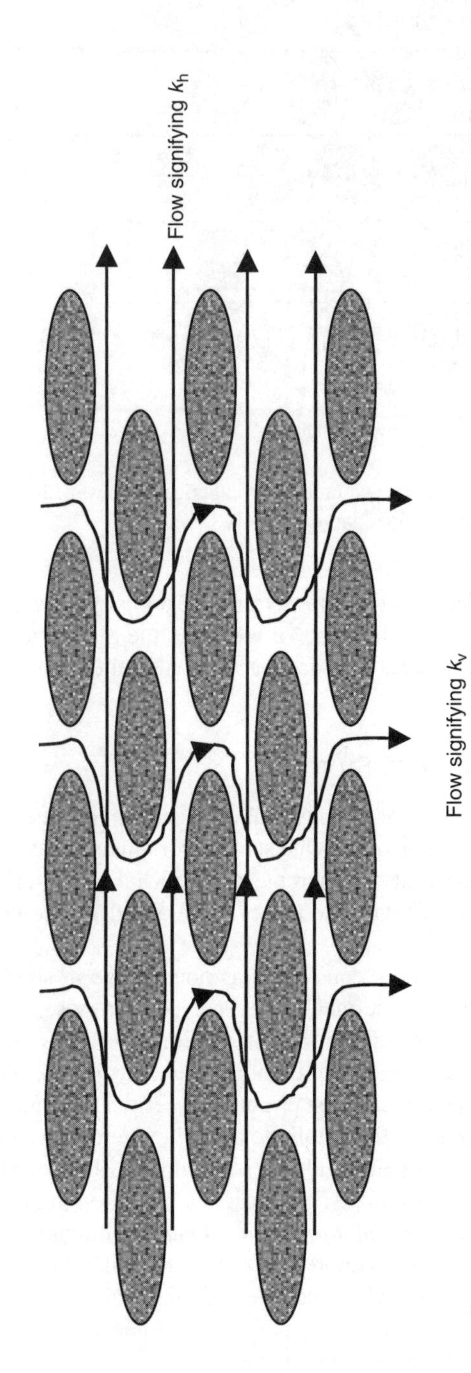

FIGURE 4.11 Schematic representation of a hypothetical porous medium consisting of uniformly arranged, identically shaped, large, flat grains ($k_h > k_v$).

waters and waters of different salinities, such as seawater, may result in salt precipitation as solubility limits of some of the salt components are exceeded when certain pressures and temperatures are reached. If salt deposition takes place in the pore spaces it may again alter the internal geometry of the porous medium usually resulting in a reduction in permeability. This type of permeability reduction is also termed as formation damage, in the usual petroleum engineering terminology.

This formation damage is greatly significant when water injection (e.g., seawater) is considered as a potential enhanced oil recovery method. Review of existing literature on formation damage indicates a value of $k_{final}/k_{initial}$ ($k_{seawater}/k_{brine}$) as 0.1 in Wojtanowicz et al.,[7] 0.001 in Sarkar and Sharma,[8] and as low as in the range of 0.01 to 4.0 E^{-05} in Gruesbeck and Collins.[9] In the core flooding experiments performed in Bertero et al.[10] on three different sandstone, cores with porosities in the range of 12.6 to 15.0% and air permeabilities in the range of 4 to 262 mD, permeability reduced by 50% when the scale precipitation volume was more than 1% of the pore volume.

Care should be taken when using degassed crude oil to ensure that core flooding tests for permeability measurement are carried out at high temperatures (preferably reservoir temperatures) because paraffin (wax) deposition may also take place in the pore spaces if a very waxy crude oil is used as a fluid medium at room temperatures. However, if wax deposition does take place in the pore spaces, this might alter the internal geometry of the pore network temporarily because the deposited wax can always be removed by using high temperatures.

4.9.3 Thermodynamic Factors

This review of the effect of thermodynamics factors on absolute permeability is restricted to a discussion on literature data based on the investigation of temperature effects on absolute permeability. Although, in an ordinary sense, if the same liquid is used in the experiments but at varying temperatures, ideally temperature should not have any effect on the absolute permeability because varying temperature only affects liquid viscosity (increase in viscosity when temperature decreases and vice versa), which in turn affects the differential pressure. Note the ratio of $\Delta P/\mu$ is always nearly constant (see Equation 4.7). However, some researchers[11,12] have indicated that the absolute permeability to water for confined sandstones is strongly temperature dependent.

Grunberg and Nissan[11] reported that core temperatures varied from 6°C to 30°C, in which case absolute permeability decreased by a ratio of 0.8 mD/°C. Aruna[12] reported a reduction in absolute permeability of up to 60% over a temperature range of 21.1°C to 149°C. However, it should be noted that the absolute permeability of sandstones to other fluid mediums (nitrogen, mineral oil, octanol) were reported to have almost no effect of temperature.[12] Aruna[12] concluded that water–silica interactions were responsible for the major effects observed with water.

4.9.4 Mechanical Factors

Mechanical factors effecting absolute permeability include the magnitude of overburden or confining pressure used when flow experiments are carried out. Generally, absolute permeability is inversely proportional to overburden pressure because core

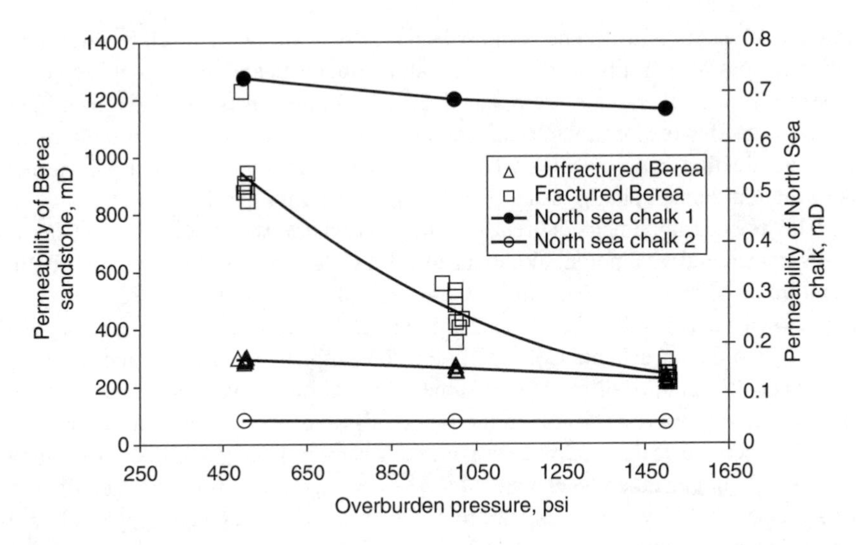

FIGURE 4.12 Effect of overburden pressure on absolute permeability. Berea sandstone data are from Putra et al.[14] and North Sea chalk data are from Dandekar.[3]

samples are compacted due to overburden and fluid flow through such samples is rather squeezed resulting in a reduction in absolute permeability.

One of the most notable outcomes in this area was first reported by Fatt and Davis[13] in 1952, which presented their results on sandstone samples from various formations in North America. Their results indicated a reduction in absolute permeability by as much as 60% for some formation samples, when comparing the values between 0 and 15,000 psi confining pressure, expressed as $k_{15,000\ \mathrm{psi}}/k_{0\ \mathrm{psi}} = 0.4$.

In addition to the work of Fatt and Davis,[13] a number of other researchers have shown similar results reviewed in Aruna.[12] Most of the results indicate that generally speaking, the higher the permeability, the higher the percentage of reduction. This is also clearly evident from results reported by Putra et al.[14] and Dandekar.[3] Putra et al. presented absolute permeability measurements for a fractured as well as unfractured Berea sandstone; Dandekar's[3] results are on absolute permeability of two North Sea chalk (carbonate) samples at overburden pressures of 500, 1000, and 1500 psi. These results are shown in Figure 4.12, which indicates a marginal reduction in the absolute permeability when overburden pressure is increased from 500 to 1500 psi for both the unfractured Berea Sandstone as well as the two chalk samples. However, a substantial reduction in the absolute permeability is evident from the results for the fractured Berea sandstone. Most routine and special core analysis tests are usually carried out by applying the representative overburden or confining pressure that is determined from the sample depth and reservoir pressures.

4.10 POROSITY AND PERMEABILITY RELATIONSHIPS

No direct relationship exists, in either practice or theory, between porosity and permeability simply because permeability depends on continuity of the pore space whereas porosity basically signifies the availability of a pore space. The only relationship between porosity and permeability is the two end points, that is, when porosity is zero, permeability is zero and when porosity is 100% (e.g., a pipe) permeability is infinite. Unfortunately, these two end points are insufficient to derive a generalized relationship between porosity and permeability.

It should; however, be noted that a reservoir rock must have a nonzero porosity in order to have a nonzero permeability. Or one can only qualitatively state that the higher the porosity, the greater the chance is for the *likelihood* of a higher permeability. On the other hand it is possible to have a very high porosity without having any permeability at all, as in the case of pumice stone or shales. But note that pumice stone has practically zero effective porosity or high isolated porosity. A similar trend is shown byMagara[15] that indicated a small increase in the shale permeability when porosity increased. Similarly, the reverse is also true, low porosity rock may have a high permeability, as in the case of a microfractured carbonate where most of the relatively unimpeded flow takes place through the microfracture.

In general, reservoir rocks do not demonstrate any theoretical relationship between porosity and permeability, therefore any practical relationship represents a best fit normally expressed in terms of a mathematical relationship, for example a straight-line fit or a second-order polynomial. When porosity and permeability plots are constructed, they usually yield a scatter or cloud merely indicating a trend of the cloud moving in the direction of increasing permeability when porosity increases. Sometimes, porosity permeability data are grouped according to the depositional environment and the rock type to enhance the clarity of the clouded data points into some meaningful trend, such as different "best fits" would be valid for particular rock types. An example of such plot(s) is shown in Altunbay et al.[16] for a Middle Eastern carbonate reservoir.

In 1927, Kozeny[17] actually developed one of the most fundamental and popular correlations expressing permeability as a function of porosity and specific surface area. Kozeny's correlation was based on the analogy between Darcy's law for flow in porous media and Poiseuille's equation for flow through n number of capillary tubes:

$$Q = \frac{kA\Delta P}{\mu L} = n\frac{\pi r^4 \Delta P}{8\mu L} \tag{4.46}$$

solving for k yields

$$k = \frac{n\pi r^4}{8A} \tag{4.47}$$

Now porosity of the capillary bundle can be defined as

$$\phi = \frac{\text{pore volume}}{\text{bulk volume}} = \frac{n\pi r^2 L}{AL} \tag{4.48}$$

where L = length of the capillary tube. From Equations 4.47 and 4.48,

$$k = \frac{\phi r^2}{8} \tag{4.49}$$

further defining the specific surface area as

$$S_p = \frac{\text{surface area}}{\text{pore volume}} = \frac{n2\pi rL}{n\pi r^2 L} \tag{4.50}$$

that gives

$$r = \frac{2}{S_p} \tag{4.51}$$

Combining Equations 4.49 and 4.51,

$$k = \left(\frac{1}{2}\right)\frac{\phi}{S_p^2} \tag{4.52}$$

if for the constant ½, $1/k_z$ is substituted, then

$$k = \frac{\phi}{k_z S_p^2} \tag{4.53}$$

Equation 4.53 is the Kozeny equation, where k_z is the Kozeny constant.

4.11 PERMEABILITIES OF DIFFERENT TYPES OF ROCKS

The absolute permeability of reservoir rocks can typically fall in a fairly wide range, however, 0.1 to 1000 mD or more is quite common. As outlined earlier, reservoir rock permeability depends on a number of inherent factors such as grain shape and size, grain arrangement, clay cementation, etc., that can substantially vary from formation to formation obviously imparting wide ranging absolute permeability values.

In general, the quality of a hydrocarbon-bearing formation is judged according to its permeability. Formations having permeabilities greater than 250 mD are considered very good; those having permeabilities less than 1 mD, typically found in chalk formations, are considered poor. However, classification of reservoir rocks on a scale of poor to very good is rather subjective and relative. Because for instance, reservoir rocks having permeabilities less than 1 mD, which are sometimes termed as *tight* formations, were once considered as too tight for economically attractive commercial production. However, today, petroleum reservoir fluids are being produced from many tight formations such as those in the Danish North Sea area. Table 4.1 shows porosity and absolute permeability data for some of the sandstone and carbonate formations in the world.

PROBLEMS

4.1 In an experiment similar to that of Darcy's, the flow rate of water was observed to be 5.0 cm^3/min. If the experiment were repeated with oil,

TABLE 4.1
Porosity and Permeability Data of Some Sandstone and Carbonate Formations in the World

Field/Formation	Type of Rock	Porosity (%)	Permeability (mD)
Prudhoe Bay, United States	Sandstone	22	265
Ghawar (Ain Dar), Saudi Arabia	Carbonate	19	617
Bombay-High, India	Carbonate	15–20	100–250
Ford Geraldine Unit, United States	Sandstone	23	64
Elk hills, United States	Sandstone	27–35	100–2000
Pullai Field, Malaysia	Sandstone	18–31	300–3000
Chincotepec, Mexico	Sandstone	5–25	0.1–900
Ekofisk, Norway	Carbonate	30–48	0.25[a]
Upper and Lower Cretaceous, Denmark	Carbonate	15–45	0.01–10
Daqing (Lamadian), China	Sandstone	24.6–26.4	200–1300
Hassi Messaoud, Algeria	Sandstone	7.4	2.5

[a] Ratio of vertical to horizontal permeability (k_v/k_h)

what would be the flow rate for oil? The difference between the upstream and downstream hydraulic gradients Δh are the same for both the experiments (measured with water for water experiment and with oil for oil experiment).

Additional data: oil viscosity = 2.5 cP, water viscosity = 0.8 cP, oil density = 0.85 g/cm^3 and water density = 1.0 g/cm^3

4.2 Brine flood in a 1.9-in-long and 1.5-in-diameter core plug from the North Sea resulted in a stabilized pressure drop of 46.05 psi. The flood was carried out at 0.05 mL/min with brine viscosity of 0.443 cP. Determine the absolute permeability of this plug in millidarcies.

4.3 Three beds of equal cross section have permeabilities of 100, 200, and 300 mD and lengths of 50, 15, and 85 ft, respectively. What is the average permeability of the beds placed in series?

4.4 Three beds of 50, 110, and 795 mD, and 5, 7, 15 ft thick, respectively, are conducting fluid in parallel flow. If all are of equal length and width, what is the average permeability?

4.5 Develop equations for radial flow in parallel and serial flow systems.

4.6 Following data were obtained during a nitrogen flood in a 1.5 cm diameter and 3.0 cm long core plug sample. Determine the Klinkenberg corrected absolute permeability of the core. Nitrogen viscosity μ_g = 0.02 cP, downstream pressure (P_2) is maintained atmospheric.

Run Number	q_g (cm^3/s)	Upstream Pressure, P_1 (atm)
1	5.11	1.95
2	18.15	2.45
3	35.61	3.11
4	62.31	3.55

REFERENCES

1. Darcy, H., *Les Fontaines Publiques de la Ville de Dyon*, Victor Dalmont, 1856.
2. Croft, H.O., *Thermodynamics, Fluid Flow and Heat Transmission*, McGraw-Hill, New York, 1938, p. 129.
3. Dandekar, A.Y., Unpublished data, 1999.
4. Klinkenberg, L.J., The permeability of porous media to liquids and gases, *Drilling and Production Practices*, American Petroleum Institute, 200, 1941.
5. Calhoun, J.R., *Fundamentals of Reservoir Engineering*, University of Oklahoma Press, Norman, OK (USA), 1976.
6. Zhou, Z., Cameron, A., Kadatz, B., and Gunter, W., Clay swelling diagrams: their applications in formation damage control, *Soc. Pet. Eng. J.*, 2, 99, 1997.
7. Wojtanowicz, A.K., Krilov, Z., and Langlinais, J.P., Study on the effect of pore blocking mechanisms on formation damage, Society of Petroleum Engineers (SPE) paper number 16233.
8. Sarkar, A.K. and Sharma, M.M., Fines migration in two-phase flow, Society of Petroleum Engineers (SPE) paper number 17437.
9. Gruesbeck, C. and Collins, R.E., Entrainment and deposition of fine particles in porous media, *Soc. Pet. Eng. J.*, 847, 1982.
10. Bertero, L., Chierici, G.L., and Mormino, G., Chemical equilibrium models: their use in simulating the injection of incompatible waters, Society of Petroleum Engineers (SPE) paper number 14126.
11. Grunberg, L. and Nissan, A.H., The permeability of porous solids to gases and liquids, *J. Inst. Pet.*, 29, 193, 1953.
12. Aruna, M., The Effects of Temperature and Pressure on Absolute Permeability of Sandstones, Ph.D. thesis, Stanford University, Stanford, California, 1976.
13. Fatt, I. and Davis, D.H., Reduction in permeability with overburden pressure, *Trans. AIME*, 329, 1952.
14. Putra, E., Muralidharan, V., and Schechter, D.S., Overburden pressure affects fracture aperture and fracture permeability in a fractured reservoir, *Saudi Aramco J. Technol.*, 57, 2003.
15. Magara, K., Porosity-permeability relationship of shale, Society of Petroleum Engineers (SPE) paper number 2430.
16. Altunbay, M., Georgi, D. and Takezaki, H.M., Permeability prediction for carbonates: still a challenge?, Society of Petroleum Engineers (SPE) paper number 37753.
17. Kozeny, J., Uber kapillare leitung, des wassers im boden (Aufstieg versikeung und anwendung auf die bemasserung), *Sitzungsber Akad., Wiss, Wein, Math-Naturwiss, KL*, 136 (IIa), 271, 1927.

5 Mechanical and Electrical Properties of Reservoir Rocks

5.1 INTRODUCTION

Petroleum reservoirs are dynamic systems that are constantly changing during the depletion of reservoir fluids. The production of reservoir fluids from petroleum reservoirs causes the reduction in reservoir pore pressure resulting in an increase in net effective stress. Similarly, during water flooding or gas injection the equilibrium rock stresses can also be altered in a dynamic manner. Injection of external fluids results in an increase in pore pressure and a decrease in net effective stress. The knowledge of changes of net effective stress is an important element of reservoir management because the alteration of the net effective stress during production can have significant impact especially on stress-sensitive reservoirs. These types of rock characteristics are normally evaluated by measuring various mechanical properties of reservoir rocks.

Petrophysical properties such as porosity, permeability, capillary pressure, resistivity, and relative permeability are influenced by the state of stress acting on a rock. These properties should be measured at a stress state that resembles the *in situ* stress. Therefore, in general, the practice is to conduct core analyses for petrophysical properties under elevated confining pressures that are usually generated by hydrostatic, or Hassler-type, cells.

For electrical properties of reservoir rocks, the most significant property is electrical resistivity that is generally dependent on the geometry of the pore space and the fluids that occupy the pore space. The reservoir rock pore space is normally occupied by gas, oil, and water, out of which gas and oil are nonconductors and water is the only conductive fluid if it contains dissolved salts. These factors indicate that electrical resistivity generally decreases with increasing porosity and increases with increasing hydrocarbon saturation. Therefore, resistivity of reservoir rocks saturated with interstitial fluids such as gas, oil, and water serves as an indicator of fluid saturations making it a valuable tool for evaluating the producibility of a formation.

Clearly it is very important to know the mechanical properties and electrical properties of reservoir rocks. The primary purpose of this chapter is to introduce the various mechanical and electrical properties of reservoir rocks. It should, however, be noted that this chapter provides a very basic discussion of the mechanical and electrical properties of reservoir rocks. For a more comprehensive discussion of these properties refer to other specialized texts on this subject.

5.2 MECHANICAL PROPERTIES

The determination of mechanical properties of reservoir rocks falls under a specialized area called *rock mechanics*, which includes the study of the strength properties of rocks. However, in order to understand the strength properties of rocks, it is first necessary to review fundamental concepts such as stress and strain because the mechanical properties of rocks are evaluated on the basis of stress–strain relationships.

5.2.1 STRESS

Stress, commonly denoted by σ, refers to the force applied to a rock that tends to change its dimensions. The external force applied to a rock is normally referred to as *load.* Mathematically, stress is the concentration of force per unit area, defined as

$$\sigma = \frac{F}{A} \tag{5.1}$$

where σ is the stress, usually expressed in Pa(1 Pa = 1 N/m^2), F the force in N, and A the area in m^2.

Reservoir rock stresses are usually in the range of megapascals (MPa = 10^6 Pa). The three basic recognized stress conditions are

1. Tensile
2. Compressive
3. Shear

as illustrated in Figure 5.1. Tensile stress is a type of stress in which the two sections of material on either side of a stress plane tend to pull apart or elongate. Compressive stress is exactly the opposite of tensile stress; adjacent parts of the material tend to press against each other or external forces are directed toward each other along the same plane. Shear stress occurs when the external forces are parallel and directed in opposite directions but in different planes.

Assessing mechanical properties of reservoir rocks is made by addressing the three basic stress types. The ability of a rock material to react to compressive stress or pressure is called *rock compressibility* (discussed later).

5.2.2 STRAIN

The effect of stress applied to rocks is studied by measuring the strain produced by the application of the stress. Strain, commonly denoted by ε, is the relative change in shape or size of a rock due to externally applied forces (i.e., stress). In other words, strain is a measure of the deformation of a material when a load is applied. Figure 5.2 shows the effect of stress on the length and diameter of a cylindrical core

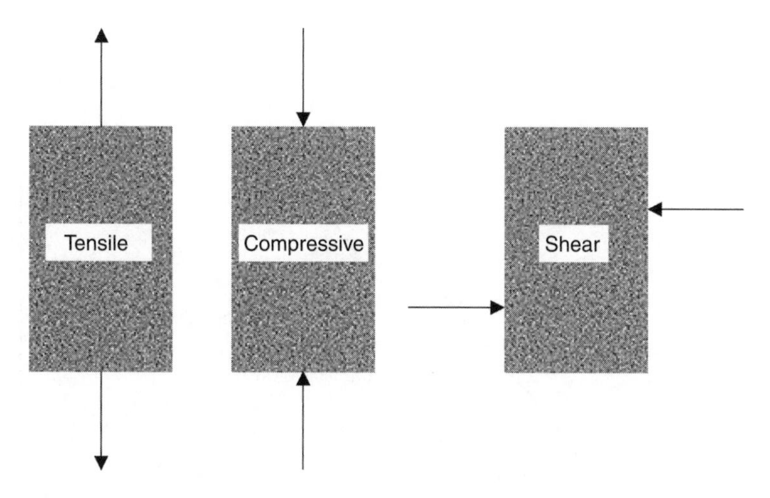

FIGURE 5.1 Schematic representation of the three stresses.

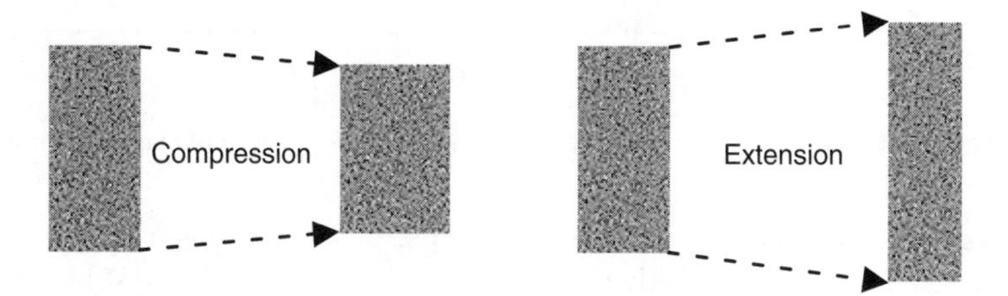

FIGURE 5.2 Deformation of a core sample on application of stress.

sample. As seen in Figure 5.2, compression makes the core sample shorter and wider; extension makes the core sample narrower and longer. Strain is calculated as a ratio of change in length to original length or as a ratio of change in diameter to original diameter and is therefore dimensionless. For example, consider a core plug of original length L_o that has been subjected to tensional stress. After applying the stress, if the original length is increased to L, then the axial strain is defined as:

$$\varepsilon = \frac{L - L_o}{L_o} \tag{5.2}$$

The strain defined by Equation 5.2 can also be expressed in terms of percentage.

5.2.3 The Stress–Strain Relationship

In most materials as stress increases, for a time strain also increases and is proportional to the stress. If the stress is removed, the strain goes back to zero. This is called *elastic deformation* or *elastic strain*. However, if the stress continues to increase, it reaches the yield point; the strength of the rock is overcome and the deformation is permanent. This is called *plastic deformation* or *plastic strain*; that is if the stress is

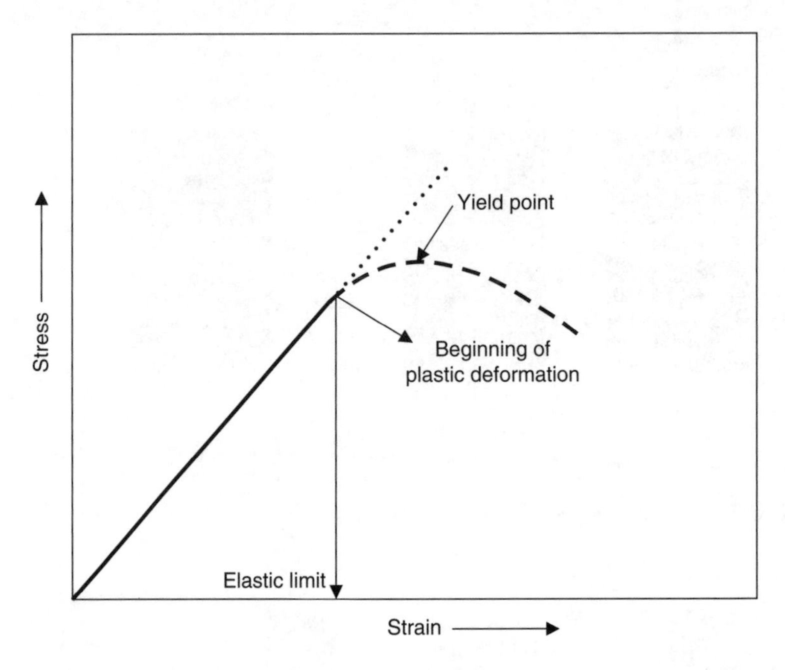

FIGURE 5.3 Ilustration of the stress–strain relationship for elastic and plastic deformations.

removed, the strain does not go back to zero. The plastic strain is usually accompanied by some elastic strain initially. Figure 5.3 illustrates the stress–strain relationship for the elastic and plastic deformations, respectively. As seen in Figure 5.3, beyond the elastic limit, the material starts to behave irreversibly in the plastic deformation region where the stress–strain curve deviates from linearity. The exact nature of the stress–strain relationship will, however, depend on the characteristics of the rock, i.e., its ductility, brittleness, and so on.

5.2.3.1 Factors Affecting the Stress–Strain Relationship

The stress–strain relationship of petroleum reservoir rocks is affected by a variety of factors that include mineralogy and fluids contained in those rocks. Therefore, it is important to have an understanding of these factors (1) when virgin reservoir conditions exist and (2) to know how they will change the mechanical behavior during primary depletion and secondary or tertiary flood. These factors require an understanding of the following several reservoir conditions:

- Confining pressure
- Temperature
- Time
- Reservoir fluids that occupy the pore space
- Anisotropy
- Porosity
- Permeability

- Degree of cementation
- Types of cementing material

To study the effect of these factors on reservoir rock stress–strain relationship, many of these factors are recreated in the laboratory based on which the resulting deformation is measured by a variety of laboratory techniques. Since many of these factors are interdependent, their separate and combined effect on the stress–strain relationship can be measured in the laboratory using an actual rock sample from the reservoir and controlling the test parameters to accurately simulate the *in situ* conditions.[1] Two loading and measuring techniques commonly used to obtain the stress–strain relationship of reservoir rocks are *uniaxial* and *triaxial* (sometimes referred to as biaxial). These techniques, which are discussed later, essentially involve the application of a specified load and measuring the corresponding strain.

5.2.4 Rock Strength

The strength of a solid material is its ability to resist the stress without yielding or to resist the deformation. Considering the three basic stress conditions, the strength of a material is always specified by the type of stress, such as the tensile strength (resistance against pulling apart), compressive strength (resistance to compression), and shear strength (resistance to shear stress). For reservoir rocks, strength is a result of the various depositional processes that formed the rock in the first place. Rock strength basically reflects geological origin.[1]

Compressive strength is the significant strength for reservoir rocks. For example, the determination of compressive rock strength is critical to understanding wellbore stability during drilling and production. The compressive strength is determined from two common laboratory techniques: uniaxial compressive strength test and triaxial compressive strength test. These tests basically determine the ultimate strength of a given reservoir rock, that is, the maximum value of stress attained before failure. As an example, the compressive strengths of different types of reservoir rocks are shown in Figure 5.4. As seen in this figure, although compressive strengths are generally very high, they do vary significantly for different rocks. Figure 5.4 also shows a comparison between the dry and saturated compressive strengths. Dry compressive strength is generally higher than the saturated compressive strength because saturated fluids normally tend to weaken the rock.

5.2.5 Rock Mechanics Parameters

The following rock mechanics parameters are generally used to characterize the mechanical properties of reservoir rocks.

5.2.5.1 Poisson's Ratio

Poisson's ratio, denoted by v, is an elastic constant that is a measure of the compressibility of material perpendicular to applied stress or defined as the ratio of latitudinal

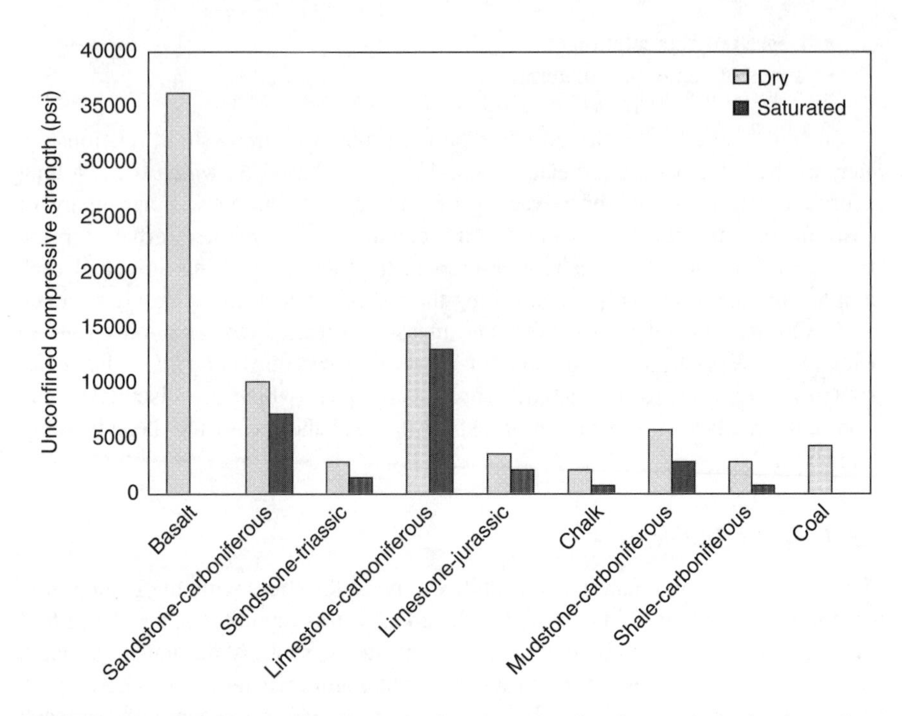

FIGURE 5.4 Typical unconfined compressive strengths of various rocks under dry and saturated conditions.

to longitudinal strain. Mathematically, Poisson's ratio is expressed by the following equation:

$$\nu = \frac{\varepsilon_{\text{latitudinal}}}{\varepsilon_{\text{longitudinal}}} = \frac{\Delta d / d_o}{\Delta L / L_o} \tag{5.3}$$

where $\varepsilon_{\text{latitudinal}}$ is the latitudinal strain, $\varepsilon_{\text{longitudinal}}$ the longitudinal strain, d_o and L_o are the original diameter and length of the cylindrical core sample respectively, and Δd and ΔL the change in the diameter and length, respectively.

Poisson's ratio is usually lower for stiffer materials than softer materials; for example carbonate rocks that are usually softer have a Poisson's ratio of the order of 0.3; sandstones, which are usually stiffer, have a Poisson's ratio of the order of 0.2.

5.2.5.2 Young's Modulus

Young's modulus, denoted by E, is defined as the ratio of longitudinal stress to longitudinal strain. Longitudinal stress is defined as the force per unit area of cross section; longitudinal strain is the increase in length per unit length, $\Delta L/L_o$.

Mathematically, Young's modulus is expressed by the following equation:

$$E = \frac{F/A}{\Delta L / L_o} \tag{5.4}$$

where F/A is the total load per unit area.

5.2.5.3 Modulus of Rigidity

Modulus of rigidity or shear modulus, denoted by G, is an important elastic constant that is a measure of the resistance of a body to change in shape and is mathematically expressed by

$$G = \frac{\text{shear stress}}{\text{shear strain}} \tag{5.5}$$

5.2.5.4 Bulk Modulus

The bulk modulus, denoted by K, gives the change in volume of a solid substance as the pressure on it is changed and is mathematically expressed as

$$K = \frac{\Delta P}{\Delta V / V_o} \tag{5.6}$$

where ΔP is the change in hydrostatic pressure, ΔV the change in volume, and V_o the original volume.

The bulk modulus is the reciprocal of matrix compressibility, C_r:

$$K = \frac{1}{C_r} \tag{5.7}$$

5.2.6 Laboratory Measurement of Rock Strength

The strength of a rock is measured by laboratory testing. The two different types of strength measurements are compressive strength and tensile strength. The two common laboratory tests to determine the compressive strength of rocks are:

1. Uniaxial compression test
2. Triaxial compression test

In the uniaxial test, a cylindrical rock core is loaded axially until it fails; in the triaxial test, a cylindrical rock core is placed in a cell, subjected to all-around (confining) pressure by a confining fluid (typically hydraulic oil or water) acting through an impermeable membrane, and loaded axially to failure.

The tests commonly carried out to determine the tensile strength are:

1. Direct pull test
2. Brazilian test
3. Beam flexure test

In the direct pull test, a cylindrical rock sample is anchored at both ends and stretched. In the Brazilian test, a relatively thin disk of the rock is loaded across the diameter until it splits. In the beam flexure test, a thin slab of rock is loaded vertically when supported at three or four points (known as four point flexure) along its length.

5.2.6.1 Triaxial Cell

The most common laboratory test used to determine the compressive strength properties of reservoir rock core samples is the triaxial test. The triaxial test provides by far the most, and most important, data for characterizing the strength of reservoir rock samples. Figure 5.5 shows the layout of a triaxial test setup. In this test, a cylindrical rock specimen (generally saturated with water or brine and normally encased in a sleeve) is placed inside a chamber (triaxial cell) that is usually filled with water or a hydraulic oil. Initially, the core specimen is confined by compressing the water or hydraulic oil in the cell; then, the specimen is subjected to axial stress until failure.

The application of axial stress can be performed in the following manner:

1. By applying hydraulic pressure in equal increments until sample failure or
2. By applying axial deformation at a constant rate by means of a geared loading press until sample failure.

A triaxial test can be conducted under either one of the following conditions: drained or undrained. If a drained test is performed, no excess pore pressures are developed during testing. If an undrained test is performed, the changes in pore fluid pressure inside the core sample are measured because pore fluid is retained within the sample and is not allowed to drain.

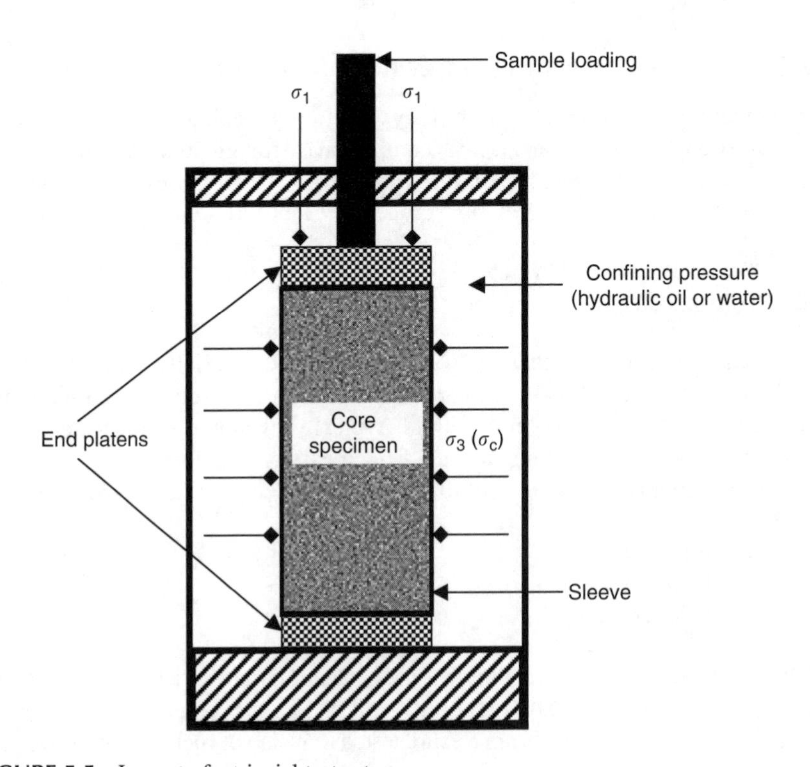

FIGURE 5.5 Layout of a triaxial test setup.

The common form of triaxial test is the conventional triaxial compression (CTC) test. This test involves loading the core sample in the axial direction while maintaining a constant confining pressure, σ_c (radial), as shown in Figure 5.5. At the peak load when failure occurs, the stress conditions are given by $\sigma_1 = F/A$ and $\sigma_3 = \sigma_c = P$. The highest load acting on area A, supportable parallel to the cylindrical axis is given by F, where P is the pressure in the confining medium. Based on the assumption that no shear stresses occur at the end platens of the geared loading press, σ_1 and σ_3 can be taken as the major and minor principal stresses, respectively. In an unconfined test, σ_3 is zero as no confining pressure is applied.

The relationship between σ_1 and σ_3 can be shown by the following equation[2]:

$$\sigma_1 = (\tan \Psi)\sigma_3 + \sigma_{uc} \tag{5.8}$$

where Ψ is the angle between principal stress and the radial stress and σ_{uc} the unconfined compressive strength.

A plot of σ_1 and σ_3 shows a regression line that follows the equation of a straight line. In a triaxial test, different core samples of the same rock type are tested for loading and each sample is subjected to loading until failure at a given confining pressure. The value of σ_1 is calculated from the ratio of peak load and the cross-sectional area of the sample. The values of σ_3 are known, which is the applied confining pressure.

Figure 5.6 shows the plot of σ_1 and σ_3 for seven granodioritic rock samples[2] from Alaska. The intercept (zero confining pressure) of the straight-line fit gives an

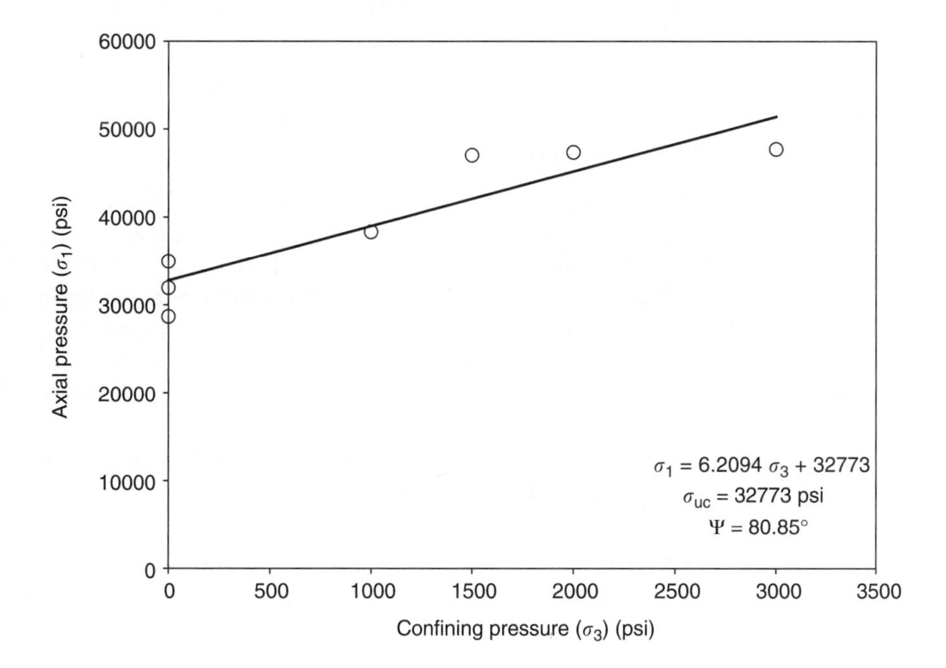

FIGURE 5.6 Analysis of triaxial test data measured on a granodioritic rock sample from Alaska. (Data from chen, G., personal communication, University of Alaska, Fairbanks, 2005.)

unconfined compressive strength, σ_{uc}, of 32,773 psi and the angle between the principal stress and the radial stress, Ψ, of 80.85° for this particular rock. The plot of σ_1 and σ_3 for this particular rock thus represents its failure conditions.

Based on the data collected during the triaxial compression test, a plot of load vs. the axial and lateral strain can also be constructed (normally a V-shaped curve) for each sample from which the two elastic constants, Young's modulus and Poisson's ratio, are determined. For example, the slope of the linear portion of the axial strain and load curve represents the linear elastic constant, Young's modulus, whereas Poisson's ratio is determined by a ratio of average lateral strain over average axial strain.

In undrained triaxial compression tests, the presence of pore fluids tends to affect the rock strength, that is rock strength depends on effective pressure defined as the difference between the confining pressure and the pore pressure. In other words, increased pore pressure tends to offset effects of confining pressure. For undrained tests, Equation 5.8 is modified as

$$\sigma_1 = (\tan\Psi)(\sigma_3 - P_{p)} + \sigma_{uuc} \tag{5.9}$$

where P_p is the pore pressure and σ_{uuc} the undrained, unconfined compressive strength.

As seen by Equation 5.9, the inclusion of pore pressure indicates a shift in the failure conditions of a given rock, or in other words, a decrease in the rock strength with increasing pore pressure. It should, however, be noted that the effect is more pronounced if the rock is saturated with an incompressible liquid rather than with a compressible gas. Nevertheless, this is an important aspect in deeply buried sedimentary rocks.

5.2.7 Reservoir Rock Compressibility

As fluids are depleted from reservoir rocks, a change in the internal stress in the formation takes place that causes the rocks to be subjected to an increased and variable overburden load. This change in the overburden load results in the compaction of the rock structure due to an increased effective stress. This compaction results in changes in the grain, pore, and bulk volume of the rock.[3] Out of these three volume changes, of principal interest to the reservoir engineer is pore compressibility. The change in the bulk volume may be important in areas where surface subsidence could cause appreciable property damage.[4]

The three compressibilities, rock matrix compressibility, rock bulk compressibility, and pore compressibility are defined as follows:

1. Rock matrix (grains) compressibility is the fractional change in the volume of the solid rock material with a unit change in pressure, and is mathematically expressed as

$$C_r = -\frac{1}{V_r}\left(\frac{\partial V_r}{\partial P}\right)_T \tag{5.10}$$

2. Rock bulk compressibility is the fractional change in volume of the bulk of the rock with a unit change in pressure and is mathematically expressed as

$$C_b = -\frac{1}{V_b}\left(\frac{\partial V_b}{\partial P}\right)_T \tag{5.11}$$

3. Pore compressibility is the fractional change in the pore volume of the rock with a unit change in pressure and is mathematically expressed as

$$C_p = -\frac{1}{V_p}\left(\frac{\partial V_p}{\partial P}\right)_T \tag{5.12}$$

where C_r, C_b, and C_p are the rock, bulk, and pore compressibilities, generally expressed in psi^{-1}; and V_r, V_b, and V_p the grain, bulk, and pore volumes, respectively.

In Equations 5.10 to 5.12, the subscript T indicates that the derivatives are taken at constant temperature. Ahmed[5] suggested an alternative expression of Equation 5.12 in terms of porosity, ϕ, by noting that porosity increases with the increase in the pore pressure:

$$C_p = \frac{1}{\phi}\left(\frac{\partial \phi}{\partial P}\right)_T \tag{5.13}$$

Since the rock and bulk compressibilities are considered small in comparison with the pore compressibility, the formation compressibility C_f is the term commonly used to describe the total compressibility of the formation and is equated to C_p[5]:

$$C_f = C_p = \frac{1}{\phi}\left(\frac{\partial \phi}{\partial P}\right)_T \tag{5.14}$$

Typical values[5] of formation compressibilities range from 3×10^{-6} to 25×10^{-6} psi^{-1}.

Based on Equations 5.11 and 5.12 and the relationship between bulk volume and pore volume ($V_p = \phi V_b$), Geertsma[3] suggested that bulk compressibility is related to the pore compressibility by the following expression:

$$C_b \cong \phi C_p \tag{5.15}$$

Geertsma[3] also states that in a reservoir, only the vertical component of hydrostatic stress is constant and stress components in the horizontal plane are characterized by the boundary condition and that there is no bulk deformation in those directions. For those boundary conditions, he suggested that the reservoir pore compressibility is half the laboratory measured pore compressibility.

The total reservoir compressibility, denoted by C_t, is extensively used in reservoir engineering calculations and reservoir simulations defined by the following expression:

$$C_t = S_g C_g + S_o C_o + S_w C_w + C_f \tag{5.16}$$

In the absence of a gas cap (i.e., undersaturated oil reservoirs, discussed later in fluid properties), Equation 5.16 reduces to

$$C_t = S_o C_o + S_w C_w + C_f \tag{5.17}$$

where S_g, S_o, and S_w are the gas, oil, and water saturations, respectively (see Chapter 6 on fluid saturations); C_g, C_o, and C_w the gas, oil, and water compressibilities, respectively, in psi^{-1} (discussed in Sections 15.2.5.2, 15.2.8.4 and 17.9); and C_t is the total reservoir compressibility in psi^{-1}.

5.2.7.1 Empirical Correlations of Formation Compressibility

Several authors have attempted to correlate formation compressibility with various parameters including the formation porosity. In 1953, Hall[6] correlated formation compressibility with porosity given by the following relationship:

$$C_f = \left(\frac{1.782}{\phi^{0.438}}\right)10^{-6} \tag{5.18}$$

where C_f is the formation compressibility in psi^{-1}, and ϕ the porosity in fraction.

In 1973, Newman[7] presented a correlation for consolidated sandstones and limestones based on 79 samples. The formation compressibility and porosity is correlated by the following generalized hyperbolic equation:

$$C_f = \frac{a}{(1+bc\phi)^{1/b}} \tag{5.19}$$

The parameters in Equation 5.19 have the following values:

For consolidated sandstone:

$a = 97.32 \times 10^{-6}$
$b = 0.699993$
$c = 79.8181$

For limestone:

$a = 0.8535$
$b = 1.075$
$c = 2.202 \times 10^6$

5.3 ELECTRICAL PROPERTIES

All reservoir rocks are comprised of solid grains and void spaces that are occupied by the fluids of interest in petroleum reservoirs (i.e., hydrocarbon gas and oil, and water). The solids that make up the reservoir rocks, with the exception of certain clay minerals, are nonconductors. Similarly, the two hydrocarbon phases, gas and oil, are also nonconductors. However, water is a conductor when it contains dissolved salts such as NaCl, $MgCl_2$, and KCl normally found in formation reservoir water. Electrical current is conducted in water by movement of ions and can therefore be termed *electrolytic conduction*. The electrical properties of reservoir rocks depend on the geometry of the voids and the fluids with which those voids are filled. Due to the electrical properties of reservoir formation water, the electrical well-log technique has become an important tool in the determination of water saturation vs. depth and thereby a reliable resource for *in situ* hydrocarbon evaluation. The remainder of this chapter presents important aspects related to the electrical properties of reservoir rocks.

5.3.1 Fundamental Concepts and the Archie Equation

The *resistivity* of a porous material can be defined by the following simple generalized equation:

$$R = \frac{rA}{L} \tag{5.20}$$

where R is the resistivity expressed in Ωm, r the resistance in Ω, A the cross-sectional area in m^2, and L the length in m.

In Equation 5.20, the electrical resistance r of a circuit component or device is defined as the ratio of the voltage difference ΔV to the electric current I which flows through it. However, for a complex system like a reservoir rock containing hydrocarbons and water, the resistivity of the rock depends on factors such as the salinity of water, temperature, porosity, geometry of the pores, formation stress, and composition of rock.

The theory of the electrical resistivity log technique that is applied in petroleum engineering was developed by Archie[8] in 1942 and is called the famous Archie equation. This empirical equation was derived for clean water-wet sandstones over a reasonable range of water saturation and porosities. In practice, Archie equation should be modified according to the rock properties: clay contents, wettability, pore distribution, and so on. The following is a brief presentation of the main electrical properties of reservoir rocks and related parameters.

5.3.1.1 Formation Factor

The most fundamental concept considering electrical properties of rocks is the formation factor F, defined by Archie as

$$F = \frac{R_o}{R_w} \tag{5.21}$$

where R_o is the resistivity (opposite of conductivity) of the rock when saturated 100% with brine expressed in Ωm and R_w the resistivity (opposite of conductivity) of the saturating brine in Ωm.

As seen in Equation 5.21, the formation factor shows a relationship between water- or brine- saturated rock conductivity and bulk water conductivity. However, considering the complexity of the reservoir rock pore space, the formation factor defined by Equation 5.21 is not readily applicable to reservoir rocks.

5.3.1.2. Tortuosity

Wyllie and Spangler[9] developed the relationship between the formation factor and other properties of rocks, such as porosity and tortuosity. A relationship among formation factor, porosity, and tortuosity can be developed on the basis of simple pore (capillary) models:

$$F = \frac{\tau}{\phi} \tag{5.22}$$

where τ is the tortuosity (dimensionless) and is defined by $(L_a/L)^2$, L_a the effective path length through the pores, L the length of the core, and ϕ the porosity.

5.3.1.3 Cementation Factor

A different form of Equation 5.22 is generally suggested to describe the relationship between the formation factor and porosity, by introducing the cementation factor m where

$$F = \frac{R_o}{R_w} = a\phi^{-m} \tag{5.23}$$

In Equation 5.23, $a \approx 1$ (taken as 0.81 for sandstones and 1 for carbonates) and $m \approx 2$. Clearly, Equation 5.23 results in an infinite formation resistivity factor when $\phi = 0$ and 1 when $\phi = 1$. Alternatively, if formation factor values and porosity values are known, a plot of log (F) vs. log (ϕ) can be used to estimate the parameters a and m, for a given rock type, of the Archie's formation factor equation. Porosity values can be measured by any of the techniques described in Chapter 3. The resistivity of the core plug saturated with 100% brine can be measured using a conductivity bridge, such as the one shown in Figure 5.7. The saturated core plug (under appropriate confining pressure) is held between electrodes in the bridge circuit. The resistivity of the brine can be determined by a platinum electrode dipped into brine, forming an element of the bridge circuit. As an example, the general nature of the log–log plot of formation factor vs. porosity and the subsequently determined Archie equation parameters for various carbonate cores are shown in Figure 5.8.

5.3.1.4 Resistivity Index

In a pore space containing hydrocarbons (gas or oil), both of which are nonconductors of electricity, with a certain amount of water, resistivity is a function of water or brine saturation S_w. For the given porosity, at partial brine saturations, the resistivity of a rock is higher than when the same rock is 100% saturated with brine. Archie determined experimentally that the resistivity factor of a formation partially saturated with brine can be expressed by

$$\frac{R_o}{R_t} = (S_w)^n \tag{5.24}$$

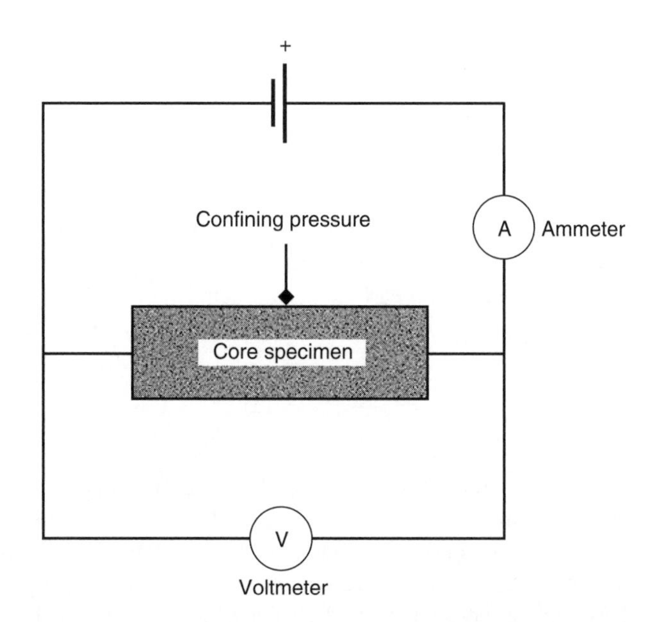

FIGURE 5.7 Core sample resistivity measurement using a conductivity bridge.

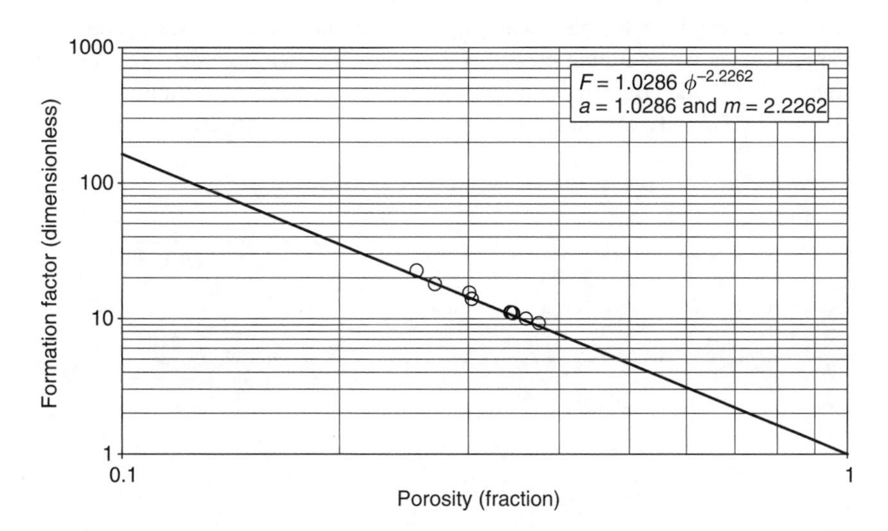

FIGURE 5.8 Log–log plot of formation factor vs. porosity for various carbonate cores for the determination of Archie equation parameters *a* and *m*.

where R_o is the resistivity of the same rock when fully saturated with brine expressed in Ωm, R_t the resistivity of the rock when partially saturated with brine in Ωm, and n the saturation exponent.

The resistivity of the rock partially saturated with brine, R_t, is also referred to as true resistivity of formation containing hydrocarbons and formation water.

Comparing Equation 5.23 and 5.24, R_o can be eliminated to obtain a generalized relationship for water saturation:

$$S_w = \left(\frac{R_o}{R_t}\right)^{1/n} = \left(\frac{FR_w}{R_t}\right)^{1/n} = \left(\frac{aR_w}{\phi^m R_t}\right)^{1/n} \tag{5.25}$$

The ratio of R_t/R_o is commonly referred to as the resistivity index, I. The resistivity index is equal to 1 for a fully brine saturated rock, whereas $I > 1$ when the rock is partially saturated with brine or hydrocarbons are present.

Equation 5.25 can also be expressed in terms of the resistivity index

$$S_w = \left(\frac{R_t}{R_o}\right)^{-1/n} = (I)^{-1/n} \tag{5.26}$$

As shown in Figure 5.9, a plot of log(I) vs. log(S_w) gives a straight line of slope $-n$. For a given core plug sample, after measurement of R_o, at 100% brine saturation, the core plug can be desaturated in several steps by displacing the brine with oil. At each step, voltage drop and water saturation can be measured. The measured voltage drop, current, and the sample dimensions yield the value of R_t at that particular water saturation. These measurements on each core plug typically continue up to the irreducible water saturation (for definition see Chapter 6). It should also be noted here that, since overburden also affects the electrical properties, all measurements should be carried out at representative confining pressures. Based on the measured data, the saturation exponent can be determined by a straight-line fit for each core plug. An average value of n is normally calculated for a particular rock type on the basis of n values

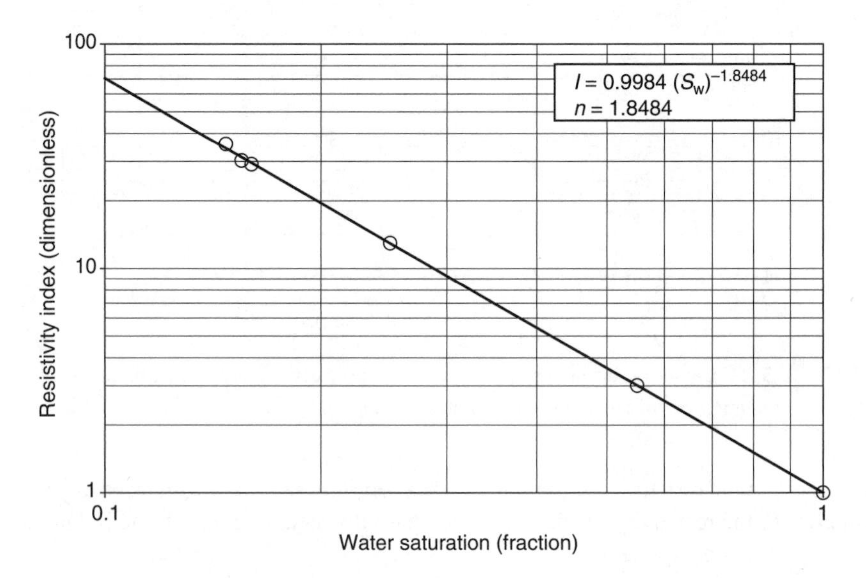

FIGURE 5.9 Log–log plot of resistivity index vs. water saturation for a carbonate core sample for the determination of saturation exponent n.

determined for multiple core plug samples. In summary, the saturation exponent and R_o are experimentally determined in the laboratory, whereas the true resistivity can be obtained from the well logs. Therefore, the *in situ* water saturation can be calculated using Equation 5.26. Finally, based on the material balance equation for the formation; $S_w + S_o + S_g = 1.0$, the in place hydrocarbons can be estimated.

5.3.2 Effect of Wettability on Electrical Properties

The electrical resistivity of a porous medium can be significantly affected by important factors such as wettability and saturation history because they control the location and distribution of fluids. The most comprehensive review of the effect of wettability on electrical properties of porous media was presented by Anderson[10] as part of the series of review papers published on the effect of wettability on various rock properties. In fact, the parameter most significantly affected by wettability is the saturation exponent n because of its dependence on the distribution of the conducting phase in the porous medium, which in turn depends on the wettability of the system (see Chapter 7 for discussion on wettability). The uncertainty in the saturation exponent can directly impact the calculated water saturation (Equation 5.25 or 5.26) and will obviously lead to errors in the calculation of hydrocarbons in place.

Anderson's[10] examination of the effect of wettability on saturation exponent basically resulted in the following major conclusions:

1. The saturation exponent is essentially independent of the system wettability when the brine saturation is sufficiently high to form a continuous film on the grain surfaces of the porous medium. The film provides a continuous path for a current flow.
2. This type of film continuity is common in clean and uniformly water-wet systems. The saturation exponent in such systems is close to 2 and remains essentially constant as the core sample is desaturated to its irreducible water saturation.
3. These two observations, however, do not apply to uniformly oil-wet systems. The saturation exponent remains close to a value of 2 upto a certain minimum water saturation. However, as the core is desaturated further from this minimum water saturation to its irreducible water saturation, rapid increase in the saturation exponent is observed. Values of n as high as 9 at irreducible water saturation are not uncommon.
4. The rapid increase in the saturation exponent with decreasing brine saturation for oil-wet systems is attributed to an increase in the resistivity of the system. The increase in resistivity is due to the disconnection and trapping of the portion of the brine (nonwetting but conducting phase) by oil (wetting but nonconducting phase). The disconnected portion of the brine obviously no longer contributes to the flow of current because it is surrounded by oil that is the nonconducting phase, eventually resulting in an increase in the resistivity of the system.

Mungan and Moore[11] studied the effects of wettability on resistivity using a Teflon® (DuPont Dow Elastomers L.L.C., Wilmington, DE) core. The two fluid pairs they used were air–brine and oil–brine. The brine is then the conducting, non-wetting phase, behaving in a fashion similar to brine in an oil-wet core. The saturation exponents for the two systems are shown in Figure 5.10. An examination of Mungan and Moore's[11] data shown in Figure 5.10 demonstrates what typically happens in an oil-wet system as the brine saturation is decreased. Above a certain conducting phase saturation, the saturation exponent is fairly constant and is near 2. However, below this saturation, the exponent begins to increase rapidly by a small decrease in the water saturation. For example, the data of Mungan and Moore indicate that for a reduction of water saturation from 34.3 to 33.9% in the case of oil–brine system; the saturation exponent jumps from 4 to 7.15 and eventually reaches a value of 9 at a water saturation of 31%. A similar behavior is observed in the case of air–brine system.

Anderson[10] also recommends that unless the reservoir is known to be strongly water-wet, the saturation exponent should be measured on native or restored state cores. Anderson also states that if a clean core is used to measure the saturation exponent and the reservoir is actually oil-wet; the water saturation can be underestimated when logging. These conclusions by Anderson were based on the work of Moore[12] in which the effects of cleaning on the Archie saturation exponent of the Bradford third sand, known to be oil-wet, were examined. Moore studied six pairs of adjacent core plugs; one core plug from each set was extracted with toluene, making it more water-wet, while the other core was unextracted and left oil-wet.

In case of each core plug, extraction was found to significantly lower the saturation exponent. In the case of unextracted core plugs the saturation exponent was observed to be higher than the extracted samples. The saturation exponent data for the

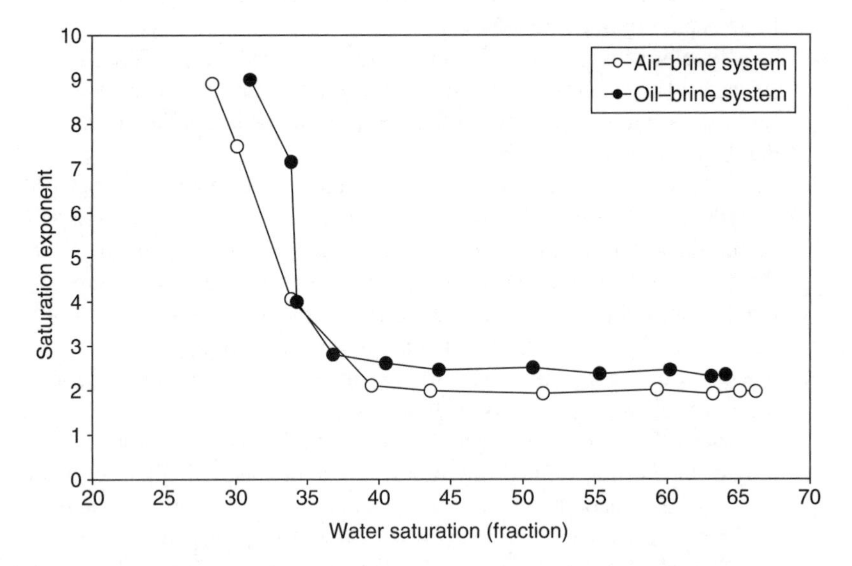

FIGURE 5.10 The variation in saturation exponent n, as a function of brine (conducting, nonwetting phase) saturation for an oil-wet system. (Data from Mungan, N. and Moore E.J., *J. Can. Pet. Technol.*, 7, 1968.)

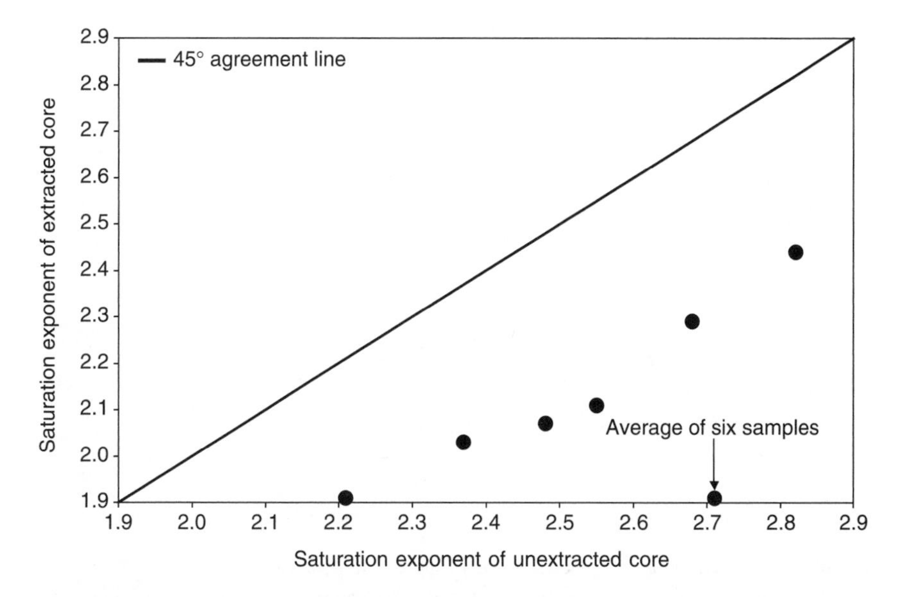

FIGURE 5.11 The alteration in the values of saturation exponents caused by core extraction with toluene. (Data from Moore, J., *Producers Monthly*, 22, 1958.)

six extracted and unextracted core plugs reported by Moore are shown in Figure 5.11, which clearly shows the differences in the saturation exponent. The differences in the calculated water saturations for various iso-resistivity values using the extracted and unextracted saturation exponents for one of the core plugs are shown in Figure 5.12. Clearly such differences in the calculated water saturation would obviously impact the determination of hydrocarbon saturation.

Moore, however, measured the resistivity of the unextracted cores only for brine saturations greater than 35%. Therefore, it is quite plausible that the saturation exponent would have shown a rapid increase at lower brine saturations, as observed by Mungan and Moore.

5.3.3 Effect of Clay on Electrical Properties

The clay minerals present in a reservoir rock act as separate conductors and are referred to as *conductive solids*. As a matter of fact, the water in the clay and the ions in the water act as the conducting materials. The effect of the clay on the resistivity of the rock is dependent upon the amount, type, and manner of distribution of the clay in the rock. The presence of conductive solids or clays in reservoir rocks requires a different approach for calculation of the formation factor.

Regarding the effect of clays on electrical properties, investigations by Wyllie[13] indicated that clays contribute significantly to the conductivity of a rock when the rock is saturated with a low conductivity water. The formation factor of a clayey sand increases with decreasing water resistivity and approaches a constant value at

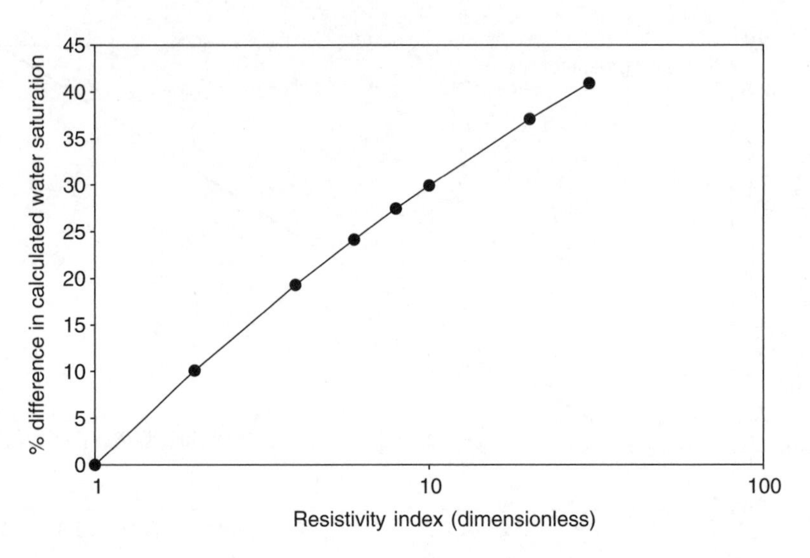

FIGURE 5.12 The percentage difference in the calculated water saturations based on the altered saturation exponents due to core extraction. Water saturations are calcuated from Equation 5.26 using $n = 1.91$ (extracted) and $n = 2.71$ (unextracted). (Data from Moore, J., *Producers Monthly*, 22, 1958.)

a water resistivity of about 0.1 Ωm, whereas the formation factor of a clean (clay-free) sand remains constant throughout the wide range of water resistivities.[4]

For the determination of formation factor of clay-laden rocks, Wyllie proposed that the observed effect of clay minerals was similar to having two electrical circuits in parallel, that is, the conducting clay minerals and the water-filled pores. Therefore

$$\frac{1}{R_{oa}} = \frac{1}{R_c} + \frac{1}{R_o} \tag{5.27}$$

where R_{oa} is the resistivity of the clayey rock 100% saturated with water of resistivity and R_w and R_c are the resistivities due to the clay minerals.

Substituting the value of R_o from Equation 5.21,

$$\frac{1}{R_{oa}} = \frac{1}{R_c} + \frac{1}{FR_w} \tag{5.28}$$

The apparent formation factor F_a for clayey rock by definition is given by

$$F_a = \frac{R_{oa}}{R_w} \tag{5.29}$$

A plot of $1/R_{oa}$ vs. $1/R_w$ thus results in a straight line having a slope of $1/F$ and an intercept of $1/R_c$. Equation 5.28 will reduce to Equation 5.21 for a clay-free clean rock because the intercept will be zero.

Equation 5.28 can be rearranged to express R_{oa} for developing the expression for F_a:

$$R_{oa} = \frac{R_c R_w}{[R_w + (R_c/F)]} \tag{5.30}$$

$$F_a = \frac{R_c}{[R_w + (R_c/F)]} \tag{5.31}$$

As R_w approaches zero, F_a equals F as shown in Equation 5.31.

PROBLEMS

5.1. A 10-mm-diameter and 50-mm-long sandstone core plug is pulled with 1500 N force. The final reading of the extensometer (an instrument used to measure deformations) is 50.07 mm. Calculate the stress and strain under this load.

5.2 An unconfined triaxial test was carried out on a 1.0-in.-diameter and 2.5-in.-long core sample from an Australian field. The sample was axially loaded at a rate of 210 lb_f/sec up to 80 sec, the time at which it failed. The triaxial test resulted in a deformation in both the axial and lateral directions. The change in the diameter is 0.0005 in.; the change in the length is 0.004 in. Calculate the latitudinal and longitudinal strains, Poisson's ratio, Young's modulus, and the ultimate strength of the sample.

5.3 Seven core plug samples were drilled from a whole core recovered from a North Sea reservoir. All samples were tested in a triaxial cell under dry conditions, at various confining pressures. The axial stress at failure was measured for each sample. Given the following data, calculate the unconfined compressive strength and the angle between principle stress and the radial stress, for this North Sea rock.

Core Plug	Axial Stress at Failure (psi)	Confining Pressure (psi)
1	19000	500
2	22200	750
3	26000	1000
4	29500	1250
5	32500	1500
6	36000	1750
7	40000	2000

5.4 Estimate the total reservoir compressibility for a sandstone formation that is characterized by a porosity of 25%. The reservoir is undersaturated (i.e., no initial gas cap is present) and the oil and water saturations are 70 and 30%, respectively. The compressibility of oil and water are 7.5×10^{-5} psi^{-1} and 2.5×10^{-5} psi^{-1}, respectively.

5.5. Six cylindrical core plugs of 2.54 cm diameter and 3.81 cm length were taken from an Alaskan North Slope reservoir. After cleaning, porosities of all the plugs were measured by a helium porosimeter. Subsequently, all samples were fully saturated with a 0.07 Ωm brine such that, $S_w = 1$. Each sample was placed in a resistivity apparatus and ΔV values were measured for current flow of 0.01 A. Determine the formation factor F, for each core plug and estimate parameters a and m for Archie's formation factor equation.

Core Plug	ϕ (fraction)	ΔV (V)
1	0.175	1.720
2	0.190	1.450
3	0.165	1.990
4	0.230	1.100
5	0.160	2.150
6	0.150	2.450

5.6 Sample 4 from the previous data set was flooded with crude oil, in several steps, in order to displace the brine. The remaining water saturation and ΔV values were measured at each of these displacement steps. Based on the measured data given below and other data from the problem 5.5, calculate the true formation resistivity R_t as a function of water saturation S_w and subsequently determine the saturation exponent n of the Archie's saturation equation.

Water Saturation (fraction)	ΔV (V)
1.0	1.1
0.8	1.8
0.6	3.6
0.4	9.0
0.3	24.0

5.7 Based on the results from problems 5.5 and 5.6, estimate the hydrocarbon saturation in the reservoir if the log analysis indicates that the porosity is 20% and the true formation resistivity is 5.25 Ωm.

5.8 The following table gives the values of the resistivity, R_{oa}, of clay-laden rocks when 100% saturated with water of resistivity, R_w. Based on the given data, calculate the values of R_c and F. Subsequently, calculate and plot the apparent formation factor vs. the water resistivity in the range of 0.01 to 20 Ωm.

R_{oa} (Ωm)	R_w (Ωm)
0.95	0.05
1.43	0.07
2.22	0.11
3.33	0.20

REFERENCES

1. Tiab, D. and Donaldson, E.C., *Theory and Practice of Measuring Reservoir Rocks and Fluid Transport Properties,* Gulf Publishing Company, Houston, TX, 1996.
2. Chen, G., personal communication, University of Alaska, Fairbanks, 2005.
3. Geertsma, J., The effect of fluid pressure decline on volumetric changes of porous rocks, *Trans. AIME*, 210, 331–340, 1957.
4. Amyx, J.W., Bass, Jr., D.M., and Whiting, R.L., *Petroleum Reservoir Engineering*, McGraw-Hill, New York, 1960.
5. Ahmed, T., *Reservoir Engineering Handbook*, Butterworth-Heinemann, Woburn, MA, 2001.
6. Hall, H.N., Compressibility of reservoir rocks, *Trans. AIME,* 198, 309, 1953.
7. Newman, G.H., Pore volume compressibility, *J. Pet. Technol.*, 129–134, 1973.
8. Archie, G.E., The electrical resistivity log as an aid in determining some reservoir characteristics, *Trans. AIME,* 146, 54–62, 1942.
9. Wyllie, M.R.J. and Spangler, M.B., Application of electrical resistivity measurements to problem of fluid flow in porous media, *Bull. Am. Assoc. Pet. Geologists,* 159, 1952.
10. Anderson, W.G., Wettability literature survey - Part 3: the effects of wettability on the electrical properties of porous media, *J. Pet. Technol.*, 1371–1378, 1986.
11. Mungan, N. and Moore, E.J., Certain wettability effects on electrical resistivity in porous media, *J. Can. Pet. Technol.*, 7, 20–25, 1968.
12. Moore, J., Laboratory determined electric logging parameters of the Bradford Third sand, *Producers Monthly*, 22, 30–39, 1958.
13. Wyllie, M.R.J., Formation factors of unconsolidated porous media: influence of particle shape and effect of cementation, *Trans. AIME,* 198, 103–110, 1953.

6 Fluid Saturation

6.1 SIGNIFICANCE AND DEFINITION

In the previous chapters on porosity and permeability, the storage capacity of a rock and the conductive capacity of a rock were discussed. However, for the reservoir engineer, yet another very important factor needs to be determined, apart from porosity and permeability: the amount of hydrocarbon fluids present in the reservoir rock. While porosity represents the maximum capacity of a reservoir rock to store fluids, fluid saturation or pore space saturation actually quantifies how much of this available capacity actually does contain various fluid phases; in other words, how is that storage capacity, pore volume, or pore space distributed or partitioned among the three typical reservoir fluid phases: gas, oil, and water (usually referred to as brine or formation water). Therefore, initial fluid saturations defined as fractions of the pore space occupied by gas, oil, and water are key factors in the determination of initial reserves of actual and recoverable hydrocarbons in place.

Fluid saturation also dominates important flow properties due to the strong influence it has on relative permeability functions. In the case of many reservoirs, initial fluid saturation is virtually unknown or is inaccurately measured, resulting in gross over- or under estimation of hydrocarbon reserves in place. For example, inaccurate determination of initial fluid saturation existing in porous media often leads to expensive mistakes in the development of a field, for example, large amounts of capital are invested where minimal reserves are present and in other cases, viable pay is overlooked due to a perceived belief, from improper saturation evaluations, that the pay will be nonproductive.

The importance of accurate fluid saturation information can also be highlighted because hydrocarbons in place (gas or oil) are calculated on the basis of a simple volumetric balance of hydrocarbons present in the effective pore space of the system. For example, if a reservoir is 50% saturated with water, this means that half of the available pore space in the reservoir actually contains oil. If this figure of 50% is erroneous, that is, an over- or underestimation of the initial water saturation, it can lead to an incorrect estimate of initial gas or oil in place. An underestimation of water saturation can result in the development of a field that may not be worth developing because of less gas or oil in place. However, the converse is also true, that is, if the water saturation is overestimated, the development of a potentially viable field may be wrongfully abandoned.

Considering the importance of water saturation in determining the original hydrocarbons in place by volumetric balance, in most situations, the central-most objective is to obtain accurate initial water saturation that exists in the porous media. Similar to the initial water saturation, fluid saturation measurements may also be used to determine the target oil in place for secondary- or tertiary-enhanced oil recovery (EOR) projects. This target oil saturation in most cases is basically the remaining oil saturation that may be mobile and hence become the goal of a particular type of EOR process.

The method frequently used to obtain such types of fluid saturation data (initial water saturation or mobile remaining oil saturation) is direct measurement on core material taken from the interval of interest. Obtaining samples from the formation of interest in their original state and measure saturations directly is ideal. In summary, the reservoir engineer uses the fluid saturation data along with porosity, permeability, and other data to determine the feasibility and estimate profitability of completing a well or developing a particular reservoir.

6.2 DISTRIBUTION OF FLUID SATURATION IN A PETROLEUM RESERVOIR

It was believed that initially the reservoir rock in most hydrocarbon bearing formations is completely saturated with water (even though petroleum engineering literature many times refers to it as simply *water*, more precisely it means formation water or reservoir water or brine that usually contain a higher concentration of salts as compared to potable water). Subsequently, when the hydrocarbon invasion took place as part of the migration process; gas, oil, and water distributed in the pore spaces of the reservoir in a manner dictated primarily by a balance between the gravitational and capillary forces. The less dense hydrocarbon phases (gas and oil) migrated to the structurally high part of the reservoir rock due to gravity. However, complete gravity segregation into three distinct layers of gas, oil, and water was not possible because of the resistance due to capillary forces. Therefore, reservoir rocks normally contain both hydrocarbon fluid phases as well as water occupying the same or adjacent pore spaces.

In order to determine the quantity of hydrocarbons accumulated in a porous rock formation, it is necessary to determine the individual fluid-phase saturation, i.e., gas, oil, and water. Herein lies the importance and significance of fluid saturations in reservoir engineering. This aspect of fluid saturation is addressed later in this chapter when the three different types or classes of fluid saturations are discussed.

6.3 DEFINITION AND MATHEMATICAL EXPRESSIONS FOR FLUID SATURATION

Generally, fluid saturation is defined as the ratio of the volume of a fluid phase in a given reservoir rock sample to the pore volume of the sample. In other words, fluid saturation is defined as that fraction, or percent of the pore volume occupied by a particular fluid phase (gas, oil, or water) expressed by a generalized mathematical expression:

$$\text{fluid saturation} = \frac{\text{total volume of the fluid phase}}{\text{pore volume}} \tag{6.1}$$

It should be noted that the fluid saturation may be reported either as a fraction of the total pore volume or the effective (interconnected as well as dead-end or cul-de-sac type pores) pore volume. However, fluid saturation is generally reported as a fraction of the effective pore volume rather than the total pore volume because it is more meaningful as fluids present in the completely isolated pore space cannot be produced. Therefore, Equation 6.1 assumes that the pore volume is effective pore volume.

Equation 6.1 can now be applied to the specific fluid phases

$$S_g = \frac{\text{volume of gas}}{\text{pore volume}} \tag{6.2}$$

$$S_o = \frac{\text{volume of oil}}{\text{pore volume}} \tag{6.3}$$

$$S_w = \frac{\text{volume of water}}{\text{pore volume}} \tag{6.4}$$

where S_g, S_o, and S_w are the gas, oil, and water saturations respectively.

Fluid saturation can be expressed as a fraction or percentage (by multiplying the values in Equations 6.2 to 6.4 by 100) of the pore volume. Equations 6.2 to 6.4 clearly indicate that saturations can range from 0 to 100% or 0 to 1, and since all saturations are scaled down to the pore volume their summation should always equal 100% or 1 leading to

$$S_g + S_o + S_w = 1.0 \tag{6.5}$$

Equation 6.5 is probably the most simple, yet fundamental equation in reservoir engineering, and is used almost everywhere in reservoir engineering calculations. Moreover, many important reservoir rock properties, such as capillary pressure and relative permeability, are actually related or linked with individual fluid-phase saturations. The definition of properties such as relative permeabilities or capillary pressures without relating them to fluid-phase saturations is basically meaningless as is shown in the pertinent chapters of this book.

It can also be seen from Equations 6.2 to 6.5 that

$$\text{volume of gas} + \text{volume of oil} + \text{volume of water} = \text{pore volume} \tag{6.6}$$

So if fluid saturations are accurately measured on a reservoir rock sample, the summation of volumes of individual fluid phases can also be used to determine the pore volume (or porosity if the bulk volume is also known) of that particular sample because fluid phases originated from the pore spaces of that very sample. In order to illustrate the significance of Equation 6.5, the fluid saturation distribution for a hypothetical core plug sample is shown in Figure 6.1.

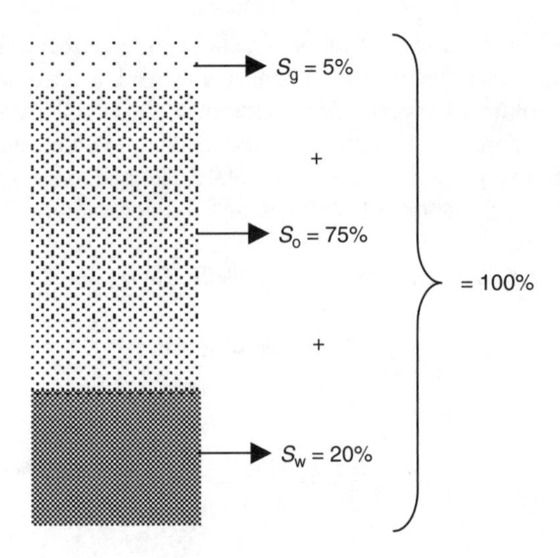

FIGURE 6.1 Fluid saturation distribution in a hypothetical reservoir rock sample.

6.4 RESERVOIR ROCK SAMPLES USED FOR FLUID SATURATION DETERMINATION

In fluid saturation determination of reservoir rock samples, the very first issue to address is: which rock samples are going to be used? Generally, fluid saturations are determined on small core plug samples that are drilled from the large whole core because ideally these core plug samples are assumed to contain the original *in situ* gas, oil, and water phase. These core plug samples are sometimes also called *native state* or *preserved state* samples.

The use of these preserved state samples for the determination for fluid saturation actually brings up two important issues:

1. Before measuring porosity and permeability, the core samples must be cleaned of residual fluids thoroughly. (This cleaning process may also be part of the fluid saturation determination.)
2. The use of preserved state samples is highly recommended in special core analysis tests, such as relative permeability measurement, so that alterations in wettability do not influence the measured relative permeabilities. If cleaned core plug samples are used in relative permeability measurements, the original wettability of the reservoir rock may be altered or changed thereby resulting in *nonrepresentative* relative permeability values that may further affect the reservoir engineering calculations.

Therefore, clearly these two issues pose a “Catch 22” that is, initial fluid saturation, porosity, and absolute permeability data are required in relative permeability determination and yet such data are not available or cannot be obtained because preserved state samples are being used.

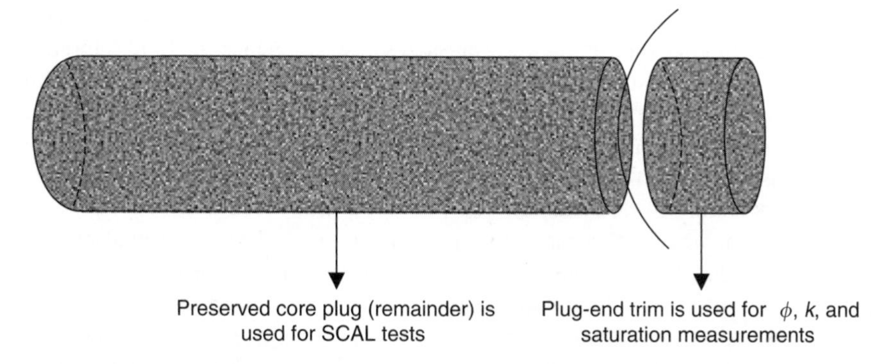

FIGURE 6.2 Plug-end trims from a preserved core plug sample.

In order to circumvent this problem or address these issues, sometimes subsamples from the preserved state core plugs are taken. These subsamples are also called *plug-end trims* (see Figure 6.2). The basic routine core analysis data are then measured on the plug-end trims while the remainder of the preserved core plug is employed in the special core analysis (SCAL) testing such as relative permeability measurements. The routine core analysis data (porosity, absolute permeability, and fluid saturations) measured on the plug-end trims are then considered as representative of the entire preserved core plug and are also used in the relative permeability testing.

It may, however, be argued, why not drill one small subsample from the entire whole core and measure the routine core analysis data and consider that as representative for the entire whole core? This procedure does not give representative results because the small subsample is a fraction of the much larger whole core. In comparison, if a plug-end trim is sliced from a small core plug sample, the likelihood of representative basic data is much greater because the plug-end trim is typically a quarter of the entire core plug sample. This still is a highly simplified assumption as reservoir rock heterogeneities may affect the representativity of the plug-end trim data.

In summary, plug-end trim data provide the basic information on porosity, absolute permeability, and fluid saturations for starting relative permeability testing. However, the same data can also be obtained for the very core plug that was used for relative permeability testing. After termination of SCAL studies, the core plug is cleaned (fluid saturations are determined) and porosity and absolute permeability are measured. The measured fluid saturations are then used to back calculate the original saturation conditions of the core plug. Finally, the two data sets on the plug-end trims and the actual core plug can be compared to validate the assumption that the former is representative of the latter.

6.5 LABORATORY MEASUREMENT OF FLUID SATURATION

Fluid saturation in reservoir rocks can be determined by essentially two different approaches: direct and indirect. The direct approach involves using preserved core plug samples or rather plug-end trims of the core plug, that is, a rock sample removed from a petroleum reservoir, for fluid saturation determination. The indirect method is further divided into two categories: (1) use of some other measurements on core plug samples such as capillary pressures based on which the fluid saturations are determined and (2) use based on traditional well-logging techniques where fluid saturations are not measured on core plugs but are measured *in situ*, the entire formation itself at various depths. Even with improved well-logging tool technology and increasing experience, accurate definition of fluid saturation and prediction of productivity could be often elusive.

The best approach is to measure the fluid saturation from the actual physical core sample. Discussion in this section is restricted to only those methods that are actually used to directly determine saturation values from preserved reservoir rock core plug samples. Additionally, x-ray CT scanning technique, also considered as a direct approach for measuring the fluid saturations, can also be used. However, the x-ray CT scanning technique is not included in this discussion.

All the methods for measurement of original reservoir rock saturation are based on the principle of *leaching* that basically refers to the process of removal of liquids from a solid (rock sample in this case). Based on the principle of leaching, two methods are devised for the determination of fluid saturation. The first methods involves using heat to extract the fluids present in the pore spaces of the rock and is termed *retort distillation*. The second method involves using both heat as well as an organic solvent to extract the pore fluids and is called Dean–Stark extraction. Both these methods are discussed in the following two sections.

6.5.1 Retort Distillation

Figure 6.3 shows the retort distillation apparatus consisting of three principal components: a heating unit, a condenser, and a receiver. The heating element or unit is used to apply very high heat to a given reservoir rock sample. The rock samples can be small cylindrical core plugs (or end trims) or crushed core samples. These rock samples, either intact or crushed, are usually weighed before placing them in the retort. The application of heat is carried out either in stages or directly to temperatures as high as 650°C (1200°F) resulting in the vaporization of oil, and water. This vaporized oil and water is then condensed in the condenser and collected in a small receiving vessel. The volumes of oil and water are measured directly. A horizontal or a plateau in the plot of collected oil and water volume vs. the heating time indicates no further extraction of pore fluids.

The oil and water saturations are subsequently determined by applying Equations 6.3 and 6.4; Equation 6.5 is used to calculate the gas saturation. It should, however, be noted here that if crushed rock sample is used in the retort distillation, the applied heat removes the fluids from interconnected and cul-de-sac pores as well as from the completely isolated or disconnected pore spaces, because these are destroyed as part of the crushing process anyway.

Despite being a very simple and rapid technique, the retort distillation method has certain drawbacks or disadvantages. First, the rock sample is completely destroyed and second, high temperatures are required. However, the application of very high heat such as 650°C, is in fact unavoidable because the oil in reservoir rock samples almost always contains very high molecular weight or high boiling point components. The application of very high temperatures becomes essential to ensure that all the oil is completely extracted from the rock sample.

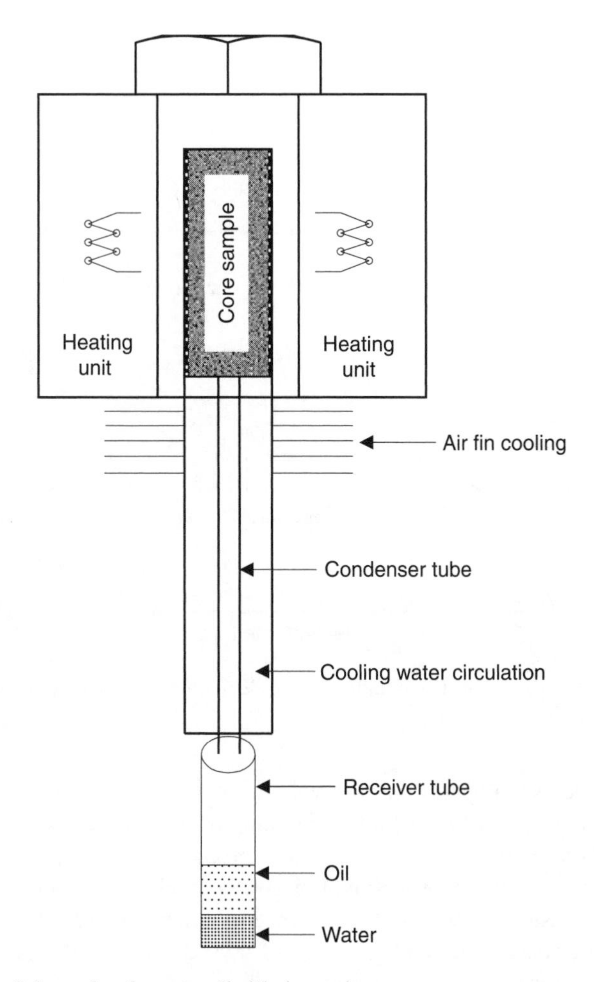

FIGURE 6.3 Schematic of a retort distillation unit.

Using temperatures of this magnitude results in a twofold problem or error — at such high temperature the water of crystallization within the rock is driven off, causing the water recovery values to be greater than the pore water (see Figure 6.4). Second high temperatures may crack and coke the oil causing the collected oil volume to not correspond to the volume of oil initially in the rock sample. The cracking and coking of the hydrocarbon molecules, in fact, tends to decrease the liquid volume and also in some cases may coat the internal walls of the rock sample itself.

The information detailing the water of crystallization and the cracking and coking of hydrocarbons is quantified in Emdahl,[1] based on the core analysis of Wilcox sands in which fluid saturations were measured by the retort distillation method, indicating an error of around 33% in the water saturation with the volume of oil recovered and the volume of oil in the sample varied due to

$$V_{\text{oil actually in sample}} = 1.2198\,(V_{\text{oil collected in receiver}}^{0.859}) \tag{6.7}$$

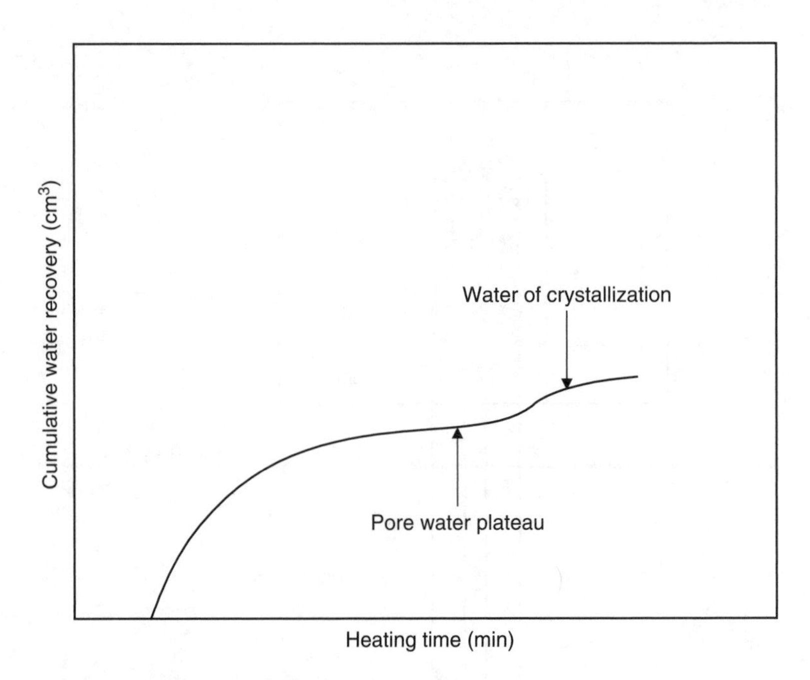

FIGURE 6.4 A typical retort distillation curve depicting the recovery of pore water and removal of water crystallization.

Equation 6.7 clearly indicates that the volume of oil recovered or collected in the receiver is decreased due to cracking and coking of the hydrocarbon molecules.

In addition to these errors, other practical problems can also occur in the retort distillation method, such as formation of oil–water emulsions that do not allow accurate volume measurements and the absence of clear demarcation between the plateaus of pore space water and the water of crystallization introducing uncertainties in the measurement of water volume.

6.5.2 Dean–Stark Extraction

In the Dean–Stark extraction technique, fluid saturation is measured by a process of distillation extraction. The setup basically consists of a long-neck round-bottom flask that contains a suitable hydrocarbon solvent such toluene, a heating element or electric heater to boil the solvent, a condenser, and a graduated tube receiver to measure the volume of extracted fluids. In the Dean–Stark extraction apparatus, shown in Figure 6.5, toluene is heated to its boiling point of 110°C; its vapors move upward, and the rock sample (typically end trim) becomes engulfed in the toluene vapors that begin to extract or leach the oil and water present in the rock sample. The rising vapor is condensed in the condenser and eventually collected in the graduated tube. Since toluene is completely miscible with the extracted oil, the condensed liquid in the graduated tube consists of two liquid phases: water and a mixed hydrocarbon phase containing toluene and oil from the rock sample. The water phase, due to its higher density, settles at the bottom of the graduated tube; the solvent (mixed)

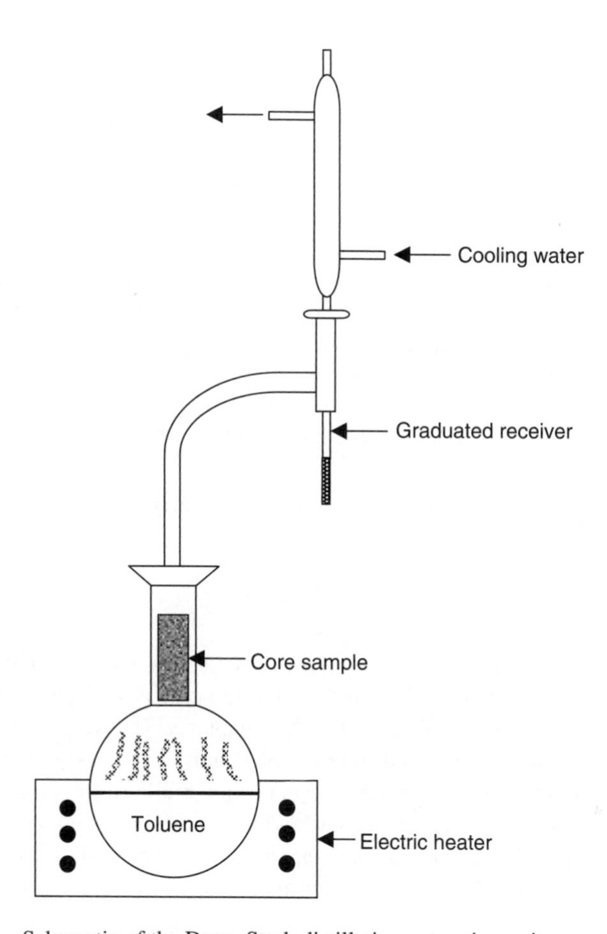

FIGURE 6.5 Schematic of the Dean–Stark distillation extraction unit.

overflows and drips back over the rock sample. The process is continued until no more water is collected in the receiving tube.

One major disadvantage of the Dean–Stark method is an inordinate amount of time required to extract all the water from very tight formations, such as low permeability chalk, in which case the production rate of water may be very slow causing the process to continue for days to ensure complete recovery. Additionally, due to the very low rate of water recovery, in some cases, a tendency to prematurely terminate the distillation process may occur under the assumption that a plateau in the cumulative water volume vs. time has been reached.

Unlike the retort distillation method, only the water saturation can be directly determined (Equation 6.4) using the Dean–Stark extraction because this is the only directly measured quantity, whereas the gas and oil saturations are determined indirectly. However, to indirectly determine the gas and oil saturations, it is necessary to record the weight of the rock sample prior to the extraction process. Then, after the rock sample has been cleaned and dried, it is again weighed. The following equations describe the procedure for indirectly calculating the gas and oil saturations from the Dean–Stark data and the rock sample weight data:

Let WW be the wet weight of the rock sample (or as received sample), DW the dry weight of the rock sample after Dean–Stark extraction, cleaning, and drying, M_g the weight of the gas (unknown, to be determined), M_o the weight of the oil (unknown, to be determined), M_w the weight of the water recovered from Dean–Stark, V_g the volume of the gas (unknown, to be determined), V_o the volume of the oil (unknown, to be determined), V_w the volume of the water recovered from Dean–Stark, ρ_g the density of the gas, ρ_o the density of the oil, and ρ_w the density of the water.

The difference between the wet weight and the dry weight of the rock sample is equal to the weight of fluids in the rock sample

$$\mathrm{WW} - \mathrm{DW} = M_g + M_o + M_w \tag{6.8}$$

Similarly, pore volume (PV) of the rock sample is equal to the summation of the volumes of gas, oil, and water

$$\mathrm{PV} = V_g + V_o + V_w \tag{6.9}$$

WW, DW, and V_w are directly measured quantities whereas PV (generally after cleaning and drying) can be obtained from the sample porosity and bulk volume determined by any of the methods discussed in the Chapter 3. The weights of the gas, oil, and water phases can be expressed in terms of volume and density, $V_g\rho_g$, $V_o\rho_o$, and $V_w\rho_w$. The density of water recovered from the Dean–Stark extraction can be directly measured, whereas the density data of gas and oil are normally at the surface conditions and are usually taken from the accompanying reservoir fluid studies report.

Equations 6.8 and 6.9 can now be rearranged as

$$V_g\rho_g + V_o\rho_o = \mathrm{WW} - \mathrm{DW} - V_w\rho_w \tag{6.10}$$

$$V_g + V_o = \mathrm{PV} - V_w \tag{6.11}$$

Equations 6.10 and 6.11 can be solved to obtain the values of V_g and V_o from which fluid saturations are calculated by

$$S_g = \frac{V_g}{\mathrm{PV}}; \quad S_o = \frac{V_o}{\mathrm{PV}}; \quad S_w = \frac{V_w}{\mathrm{PV}} \tag{6.12}$$

6.6 ASSESSING THE VALIDITY OF FLUID SATURATION DATA MEASURED ON THE PLUG-END TRIM FOR THE CORE PLUG SAMPLE

Section 6.4 discussed in detail the various issues related to the consideration of routine core analysis data (particularly fluid saturations) of plug-end trims as valid for the entire core plug sample from which the end trim originates. However, a way exists by which this assumption can be assessed or verified. Consider a preserved core plug sample from which an end trim was taken off, as shown in Figure 6.2. All the routine core analysis data are measured on this plug-end trim, and SCAL tests are planned on the remainder of this preserved core plug sample. However, before beginning the SCAL tests, the representativity of the end trim data especially on fluid saturations, as that valid for the entire core plug should be evaluated. The required assessment can be easily carried out in the following manner.

Let W_c be the recorded (from a balance) weight of the trimless core plug sample (the one used for the SCAL test) and W_{ct} the theoretical or calculated weight of the trimless core plug sample where

$$W_{ct} = \text{weight of grain} + \text{weight of fluids} \tag{6.13}$$

$$\text{weight of grain} = \rho_{grain} \times (\text{BV} - \text{PV}) \tag{6.14}$$

$$\text{weight of fluids} = \text{PV}\,S_g\rho_g + \text{PV}\,S_o\rho_o + \text{PV}\,S_w\rho_w \tag{6.15}$$

Equation 6.13 now becomes

$$W_{ct} = \rho_{grain}(\text{BV} - \text{PV}) + \text{PV}\,S_g\rho_g + \text{PV}\,S_o\rho_o + \text{PV}\,S_w\rho_w \tag{6.16}$$

where ρ_{grain} is the grain density measured from plug-end trim grains (and is assumed to also represent the trimless core plug because at this stage trimless core plug value is unknown), BV the bulk volume of the trimless core plug sample (e.g., measured from dimensions), PV the pore volume of the trimless core plug sample, calculated from $\phi \times$ BV, ϕ the porosity measured on the plug-end trim (and is assumed to also represent the trimless core plug because at this stage the trimless core plug value is unknown), and S_g, S_o, and S_w the gas, oil, and water saturations measured on the plug-end trim from Dean–Stark or retort distillation techniques (and are assumed to also represent the trimless core plug because at this stage the trimless core plug value is unknown).

Generally, if the recorded and calculated or theoretical weights of the trimless core plug, W_c and W_{ct} are equal then it can be assumed that the plug-end trim porosity, grain density, and fluid saturation data are also valid or can be considered as representative for the trimless core plug sample. Even though the equality of W_c and W_{ct} might be a rare occurrence in practical core analysis, agreement within few percent generally confirms the validity of the assumption.

6.7 SPECIAL TYPES OF FLUID SATURATIONS

Section 6.6 provided the basic definitions and the measurements of gas, oil, and water saturations in a reservoir rock. These definitions allowed us to define the distribution of pore space or pore volume of a reservoir rock into the individual fluid phases of gas, oil, and water. However, three special types of fluid saturations are important or rather the magnitude of these saturations associated with gas, oil, and water phases that are of particular importance and interest in reservoir engineering. These are called

1. Critical gas saturation
2. Residual oil saturation
3. Irreducible water saturation

The three saturations play a key role in understanding the flow of multiphase fluids in porous media and the recovery of hydrocarbon fluids from petroleum reservoirs. As addressed in Chaper 9, these three saturations in fact constitute the end points of the relative permeability curves. Therefore, it is appropriate and logical to discuss these saturations here and hence set the groundwork for relative permeability.

6.7.1 Critical Gas Saturation

Hydrocarbon fluids in a petroleum reservoir normally exist at high pressure and high temperature conditions. Due to this high pressure, hydrocarbon gas is normally dissolved in the liquid phase. Because production from a petroleum reservoir is initiated, the reservoir pressure begins to decrease while the reservoir temperature generally remains constant. The steadily declining reservoir pressure results in the evolution of a gas phase (gas saturation increase from 0) when pressure falls below a certain solubility limit, known as *bubble point pressure.* Subsequently, the saturation of the gas phase increases as the depletion of reservoir pressure continues. This gas phase, however, remains immobile or is trapped until its saturation exceeds a certain saturation value, called *critical gas saturation* and denoted by S_{gc}. The gas phase then begins to move above this critical gas saturation. The entire process is attributed to the physical process of the gas phase becoming continuous through the system in order to flow.

A typical sequence of events related to critical gas saturation is depicted in Figure 6.6. Critical gas saturation can significantly impact the production of oil from petroleum reservoirs. In primary oil production, solution gas drive (hydrocarbon gas phase dissolved in the hydrocarbon liquid phase) is the chief mechanism of oil production because gas comes out of solution and expels the oil. Since gas is quite compressible, it maintains the reservoir pressure high enough to cause recovery. However, when the gas saturation reaches a critical value it begins to flow thus reducing the reservoir pressure, hampering oil production.

6.7.2 Residual Oil Saturation

Residual oil saturation generally is denoted by S_{or}, and can be construed in two different ways by reservoir engineers. To some engineers, it is the oil saturation remaining in the reservoir at the conclusion of primary production or after either the gas or water displacement process. However, S_{or} defined in this manner is normally the target for EOR. To others, residual oil saturation is the final oil saturation in a reservoir rock core sample at the end of a laboratory gas displacement or water displacement process. The concept of residual oil saturation from a laboratory core flood view point

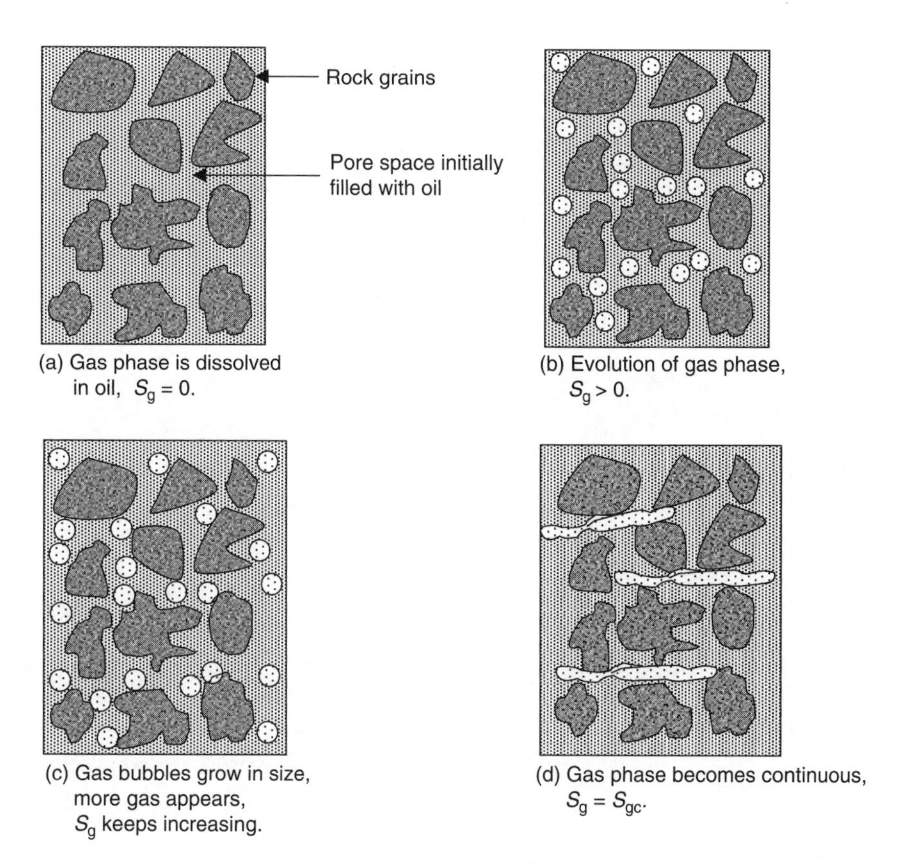

FIGURE 6.6 Schematic representation of events leading to critical gas saturation.

can be best described by a simple core plug displacement experiment discussed in the following text.

Figure 6.7a considers a core plug that is initially 100% saturated with a hydrocarbon liquid or oil phase and into which either gas or water is injected. As soon as the displacing phase, either gas or water, is injected in the core it will start replacing the oil phase from the pore spaces, and oil will be produced from the opposite end of the core plug. As the process continues, more and more oil is produced, however, at a certain point in time the oil production declines (as the displacing phase is also produced) and eventually cease and only the displacing phase is produced from the opposite end. If cumulative oil production is now plotted as a function of time; the plot shows a horizontal or a plateau after a certain time, which basically signifies the maximum amount of oil that can be produced from this core plug by either gas or water injection (see Figure 6.7b). However, as seen in Figure 6.7c, a 100% recovery of oil from this core plug is not possible by injection of either gas or water, since some oil still remains trapped inside the pore spaces of this core plug sample. This particular trapped oil or remaining oil is nothing but the residual oil saturation. In summary, if gas or water injection is continued further it simply

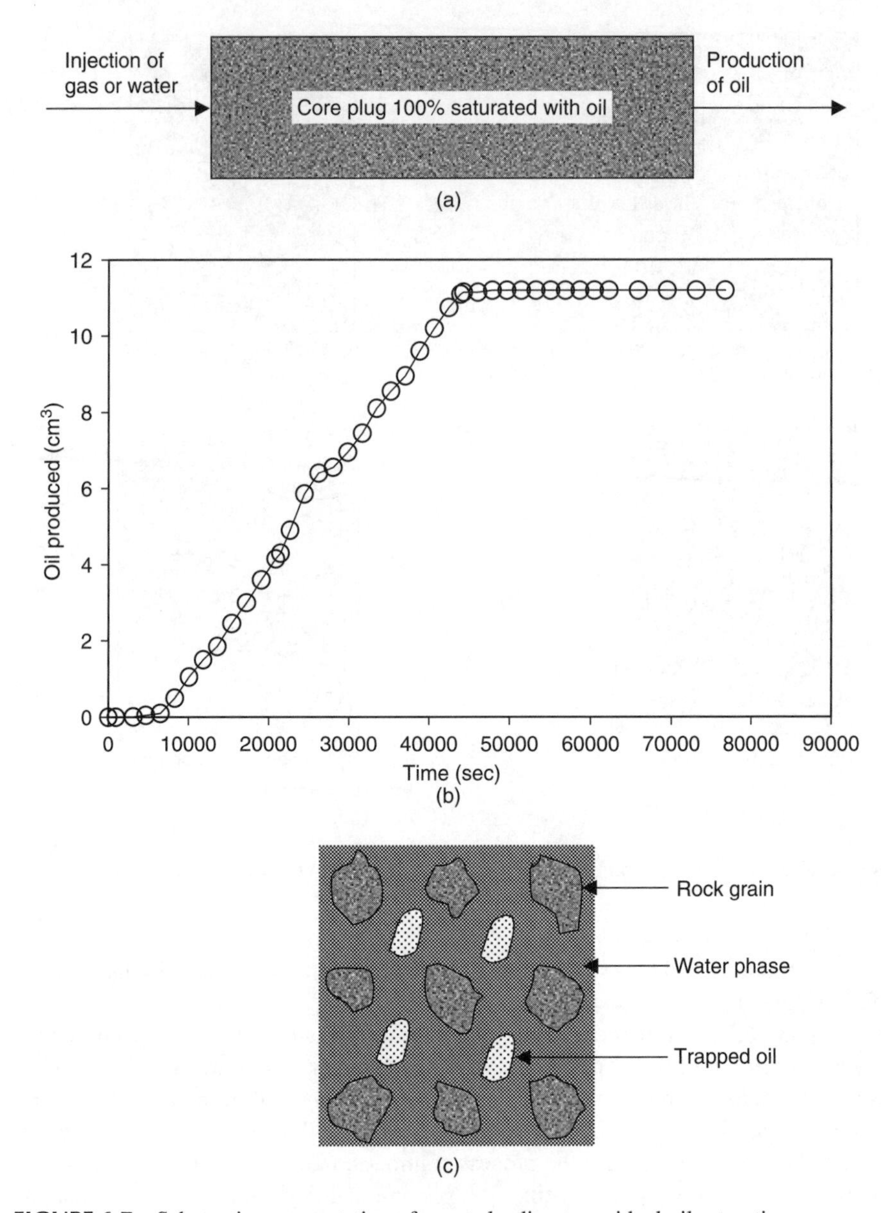

FIGURE 6.7 Schematic representation of events leading to residual oil saturation.

bypasses this trapped oil and only the displacing phase is produced at the opposite end of the sample. For a simple experiment of this nature, the residual oil saturation can be easily determined from the following equation:

$$S_{or} = \frac{(\text{PV} - \text{cumulative vol. of oil produced})}{\text{PV}} = \frac{\text{trapped oil in the sample}}{\text{PV}} \tag{6.17}$$

Depending on the type of displacing phase used (i.e., gas or oil) the S_{or} in Equation 6.17 is further categorized as S_{org} (gas flood residual oil saturation) or S_{orw} (water flood residual oil saturation). As outlined in Chapter 9, the S_{or} defined by Equation 6.17 is in fact somewhat analogous to the end-point saturation of the relative permeability curves. Finally, whichever manner one looks at residual oil saturation; probably it is the most important term in the petroleum industry as this signifies how much oil can be ultimately recovered or how much is left behind. Despite that laboratory core floods give a fairly reasonable indication of S_{or} for a particular formation, remember that these tests are affected by a number of factors such as the type of test conducted, test conditions and procedures, rock types, and properties of the displaced oil and the displacing phases.

6.7.3 Irreducible Water Saturation

The term *irreducible water saturation*, denoted by S_{wi}, is defined as the minimum water saturation or the least value of water saturation that is present in a porous medium. The other commonly used terms for irreducible water saturation are: interstitial water saturation, initial water saturation, connate water saturation, or capillary bound water. Frequently, these terms are used interchangeably in petroleum engineering literature.

To understand the concept of irreducible water saturation, first consider an idealized petroleum reservoir showing gas, oil, and water distribution, as shown in Figure 6.8. The fluids in most petroleum reservoirs, shown in Figure 6.8, have reached a state of equilibrium and have become somewhat separated as per their densities (i.e., gas on top) followed by the oil phase, and underlain by water. Prior to reaching this state of equilibrium, tiny pore spaces of the petroleum reservoir rocks are completely saturated with water in which the hydrocarbons migrated from a source rock. However, due to the competition between capillary and gravity forces, during this migration process, complete gravity segregation between the fluid phases never takes place, and the connate water is distributed throughout the gas and oil zones, as shown in Figure 6.8. The water in these zones is reduced to some irreducible minimum that is nothing but the irreducible water saturation, S_{wi}.

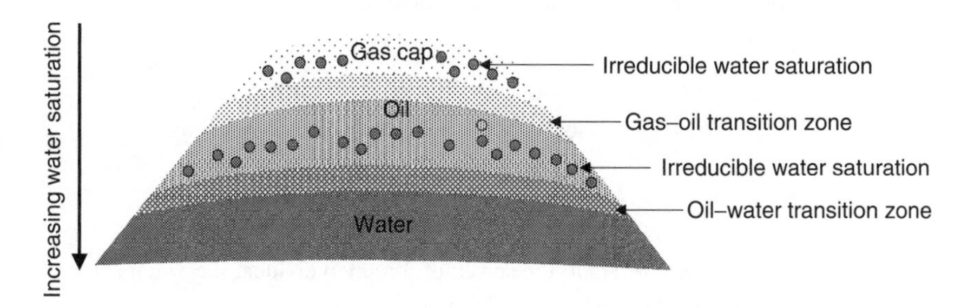

FIGURE 6.8 Schematic representation of irreducible water saturation in an idealized gravity–capillary equilibriated petroleum reservoir.

The forces retaining the water in the gas and oil zones are referred to as capillary forces because they are important only because of the tiny pore spaces of capillary size. Once this entire process was complete the petroleum reservoirs reached the state of equilibrium known as *capillary–gravity equilibrium*. The irreducible water saturation is generally not uniformly distributed throughout the reservoir but varies with permeability, lithology, height above the free water table, and most importantly with the capillary and gravity forces in a petroleum reservoir.

Irreducible water saturation can be addressed in a laboratory scenario with a core flooding experiment similar to the one discussed in the Section 6.7.2 but the fluids used must be switched. Initially the core plug sample is 100% water saturated in which either gas or oil is injected until no more water is produced; that is, the core plug sample is flooded down to irreducible water saturation. The numerical value of S_{wi} can be determined by changing the terms in Equations 6.17 to cumulative volume of water produced or water remaining in the sample. However, one remarkably distinguishing feature between the laboratory-obtained S_{wi} and one found in gas and oil zones in the petroleum reservoirs, is that the former is capillary–viscous-based while the latter is capillary–gravity-based. Therefore, S_{wi} achieved in the laboratory is relative and perhaps the term *irreducible water saturation* is somewhat imprecise because it depends on the final drive pressure or viscous pressure drop when flowing gas or oil.

In summary, whichever manner is used, S_{wi} is a very important parameter because it reduces the amount of space available for the hydrocarbon phase of either gas or oil. Additionally, unlike critical gas saturation or residual oil saturation that can be construed as an artificially created saturation, irreducible water saturation in petroleum reservoirs, on the other hand is an entirely nature-driven process and is influenced only by the competition between the capillary and gravity forces. A range of 20 to 40% in irreducible water saturation in petroleum reservoirs is rather common, however, values ranging from as low as 5% to those as high as 60% (depending on the capillary properties of rocks) have also been reported for some North Sea chalk reservoirs.[2]

6.8 SATURATION AVERAGING

The fluid saturation data for various intervals of a given thickness and porosity can be averaged as per the following generalized equation:

$$S_{l\mathrm{avg}} = \frac{\sum_{i=1}^{n} S_{li}\phi_i h_i}{\sum_{i=1}^{n} \phi_i h_i} \tag{6.18}$$

where l is gas, oil, or water, i refers to the subscript for any individual measurement, and h_i represents the depth interval to which ϕ_i and S_{li} apply.

The averaged saturation values in Equation 6.18 are weighted by both the interval thickness and the interval porosity.

6.9 FACTORS AFFECTING FLUID SATURATION DETERMINATION

The determination of fluid saturation in the laboratory on core samples is probably one of the least reliable reservoir property measurements. Other than inaccuracies associated with the various measurement techniques discussed in Sections 6.5.1 and 6.5.2 factors that are more likely to introduce errors and uncertainties in fluid saturation measurements are improper drilling, handling, preservation, and analysis of the core material in an improper fashion.

In principle, two different processes are likely to introduce uncertainties in the fluid saturations or alter the fluid saturations: one related to the invasion of the core sample by the mud or the mud filtrate during the coring process and the other related to the shrinkage and expulsion of fluids from the core material as the core is brought to the surface from the reservoir. These two processes of mud filtrate invasion and the shrinkage and expansion alter the initial fluid content of the core material. Although these two factors initially affect the whole core sample, it is also consequential for subsamples (core plugs) and plug-end trims. Each of these factors will now be examined individually.

6.9.1 EFFECT OF DRILLING MUDS ON FLUID SATURATION

In all drilling operations, usually a particle suspension mixture of finely divided heavy material, such as barite or bentonite blended with a liquid, is pumped down the hole through a drill pipe primarily to cool and lubricate the rotating bit and to hinder the penetration of reservoir fluids into the well bore. These drilling muds can be generally categorized according to the liquid used (i.e., water-based, non-water-based, or oil-based). Additionally, gaseous or pneumatic drilling muds are used. In the case of rotary drilling, the formation is under greater pressure from the mud column in the well than from the fluid in the formation. Thus, the differential pressure across the well face causes mud and mud filtrate to invade the formation resulting in flushing the formation with mud and its filtrate, thereby altering the fluid saturations.

This invasion process is rather complex. It is generally considered to start with a short initial spurt loss when the drill bit penetrates the formation and continues in a dynamic fashion when the mud is circulated. The mud filtrate invasion can also be characterized by *diameter of invasion*. This term assumes equal invasion on all sides, basically forming a circle for which the center is the center of the borehole. When core samples are obtained, the process results in two diameters of invasion; one with respect to formation and the other with respect to core sample that can be categorized as outward and inward invasions, respectively.

In a normal drilling operation when core samples are not taken, the mud filtrate invasion occurs only in the formation resulting in only one diameter of invasion. When the mud filtrate invasion occurs only in the formation, the process usually starts with a build-up of mud cake on the walls of the formation and subsequently the filtrate from the cake penetrates the formation resulting in formation damage in the invaded zone. Only the invasion that occurs in the core sample recovered from the formation affecting fluid saturations is discussed here. Figure 6.9 illustrates the concept of diameters of invasion.

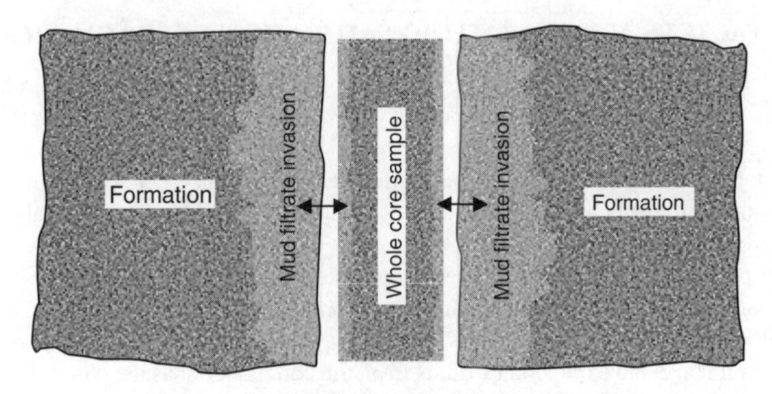

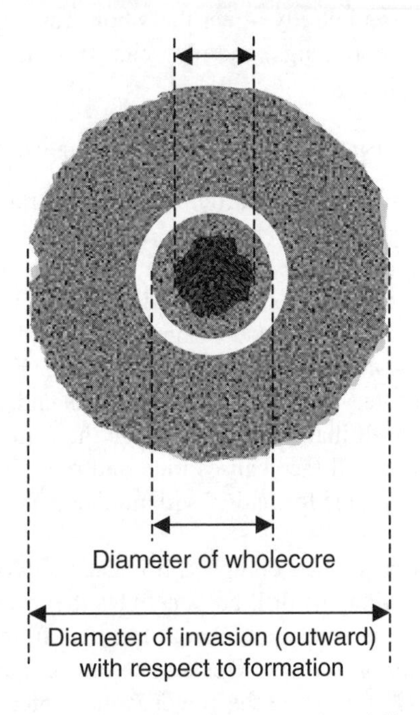

FIGURE 6.9 Conceptual illustration of the diameters of mud filtration invasion.

Therefore, any type of drilling mud almost always affects the initial fluid saturations of the core material, however, the use of a particular type of drilling mud (water or oil-based) actually dictates the alteration of the saturation of a particular fluid phase for water or oil. If the drilling mud is water-based, most likely this results in an increase in the water saturation because the water filtrate may invade the core material and displace some of the oil leading to higher water saturation. On the other hand, if the drilling mud is oil-based, the initial water saturation is almost likely unaltered if it is at an irreducible (immobile) value.

However, this situation may not exist for transition zones or water zones because water is mobile and the oil-based mud filtrate may flush the core and reduce the water saturation. Oil-based drilling muds, usually provide a reasonably good estimate of irreducible water saturation but not necessarily the water saturation existing in a transition zone or water zone in a reservoir. Therefore, the connate water saturation determined from cores drilled with oil-based muds has long been an industry standard, whereas with water-based drilling muds the contamination from the mud filtrate becomes much more an issue and thus the initial water saturation values are considered as questionable or unreliable.

Ringen et al.[3] presented data for a number of North Sea wells that were analyzed with respect to water saturation and mud filtrate invasion from water-based muds indicating very high levels of invasion that averaged about 22% of pore volume. The invasion of mud filtrate is generally controlled by parameters such as rate of penetration, overbalance or the pressure differential across the well face, pumping rate, stability of the filter cake, reservoir fluid properties such as oil viscosity, and formation permeability.[3] For example, higher oil viscosity inhibits invasion, while higher permeability increases invasion at the core bit.

Pneumatic-based drilling muds also experience problems similar to the ones associated with water- and oil-based drilling muds on initial fluid saturation. In gaseous or pneumatic-drilling muds, due to the poor heat transfer capacity of gas, a large amount of heat is generated during the coring process. This heat, combined with the dehydrated nature and high rate of gas circulation required to clean the hole often results in desiccation of the core material and artificially low water saturations.

6.9.2 Effect of Fluid Expansion on Fluid Saturation

The second factor that contributes to the errors or uncertainties in core-measured fluid saturation is related to pressure and temperature changes that in turn affect the properties or characteristics of the hydrocarbons present in the core material. This section discusses this effect on fluid saturation.

At reservoir pressure and temperature (elevated pressure as well as elevated temperature) most liquid hydrocarbons contain dissolved solution gas. It is this particular liquid hydrocarbon (oil saturation) that is present in the reservoir rock along with the formation water (water saturation). When cores are cut from such reservoir rocks and are brought to the surface, the core is subjected to pressure and temperature reduction. The pressure depletion experienced by the core results in the release of gases that are initially dissolved in the liquid hydrocarbons, causing a shrinkage in oil volume, and, as the gases expand and escape from the core, an expulsion of some of the mobile oil and water from the pore system.

In addition to the pressure effect, thermal contraction of any oil and water present in the pore system may also be significant as the core material cools from reservoir temperature to surface temperature. Therefore, these pressure and temperature effects result in completely altered fluid saturation (mostly the hydrocarbon phase) in the core sample as compared to the actual reservoir. Hence, fluid saturations measured in a core sample at the surface do not necessarily reflect the true saturation that exist in the reservoir.

Bennion et al.[4] illustrated a reduction in oil saturation by almost 50% due to the expansion effects while water saturation remained almost unchanged during comparison of the core data and the reservoir data. Similarly, Koepf,[5] reported that the oil saturation of a core sample on its trip from the reservoir to the surface, for a virgin oil productive formation, reduced from 70 to 20% due to shrinkage and expulsion, while water saturation remained the same. However, in certain cases core fluid saturations found at the surface may be quite similar to those found in the reservoir; for instance, in a pressure-depleted reservoir the oil may contain little or no gas in solution, thus preserving the reservoir oil saturations in the recovered cores.[6]

6.9.3 Combined Effects of Mud Filtrate Invasion and Fluid Expansion on Fluid Saturation

To better understand the overall or combined effect of mud filtrate invasion and fluid expansion on fluid saturation, Kennedy et al.[7] actually simulated rotary coring techniques. In a specially designed test cell, a cylindrical core sample with original oil and water saturations simulating reservoir saturations was used. It also had a hole drilled in the middle to represent the well bore. In this middle hole, mud under pressure was pumped to allow the filtrate to invade the core sample. The oil phase and the water phase flushed from the core sample were measured at the outer boundary. The values gave the change in saturation caused by the flushing action of the mud filtrate. Subsequently, the pressure on the core sample was reduced to atmospheric pressure and the amount of oil and water phase remaining in the sample was determined. These two steps thus allowed the quantification of the combined effects of mud filtrate invasion and fluid expansion on fluid saturations.

In their experiments, Kennedy et al.[7] used both water-based as well as oil-based mud to compare the magnitude of invasion a particular type of mud would cause. Typical alterations in fluid saturations of the core sample flushed with water- and oil-based mud, and fluid expansion, are illustrated in Figures 6.10 and 6.11, respectively. For water-based mud, about 14% of the original oil is displaced by the mud filtrate increasing the water saturation from 32.4 to 46.6% and subsequently reducing to 38.5% as some oil and water is expelled due to the pressure reduction (expanding gas displacing the oil and water). This indicates that the final water saturation was greater than the water saturation prior to coring. However, when oil-based mud is used, the mud filtrate invasion did not alter the initial water saturation but did result in the replacement of about 20% of the initial oil. In the subsequent step of pressure depletion, water saturation reduced by less than 2% from its original value, while oil saturation decreased even further to about 27%. These results indicate that water saturation values obtained with oil-based muds may be considered as representative of the initial water saturation in the reservoir. Hence, by judicious selection of the drilling fluid, it is feasible to obtain fairly representative values of in-place water saturations.

Kennedy et al.[7] tested core samples (mostly limestones) with wide-ranging porosities and permeabilities to evaluate the combined effect of drilling fluids and fluid expansion on fluid saturation. Based on their experiments, they correlated the hydrocarbon saturations before and after coring. The initial and final hydrocarbon saturations could be correlated by a straight-line fit for initial saturations greater than 15%. These type of correlations can be used to correct the saturations measured from core sample to original conditions, however, additional data on a wide variety of core

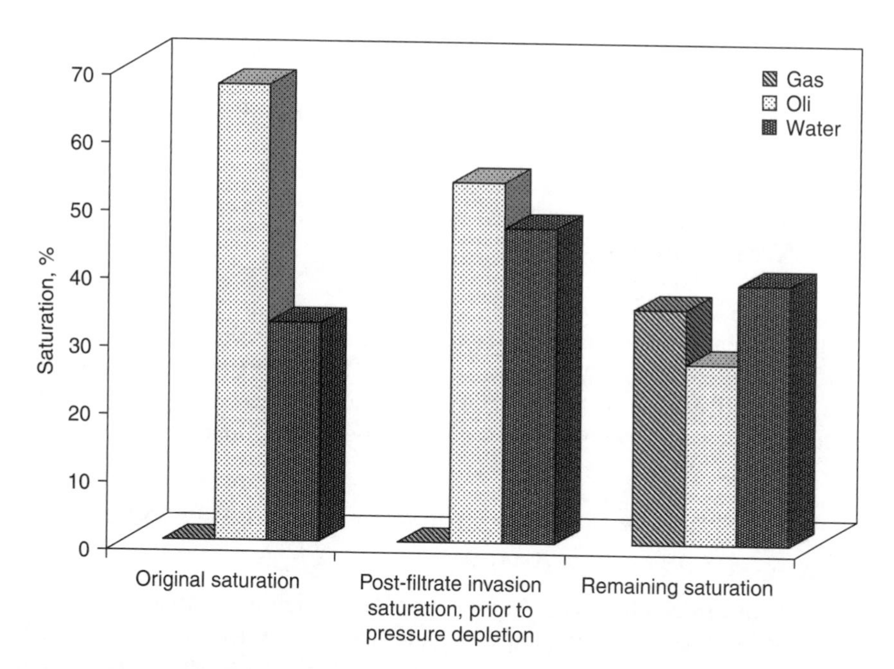

FIGURE 6.10 Typical alterations in the fluid saturations of a core sample flushed with water-based mud, from Kennedy et al.[7]

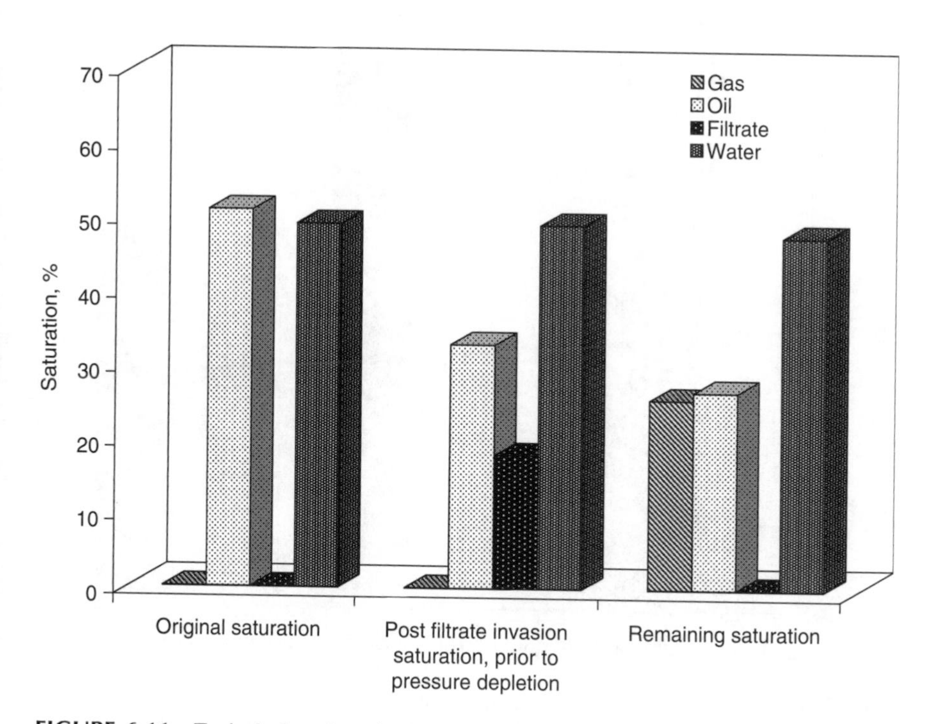

FIGURE 6.11 Typical alterations in the fluid saturations of a core sample flushed with oil-based mud, from Kennedy et al.[7]

samples are required before generalized or universally applicable correlations are established.

The results similar to the ones reported by Kennedy et al.[7] were also presented by Koepf[5] for three different cases:

1. A virgin productive formation
2. A water-flooded reservoir
3. A pressure-depleted reservoir

Fluid saturations were compared at all three levels: in the reservoir, in the core barrel, and at the surface. The results presented by Koepf[5] for case 1 and case 2 are summarized in Figures 6.12 and 6.13, respectively. It should, however, be noted that all the results in Figures 6.12 and 6.13 are for a water-based drilling mud.

6.9.4 Mitigation of Mud Filtrate Invasion and Fluid Expansion Effects on Fluid Saturation

Accurate knowledge of initial fluid saturation distribution in a formation is of significant importance in determining the initial reserves of actual and recoverable hydrocarbons in place. The most common source of fluid saturation data is based on the routine core analysis of the core material recovered from a given formation.

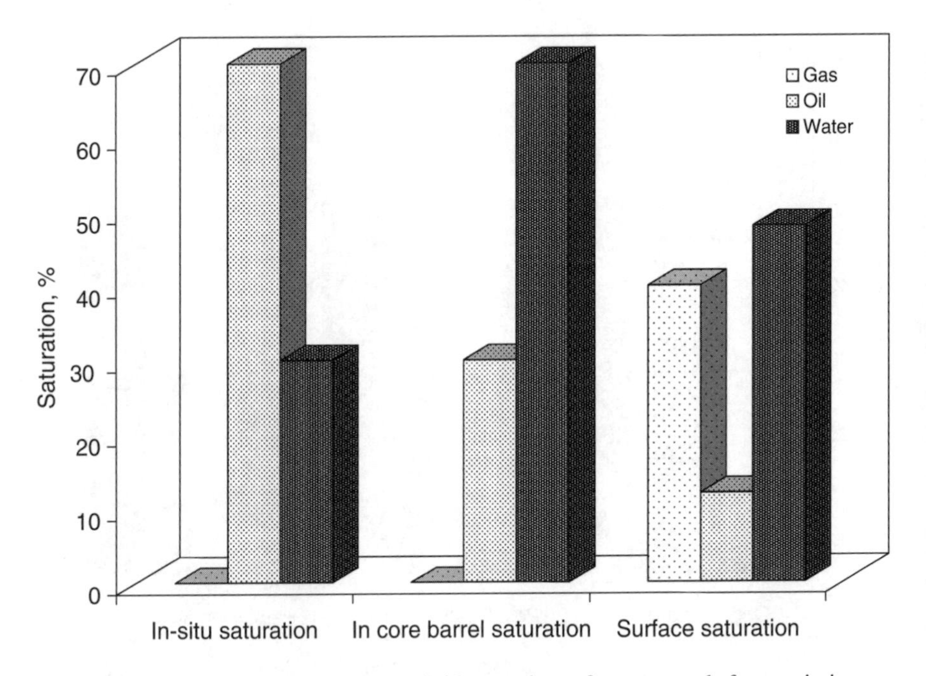

FIGURE 6.12 Typical alterations in the fluid saturations of a core sample from a virgin productive formation that was badly flushed with water-based drilling mud, from Koepf.[5]

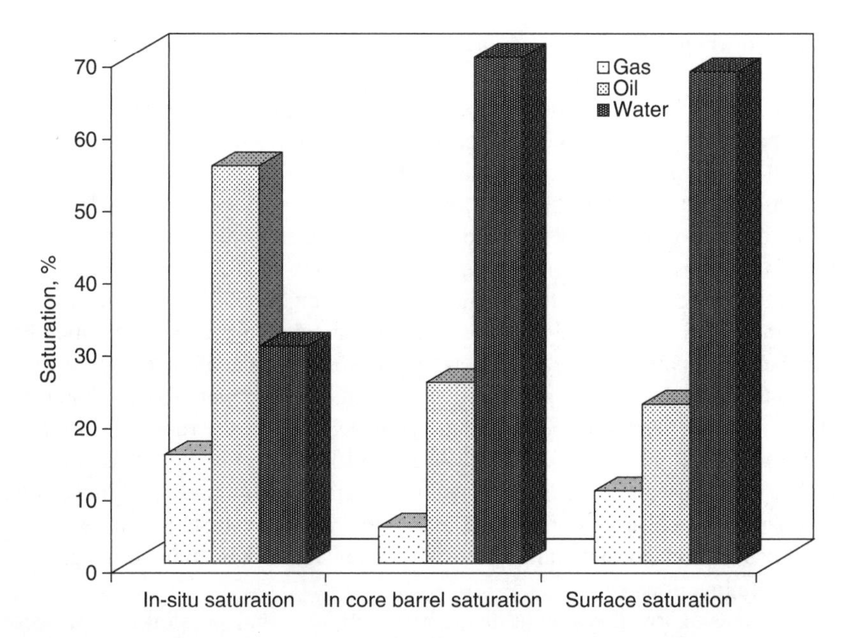

FIGURE 6.13 Typical alterations in the fluid saturations of a core sample from a pressure-depleted formation that was badly flushed with water-based drilling mud, from Koepf.[5]

But, as discussed in Section 6.9, these render inaccurate results due to inherent mud filtrate invasion during coring as well as oil shrinkage and expulsion due to gas expansion during the core trip from the reservoir to the surface. Therefore, attempting to infer accurate *in situ* fluid saturations from core material recovered from hydrocarbon formations can be fraught with problems and is indeed a very challenging task. The accuracy of the results may be seriously compromised if the core material is not drilled, handled, preserved, and analyzed in the proper fashion.

The coring of a given formation and its subsequent analysis in the laboratory is generally a very resource-intensive activity that if planned properly by all the involved parties, yields maximum amount of accurate and useful data (fluid saturation being one) from the recovered core material. It is generally through this careful planning process, attempts are made by various parties involved to actually mitigate the problems associated with mud filtrate invasion and gas expansion so that accurate fluid saturation data are obtained. The following two sections discuss the various measures taken to minimize these effects and correct the measured data.

6.9.4.1 Measures That Avoid or Account for Mud Filtrate Invasion

The measures that can be taken to avoid or correct the mud filtrate invasion are usually dictated by the fluid saturation specifically desired, that is, initial water saturation or oil saturation. However, the former is almost always the desired key parameter, based on which the hydrocarbon pore volume or oil saturation is calculated. Hence if oil-based drilling muds are used then the core-measured water saturations (irreducible) are generally reliable. Proper selection of the drilling mud also helps mitigate mud filtrate invasion, yielding accurate water saturations. Holstein and Warner[8] and Richardson et al.[9] reported that the as-received (core sample) S_{wobc} (cored using oil-based mud) values represent reliable measurements of *in situ* S_w values for the reservoir (Ivishak, Prudhoe Bay Field) interval above the oil–water transition zone.

The other most common measure that can be taken to avoid or minimize mud filtrate invasion is use of bland coring fluid. This type of coring or drilling fluid is formulated with components that are not likely to alter wettability in the pores of the rock sample and have low dynamic-filtration characteristics. These qualities help retain the core's native properties and can retain some (or all) of the reservoir fluids, i.e., gas (if maintained under pressure), oil and water. Bland water-based fluid is formulated to make the filtrate resemble the connate water in the reservoir. Similarly, bland oil-based fluids should be water-free, and the base oil should resemble the reservoir oil (reservoir crude is used in some cases). The data presented by Egbogah and Amar[10] for a coring program for Dulang field in Malaysia demonstrate the successful use of bland water- and oil-based drilling fluids, resulting in insignificant mud filtrate invasion.

Despite using low invasion or bland drilling fluids, some possibility of invasion, particularly in zones of high permeability, may still exist. Saturations evaluated from a low-invasion coring process may still be questionable. This can be remedied by doping or tagging the drilling or coring fluid with some kind of a tracer material. The tracer usually consists of material, not present in the naturally occurring formation fluids, that can be readily analyzed. The occurrence of invasion is evidenced or ascertained by the presence of this tracer in the fluids removed from the core samples. Tracers can be used in both oil- and water-based systems (although water-based tracer systems tend to be much more common). The materials used as tracers include deuterium and tritium (isotopic forms of water), calcium chloride, hexachloroethane, and so on. These tracers can also be used for detecting the mud filtrate invasion in cores recovered using ordinary water- or oil-based drilling muds. Based on the concentration of tracer found in the extracted fluids (oil and water by either retort distillation or Dean–Stark method) from the core samples, the measured saturation data are corrected to the reservoir saturation data. An example is shown in Figure 6.14, based on the water saturation data reported by Bennion et al.[4] As seen in this example, as the tracer (tritium) concentration increases, the difference between the measured and corrected saturation increases, or the higher the tracer concentration, the greater the applied correction.

Other measures taken to avoid or minimize the mud filtrate invasion include the use of gels[11] that throughout the coring process encapsulate the core downhole in a viscous, noninvasive, and protective medium. These gels are designed to eliminate mud filtrate invasion thus enhancing the representation of the core sample with respect to *in situ* wettability and water saturation. Additionally, by controlling mechanical factors such as maintaining low overbalance, low pump rate, and high rate of penetration, may also result in coring with minimum mud filtrate invasion.

Another very important issue that needs to be addressed regarding mud filtrate invasion is the consequence this invasion has on core plugs or subsamples that are

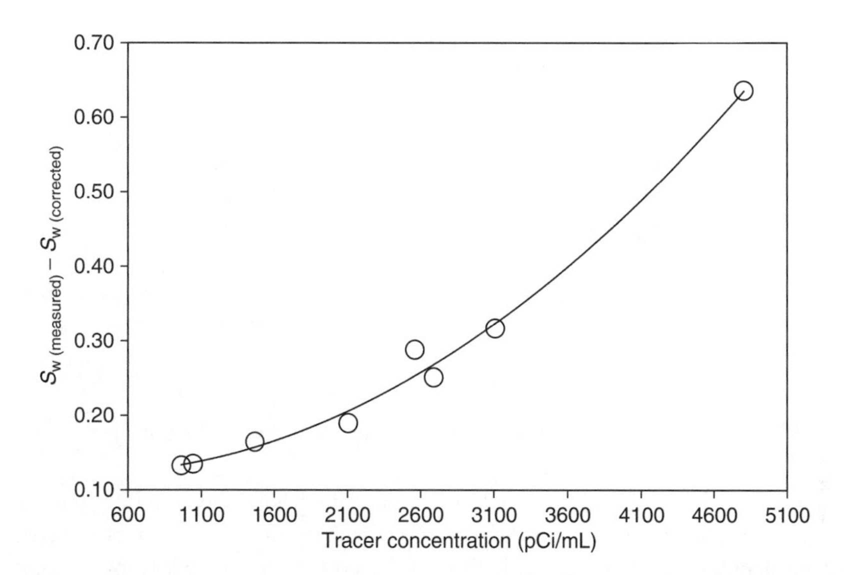

FIGURE 6.14 Correction of measured water saturation based on tracer data, from Bennion et al.[4] Unusually high water saturations indicate severe invasion from mud filtrate of the water-based drilling fluid has occurred and can be corrected using the tracer concentration.

drilled from the whole core. Core plug samples are routinely used in special core analysis studies during which accurate fluid saturations (especially initial or connate water) are equally as important as they are determining hydrocarbon pore volume in a given formation. To understand the implications of mud filtrate invasion on water saturations of core plug samples draw a top view of a water-based mud filtrate invaded core sample, as illustrated in Figure 6.15, where the invaded zone is shown by the shaded area formed by a concentric circle with respect to the core diameter. This is a vertical whole core from which core plug samples are drilled horizontally (parallel to the bedding plane or perpendicular to the long axis of the whole core). The specific effect the invasion may have on water saturation is qualitatively described by the following points:

- The overall water saturation in the core plugs are not representative of the true reservoir water saturation.
- The outer section of the core plug has high water saturation, while the inner section of the core plug has a representative water saturation; see Figure 6.15.
- Subsequently, if plug-end trims are sliced from such a core plug sample for routine core analysis, then depending on which side the end trim is taken, it will either have a very high water saturation or a representative water saturation (see the discussion on plug-end trim in Sections 6.4 and 6.6).
- If these core plugs are used in special core analysis studies, such as relative permeability, then the measured data are meaningless because the starting water saturation is not representative of true reservoir water saturation.

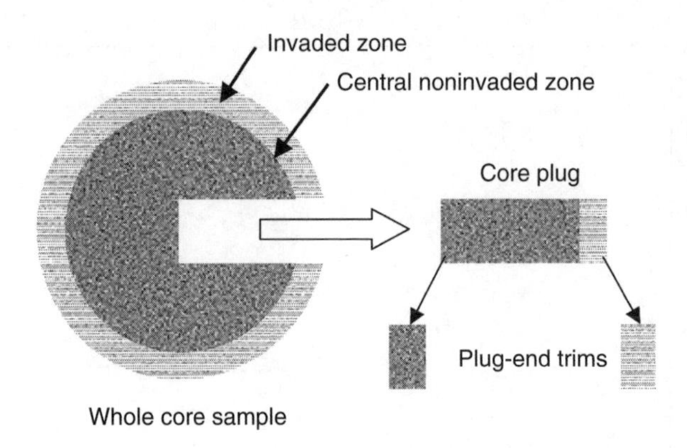

FIGURE 6.15 Effect of water-based mud filtrate invasion on core plugs and end trims.

Therefore in situations summarized by the preceding points, where only the initial water saturation is desired and the degree of flushing of the exterior of the core is inconsequential, the best option is to drill the core plug samples from the central noninvaded section of the whole core.

6.9.4.2 Measures That Avoid or Account for Fluid Expansion

Even if the mud filtrate invasion factor is minimized or corrected for fluid saturation, another issue to address is the fluid saturation changes that take place in the core sample due to the pressure and temperature reduction as the sample is lifted from the reservoir to the surface. Section 6.9.2 explains that oil saturation changes taking place can be quite significant. The most affected fluid saturation is in fact the oil saturation because the gas that expands is basically released from the reservoir oil, thereby causing the oil to shrink in volume as well as be expelled from the core. Conversely, gas solubility is rather low in formation waters and water has very low compressibility, actually resulting in minimal water shrinkage. However, if the determination of oil saturation (e.g., for the target oil in place for secondary or tertiary EOR projects) is the primary objective then fluid expansion effects assume utmost importance. Therefore, the fluid shrinkage and expulsion effects ought to be either precluded or corrected for determining meaningful fluid saturation data from the core samples. The various measures that are taken either avoid or account for the fluid expansion effects and are discussed in the following text.

Under some conditions, if flushing by mud filtrate is minor, valid fluid saturations can often be obtained by using a pressure-retaining core barrel to prevent loss of fluids by gas expansion. Pressure coring actually eliminates the expulsion of fluids from the core by maintaining the reservoir pressure until the samples and the fluids contained therein can be subjected to cryogenic freezing. However, core handling and analysis procedures for pressure cores are somewhat more complicated and much more demanding than those for conventional cores. Although, the use of a pressure core barrel can prevent loss of fluids by gas expansion, it does not solve the problem of mud filtrate invasion during coring. However, invasion effects can either be minimized or accounted for by adherence to certain operating procedures, as discussed in the previous section.

The determination of fluid saturations from pressure cores is in fact a two-step process involving distillation and extraction. The process has been discussed in great detail by Treinen et al.[12] The frozen pressure cores are first subjected to an overnight thawing period at ambient temperature that is followed by vacuum distillation at about 140°C. The first step effectively removes water and some of the hydrocarbon components completely while some other heavy components are partially distilled. In the next step, the core samples are then extracted using a two-solvent system with an equal volume blend of methylene chloride and carbon disulfide. The measured oil and water saturations can then be adjusted using a formation volume factor to yield *in situ* fluid saturations:

$$S_{\mathrm{o},in\ situ} = S_{\mathrm{o,\ distillation+extraction}} B_{\mathrm{o}} \tag{6.19}$$

$$S_{\mathrm{w},in\ situ} = S_{\mathrm{w,\ distillation}} B_{\mathrm{w}} \tag{6.20}$$

where B_o and B_w are the formation volume factors (FVF) for reservoir oil and formation water, respectively. Both formation volume factors represent the ratio of volume of oil or water at reservoir conditions and at surface conditions. The numerical values of both B_o and B_w are not constants; they keep changing and follow a well-defined pattern as reservoir pressures change. Therefore, oil and water saturations are corrected on the basis of the FVFs corresponding to the reservoir pressure at which pressure cores are recovered.

The FVF for oil is always >1 because the volume of oil at reservoir conditions is always higher than that at surface conditions. This is because at reservoir conditions the oil is *swollen* due to the dissolved gas and usually high reservoir temperatures that subsequently shrinks due to temperature reduction and evolution of gas due to pressure reduction as the oil travels to the surface. However, water FVF is usually quite close to unity because of substantially low gas solubility in water and contraction and expansion due to pressure and temperature reduction, as water is transported from the reservoir conditions to the surface conditions, are small and offsetting. Therefore, the adjusted water saturation to *in situ* conditions in Equation 6.20 basically reflects a small correction, while oil saturation corrections can be substantial due to the contributing factors as explained here.

The logic behind the application of a shrinkage correction factor, such as the one defined in Equations 6.19 and 6.20, is based on the analogy between the producing of oil (and the associated water that is also produced) and the lifting of the core sample from the reservoir to the surface because both are considered to experience a similar magnitude of shrinkage. This type of shrinkage correction is applied to all methods in an attempt to calculate the *in situ* fluid saturations.

Hagedorn and Blackwell[13] provide a good summary of experience with pressure coring. Since the use of pressurized core barrels has practical limitations, implementating reduced tripping time from the reservoir to the surface has also been suggested as an alterative to minimize the gas expansion effects.

The other technique quite commonly used is *sponge coring* and is designed to essentially capture any expelled fluids in the surrounding sponge material for subsequent analysis to determine the volume of the expelled fluids and institute these saturations back into the pore volume of the adjacent whole core sample. The sponge coring technique can be used for accurate determination of oil as well as water saturations. The sponge coring technique reports three saturations: the saturation in the core (by Dean–Stark or retort method), the saturation in the sponge (accounting for expulsion), and the total saturation (sponge and core). Assuming minimal mud filtrate invasion of the core during coring and correcting the fluid volumes for shrinkage (by applying corrections similar to Equations 6.19 and 6.20), the corrected fluid saturations may approach true reservoir saturations.

Rathmell et al.[14] pointed out that oil saturations routinely determined by retort distillation method (or perhaps the Dean–Stark method) can be used to evaluate water flooding residual oil saturation, provided the surface oil saturations of the core samples are adjusted for the shrinkage and expulsion that occurs during lifting of the core material. The corrections suggested by Rathmell et al.[14] are expressed as

$$S_{\text{o},\mathit{in\ situ}} = S_{\text{o,core}} B_{\text{o}} E \tag{6.21}$$

where $S_{\text{o},\mathit{in\ situ}}$ is the water flooding residual oil saturation and $S_{\text{o,core}}$ the oil saturation measured in the core at surface conditions by either the retort method or the Dean–Stark method. The oil FVF (B_{o}) is used to correct for shrinkage whereas the constant E is used to correct for the expulsion losses (also referred to as bleeding). In the absence of bleeding measurements on the specific reservoir they used a value of $E = 1.11$. This factor was determined on the basis of their simulated lifting experiments that resulted in about 10 cm^3 of a residual oil volume (expelled volume or bleeding volume) of approximately 100 cm^3 at atmospheric pressure (determined in the core). They computed the E factor as

$$E = \frac{1}{\left[1 - \frac{10}{100}\right]} = 1.111 \tag{6.22}$$

However, Egbogah and Amaefule[15] addressed correction factor for residual oil saturation from conventional cores by stating that after correcting the oil saturations for shrinkage, the resulting value is arbitrarily increased again by a factor of 10 to 15% in order to account for the bleeding factor. Unlike the single value proposed by Rathmell et al.,[14] Egbogah and Amaefule[15] determined the value of the E factor based on the data set of 29 full diameter cores recovered by a bland water-based sponge coring program. Their E factor values ranged from 1.004 to 1.209.

PROBLEMS

6.1 A petroleum reservoir has an areal extent of 55,000 ft^2 and a oil pay zone thickness of 100 ft. The reservoir rock has a uniform porosity of 25% and the connate water saturation is 30%. Calculate the initial oil in place.

6.2 A chalk core plug having a pore volume of 17.0307 cm^3 is fully saturated with reservoir brine. A synthetic oil (Isopar-L) flood is conducted on this plug. It is found that 12.25 cm^3 of reservoir brine was displaced from this plug by the Isopar-L. After reaching this value, no further reservoir brine could be displaced from the core plug. What is the connate water or irreducible water saturation of this core plug?

6.3 For the following core plugs, gas floods were carried out using nitrogen. The oil produced from plugs 1 and 2 for the gas floods was 9.0 and 6.9 cm^3, respectively. What is the residual oil saturation (S_{org}) in these two plugs?

Plug No.	Initial Saturations (%)			Ø (%)	BV (cm^3)	ρ_{grain} (g/ cm^3)	ρ_o (g/ cm^3)	ρ_w (g/ cm^3)	ρ_g (g/ cm^3)
	S_o	S_w	S_g						
1	64.64	35.36	0	38.27	63.05	2.719	0.723	1.0216	0.001
2	71.93	28.07	0	34.63	51.05	2.724	0.723	1.0216	0.001

6.4 A Dean–Stark extraction is performed on a North Sea chalk core plug sample, which extracted 5.77 cm^3 of water. The core plug has a porosity of 36.1% and a bulk volume of 24.5 cm^3. The wet and dry weights of the sample are 50.64 and 42.33 g, respectively. The gas, oil, and water densities are 0.001, 0.85, and 1.035 g/cm^3, respectively. Calculate the gas, oil, and water saturations in the core plug sample.

6.5 The following data is available for the end trim of a chalk core plug sample:
$S_g = 5\%$, $S_o = 48\%$ and $S_w = 47\%$
$\rho_{grain} = 2.713$ g/cm^3
$\phi = 38.31\%$

Additional data include gas, oil, and water of densities 0.001, 0.8532 and 1.0351 g/cm^3, respectively.

The bulk volume of the core plug sample (in a preserved state), from which the end trim was taken, is 65.91 cm^3, and its measured weight (trimless) is 133.0 g. No additional data are available for the core plug sample. Perform an assessment check to evaluate if the end trim data is also valid for the core plug, so that the core plug sample can either be used or discarded for SCAL tests.

REFERENCES

1. Emdahl, B.A., Core analysis of Wilcox sands, *World Oil,* 1952.
2. Larsen J.K. and Fabricius, I.L., Interpretation of water saturation above the transitional zone in chalk reservoirs, *SPE Reservoir Evaluation Eng.,* 155–163, 2004.
3. Ringen, J.K., Halvorsen, C., Lehne, K.A., Rueslaatten, H., and Holand, H., Reservoir water saturation measured on cores; case histories and recommendations, *Proceedings of the 6th Nordic Symposium on Petrophysics, Trondheim,* Norway, 2001.
4. Bennion, D.B., Thomas, F.B., and Ma, T., Determination of initial fluid saturations using traced drilling media, paper number 99-08, Hycal Energy Research Laboratories Ltd., www.hycal.com.

5. Koepf, E.H., Coring Methods in Determination of Oil Saturation, Interstate Oil Compact Commissions, 1978.
6. Murphy, R.P. and Owens, W.W., The use of special coring and logging procedures for defining reservoir residual oil saturations, *Trans. AIME*, 255, 841, 1973.
7. Kennedy, H.T., Van Meter, O.E., and Jones, R.G., Saturation determination of rotary cores, *Pet. Eng.*, 1954.
8. Holstein, E.D. and Warner, H.R., Jr., Overview of water saturation determination for the Ivishak (Sadlerochit) reservoir, Prudhoe Bay field, Society of Petroleum Engineers (SPE) paper number 28573.
9. Richardson, J.G., Holstein, E.D., Rathmell, J.J., and Warner, H.R., Jr., Validation of as-received oil-based-core water saturations from Prudhoe Bay, *SPE Reservoir Eng.*, 31–36, 1997.
10. Egbogah, E.O. and Amar, Z.H.B.T., Accurate initial/residual saturation determination reduces uncertainty in further development and reservoir management of the Dulang field, offshore peninsular Malaysia, Society of Petroleum Engineers (SPE) paper number 38024.
11. Whitebay, L., Ringen, J.K., Lund, T., van Puymbroeck, L., Hall, L.M., and Evans, R.J., Increasing core quality and coring performance through the use of gel coring and telescoping inner barrels, Society of Petroleum Engineers (SPE) paper number 38687.
12. Treinen, R.J., Bone, R.L., Vogel, H.S., and Rathmell, J.J., Hydrocarbon composition and saturation from pressure core analysis, Society of Petroleum Engineers (SPE) paper number 27802.
13. Hagedorn, A.R. and Blackwell, R.J., Summary of experience with pressure coring, Society of Petroleum Engineers (SPE) paper number 3962.
14. Rathmell, J.J., Braun, P.H., and Perkins, T.K., Reservoir waterflood residual oil saturation from laboratory tests, *J. Pet. Technol.*, 175–185, 1973.
15. Egbogah, E.O. and Amaefule, J.O., Correction factor for residual oil saturation from conventional cores, Society of Petroleum Engineers (SPE) paper number 38692.

7 Interfacial Tension and Wettability

7.1 INTRODUCTION AND FUNDAMENTAL CONCEPTS

Chapters 3 and 4 define porosity and absolute permeability in terms of single-fluid systems–, i.e., a single fluid phase occupying the pore space of the reservoir rock signifies its porosity or storage capacity, whereas the flow behavior of that particular phase in the pore space characterized its absolute permeability. However, such simplified single-fluid systems are seldom found in the actual petroleum reservoirs. As outlined in Chapter 6, the pore space or the pore volume of a reservoir rock does not contain one (single phase) but multiple fluid saturations of gas, oil, and water, and the magnitude of these determines the distribution of the pore space among these multiple fluid saturations.

Therefore, in petroleum reservoirs, two fluids are present, and many times, three fluid phases are involved; in some rare cases, three fluid phases and a solid phase such as asphaltenes can also exist at certain reservoir conditions. Asphaltenes are the heaviest fractions found in crude oil, having a very complex molecular structure. Hence, considering the existence of multiple fluid saturations in a petroleum reservoir, other definitions must be added for a complete classification of the properties of a petroleum reservoir. All the chapters from this point onward deal with properties of petroleum reservoir rocks permeated with multiple fluid saturations.

When only one fluid exists in the pore spaces of a reservoir rock, only one set of forces is considered and that is the attraction between the rock and the fluid. In any reservoir where a single fluid is present, such as an aquifer, these forces may not be that important because porosity and absolute permeability are to some extent adequate to define the characteristics of such reservoirs. However, when more than one fluid phase is present, at least three sets of active forces need to be considered; thus for a two-fluid system the forces for consideration are:

Fluid 1 ⇔ Fluid 2
Fluid 1 ⇔ Rock
Fluid 2 ⇔ Rock

It is the existence of these forces that gives rise to fundamental properties such as interfacial tension and wettability. In addition to these two fundamental properties, the simultaneous existence of two or more fluids in a pore space also necessitates the introduction of other properties such as capillary pressure and relative permeability.

All these properties of petroleum reservoir rocks permeated with multiple fluid saturations should be determined to accurately describe the characteristics or production potential of a given petroleum reservoir.

The first set of forces to be considered is the surface forces or the interfacial tension. Because wettability depends on interfacial tension, capillary pressure depends on interfacial tension and wettability, whereas relative permeabilities are dependent on interfacial tension, wettability, and capillary pressure along with some other properties. This dependence can be summarized as:

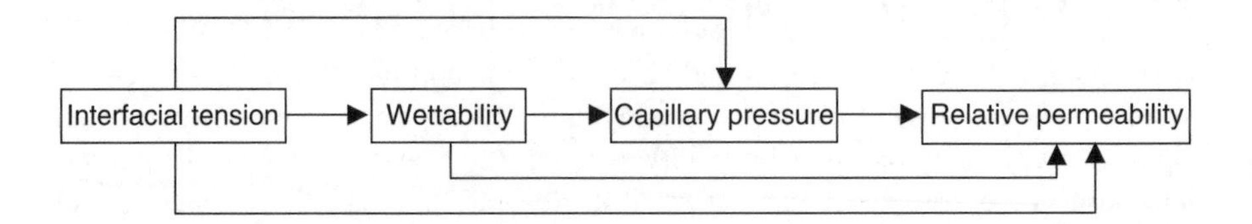

The remainder of this book discusses these properties in the most logical manner or sequence possible. This chapter focuses on interfacial tension and wettability; considering the vastness of the topic of capillary pressure and relative permeabilities, these two properties are discussed in two separate chapters, Chapter 8 and 9, respectively. This chapter begins with the discussion on interfacial tension and concludes with wettability. The fundamental definitions of these two properties from reservoir engineering perspectives, various methods of measurement, factors affecting these properties, and their significance in the recovery of hydrocarbons are some of the topics presented and discussed in the following text.

7.2 INTERFACIAL AND SURFACE TENSION

In petroleum reservoirs, up to three fluid phases, gas, oil, and water, may coexist. All these fluid phases are immiscible at the pertinent reservoir conditions. When these immiscible fluid phases in a petroleum reservoir are in contact, these fluids are separated by a well-defined interface between gas–oil, gas–water, and oil–water pairs. This particular interface is only a few molecular diameters in thickness. In dealing with multiphase systems such as those encountered in petroleum reservoirs, it is necessary to consider the effect of the forces that exist at the interface when two immiscible fluids are in contact.

The term *surface tension* (ST) is normally used when characterizing the gas–liquid surface forces, simply because this interface is the liquid surface. However, in the case of two immiscible liquids, the term *interfacial tension* (IFT) is used when describing the liquid–liquid interfacial forces. However, regardless of the terminology used, the physical forces that cause the boundary or surface or interface are the same. Frequently, in petroleum engineering literature, these terms are used interchangeably. Technically, in a petroleum reservoir that contains all the three phases — gas, oil and water — three different IFT or ST values are of significance: gas–oil ST, gas–water ST, and oil–water IFT.

To understand the concept of interfacial tension or surface tension, consider a system of two immiscible fluids, oil and water, as shown in Figure 7.1. An oil or

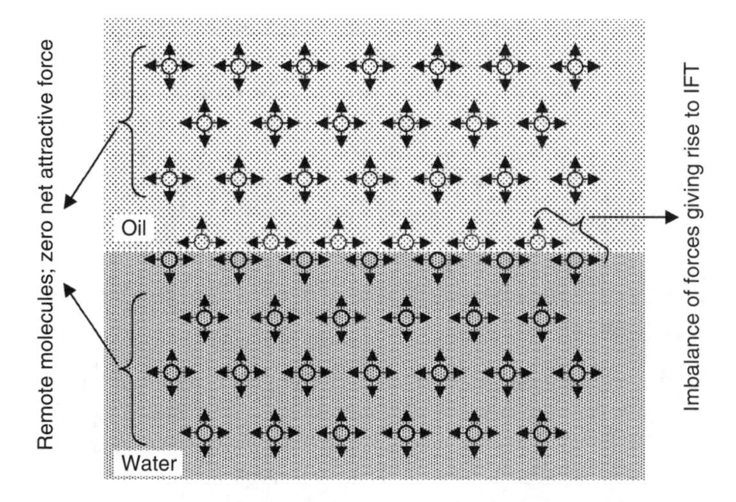

FIGURE 7.1 The concept of interfacial tension (IFT) between two immiscible liquids.

water molecule, remote from the interface, is surrounded by other oil or water molecules, thus having a resulting net attractive force on the molecule of 0 as it is pulled in all directions. However, a molecule at the interface has a force acting upon it from the oil lying immediately above the interface and water molecules lying below the interface. The resulting forces are not balanced because the magnitude of forces are different (i.e., forces from above and below) and gives rise to interfacial tension. A system similar to the one illustrated in Figure 7.1 can also be defined in terms of an immiscible pair of a gas and a liquid, in which case the resulting unbalanced forces at the surface give rise to surface tension. For oil–water or gas–oil/water, the unbalanced attraction force creates a membrane-like surface with a measurable tension such as interfacial or surface tension. Generally the interfacial tension of two liquids is less than the highest individual surface tension of one of the liquids because the mutual attraction is moderated by all molecules involved. Therefore, the interfacial or surface tension has the dimensions of force per unit length usually expressed as mN/m or 10^{-3}N/m (dyn/cm) and commonly denoted by the Greek symbol σ. Some other commonly used definitions of interfacial or surface tension include:

- A quantitative index of the molecular behavior at the interface
- Measure of the specific surface free energy between two phases having different composition
- The boundary tension at an interface between a gas and a liquid or between two immiscible liquids
- The measure of free energy of a fluid interface

Unlike other common specific properties of fluids, such as density, boiling and freezing points, viscosity, and thermal conductivity that are properties of the main body of the fluids, interfacial tension or surface tension is the best known property of fluid interfaces. Despite the fact that interfacial tension or surface tension is an entirely fluid- or interface-related property and not a petroleum reservoir rock property, it significantly influences other important rock properties such as wettability, capillary pressure, and relative permeabilities, all of which in turn affect the recovery of hydrocarbon fluids from petroleum reservoirs. A discussion on interfacial tension or surface tension as part of the discussion on petroleum reservoir rock properties is thus indispensable and appropriate.

7.2.1 Effect of Pressure and Temperature on Interfacial Tension and Surface Tension

Since gas, oil, and water coexist at a variety of high-pressure and high-temperature conditions in petroleum reservoirs, it is important to understand the effect of these variables on IFT or ST. The variation of IFT or ST with temperature and pressure strongly influences the transport of fluids in a reservoir and therefore are fundamental to the understanding of the role of interfacial forces in oil recovery. Generally speaking, even though both IFT and ST are significant from an oil recovery point of view, properties such as wettability (discussed later in this chapter) are more closely related to the IFT between oil and water. It is imperative, however, to address the effects of pressure and temperature on IFT and ST separately. First, the behavior of ST with respect to pressure and temperature is studied because these effects are rather clearly understood in comparison to IFT.

Surface tension generally decreases with an increase in pressure and temperature. As temperature increases, the kinetic agitation of the molecules and the tendency of the molecules to fly outward increases, resulting in a decrease in ST values. Katz[1] presented ST data as a function of temperature, on a number of pure hydrocarbon components showing the decline in ST values with an increase in the temperature. As an example, ST data of pure water,[2] reservoir brine,[3] and a flashed North Sea crude oil[3] are shown in Figure 7.2, that also shows a similar trend of decreasing ST values with increasing temperature. The effect of pressure on ST is also somewhat similar to the temperature effect. When considering the gas–liquid ST at high pressures, in most cases, the high-pressure vapor over the surface of a liquid would result in a low ST by bringing a fairly large number of gaseous molecules within reach of the surface. The attractions of these molecules to the surface molecules of the liquid would neutralize, to some extent, the inward attraction on the surface molecules diminishing the ST. The high-pressure of the gas above the liquid is somewhat analogous or equivalent to placing a second liquid of rather small attraction for the first, in place of gas. In other words, the high-pressure gas phase tends to develop miscibility toward the companion liquid phase thereby reducing ST as pressure increases.

Kundt's[4] measurements showed a decrease in ST values of several common liquids, with an increase of pressure of the gas above them. Dandekar[5] reported a large number of ST data on a wide variety of gas–liquid hydrocarbon systems at reservoir conditions showing a decrease in ST with increasing pressure. A similar behavior is also evident from ST data presented by Katz[1] for crude oils, Mulyadi and Amin[6] for gas–water (brine) and gas–oil at various pressures and reservoir temperature, Huijgens

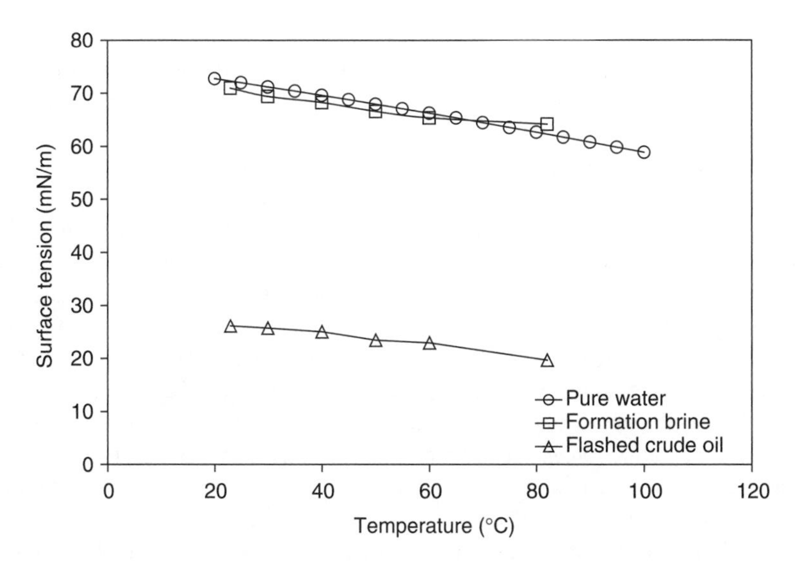

FIGURE 7.2 Pure water, formation brine, and flashed crude oil surface tension data as a function of temperature.

and Hagoort[7] for nitrogen-simulated North Sea oils at various pressures and 100°C. The ST data reported by Jennings and Newman[8] on a methane–water system in the pressure range of 14.7 to 12,000 psia and 74 to 350°F clearly aids in understanding the effect of both pressure and temperature because ST values show a decrease with increasing pressure and temperature.

The behavior of IFT values between oil and water, with regards to the pressure and temperature changes, i.e., increase or decrease is, however, not that well understood. Wang and Gupta[9] presented IFT data for crude oil and two different brine systems in a pressure and temperature range of 14.7 to 10,000 psia and 70 to 200°F, respectively. In addition to this data they also reported IFT data for mineral oil and distilled water systems in the same pressure and temperature range. When the IFT data is plotted vs. temperature (at various constant pressures) for the crude oil and the two brine systems, one shows a decrease in IFT as temperature decreases, whereas the other system shows an increase in the IFT with increasing temperature. Although this trend is indicated from the straight-line fits of trend lines of the IFT temperature data, a lot of scatter in the data is evident when individual data points are considered, clearly demonstrating the absence of any particular trend for the effect of temperature on oil–water IFT. A similar observation can also be made while considering the IFT values vs. pressure (at various temperatures) for the crude oil and the two brine systems. Trend lines fitted to the data do indicate an increase in the IFT with increasing pressure (exactly opposite of ST behavior with pressure); however, a fair amount of scatter in the plots indicate the absence of any clear trend.

Jennings and Newman[8] presented IFT data for water and simulated live crude oils at pressure ranging from 14.7 to 12,000 psia and temperatures ranging from 74 to 350°F. Their IFT data when plotted as a function of pressure at the tested temperatures are also devoid of any particular trend. The IFT data reported by Hjelmeland and Larrondo,[10] on flashed crude oil and formation brine also indicated inconsistent trends of IFT values vs. temperature. When their IFT measurements in two different experimental setups were compared, one showed an increase in the IFT with increasing temperature, whereas the other indicated a decrease in the IFT with an increase in the temperature. A similar inconsistency is also evident from the oil database presented in Ref. 11, where IFT data are presented for a number of reservoir oils and brines as well as fresh water. It has been stated[9] that the previous studies about the effects of pressure and temperature on interfacial tension indicate that observed trends would depend on the type of systems studied. This phenomenon has not been well explained and a sound theoretical explanation is still lacking. However, considering the oil–brine IFT data presented in the literature [6, 9–12] one important observation can be made, i.e., most of the IFT data seem to have an average value of about 25 mN/m in a wide variety of pressures and temperature ranges (see Figure 7.3). Therefore, considering the inconsistent trends of oil–water IFT data with respect to pressure and temperature, and the important observation from the literature data, an average value of around 25 mN/m for the oil–water IFT can be considered a reasonable assumption if experimental values for the pertinent pressure and temperature conditions are unavailable.

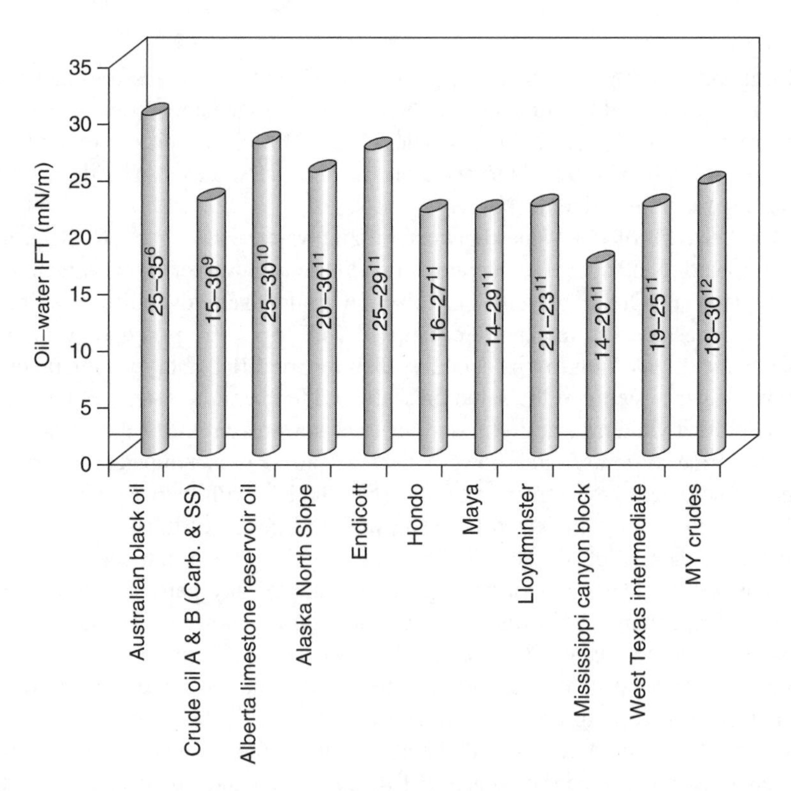

FIGURE 7.3 Oil–water (brine) interfacial tension data for different systems. Values plotted on the *y* axis are average IFT values. The IFT range and literature reference are shown on the data bars.

7.2.2 Laboratory Measurement of Interfacial Tension

The experimental techniques that are used for measuring interfacial tension or surface tension are essentially identical, i.e., an apparatus or experimental setup used for measuring the interfacial tension between two liquids, generally can also be employed for conducting surface tension measurements. A variety of experimental techniques are available for the measurement of IFT or ST values and are referred to as *tensiometers*. These tensiometers are instruments that simply measure force. Some of the commonly used tensiometers are the Wilhelmy plate method, DuNouy ring method, spinning drop method, and the maximum pull force method. These methods are described in great details in most standard physical chemistry text books and they will not be discussed here. Moreover, many of these standard methods are not even applicable at the usual reservoir conditions of high pressure and high temperature. However, in the petroleum industry, the most commonly used technique for measurement of IFT or ST of petroleum reservoir fluids is the pendant drop method. It is perhaps the most widely used method for measuring the IFT and ST of a variety of fluids with established accuracy and reliability. A survey of IFT and ST literature showed that the majority of experimental data on IFT as well as ST have been generated using the pendant drop method.[5–12] Therefore, considering the wide acceptance of this method and its routine use in the petroleum industry, only this technique is discussed in detail in this section.

The pendant drop method involves suspending a droplet of the liquid in the companion or second-liquid phase (e.g., a droplet of heaviest liquid in the surrounding light–liquid phase) for IFT determination, whereas a liquid droplet is suspended in a companion gas or vapor phase for ST measurements. The liquid droplet is allowed to hang from a narrow tube, spout, or a syringe from its tip. The shape and size of the liquid droplet is mainly a function of the prevailing IFT or ST between the given fluid pairs. With this method the IFT or ST values are determined from the profile (image) of the static pendant drop for a given density difference between the gas–liquid phases or the liquid–liquid phases. A video system is generally used to capture the image of the drop, subsequently dimensioned by image analysis.

For oil–water IFT measurements, if water droplet is formed in the surrounding oil phase, then the water droplet cannot be visualized in the case of dark oil. In such a situation, the narrow tube or the spout is bent in a "J" shape and the oil droplet is formed on the tip of the narrow tube because if the oil droplet is formed on the tip of a normal vertical narrow tube then the oil droplet almost instantly detaches due to dominant gravity forces (oil is usually lighter than water).

A typical pendant drop is shown in Figure 7.4. The pendant drop assembly is normally integrated in a high-pressure cell that is usually enclosed in a climatic air bath so that the pertinent reservoir or test pressure and temperature conditions can be maintained. The determination of IFT or ST from pendant drop analysis is based on the assumptions that the drop is symmetric about a central vertical axis (i.e., the drop can be viewed from any angle) and the drop is not in motion (i.e., surface forces (IFT or ST) and gravity are the only forces affecting the dimensional characteristics of the drop). One major advantage of the pendant drop method is that it does not require any calibration. However, considering the small dimensions of the droplet (usually a few millimeters), the image is normally enlarged or magnified by a factor

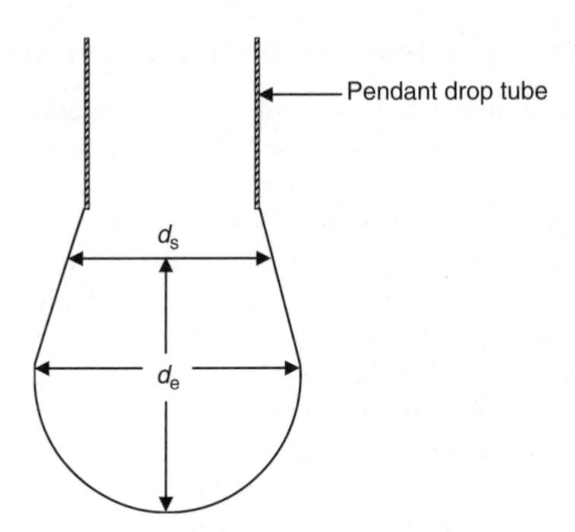

FIGURE 7.4 Schematic of a pendant drop of water or brine in a clear oil phase.

of 50 times its actual size in order to accurately dimension the droplet. Therefore, the only correction that is required is the calculation of the actual dimensions of the drop, based on the magnification factor used for enlarging the drop image.

The IFT or ST values are calculated from the following equation:

$$\sigma = \frac{\Delta \rho g d_e^2}{H} \tag{7.1}$$

where $\Delta\rho$ is the density difference between the two immiscible phases (e.g., gas–liquid or oil–water) in g/cm^3, g the acceleration due to gravity in cm/sec^2 d_e the equatorial or maximum horizontal diameter of the unmagnified (or magnification corrected) droplet in cm (see Figure 7.4), H the drop shape factor as a function of $S = d_s/d_e$, and d_s the diameter of the droplet measured at a distance d_e above the tip of the droplet (see Figure 7.4).

In Equation 7.1, the measured values of the individual fluid phase densities are normally used. After dimensioning the droplet, first the value of S is calculated from which the value of the drop shape factor is read from the tables published by Niederhauser and Bartell.[13] However, in the absence of the drop-shape factor tables, equations relating $1/H$ and S derived by Misak[14] can also be utilized, eliminating the need to use the tables and even further simplifying the IFT or ST calculations. Hence, based on these variables, IFT or ST values can be readily determined.

7.3 WETTABILITY

In dealing with petroleum reservoir fluids in a reservoir system, it is necessary to consider not only the surface forces between a gas and a liquid and the interfacial forces between two immiscible liquids (i.e., ST and IFT as we saw earlier) but also the forces that are active at the interface between the liquids and the solids. The consideration of the interface between the liquids and the solid assumes a significant importance in reservoir engineering simply because the petroleum reservoir fluids are ubiquitously in contact with the solid (reservoir rocks) until they are produced on the surface. It is the combination of all the active forces that determines the wettability of reservoir rocks. Wettability is a key parameter that affects the petrophysical properties of reservoir rocks. The knowledge of reservoir wettability is critical because it influences various important reservoir properties such as the distribution of gas, oil, and water within a reservoir rock, capillary pressure and relative permeability characteristics, and consequently the production of hydrocarbons. Therefore, an understanding of the wettability of a reservoir is crucial for determining the most efficient means of hydrocarbon recovery from petroleum reservoirs.

The subject of reservoir wettability, due to its vast nature, is one of the most widely debated and discussed areas in the petroleum engineering literature, evident from a large number of publications in this area. To clearly understand the various aspects of this important reservoir property, the topic of wettability is divided into the following four areas and discussed individually:

1. Fundamental concepts of wettability
2. Practical aspects of wettability
3. Measurement of wettability
4. Factors affecting wettability

The fundamental aspects and definitions of wettability are addressed first, followed by the discussion on wettability from a practical point of view, types of wettability and so on. The methods of laboratory measurement of wettability are presented in the third part, and the discussion ends with factors that affect wettability. Some examples on typical reservoir wettabilities are also presented in each section.

7.4 FUNDAMENTAL CONCEPTS OF WETTABILITY

To understand the fundamental concept behind wettability, various definitions of reservoir wettability must be addressed. Wettability has been defined as:

- The relative ability of a fluid to spread on a solid surface in the presence of another fluid.
- The tendency of surfaces to be preferentially wet by one fluid phase.
- The tendency of one fluid of a fluid pair to coat the surface of a solid spontaneously.
- The tendency of a fluid to spread on and preferentially adhere to or wet a solid surface in the presence of other immiscible fluids.
- Reservoir wettability is determined by complex interface boundary conditions acting within the pore space of sedimentary rocks.
- The term used to describe the relative adhesion of two fluids to a solid surface.
- When two immiscible fluids contact a solid surface, one of them tends to spread or adhere to it more than the other.

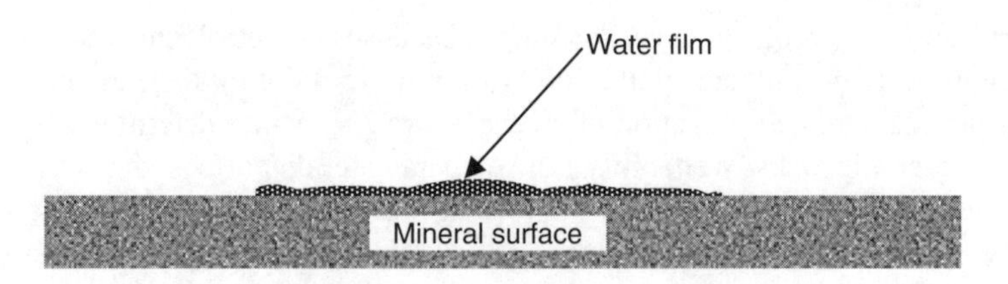

FIGURE 7.5 Schematic of a film of water spread on a mineral surface.

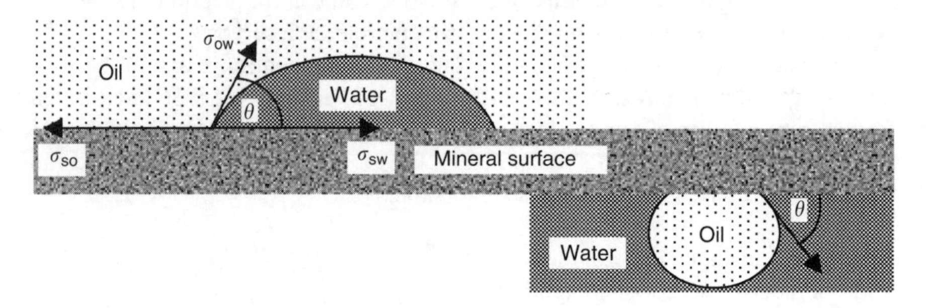

FIGURE 7.6 Schematic of a system of two immiscible liquids (oil and water) in contact with a mineral surface.

Considering all these definitions, it is clear that whichever way wettability is addressed, it basically means that in a multiphase situation; one of the fluid phases has a greater degree of affinity toward the solid surface of the reservoir rock. Thus the tendency of a fluid phase to spread over the surface of a solid is an indication of the wetting characteristics of the fluid for the solid. For example, in a system comprising of oil, water (brine), and rock, (sandstone or a carbonate), one of the phases (either the oil phase or the water phase) has a tendency to preferentially wet the rock. The concept of wettability is simple to illustrate and is shown in Figure 7.5.

The spreading tendency of a fluid can be expressed more conveniently as *adhesion tension*, A_T. Adhesion tension is a function of the interfacial tension and determines which fluid preferentially wets the solid. Understanding the concept of wettability in terms of adhesion tension requires considering a system of two liquids, such as oil and water, that are in contact with a solid, as illustrated in Figure 7.6. For the system shown in Figure 7.6, the adhesion tension is defined by

$$A_T = \sigma_{SO} - \sigma_{SW} \tag{7.2}$$

where σ_{SO} is the interfacial tension between the solid and the lighter fluid phase (oil in this case) and σ_{SW} the interfacial tension between the solid and the denser fluid phase (water in this case)

The angle of contact, θ at the liquid–solid surface is also shown in Figure 7.6. This contact angle, by convention, is measured through the denser liquid phase and can range from 0 to 180°. Therefore, by definition the cosine of the contact angle θ is

$$\cos \theta_{OW} = \frac{\sigma_{SO} - \sigma_{SW}}{\sigma_{OW}} \tag{7.3}$$

combining Equations 7.2 and 7.3,

$$A_T = \sigma_{OW} \cos\theta_{OW} \tag{7.4}$$

To determine adhesion tension for defining the wettability of the system shown in Figure 7.6, knowing the value of the oil–water interfacial tension and a measure of the contact angle is necessary. The oil–water IFT value can be obtained from a pendant drop method, whereas the contact angle can be measured by a standard contact angle meter in to which a droplet of liquid is dispensed on to the solid surface. A camera then captures the profile of the droplet on the computer screen where a software calculates the tangent to the droplet shape and contact angle. As seen earlier, the oil–water IFT values are in the range of 25 mN/m and considering Equation 7.4, the magnitude of the adhesion tension is dictated by the contact angle, making the angle of contact the predominant measure of wettability. As a matter of fact, determining the contact angle has achieved significance as a measure of reservoir wettability, details of which are discussed in Section 7.6.

According to Equation 7.4, positive adhesion tension indicates that the denser phase (water in this case) preferentially wets the solid surface, whereas a negative value of adhesion tension indicates a wetting preference by the lighter phase (oil in this case). An adhesion tension of 0 indicates that both phases (oil and water) have equal wettability or affinity for the solid surface. The wetting preferences indicated by the adhesion tension can also be expressed in terms of the contact angle; a 0° contact angle indicates a completely water-wet system, whereas a contact angle of 180° indicates an oil-wet system. A 90° contact angle indicates a neutral wet system, that is, both phases have equal affinity for the solid surface. Sometimes, the neutral-wet system is also called an *intermediate wet system*. The limits of the neutral-wetting scales are not definite, however, a range from about 70 to 110° is considered average.

Based on the contact angle and assuming a constant IFT value, an equivalent terminology called the *wetting index* (WI) is also used; for a contact angle of 0° and 180°, WI is +1.0 and −1.0, water-wet and oil-wet, respectively. For a contact angle of 90°, WI is 0, signifying a neutral or intermediate-wet system. However, the limits of the wettability scales described here for the system in Figure 7.6 are rather broad because they are for a hypothetical solid surface and fluid phases, and do not account for the mineralogy or lithology of the solid or the rock surface, and the chemistry of the fluids involved. Therefore, it is imperative to study a system that considers different fluid phases and solid surfaces so that the effect of mineralogy and fluid phase characteristics on wetting preferences can be understood.

The schematic of a hypothetical multiliquid (four different reservoir oils of varying chemical compositions and water) and varying rock lithologies (sandstone and carbonate, represented by silica and calcite respectively) system is shown in Figure 7.7. The type of system shown in Figure 7.7 provides only understanding of the effect of oil chemistry and rock lithology on wettability and does not represent any particular reservoir wettability. The contact angles in Figure 7.7 illustrate the effects that might be expected from varying the lithology of the rock and the chemical characteristics of the hydrocarbon phases. When oil A and water are used, the water phase preferentially wets both the sandstone and carbonate surfaces. In the case of oil D and water, the latter phase preferentially wets the sandstone surface with a contact angle of 35°, whereas oil

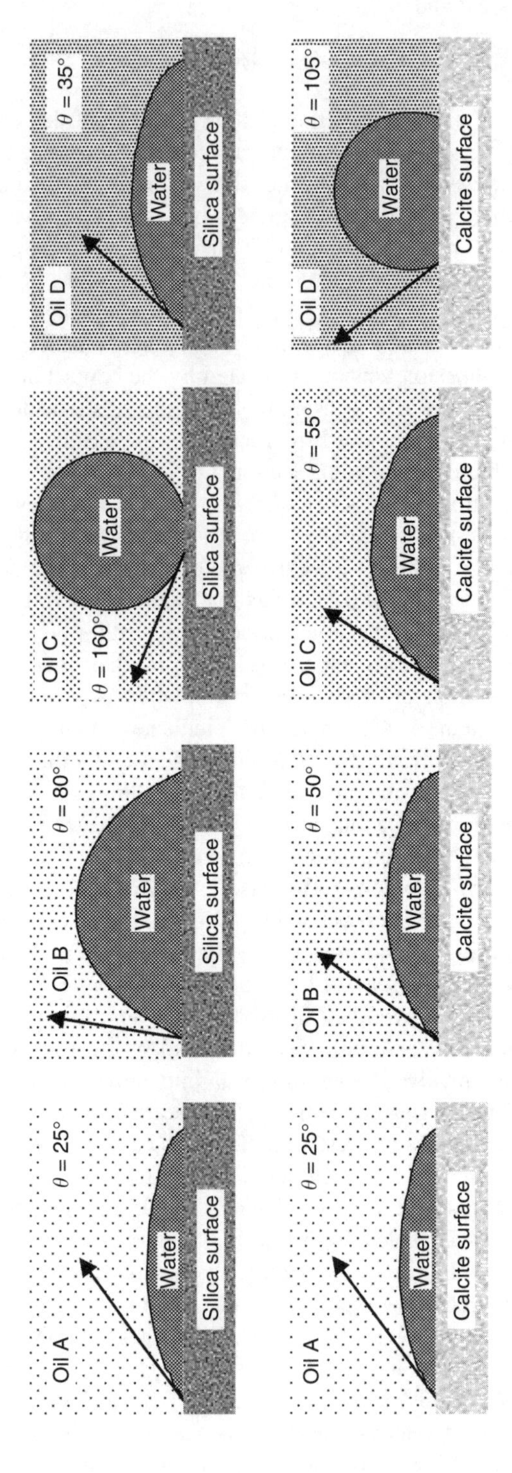

FIGURE 7.7 The effect of rock lithology and fluid characteristics on wetting tendencies.

preferentially wets the carbonate surface with a contact angle of 105°. The system of oil B for sandstone indicates slightly water-wetting characteristics, whereas the same system for carbonate exhibits a much higher degree of water-wetting. The behavior of system of oil C is reversed when comparing the behavior of system D, that is, sandstone is oil-wet and carbonate is water-wet.

In summary, depending on the chemical composition of the fluids involved and the rock, either a water-wet or oil-wet surface is possible. This discussion on wettability so far has focused only on the oil and the water phase because the wetting characteristics of the reservoir rocks are entirely dictated or competed by these two liquid phases, whereas the hydrocarbon gas phase is always the nonwetting phase.

7.5 A DISCUSSION ON PRACTICAL ASPECTS OF WETTABILITY

The wettability of a petroleum reservoir rock and its effect on various aspects related to the recovery of hydrocarbons from petroleum reservoirs have been the subject of a considerable and growing body of literature. However, this section focuses discussion on some practical aspects of reservoir wettability. Although wettability has been studied extensively by researchers for many years, much discussion still abounds and many questions remain unanswered as to the factors that actually control the wettability in the reservoir. In the early days of petroleum (reservoir) engineering it was assumed that all or at least most petroleum reservoirs were considered as strongly water-wet ($\theta = 0$) and that water completely coated the rock surface. This belief of strongly water-wet reservoirs was attributed to the saturation history of the reservoirs such that the rock was completely saturated with water prior to the migration of the hydrocarbons from the source rock. Basically, for very large time scales the reservoir rock was always in contact with water and never saw the hydrocarbons until migration occurred, and hence there was no reason to assume that this wetting condition (which was thought to have been originally established to water wet) had altered.

This assumption of strongly water-wet reservoirs led to problems that arose frequently regarding understanding of the reservoir behavior, that is, the reservoirs were expected to respond or behave in a certain fashion in congruence with the assumption of strongly water-wet characteristics. Hence, questions were raised regarding the natural wettability of hydrocarbon reservoirs and numerous examples of wettability, other than strongly water-wet reservoirs were identified.

In order to emphasize the occurrence of nonwater-wet reservoirs, Cuiec[15] evaluated the wettability of some 20 reservoirs from Europe, North America, North Africa, and the Middle East, confirming the existence of nonwater-wet reservoirs. As a matter of fact a number of reservoirs reported in Cuiec[15]were identified as oil-wet. Cuiec[15] also identified many reservoirs as partly oil-wet or intermediate oil-wet. It now appears that although strongly water-wet and truly oil-wet reservoirs do exist, many researchers, knowledgeable on the subject of wettability recognized that many if not most reservoirs are at a wettability state intermediate between water-wet and oil-wet. This basically means that reservoir wettability is not a simply defined property and classification of reservoirs as water-wet or oil-wet is a gross oversimplification.[16] It is generally agreed that preferred wettability is not a discrete-valued function (i.e., water-wet or oil-wet) but can span a continuum between these extremes.[17]

These various practical aspects also have great implications for reservoir engineering studies because wettability of a porous medium governs the relative distribution of fluids in the pores and has considerable influence on the conditions in which reservoir fluids flow, subsequently leading to the recovery of hydrocarbons. Specifically, in reservoir engineering studies of which special core analysis is a critical component; properties such as capillary pressures, relative permeabilities, irreducible water saturation, and residual oil saturation are routinely measured. The magnitude of all these SCAL properties is to a great extent dependent on a given system of reservoir fluids and the rocks. However, wettability is also a function of the characteristics of the given reservoir fluids and the rock. Therefore, wettability is an important characteristic of a rock/fluid system as it affects many of the SCAL properties that are an integral part of reservoir engineering analyses.

Another important aspect of reservoir wettability is that it is probably the only rock/fluid property that does not directly enter into any reservoir engineering calculations; therefore its sensitivity to reservoir engineering calculations cannot be mathematically established. Wettability is used in most cases to qualitatively judge, describe, or explain the behavior of a particular process in a certain fashion, such as fluid distribution in the pore spaces of a reservoir rock, capillary pressure curves and relative permeability characteristics, or the performance of water displacement or a waterflood process. However, a well-defined *pore level* classification of wettability is necessary to relate the wetting characteristics of a reservoir rock to the various SCAL properties and reservoir fluid flow processes on a *pore level.*

7.5.1 Classification/Types of Wettability

A variety of wettability states exist for petroleum reservoirs, primarily depending on both reservoir fluid and rock characteristics (as seen in the ideal example illustrated in Figure 7.7). Different types of wettability classified on a pore level are presented in the following text.

7.5.1.1 Water-Wet

In this wettability state, the rock surface (large and small) has preference for the water phase rather than the hydrocarbon phase. Therefore, hydrocarbon phases (gas and oil) are contained in the centers of the pores and hence do not cover any of the rock surfaces.

7.5.1.2 Oil-Wet

This wettability state is exactly the opposite of the water-wet state; the relative positions of the oil and water are reversed. It is believed that asphaltenic components cause this wetting state.

7.5.1.3 Intermediate Wet

In this wettability state, the rock surface has preference for both the oil and water phases. The precise nature of intermediate wetting is not defined; therefore, it includes the subclasses of both fractional and mixed wettability. In neutral wettability, the rock surface has equal tendency for oil and water to wet the rock surface (e.g. angle of contact is 0°).

7.5.1.4 Fractional Wettability

This type of wettability is also termed as *Dalmatian wettability* because some of the pores are water-wet, while others are oil-wet. Fractional wettability occurs when the surfaces of the rocks are composed of many minerals having different chemical properties leading to variations in wettability throughout the internal surfaces of the pores. Note the difference in the concept from the neutral wettability, which implies all portions of the rock have an equal preference for oil or water. Although, fractional wettability can be artificially created in the laboratory, field cases of this wettability state have not been reported before.

7.5.1.5 Mixed Wettability

This type of wettability was proposed by Salathiel.[18] In this wettability state, the smaller pores are occupied by water and are water-wet, whereas the oil preferentially wets the interconnected larger pores. It is believed that the mixed wettability conditions are associated with the original oil invasion preferentially into the larger pores, followed by the deposition of asphaltenic compounds rendering the surface oil-wet.

As an example, a pore level illustration of water-wet and oil-wet rocks is presented in Figure 7.8.

7.6 MEASUREMENT OF RESERVOIR ROCK WETTABILITY

Reservoir wettability can be evaluated by two different groups of methods: *qualitative* and *quantitative*. Qualitative methods for wettability are indirectly inferred from other measurements, for example, capillary pressure curves or relative permeability curves. However, relative permeability curve methods are all suitable only for discriminating between strongly water-wet and strongly oil-wet cores. A smaller change in wettability, for example, strongly and moderately water-wet, may not be noticed by these methods[19]. Therefore, indirect qualitative methods are not discussed here. Focus is on the direct or quantitative methods. Quantitative methods are direct measurement methods, where the wettability is measured on actual rock samples using reservoir fluid samples and wettability is reported in terms of a certain wettability index, signifying the degree of water, oil wetness, or intermediate wetness.

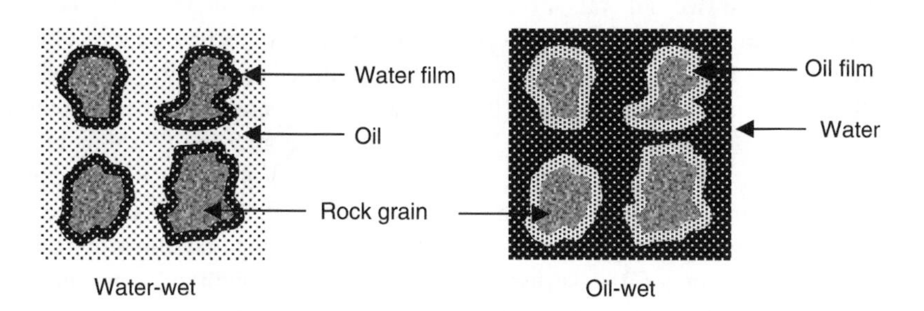

FIGURE 7.8 Schematic representation of water-wet and oil-wet pore spaces.

These direct quantitative methods include contact angle measurement, Amott's test, and the U.S. Bureau of Mines (USBM) wettability method. The contact angle measures the wettability of a specific surface, while the Amott and USBM methods measure the average wettability of a core sample. The contact angle method is the most direct method because it measures the contact angle on *representative* reservoir rock surfaces using reservoir fluids, from which the adhesion tension can be calculated if the IFT values between the oil and water are also known. For the other two methods, Amott and USBM, wettability is evaluated based on displacement characteristics of the core sample. For measurement of fractional wettability, techniques such as dye adsorption and nuclear magnetic relaxation are used, whereas mixed wettability is evaluated by the glass slide method and advanced techniques such as the atomic force microscopy. This discussion, however, is restricted to contact angle measurement, the Amott test, and the USBM method, described in the following text.

7.6.1 Contact Angle Measurement

Many methods of contact angle measurement have been used, including the tilting plate method, sessile drops or bubbles, vertical rod method, tensiometric method, cylinder method, and capillary rise method.[19] However, the method that is often used to make direct measurements of the contact angle to determine the preferential wetting characteristics of a given oil–water–rock system is called the *sessile drop method*. The determination of reservoir wettability from contact angle measurements by the sessile drop method is simple in concept. A drop of water is placed on a mineral surface in the presence of reservoir oil and the angle through the water phase is measured (see Figure 7.7). If the water drop spreads over the mineral surface, the surface is water-wet and the contact angle is low; if the water drop beads up, the contact angle is high and the surface is oil-wet. This situation can also be reversed, that is, a drop of oil placed on a mineral surface in the presence of formation water. A photograph of the system is subsequently taken for accurate measurement of the contact angle. Contact angles ranging between 0 and 70° are considered to indicate water wetness, whereas those ranging between 110 and 180° indicate oil wetness; a range of 70 to 110° demonstrates neutral or intermediate wettability of the system.

For mineral surfaces used in contact angle measurements, a large crystal of the mineral type lining the pore space of the reservoir rock is used. In general, sandstones are predominantly quartz and carbonates are predominantly calcite, so plates made out of these two minerals are chiefly used to simulate the reservoir rock surface. A modification of the sessile drop method was introduced by Leach et al.[20] to measure the water-advancing contact angle. The modified sessile drop method uses two flat, polished mineral crystals that are mounted parallel to each other on adjustable posts, as shown in Figure 7.9. The cell containing the two mineral surfaces is filled with water (brine) and subsequently an oil drop is placed between the two crystals so that it contacts a large area of each crystal. After allowing the drop of oil to age, the mobile plate is moved, shifting the oil drop and allowing brine to move over a portion of the surface previously covered with oil, thus creating the advancing contact angle. Contact angles measured in this fashion are called *water-advancing contact angles*.

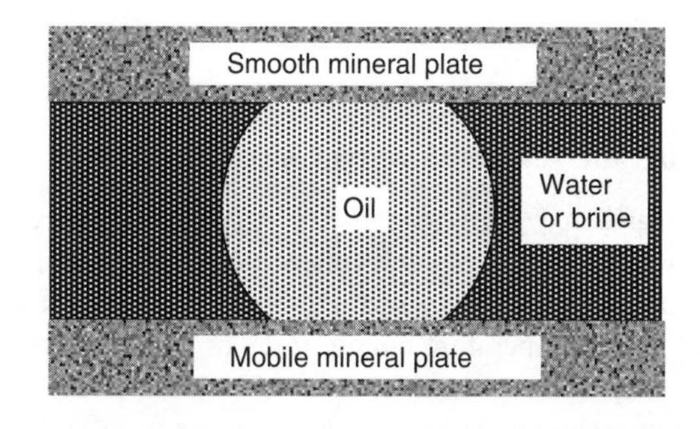

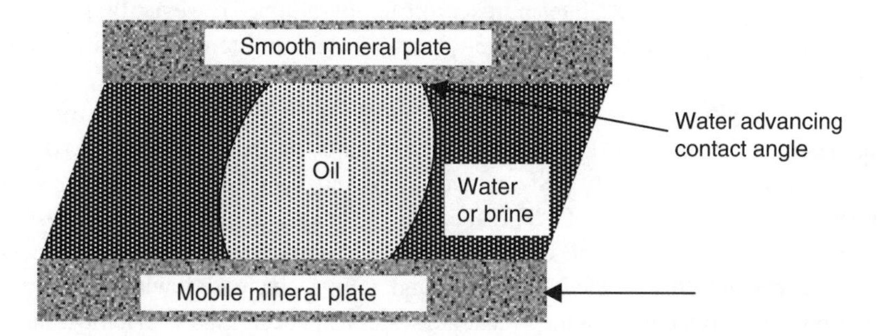

FIGURE 7.9 The measurement of contact angle for wettability determination.

The contact angle test is used as a procedure for determining whether or not the crude oil sample causes the oil to wet a reservoir rock mineral in the presence of formation water at reservoir conditions. The underlying assumption in using such a simplified method (especially the smooth-polished mineral surfaces) for wettability determination is if the oil does not wet the mineral under conditions similar to those that exist in the reservoir, it is quite possible that it does not wet the mineral in the reservoir too. Conversely, if the oil does wet the mineral under reservoir-like conditions, it very probably wets the mineral in the reservoir as well. However, in spite of this justification and that it is possible, with great care, to get exact and reproducible contact angle measurements, the question arises: how representative are these results of actual reservoir rock wettability? Since polished mineral surfaces are used, contact angle measurements do not account for factors such as roughness, heterogeneity, and complex geometry of reservoir rocks. For example, Morrow[21] has pointed out that roughness and pore geometry influences the oil–water–solid contact line and can change the apparent contact angle. On the sharp edges found in reservoir rock, a wide range of possible contact angles[21,22] exists, whereas on a smooth surface such as the one used in the contact angle measurements, the contact angle is fixed.

Although the contact angle measurements do provide interesting information considering its limitations, the tests that measure average core wettability, such as the Amott and the USBM methods, are considered the most useful and in fact preferred by many reservoir engineers. However, the main advantage of the contact angle method of determining wettability is its relatively low cost as compared with the expense of obtaining cores in their native wettability condition needed for most other types of wettability tests.

7.6.1.1 Effect of Pressure and Temperature on Contact Angles

Although wettability is considered a key parameter that affects the recovery of hydrocarbons both by conventional methods and enhanced oil recovery methods, the majority of the available data on wettability of the oil–water–rock systems are for atmospheric pressure and room temperature conditions. The effects of temperature and pressure on the wettability of reservoir rocks are not well understood.

Treiber and Owens[23] reported contact angle measurements for 50 oil-producing reservoirs from the United States at reservoir temperatures. However, their measurements did not specifically include the effects of changing temperatures on contact angles.

Wang and Gupta[9] have reported contact angle measurements at reservoir conditions. Their experimental results included contact angle measurements for two different crude oil–brine–quartz/calcite systems over a pressure range of 200 to 3000 psig and temperature range of 72.5 to 200°F, respectively. The quartz mineral surface was used to represent the sandstone and calcite was used to represent carbonate. The contact angle for the systems studied by Wang and Gupta[9] increased with pressure and increased with temperature for the sandstone system and decreased with temperature for carbonate system. However, note that although their contact angle measurements did indicate a functionality or variability with pressure and temperature, an average value of contact angles between 20° and 32° for the sandstone system and 50° and 60° for the carbonate system is observed. In other words, contact angles did not vary significantly as a function of either pressure or temperature, basically indicating a preference to water wetting for both the systems; however, the oil–brine–quartz system is more strongly water-wet as compared to the oil–brine–calcite system.

7.6.2 Core Samples Used for Amott Test and USBM Methods

While the contact angle measures the wettability of a specific surface, the Amott and USBM methods measure the average wettability of a core sample. Therefore, it is prudent to discuss the issues related to the type of core samples that are used in either the Amott test or the USBM method. Moreover, given the very delicate and sensitive nature of reservoir rock wettability, the use of appropriate core samples in wettability determination by the Amott test and the USBM method assume considerable significance.

Chapter 6 discussed at length the effect of drilling-mud filtrate invasion in the whole core sample; and as seen this invasion can have a significant impact on the initial fluid saturations. Along with the alteration in the initial fluid saturations, the invasion of mud filtrate and variations in pressure and temperature can also alter the natural wettability of the recovered core sample.

Bobeck et al.[24] reported that some drilling muds are capable of changing the rock wettability while others are not. However, some of Amott's[25] results indicated that wettability of Ohio sandstone was altered by one of the oil-based drilling mud filtrates from strongly water-wet to moderately water-wet, whereas the other oil-based drilling mud filtrate caused a marked change in the wettability; the rock was made moderately preferentially oil-wet. As a consequence, wettability measurement experiments with such cores obviously produce erroneous data, that is, the measured wettability itself may not be erroneous but it is not representative of the original or existing reservoir wettability. The natural wettability of the reservoir rock can, however, be preserved by using low-invasion drilling muds, or by using coring systems that use gels that cause only minimal alteration in the natural wettability. However, if mud filtrate invasion does occur, and if the depth of invasion is known, then the best option is to drill the core samples used for wettability measurements from the central noninvaded section of the whole core (see Figure 6.15). Otherwise the best alternative is to attempt to restore the natural wettability of the core sample, which can be carried out in the following manner. Although the procedure described below does provide the best alternative to native state or preserved state core material, there is no guarantee that this procedure yields a core sample with wettability the same as that in the reservoir.

- All fluids from the core sample are removed by extraction and subsequently the sample is dried.
- The sample is then saturated with reconstituted formation brine.
- The brine from the core sample is displaced by a synthetic oil or a laboratory oil such as Isopar-L, until the irreducible water saturation is achieved.
- The core sample is then flooded with the representative reservoir oil sample at reservoir temperature to replace the synthetic oil.
- The core sample is then aged in reservoir oil at reservoir pressure and temperature conditions for an extended period. This aging period may be up to 1000 h[26]; however, longer periods may be required. The objective of the aging process is to attempt to re-establish the wettability developed in the reservoir over a geologic timescale.

It should, however, be noted that although no specific methods or guidelines exist on precisely how the natural wettability can be restored, these steps basically summarize the procedures that can be used as an attempt to restore wettability. However, if only unpreserved core samples are available, such a procedure described here becomes indispensable and it may not be the best choice but the only option available. These procedural steps can, however, be altered or adapted to address a specific type of oil–brine–rock system to account for peculiarities in the oil or brine samples, degree of mud filtrate invasion, and so on. Moreover, the wettability measured on the restored core sample cannot really be compared with the actual reservoir wettability because of the fact that the natural wettability is not known (e.g., under the assumption that it has been altered by mud filtrate invasion). Therefore, wettability measured on the restored core sample is assumed as *the* wettability of the reservoir rock. Or, the only true test of the validity of a test on a restored core sample can be established by comparing the wettability results from the test on the same rock from a native state core sample, if such data are available.

Gant and Anderson's[26] exhaustive research on core cleaning for restoration of native wettability discusses various issues related to the cleaning of contaminated cores for restoration of natural wettability. In summary, as it has been stated by Cuiec[15], it is obvious that as long as no reliable way of determining wettability *in situ* is available, doubts will continue as to the representativity of the surface state no matter what solution is adopted.

7.6.3 Amott Test

The Amott[25] wettability test is the most commonly and routinely used test in core analysis for the determination of average wettability of core samples. The determination of wettability is based on the displacement properties of the oil–water–rock system. The rock samples typically used in the Amott test are core plugs either 1 or 1.5 in. in diameter and lengths ranging from 2 to 3 in. The Amott test basically comprises of natural and forced displacement of oil and water from a given core sample. The test begins with a residual oil saturation in the core sample obtained by forced displacement of the oil. Subsequently, the test measures the average wetting characteristics of the core sample using a procedure that involves four displacement operations:

1. Immersion of the core sample in oil to observe the spontaneous displacement of water by oil (see Figure 7.10).
2. Forced displacement of water by oil in the same system by applying a high displacement pressure.
3. Immersion of the core sample in water to observe the spontaneous displacement of oil by water (see Figure 7.10).
4. Forced displacement of oil by water. The volume of water and oil released in the spontaneous and forced displacement steps are recorded. The forced displacement of water by oil and oil by water can be carried out by using either a centrifuge procedure or by mounting the core sample in a displacement apparatus (see Figure 4.7).

Amott, in his experiments, allowed a time period of 20 h for the spontaneous displacement and a centrifugal force of 1800 times gravity for the centrifuge procedure (selected on the basis of its convenience and rapidity) for the forced displacement of oil and water. Although, in Amott's experiments, a time limit of 20 h was chosen for the spontaneous displacement, the best approach should be to plot the results of periodically measured amount of fluids displaced and wait until a stable equilibrium value is observed on the graphs.

Spontaneous displacement tests can be conveniently carried out in equipment already available in most laboratories or easily obtained (see Figure 7.10). For spontaneous displacement, the majority of authors call this *spontaneous imbibition* rather than *displacement*. In an ordinary sense, imbibition does mean "absorption of a liquid by a solid," however, the term imbibition should not be used to describe spontaneous displacement. Its use is inappropriate and not justified because when dealing with the displacement of fluids in the porous media, the term imbibition is specifically used to describe the displacement of a nonwetting phase by a wetting phase, for example, the displacement of oil by water in a water-wet rock, under the assumption that the wettability is already known. Therefore, the use of the term spontaneous imbibition instead of displacement is incorrect because assumption is made that the sample is either oil-wet or water-wet, an unknown at this stage and the very purpose of the Amott test is to determine the wettability.

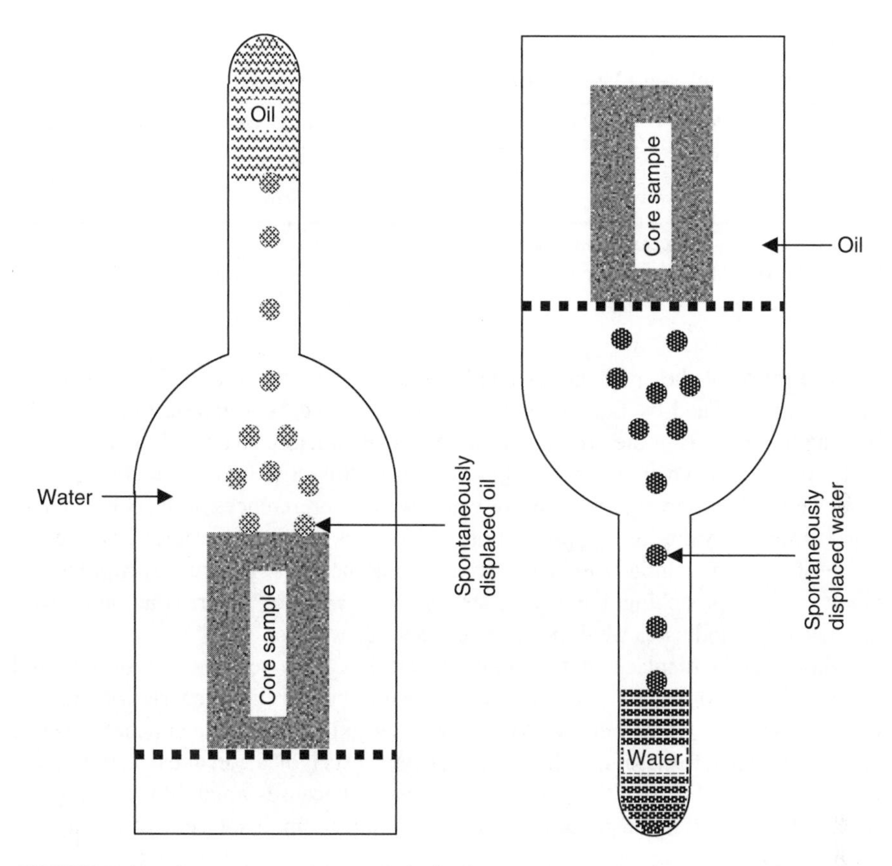

FIGURE 7.10 Spontaneous water and oil displacement setup for the Amott wettability test.

The core sample wettability from the previous steps of the Amott test is determined as follows: Let V_{ws} be the volume of water spontaneously displaced by oil, V_{wf} the volume of water released by forced displacement of water by oil, $V_{wt} = V_{ws} + V_{wf}$ the volume of water from spontaneous and forced displacements, V_{os} the volume of oil spontaneously displaced by water, V_{of} the volume of oil released by forced displacement of oil by water, and $V_{ot} = V_{os} + V_{of}$ the volume of oil from spontaneous and forced displacements.

Based on these steps, the test results are expressed as *displacement by oil ratio*, δ_o and *displacement by water ratio*, δ_w, respectively, defined by the following equations:

$$\delta_o = \frac{V_{ws}}{V_{wt}} \tag{7.5}$$

$$\delta_w = \frac{V_{os}}{V_{ot}} \tag{7.6}$$

TABLE 7.1
Relationships between Wettability and Amott[25] Wettability Indices

Diplacement Ratio	Water-Wet[a]	Neutral-Wet	Oil-Wet[a]
δ_o	Zero	Zero	Positive[b]
δ_w	Positive[b]	Zero	Zero

[a] Amott refers to this as preferentially water-wet and preferenitally oil-wet.

[b] Although this is characterized as positive in Amott's paper, what it actually means is a value approaching one.

The ratios of the spontaneous displacement volumes to the total displacement volumes, as defined by Equations 7.5 and 7.6 are used as *wettability indices*. The wetting preferences of the tested core sample are characterized according to the general criteria[25] shown in Table 7.1. In addition to this criteria, a distinction is also made between strong and weak oil- or water-wetting preferences of the core sample. For example, a value of δ_o approaching 1 indicates strong oil wetness, whereas a value of δ_o approaching 0 indicates weak preference for oil wetness. Similarly, a value of δ_w approaching 1 indicates strong water wetness, whereas a value of δ_w approaching 0 indicates weak preference for water wetness.

Based on the displacement tests and the criteria discussed here, Amott reported the wettabilities of a variety of core samples that fall under the categories of strongly oil-wet, strongly water-wet, weakly oil-wet, weakly water-wet, and neutral-wet. Amott stated that these simple displacement-type tests indicate in a reasonably direct manner the wettability of the porous rock surfaces because wettability is one of the factors that control the displacement rates and equilibrium displacement volumes in porous rocks.

7.6.3.1 Modification of the Amott Test (Amott–Harvey Test)

Other investigators[27,28] used a modification of the Amott wettability test called the *Amott–Harvey relative displacement or wettability index*. Unlike the Amott test, this procedure begins with oilflooding of the core sample to achieve irreducible water saturation, which is generally carried out in a centrifuge. The sequence of displacements used in the original Amott test are reversed in this case; the core sample containing irreducible water saturation is first subjected to the spontaneous and forced displacement of oil by water and then is followed by the spontaneous and forced displacement of water by oil. Based on the recorded volumes, the displacement by water and displacement by oil ratios are then calculated by the Amott method as shown in Equations 7.5 and 7.6. Using these ratios, the Amott–Harvey wettability is calculated as

$$I_{AH} = \delta_w - \delta_o \tag{7.7}$$

Equation 7.7 combines the two displacement ratios into a single wettability index that varies from +1 for complete water wetness to −1 for complete oil wetness. As shown in Table 7.2, Cuiec[15], however, refined the wettability scale and proposed

TABLE 7.2
Cuiec's[15] Wettability Classification Based on the Amott–Harvey Wettability Index, I_{AH}

I_{AH} Range	Wettability
+0.3 to +1.0	Water-wet
+0.1 to +0.3	Slightly water-wet
−0.1 to +0.1	Neutral
−0.3 to −0.1	Slightly oil-wet
−1.0 to −0.3	Oil-wet

a wettability classification based on the range of Amott–Harvey wettability index (I_{AH}) values.

7.6.4 USBM Method

The USBM method, developed by Donaldson et al.,[29] is also designed to measure the average wettability of a core sample. The USBM test is also one of the most popularly used methods to determine the wettability of a core sample. The entire wettability test is conducted in a centrifuge apparatus. Figure 7.11 shows a cross section of the arm of a centrifuge or the centrifuge tube setup used to house the core sample, displacing fluid, and the collection of displaced fluid.

The test begins by establishing the irreducible water saturation in the core plug sample. Irreducible water saturation in the core sample is obtained by centrifuging the water-saturated sample under the displacing oil phase at high speeds. The displacement of water by oil is monitored and centrifugation is continued until equilibrium is achieved, which is indicated by zero fractional water production or by a plateau in the cumulative water production vs. time curve. The value of irreducible water saturation is calculated by either the volume balance or mass balance.

Once the sample is prepared at irreducible water saturation, wettability determination begins with the first step in which cores are placed in brine and centrifuged at incrementally increasing speeds until an effective pressure (difference between the two phase pressures) of –10 psi is reached. This step is also known as the brine drive because brine displaces oil from the core. During the course of this first step, effective pressure and water saturation is determined at each constant speed of the centrifuge. Effective pressure is calculated from the equation suggested by Slobod et al.,[30] whereas the average water saturation is computed from the amount of oil displaced.

It should be noted that the effective pressures for the brine drive are indicated by negative pressures. This distinction is made by considering that these effective pressures are the phase pressure differences between the nonwetting phase and the wetting phase, that is, in the case of brine drive, if water is considered as a wetting phase then, $-P_{effective} = P_{water} - P_{oil}$, yielding a negative value. In the second and final step, the core is placed in oil and centrifuged. During this oil-drive step, oil displaces brine from the core. The water saturation and the effective pressures are calculated at each incremental centrifuge speed in a manner

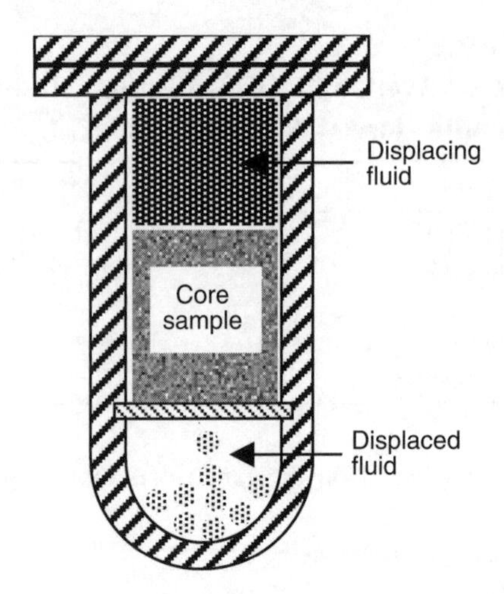

FIGURE 7.11 Conceptual diagram for a centrifuge tube setup for the USBM wettability measurement method.

similar to the first step. The second step is terminated when effective pressure of +10 psi is reached. The effective pressure in this case is indicated by positive values because $P_{effective} = P_{oil} - P_{water}$, again considering water as the wetting phase.

After completion of these two steps, the effective pressures for both the brine drive and the oil drive are then plotted against the water saturation, identified as curve I and curve II, respectively, in Figure 7.12. In each case, the curves are linearly extrapolated or truncated if the last pressure is not exactly 10 psi. The USBM wettability index is then calculated from the ratio of the area under the two effective pressure curves according to the following equation:

$$I_{USBM} = \log\left[\frac{A_1}{A_2}\right] \tag{7.8}$$

where I_{USBM} is the USBM wettability index, A_1 the area under the oil curve, and A_2 the area under the brine curve.

The areas under the oil and the brine curves represent the thermodynamic work required for the respective fluid displacements. For instance, the displacement of a nonwetting phase by a wetting phase requires less energy than displacement of a wetting phase by a nonwetting phase. Therefore, the ratio of the areas under the two curves is considered as a direct indicator of the degree of wettability.

The wettability of the core sample is determined as per the following criteria: If $I_{USBM} > 0$, the core is water-wet, and when $I_{USBM} < 0$, the core is oil-wet, whereas a near zero value of I_{USBM} indicates neutral wettability. The strong wetting preferences are indicated by larger absolute values of I_{USBM}. Donaldson et al.[29] stated that the area under the curve is considered representative of the overall wettability of the system because it is an integrated value over the practical range of saturations.

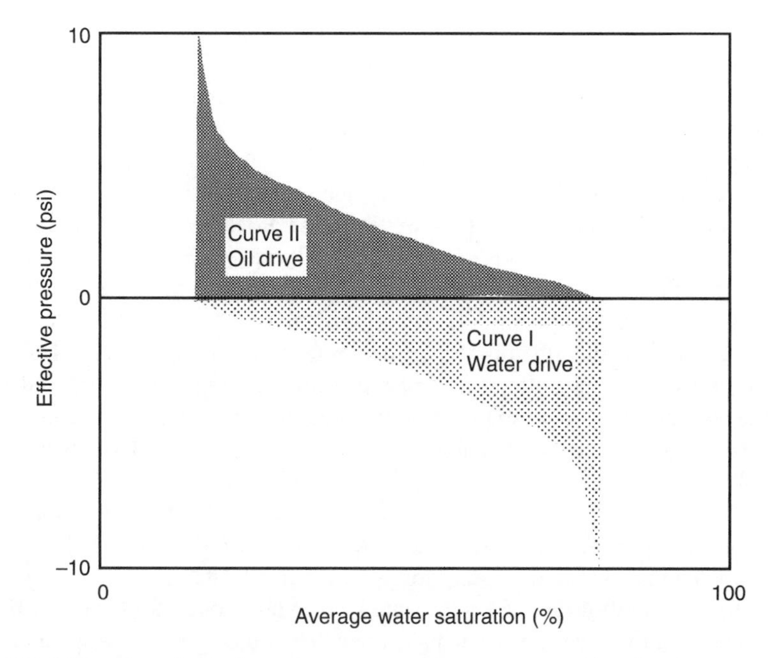

FIGURE 7.12 Plot of effective pressure vs. average water saturation for the water (brine) and oil drive used in the determination of wettability by the USBM method.

7.7 FACTORS AFFECTING WETTABILITY

Reservoir wettability is almost entirely dependent on the characteristics of the fluids involved and the lithology of the rock in question. Obviously these factors primarily affect the reservoir wettability. Additionally, reservoir pressure and temperature, locations of fluid contacts in the reservoir, effect of drilling-mud filtrate invasion are also some of the factors that play a role in dictating the reservoir wettability. However, many uncertainties exist as to the relative importance of these various factors in affecting the wettability in reservoirs. Specifically, this discussion focuses on the following factors that affect the reservoir wettability:

- Composition of the reservoir oil
- Composition of the brine
- Reservoir pressure and temperature
- Depth of the reservoir structure

7.7.1 Composition of the Reservoir Oil

While reservoir wettability is clearly affected by the composition of the reservoir oil, precisely which components of the reservoir oil are the most important is not clear. However, it is widely agreed that the presence and amount of asphaltenic components in a reservoir oil is important, such that reservoir wettability characteristics, among other factors, are often attributed to the adsorption of asphaltenes onto the mineral surfaces of reservoir rocks. Asphaltene is operationally defined as the precipitate resulting from addition of low-molecular-weight alkane to crude oil. Although numerous studies address different aspects of asphaltenes and asphaltene-containing reservoir oils on reservoir rock wettability, the results/arguments and conclusions seem to be dependent on the nature of the individual oil–water–rock systems.[31,32]

Despite the significance of asphaltenes on wettability alteration in reservoir rocks, it is particularly difficult to evaluate the underlying mechanisms from core tests because of the coupled effects of wetting and pore morphology.[32] Many studies of the evaluation of oil composition on wettability have focused on observing the interactions of crude oils and their components with smooth solid surfaces, typically via contact angle measurements.

Rayes et al.[31] studied the oil–water–rock systems for a Libyan and a Hungarian oil field to understand the effects of asphaltenes on wettability. They concluded that asphaltenes can cause substantial modification in the wetting characteristics of rocks. Their results indicated a change in the contact angle from 40–60° to 120°, i.e., a complete reversal in wettability from water-wet to oil-wet.

Liu and Buckley[33] studied the evolution of wetting alteration by adsorption of asphaltenic components from crude oil. They used four different asphaltenic crude oils to evaluate the alteration in wettability of a borosilicate glass microscope slide used as the test solid surface. They measured the contact angles for the test solid surface, using a normal decane water pair, after they were aged in these asphaltenic crude oils, indicating a change in the contact angle from about 50–70° to as high as 170°. Liu and Buckley's[33] results are somewhat similar to those of Rayes et al.[31]

Al-Maamari and Buckley[34] used freshly cleaved Muscovite mica as the solid surface to observe wettability alterations induced by ashphaltene precipitation from five different crude oils. To observe wettability alterations, they aged the test solid surface in crude oil and heptane mixture (heptane is used as asphaltene precipitant) and subsequently measured the contact angles for the decane–water–aged mica systems. The results obtained were similar to the earlier discussed studies, wettability altered from water-wet to oil-wet.

Tang and Morrow,[35] among other things, studied the effect of asphaltenes on wettability. Their experiments were based on a Berea sandstone in which displacement tests were conducted using three different crude oils. One of the dead crude oil composition was varied by removal of lighter components or by addition of alkanes such as pentane, hexane, and decane, to investigate the effect of changing oil composition on wettability. They concluded that the removal of light components from the crude oil resulted in increased water-wetness, whereas addition of alkanes to the crude oil reduced water wetness.

7.7.2 Composition of the Brine

Even though, crude oil composition or chemistry is the most important factor governing the wettability, brine composition or chemistry has also been shown to influence the wettability of the oil–water–rock systems.

Vijapurapu and Rao[36] evaluated the effects of brine dilution on the wettability of oil–water–rock (dolomite surface) system. Their results indicated that the initial oil-wet nature of the system was changed to intermediate wettability simply by diluting the reservoir brine with deionized water.

Tang and Morrow[35] carried out displacement experiments on Berea sandstone cores and various oil–brine systems, to evaluate the effect of brine concentration on oil recovery and wettability. Their results showed that salinity of the connate and invading brines greatly influences wettability and oil recovery at reservoir temperature. They observed that the oil recovery increased with dilution of the connate brine and invading brine. However, instead of directly correlating wettability changes (by a single parameter such as the Amott wettability index) occurring due to the dilution of the invading brine, they chose to characterize wettability by a dimensionless time. The recoveries obtained from the various displacement tests with different dilution rates were therefore compared with this dimensionless time for a strongly water-wet case.

7.7.3 Reservoir Pressure and Temperature

The effect of reservoir pressure and temperature on wettability can actually be perceived in many different ways. The majority of pressure and temperature effects on wettability are reflected through the changes that occur in the fluid (oil or water) characteristics with varying pressure and temperature conditions. Reservoir pressures and temperatures can cause changes in the crude oil composition that can in turn influence the precipitation of asphaltenes from crude oils. Precipitated asphaltenes can impact the wettability of the reservoir rocks, as outlined in Section 7.7.1. Similarly, wettability may change as a function of pressure and temperature through changes that occur in the oil–water interfacial tension values. However, Section 7.2.1 shows that neither pressure nor temperature seemed to significantly influence oil–water IFT values. Also, when the contact angle measurement is considered as a measure of wettability; the influence of pressure and temperature on wettability can also be evaluated on the basis of contact angle measurements at various pressures and temperatures. However, Wang and Gupta's[9] measurements of contact angles for the calcite and quartz surfaces did not indicate a very strong correlation with either pressure or temperature.

The effect of temperature on contact angle has also been studied by other investigators.[37,38] The measurements of Poston et al.[37] showed an average temperature coefficient for the contact angle of –0.27°/°C, while Phillips and Riddiford[38] reported a coefficient of –0.29°/ °C. Similarly, Lo and Mungan[39], have reported contact angle measurements for tetradecane–brine–quartz system in the temperature range of 25 to 149°C. Their results also indicate a coefficient of approximately –0.20°/°C.

It should, however, be noted that all these studies reported contact angles of about 40° for the lowest temperature studied, which reduced as test temperatures were increased. As per the first principles of contact angle and wettability, to begin with, the systems were water-wet anyway, the degree of which increased slightly with temperature. In summary, these reported results on the specific systems illustrate that the decrease in contact angle with increasing temperature is real, but small, that is, the systems becoming progressively more water-wet. However, that increase in the degree of water wetness with increasing temperature does not appear to be very significant.

Jadhunandan and Morrow[40] presented data on the wettability index as a function of aging temperature for two different oil–brine–rock systems. Two oils were from

West Texas and the North Sea, respectively, whereas the core samples were cut from blocks of Berea sandstone. These oil–water–rock systems were aged in the respective crude oils in a temperature range of 20 to 80°C. Subsequently, when the wettability indices were plotted against the aging temperature for the two systems, a trend of decreasing water wetness with increasing temperature was revealed. Thus, the aging temperature showed a dominant effect on the wetting behavior.

7.7.4 Depth of the Reservoir Structure

Another important variable that can potentially control the wettability of a reservoir rock is its location relative to the oil–water contact. For example, in the case of a gravity-capillary equilibrated reservoir, if the Amott–Harvey wettability indices are plotted vs. the height above the oil–water contact, as shown in Figure 7.13, the wettability may trend from slightly oil-wet in the oil column to water-wet near the oil–water contact.

Jerauld and Rathmell[41] documented the wettability of the Prudhoe Bay reservoir as a function of the depth of the reservoir structure. Core samples were collected at different depths of the reservoir structure and their wettability was determined using the Amott test. A plot of the Amott indices of cores vs. their subsequent depths revealed the existence of completely water-wet rocks near the oil–water contact (down structure), and mixed-wet rocks near the oil–water contact (up structure). However, the samples taken at various depths above the oil–water contact clearly indicate progressively more oil-wet behavior with height into the oil column.

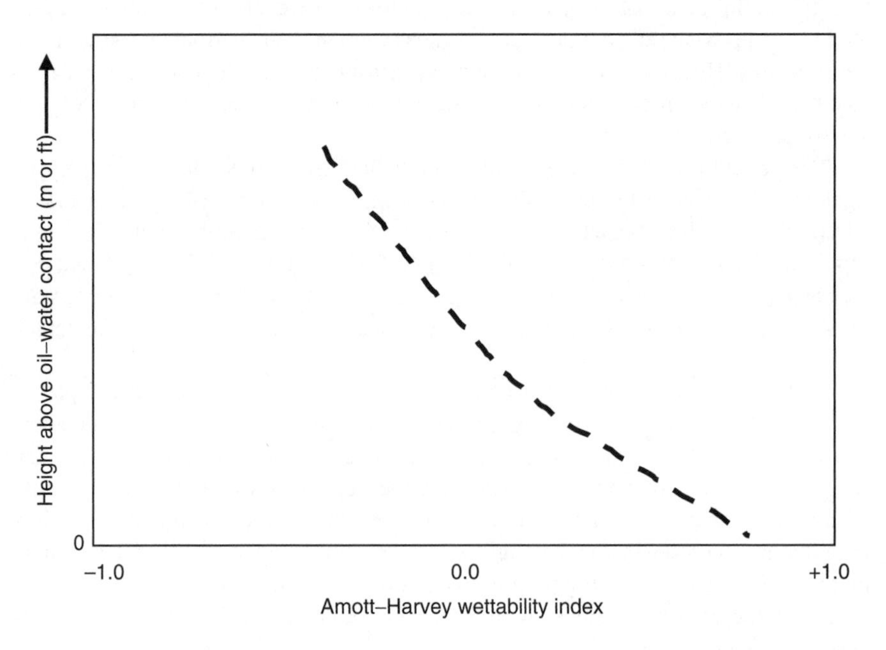

FIGURE 7.13 The effect of the depth of reservoir structure on wettability, which is a likely scenario for a gravity-capillary equilibrated reservoir.

Such data, therefore, demonstrate the existence of a wettability transition within the field, with water-wet behavior down structure, and mixed-wet behavior upstructure tending toward oil wetness in relation to the oil–water contact.

7.8 RELATIONSHIP BETWEEN WETTABILITY AND IRREDUCIBLE WATER SATURATION AND RESIDUAL OIL SATURATION

It is widely recognized that the reservoir wettability affects the relative distribution of fluids within a porous medium, which in turn strongly affects the displacement behavior, relative permeability characteristics, and consequently, the production of hydrocarbons from petroleum reservoirs. Therefore, considering the importance of wettability, this discussion focuses on the impact of wettability, on irreducible water saturation (S_{wi}) and residual oil saturation (S_{or}). The primary reason behind considering the effect of wettability on S_{wi} and S_{or} is based on the fact that these two saturations basically represent the two end points in the recovery of hydrocarbons; $1 - S_{wi}$ indicates the amount of initial hydrocarbons in place that could be potentially recovered, whereas S_{or} indicates the amount of oil left in the pore space after the termination of primary production or gasflood or waterflood.

7.8.1 WETTABILITY AND IRREDUCIBLE WATER SATURATION

The relationship between wettability and initial water saturation or irreducible water saturation is sometimes stated as a rule of thumb that water-wet rocks have connate water saturation greater than 20 to 25% of pore volume, whereas in oil-wet rocks connate water is generally less than 15% of pore volume and frequently less than 10%.[40]

Based on the data presented by Jadhunandan and Morrow,[40] a plot of wettability index vs. initial water saturation for tested systems is shown in Figure 7.14. Initial water saturation tends to decrease with increasing oil wetness, as shown in Figure 7.14. Bennion et al.[42] also presented significant data on wettability and initial water saturation for Western Canadian Sedimentary Basin. The data reported by Bennion et al.[42] also seem to indicate a trend of decreasing initial water saturation with increasing oil wetness (see Figure 7.14).

Jerauld and Rathmell[41] also presented data on wettability indices as a function of initial water saturation for Prudhoe Bay and concluded that a strong correlation exists between initial water saturation and the Amott wettability index with more water-wet behavior at high initial water saturation. Trends in Jerauld and Rathmell's[41] data are essentially the same as those observed by Jadhunandan and Morrow.[40] The observed behavior based on these works may be considered consistent with the fact that, in water-wet rocks water covers the pore surfaces and thus exists as a continuous film rather that small discontinuous globules in oil-wet rocks. However, Jadhunandan and Morrow.[40] have stated that the wettability of the reservoir may well be a consequence of the water saturation rather than water saturation being dependent on wettability.

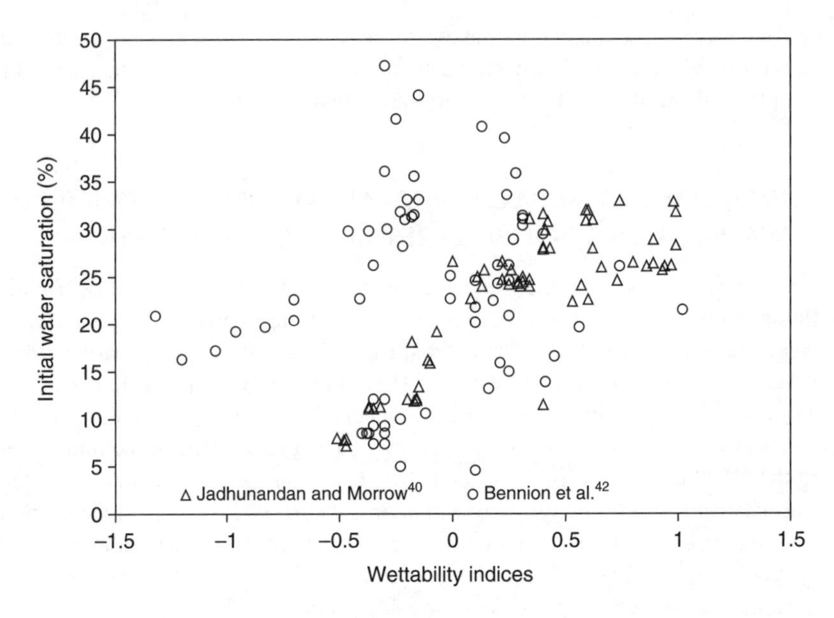

FIGURE 7.14 Relationship between wettability and initial or irreducible water saturation. The S_{wi} data of Jadhunandan and Morrow[40] are plotted against the Amott–Harvey wettability index. The data of Bennion et al.[42] are plotted against the USBM wettability index.

7.8.2 Wettability and Residual Oil Saturation

The studies on the effects of wettability on oil recovery or residual oil saturation are generally confined to waterflooding because the relationship between primary recovery (by pressure depletion) and wettability have not been developed. Therefore, this discussion focuses on the relationship between wettability and waterflood residual oil saturation, S_{or}. Amott[25] presented some of the earlier work on correlation between rock wettability and oil recovery by waterflooding. Amott conducted a group of waterflood tests on Ohio sandstone cores with the objective of comparing oil recovery with wettability. The plot of wettability (x axis in terms of δ_w and δ_o) as a function of oil recovery at 2.4 pore volume throughput indicated gradually increasing oil recoveries for δ_w from 1.0 to 0.6, a plateau of high oil recovery for δ_w from 0.6 to 0.05, and a further decrease in the oil recovery with increasing value of δ_o. Amott's results on Ohio sandstone basically indicate that low recoveries or high S_{or}'s are obtained at either wettability extremes, whereas somewhat higher recoveries or low S_{or}'s are obtained in the weakly water-wet to neutral wettability conditions.

Jadhunandan and Morrow[40] plotted results similar to those of Amott on wettability index as a function of oil recovery for two different pore volumes injected. Their results were based on waterfloods in Berea sandstone samples saturated with two different crude oils from West Texas and the North Sea. The trend seen in their plots of wettability indices vs. percentage of oil recovered were quite similar to the observations of Amott, that is, a maximum in recovery at a wettability close to but on the water-wet side of neutral (wettability index = 0.2) and generally with low oil recoveries on either wettability extremes. Therefore, the results of Amott[25] and Jadhunandan and Morrow,[40] with respect to the effect of wettability on oil recovery, provide the closest qualitative similarity. In response to the observed behavior, Jadhunandan and Morrow[40] stated that the maximum in oil recovery at near-neutral wettability has intuitive appeal because it can be argued that capillary forces are minimized.

Kennedy et al.[43] presented data on ultimate oil recovery for a synthetic silica core, East Texas crude oil and surfactant-treated brine system to depict the effect of wettability on oil recovery. They used different surfactants to vary the wettability. However, they used the sessile drop method with a smooth silica surface to determine the wettability of the system. The data presented by Kennedy et al.[43] indicate that the maximum recovery (and minimum true-residual oil saturation) occurred at slightly oil-wet conditions. The plot of oil recovery as a function of wettability reveals a trend quite similar to the one observed by Amott[25] and Jadhunandan and Morrow.[40] Kennedy et al.'s[43] data also show low oil recoveries on either wettability extremes. Despite the similarities in the observed trends, the difference between the average low recoveries at either wettability extremes and the maximum oil recovery is about 20% in Amott[25] and Jadhunandan and Morrow,[40] however, this difference in Kennedy et al.[43] is only about 5%. The change in residual oil saturation or oil recovery is small while wettability varies from one extreme to the other.

Lorenz et al.[44] presented data on average residual oil saturations vs. USBM wettability index for Squirrel oil and organochlorosilicane-treated Torpedo sandstone cores. Their residual oil saturations are based on centrifuging. The organochlorosilicane was used to vary the sample wettability ranging from strongly water-wet to strongly oil-wet. As a matter of fact, data of Lorenz et al.[44] also show a trend similar to the earlier discussed works, that is, high residual oil saturation (or low recovery) at either wettability extremes, whereas the minima in residual oil saturation (or high recovery) occurring around the neutral wetting region. Their data indicate an S_{or} of about 30% when the sample is either strongly water-wet or strongly oil-wet, whereas S_{or} reduces to about 20% when the system is in the neutrally wet regime. In summary, as illustrated in Figure 7.15, these literature results discussed with regard to the relationship between wettability and residual oil saturation or oil recovery basically represent a crest-shaped curve when oil recovery is plotted against wettability, and a trough-shaped curve when residual oil saturation is plotted against wettability.

However, when evaluating the effect of wettability on oil recovery or residual oil saturation, it should also be recognized that the obtained residual oil saturation for a particular type of wettability is also dependent on the initial oil saturation. Data presented by Masalmeh[45] from a collection of several cores from different carbonate fields in the Middle East, in fact allow the comparison of residual oil saturation for two different types of wettability and the initial oil saturation varying from about 5 to 90%. Data on residual oil saturation for the water-wet cores indicate that the residual oil saturation increases as initial oil saturation increases. However, for the mixed-wet cores, the residual oil saturation stays within a band of about 1 to 10% when initial oil saturation increases from about 10 to 90%. For example, a comparison of the residual oil saturation at 60% initial oil saturation, yields an S_{or} of about 25% for the water-wet cores and about 7.5% for the mixed-wet cores. In mixed-wet condition, the

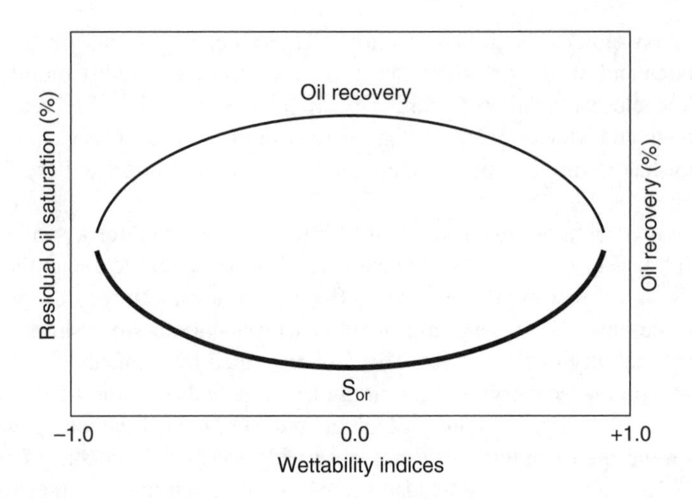

FIGURE 7.15 Relationship between wettability and residual oil saturation or oil recovery. This figure has been constructed on the basis of observations from various references discussed in Section 7.8.2.

fine pores and grain contacts are preferentially water-wet and the surfaces of the larger pores are oil-wet. However, for mixed wettability systems, if oil-wet paths were continuous through the rock (as per the definition of mixed wettability), water could displace oil from the large pores and little or no oil would be held by capillary forces in small pores or at grain contacts as stated by Salathiel.[18]

PROBLEMS

7.1 In the pendant drop technique for measuring the oil–water interfacial tension (IFT), the drop remains attached to the pendant drop tube (syringe or needle), at which point it is dimensioned. However, the drop eventually breaks away when suspended weight of the drop (due to gravity force) is no longer supported by the IFT force. For one such experiment, the following data was obtained: inside diameter of the pendant drop tube = 0.10 in.; drop diameter at the point of breaking away = 0.22 in.; $\rho_w = 1.05$ g/cm^3; and $\rho_o = 0.85$ g/cm^3. Calculate the oil–water IFT.

7.2 The following table provides the displacement data for Amott wettability test on three cores from an Alaska North Slope reservoir:

Core No.	Displacement by Oil, (mL)		Displacement by Water (mL)	
	Spontaneous	Forced	Spontaneous	Forced
1	0.05	1.25	0.81	0.85
2	0.00	1.69	0.00	0.97
3	0.47	0.53	0.01	0.59

Calculate the Amott–Harvey wettability index for each core and determine the wetting characteristics of each core.

7.3 A contact angle of 35° is measured by placing a drop of brine on a polished calcite plate submerged in Isopar-L. Subsequently, if the contact angle measurement is repeated by reversing the location of Isopar-L and brine, that is, the drop of Isopar-L on the same plate (plate on top) submerged in water; what will be the value of the contact angle?

7.4 The USBM method is commonly used for determining the average wettability of a reservoir rock core sample, based on the effective pressures and corresponding water saturations for the brine drive and oil drive, respectively. Draw hypothetical effective pressure vs. water saturation curves for three different wettability cases: water-wet, oil-wet, and neutral-wet.

REFERENCES

1. Katz, D.L., *Handbook of Natural Gas Engineering*, McGraw-Hill, New York, 1959.
2. Cooper, J.R., IAPWS release on surface tension of ordinary water substance, issued by the International Association for the Properties of Water and Steam, 1994.
3. Dandekar, A.Y., unpublished data, 1999.
4. Kundt, *Ann. Physik*, 12, 538; *International Critical Tables*, 4, 475, 1881.
5. Dandekar, A.Y., Interfacial Tension and Viscosity of Reservoir Fluids, Ph.D. thesis, Heriot-Watt University, Edinburgh, UK, 1994.
6. Mulyadi, H. and Amin, R., A new approach to 3D reservoir simulation: effect of interfacial tension on improving oil recovery, Society of Petroleum Engineers SPE paper number 68733.
7. Huijgens, R.J.M. and Hagoort, J., Interfacial tension of nitrogen/volatile oil systems, Society of Petroleum Engineers SPE paper number 26643.
8. Jennings, H.Y., Jr. and Newman, G.H., The effect of temperature and pressure on the interfacial tension of water against methane-normal decane mixtures, *Soc. Pet. Eng. J.*, 171–175, 1971.
9. Wang, W. and Gupta, A., Investigation of the effect of temperature and pressure on wettability using modified pendant drop method, Society of Petroleum Engineers SPE paper number 30544.
10. Hjelmeland, O.S. and Larrondo, L.E., Experimental investigation of the effects of temperature, pressure, and crude oil composition on interfacial properties, *SPE Reservoir Eng.*, 321–328, 1986.
11. http://www.etccte.ec.gc.ca/databases/OilProperties/Default.aspx
12. Hirasaki, G. and Zhang, D.L., Surface chemistry of oil recovery from fractured, oil-wet, carbonate formations, *Soc. Pet. Eng. J.*, 9, 151–162, 2004.
13. Niederhauser, D.O. and Bartell, F.E., A corrected table for calculation of boundary tensions by pendant drop method, *Research on Occurrence and Recovery of Petroleum, A Contribution from API Research Project 27*, 114, 1947.
14. Misak, M.D., Equations for determining $1/H$ versus S values for interfacial tension calculations by the pendant drop method, Society of Petroleum Engineers SPE paper number 2310.
15. Cuiec, L., Rock/crude-oil interactions and wettability: an attempt to understand their relation, Society of Petroleum Engineers SPE paper number 13211.

16. Morrow, N.R., Wettability and its effect on oil recovery, *J. Pet. Technol.*, 1476–1484, 1990.
17. Ehrlich, R., Hasiba, H.H., and Raimondi, P., Alkaline waterflooding for wettability alteration evaluating a potential field application, *J. Pet. Technol.*, 1335–1343, 1974.
18. Salathiel, R.A., Oil recovery by surface film drainage in mixed-wettability rocks, *J. Pet. Technol.*, 1216–1224, 1973.
19. Anderson, W.G., Wettability literature survey—Part 2: wettability measurement, *J. Pet. Technol.*, 1246–1262, 1986.
20. Leach, R.O., Wagner, O.R., Wood, H.W., and Harpke, C.F., A laboratory and field study of wettability adjustment in waterflooding, *J. Pet. Technol.*, 225, 206, 1962.
21. Morrow, N.R., Physics and thermodynamics of capillary action in porous media, *Ind. Eng. Chem.*, 62, 32, 1970.
22. Eick, J.D., Good, R.J. and Neumann, A.W., Thermodynamics of contact angles: II. Rough solid surfaces *J. Colloid Interface Sci.*, 53, 235, 1975.
23. Treiber, L.E. and Owens, W.W., A laboratory evaluation of the wettability of fifty oil-producing reservoirs, *Soc. Pet. Eng., J.*, 531–540, 1972.
24. Bobek, J.E., Mattax, C.C., and Denekas, M.O., Reservoir rock wettability —its significance and evaluation, *Trans. AIME*, 213, 155, 1958.
25. Amott, E., Observations relating to the wettability of porous rocks, *Trans. AIME*, 192, 99, 1951.
26. Gant, P.L. and Anderson, W.G., Core cleaning for restoration of native wettability, Society of Petroleum Engineers SPE paper number 14875.
27. Boenau, D.F. and Clampitt, R.L., A surfactant system for the oil-wet sandstone of the North Burbank unit, *J. Pet. Technol.*, 501–506, 1977.
28. Trantham, J.C. and Clampitt, R.L., Determination of oil saturation after waterflooding in an oil-wet reservoir - The North Burbank Unit, tract 97 project, *J. Pet. Technol.*, 491–500, 1977.
29. Donaldson, E.C., Thomas, R.D., and Lorenz, P.B., Wettability determination and its effect on recovery efficiency, *Soc. Pet. Eng. J.*, 13–20, 1969.
30. Slobod, R.L., Chambers, A., and Prehn, W.L., Use of centrifuge for determining connate water, residual oil, and capillary pressure curves of small core samples, *Trans. AIME*, 192, 127, 1951.
31. Rayes, B.H., Lakatos, I., Pernyeszi, T., and Toth, J., Adsorption of asphaltenes, on formation rocks and its effect on wettability, Society of Petroleum Engineers SPE paper number 81470.
32. Buckley, J.S., Liu, Y., and Monsterleet, S., Mechanisms of wetting alteration by crude oils, Society of Petroleum Engineers SPE paper number 37230.
33. Liu, Y. and Buckley, J.S., Evolution of wetting alteration by adsorption from crude oil, *SPE Formation Evaluation*, 12, 5–12, 1997.
34. Al-Maamari, R.S.H. and Buckley, J.S., Asphaltene precipitation and alteration of wetting: the potential for wettability changes during oil production, *SPE Reservoir Evaluation Eng.*, 6, 210–214, 2003.
35. Tang, G.Q. and Morrow, N.R., Effect of temperature, salinity and oil composition on wetting behavior and oil recovery by waterflooding, Society of Petroleum Engineers SPE paper number 36680.
36. Vijapurapu, C.S. and Rao, D.N., Effect of brine dilution and surfactant concentration on spreading and wettability, Society of Petroleum Engineers SPE paper number 80273.
37. Poston, S.W., Ysrael, S.C., Hossain, A.K.M.S., Montgomery, E.F., and Ramey, H.J., Jr., The effect of temperature on irreducible water saturation and relative permeability of unconsolidated sands, *Soc. Pet. Eng. J.*, 171–180, 1970.

38. Phillips, M.C. and Riddiford, A.C., *Nature*, 205, 1005, 1965.
39. Lo, H.Y. and Mungan, N., Effect of temperature on water-oil relative permeabilities in oil-wet and water-wet systems, Society of Petroleum Engineers SPE paper number 4505.
40. Jadhunandan, P.P. and Morrow, N.R., Effect of wettability on waterflood recovery for crude-oil/brine/rock systems, Society of Petroleum Engineers SPE paper number 22597.
41. Jerauld, G.R. and Rathmell, J.J., Wettability and relative permeability of Prudhoe Bay: a case study in mixed-wet reservoirs, Society of Petroleum Engineers SPE paper number 28576.
42. Bennion, D.B., Thomas, F.B., Schulmeister, B.E. and Ma, T., A correlation of water and gas–oil relative permeability properties for various Western Canadian sandstone and carbonate oil producing formations, www.hycal.com, Hycal PAPER 2002-066.
43. Kennedy, H.T., Burja, E.O., and Boykin, R.S., An investigation of the effects of wettability on the recovery of oil by waterflooding, *J. Phys. Chem.*, 59, 867, 1955.
44. Lorenz, P.B., Donaldson, E.C. and Thomas, R.D., Use of centrifugal measurements of wettability to predict oil recovery, U.S. Bureau of Mines, Bartlesville Energy Technology Center, Report 7873, 1974.
45. Masalmeh, S.K., The effect of wettability on saturation functions and impact on carbonate reservoirs in the Middle East, Society of Petroleum Engineers SPE paper number 78515.

8 Capillary Pressure

8.1 INTRODUCTION

Chapter 7 addressed the fact that the simultaneous existence of two or more fluid phases in a pore space of a reservoir rock requires the definition of surface and interfacial tensions, wettability, capillary pressure, and relative permeability. It now becomes necessary to accurately characterize these terms so that the production potential of a given hydrocarbon-bearing formation can be evaluated. Additionally, Chapter 7 dealt with interfacial forces and wettability. This chapter focuses on the capillary pressure characteristics of a porous medium.

Capillary phenomena occur in porous media when two or more immiscible fluids are present in the pore space. A difference in the pressure across the interface due to interfacial energy between two immiscible phases results in a curvature of the interface. The first thing that is noticeable is each immiscible fluid has a pressure that is distinct from that of the other immiscible fluids because of the curvature of the interfaces. In other words, when two immiscible fluids are in contact, a discontinuity in pressure exists between the two fluids, which depends on the curvature of the interface that separates the fluids. This particular difference in pressures is called the *capillary pressure* and is normally denoted by P_c. Capillary forces in a petroleum reservoir are the result of the combined effect of surface and interfacial tensions, pore size, geometry, and wetting characteristics of a given system.

The presence of capillary forces in a porous medium causes retention of fluids in the pore space against gravity forces, even though in large containers, such as tanks and pipes of large diameters, immiscible fluids usually completely segregate due to gravity. For instance, if a drop of oil is released from the seafloor, it would immediately rise to the sea surface; this rise being solely dependent on the density difference between the seawater and the oil. However, if the same situation is considered in a porous medium while the density difference or the gravity forces are still active, the upward movement of the oil also experiences a resistant force due to capillarity. The classic example of the retention of fluids in pore spaces against gravity forces is the migration of hydrocarbons from a source rock to the water-saturated reservoir rock, a process that resulted in the formation of petroleum reservoirs. In this particular process, complete gravity segregation into distinct layers of gas, oil, and water did not take place due to capillary forces. However, eventually equilibrium occurred when capillary forces and gravitational forces balanced and resulted in a particular fluid distribution, zonation, and fluid contacts in a given petroleum reservoir. This is the primary reason why dense fluid such as water or brine (usually represented by connate or irreducible water saturation) is found in petroleum reservoirs at higher elevations above the oil–water contact.

Capillary forces also play a major role in the process of displacement of one fluid by another in the pore space of a porous medium and this displacement process is either aided or opposed by the capillary forces. As a consequence, to maintain a porous medium partially saturated with a nonwetting fluid while the medium is also exposed to the wetting fluid, it is necessary to maintain the pressure of the nonwetting fluid at a value greater than that in the wetting fluid. Specifically, during water flooding, the capillary forces may act together with frictional forces to resist the flow of oil. It is advantageous to understand the nature of these capillary forces both from a reservoir structure (in terms of fluid contacts, transition zones, and free water level) as well as the actual hydrocarbon recovery point of view.

8.2 BASIC MATHEMATICAL EXPRESSION OF CAPILLARY PRESSURE

If the pressure in the nonwetting phase and the wetting phase is denoted by P_{nw} and P_w, respectively, then the capillary pressure is expressed as

$$\text{Capillary pressure} = \text{pressure in the nonwetting phase} - \text{pressure in the wetting phase}$$

or

$$P_c = P_{nw} - P_w \tag{8.1}$$

Equation 8.1 is the defining equation for capillary pressure in a porous medium. Basically, three types of capillary pressures exist:

- Gas–oil capillary pressure denoted by P_{cgo}
- Gas–water capillary pressure denoted by P_{cgw}
- Oil–water capillary pressure denoted by P_{cow}

However, when mathematically expressing these three capillary pressure pairs, wetting preferences of a given porous medium should be considered. For example, when considering the gas–oil and gas–water capillary pressures, gas is always the nonwetting phase. Therefore,

$$P_{cgo} = P_g - P_o \tag{8.2}$$

and

$$P_{cgw} = P_g - P_w \tag{8.3}$$

where P_g, P_o, and P_w represent the gas-, oil-, and water-phase pressures, respectively.

In the case of oil and water, either phase could preferentially wet the rock. Therefore, oil–water capillary pressure of a water-wet rock (water is a wetting phase and oil is a nonwetting phase) can be expressed as:

$$P_{cow} = P_o - P_w \tag{8.4}$$

If all the three phases are continuous, then:

$$P_{cgw} = P_{cgo} + P_{cow} \tag{8.5}$$

However, the capillary pressure is a combined effect of surface and interfacial tensions, pore size, geometry, and wetting characteristics of a given system. Therefore, a practical mathematical expression that relates capillary pressure to these properties should be developed. The development of such an equation, based on the rise of liquid in the capillaries, is described in the following section.

8.3 THE RISE OF LIQUID IN CAPILLARIES

Consider a capillary tube having an extremely small internal diameter, placed in a large open vessel containing a liquid, as shown in Figure 8.1. The liquid rises in the capillary tube above the height of the liquid in the large vessel due to the attractive forces (adhesion tension, A_T) between the tube and the liquid and the small weight

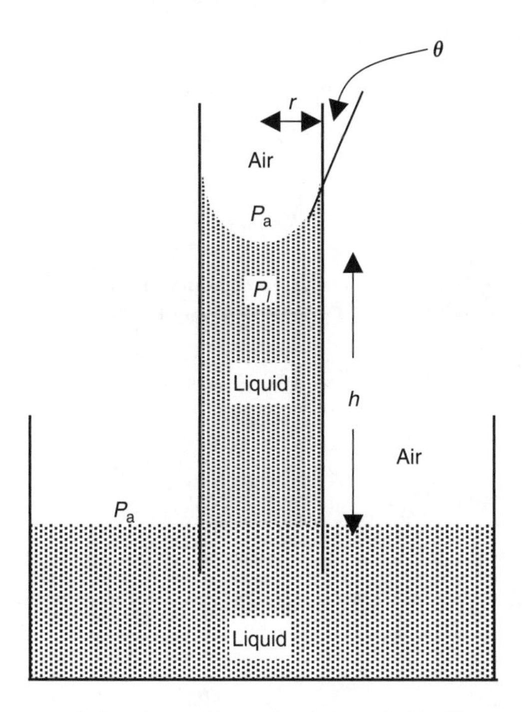

FIGURE 8.1 Pressure relations in capillary tubes for an air–liquid system.

represented by the column of liquid in the tube. The adhesion tension is the force tending to pull the liquid up the wall of the tube. The height to which the liquid rises in the capillary tube depends on the balance between the total force acting to pull the liquid upward and the weight of the column of liquid supported in the tube.

The two forces, i.e., upwards and downwards, can be expressed as

$$\text{Force up} = A_T \times 2\pi r \tag{8.6}$$

$$\text{Force down} = \pi r^2 h\rho g \tag{8.7}$$

where A_T is the adhesion tension expressed in dynes/cm, r the radius of the capillary tube in cm, h the height of capillary rise in cm, ρ the density of the liquid in tube in g/cm^3, (air density is negligible compared to liquid) and g the force of gravity in cm/sec^2.

All these variables can also be expressed in terms of any consistent set of units. Equating Equations 8.6 and 8.7 yields a force balance such that the total adhesion tension force would be just balancing the gravitational pull on the liquid column. As per the definition of capillary pressure, a difference in pressure exists across the air–liquid interface. Therefore, the pressure existing in the liquid phase beneath the air–liquid interface is less than the pressure that exists in the gaseous phase above the interface. However, for the large vessel, the air–liquid interface is essentially horizontal, and the capillary pressure is 0 in a horizontal or plane interface. Figure 8.1 shows the pressure in the liquid at the top of the liquid column is equal to the pressure in the liquid at the bottom minus the pressure due to a head of liquid h.

Therefore,

$$P_1 = P_a - \rho g h \tag{8.8}$$

or

$$P_a - P_1 = \rho g h = P_c \tag{8.9}$$

which is also the capillary pressure across the curved interface.

From force balance based on Equations 8.6 and 8.7:

$$A_T \times 2\pi r = \pi r^2 h\rho g \tag{8.10}$$

or

$$h = \frac{2\pi r A_T}{\pi r^2 \rho g} = \frac{2A_T}{r\rho g} \tag{8.11}$$

However, based on the definition of adhesion tension, $A_T = \sigma_{al} \cos\theta_{al}$. Therefore,

$$h = \frac{2\sigma_{al}\cos\theta_{al}}{r\rho g} \tag{8.12}$$

A combination of Equations 8.9 and 8.12 leads to the capillary pressure equation in terms of surface forces, wettability, and capillary size

$$P_c = \frac{2\sigma_{al}\cos\theta_{al}}{r} \tag{8.13}$$

if the liquid under consideration is water then,

$$P_{caw} = \frac{2\sigma_{aw}\cos\theta_{aw}}{r} \tag{8.14}$$

where P_{caw} is the air–water capillary pressure, σ_{aw} the air–water surface tension, θ_{aw} the contact angle, and r the capillary radius.

In Equation 8.14, if the air–water surface tension and the capillary radius are expressed in dynes/cm and cm, respectively, then the capillary pressure is in dynes/cm^2. If these variables are specified in N/m and m, respectively, then the capillary pressure is in N/m^2. The value of capillary pressure determined using Equation 8.14 can also be converted to field units by applying the appropriate conversion factors.

An expression similar to Equation 8.14 can also be developed for two immiscible liquids. Consider the capillary tube immersed in a beaker of water where oil is the other liquid, as shown in Figure 8.2.

Let P_{o1} be the pressure in oil at point 1, P_{o2} the pressure in oil at point 2, P_{w1} the pressure in water at point 1, and P_{w2} the pressure in water at point 2.

Again, considering the large vessel, the interface at 1 is a flat oil–water interface and the capillary pressure is zero, i.e., $P_{o1} = P_{w1}$ at the free water level in the large vessel.

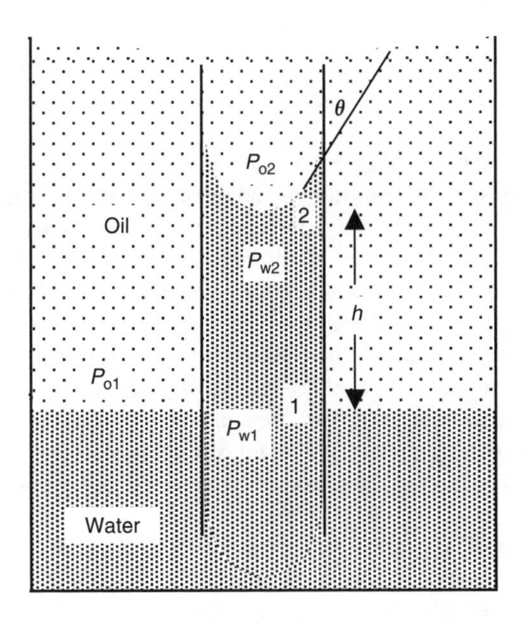

FIGURE 8.2 Pressure relations in capillary tubes for an oil–water system.

The pressure in oil and water at point 2 can now be written as

$$P_{o2} = P_{o1} - \rho_o gh \tag{8.15}$$

$$P_{w2} = P_{w1} - \rho_w gh \tag{8.16}$$

The pressure difference or the capillary pressure across the oil–water interface is

$$P_{o2} - P_{w2} = (\rho_w - \rho_o)gh = P_{cow} \tag{8.17}$$

If the oil and water is in equilibrium and not flowing, the capillary pressure must be in equilibrium with the gravitational forces. The expression for the oil–water capillary pressure in terms of interfacial forces, wettability, and capillary size can also be derived in the same manner as that for the air–water case and results in a similar equation:

$$P_{cow} = \frac{2\sigma_{ow}\cos\theta_{ow}}{r} \tag{8.18}$$

where P_{cow} is the oil–water capillary pressure, σ_{ow} the oil–water interfacial tension, and θ_{ow} the contact angle.

8.4 DEPENDENCE OF CAPILLARY PRESSURE ON ROCK AND FLUID PROPERTIES

Equations 8.14 and 8.18 show the capillary pressure of an immiscible pair of fluids expressed in terms of surface or interface forces, wettability, and capillary size. Capillary pressure is a function of the adhesion tension ($\sigma \cos\theta$) and inversely proportional to the radius of the capillary tube. Now a qualitative examination of the effect of pore size (capillary radius in this case) and the adhesion tension on capillary pressure must be made. Figure 8.3 shows the effect of varying the wetting characteristics of the system and varying the radius of the capillary tube.

In Figure 8.3a, the wetting characteristics are same, i.e., the same contact angle, but the radius of the capillary tube is different. In this case the capillary pressure is inversely proportional to the capillary tube radius, while the adhesion tension remains constant. By merely looking at the mathematical expression of capillary pressure, it is easily understood that higher the capillary tube radius, the lower the capillary pressure or vice versa. Alternatively, in terms of the weight of the liquid column; in the case of higher capillary tube radius, obviously the gravity forces are dominating because the weight of the liquid column increases and consequently the capillary pressure decreases. The opposite is true in the case of the smaller capillary tube radius.

On the other hand, when the capillary tubes of same radius but different wetting characteristics are considered, the denominator in the capillary pressure equation will be a constant and the value of capillary pressure will be directly proportional to

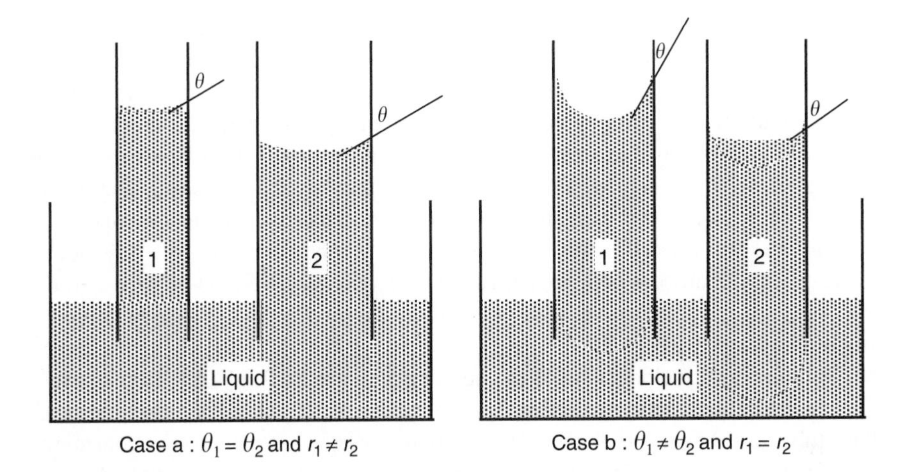

FIGURE 8.3 Dependence of capillary pressure on wetting characteristics and pore size (tube radius in this case).

the adhesion tension or the wetting characteristics of the system. Figure 8.3b shows such a system of same radius and different contact angles. In this case, the smaller the contact angle, the greater the height of liquid rise and stronger the adhesion tension, leading to higher capillary pressure, whereas the opposite is evident from the other tube having weaker wetting characteristics or adhesion tension that obviously results in lower capillary pressure.

8.5 CAPILLARY PRESSURE AND SATURATION HISTORY

Section 8.4 looked at a static case of capillary pressures, only considering the effects of varying adhesion tension and pore size that is valid for a particular fluid saturation and a fixed geometry. However, the phenomenon described earlier for a single capillary tube also exists when bundles of interconnected capillaries of varying sizes exist in a porous medium. The capillary pressure that exists within a porous medium between two immiscible fluid phases is a function of adhesion tension and the average size of the capillaries. Additionally, capillary pressure is also a function of the saturation distribution of the fluids involved and the saturation history. As the relative saturations of the phases change, the pressure differences across the fluid interfaces also change, resulting in a change in the capillary pressure. This is obviously of great significance when considering both static and dynamic problems of hydrocarbon reservoirs. Static problems primarily involve fluid distribution, fluid contacts and zonation in the reservoirs; dynamic problems consist of the transport of immiscible fluid phases in the pore spaces under the influence of forces due to gravity, capillarity, and an impressed external pressure gradient. Therefore, it is imperative to study the effect of saturation distribution and saturation history on capillary pressure. However, before saturation distribution and saturation history can be discussed, two important saturation processes must be considered, namely, imbibition and drainage.

These two saturation processes are dependent on the wetting characteristics of fluid phases; when oil and water are the two fluid phases in a porous medium, generally either oil or water will preferentially wet the pore space based on which the phases will be identified as wetting or nonwetting. Thus in a porous medium saturated with oil and water, two basic processes can occur — a wetting phase displacing a nonwetting phase and a nonwetting phase displacing a wetting phase. These two processes are called *imbibition* and *drainage*. For instance, when water displaces oil from a water-wet rock, the process is imbibition; when water displaces oil from an oil-wet rock, the process is drainage. However, when gas displaces oil or water, the process is always drainage because gas is always the nonwetting phase in comparison to oil or water.

To understand the dependence of capillary pressure on fluid saturation distribution and saturation history (resulting from imbibition or drainage), a continuous capillary tube that changes in diameter from small to large to small must be considered (see Figure 8.4). The saturation for capillary pressures of equal magnitude depends on whether the system is initially 100% saturated with a wetting fluid or it is being saturated with the wetting fluid (imbibition). Figures 8.4a and c show that the wetting-phase saturation is 100 and 0%, having correspondingly low and high capillary pressure, respectively. Forcing the entry of a nonwetting fluid (drainage) into a tube saturated with a wetting fluid causes the wetting fluid to be displaced to such a point that the capillary pressure across the interface is equivalent to the applied pressure plus the pressure due to the column of the suspended fluid (see Figure 8.4b). The capillary tube is now 90% saturated with the wetting fluid for a higher value of capillary pressure as the nonwetting fluid is displacing the wetting fluid.

Now consider the case when the capillary tube is initially saturated with a nonwetting fluid and is immersed in a container filled with a wetting fluid (see Figure 8.4c). In this case, the wetting fluid begins to imbibe because of the adhesion between the wetting fluid and the capillary tube surface. This process, however, continues as we

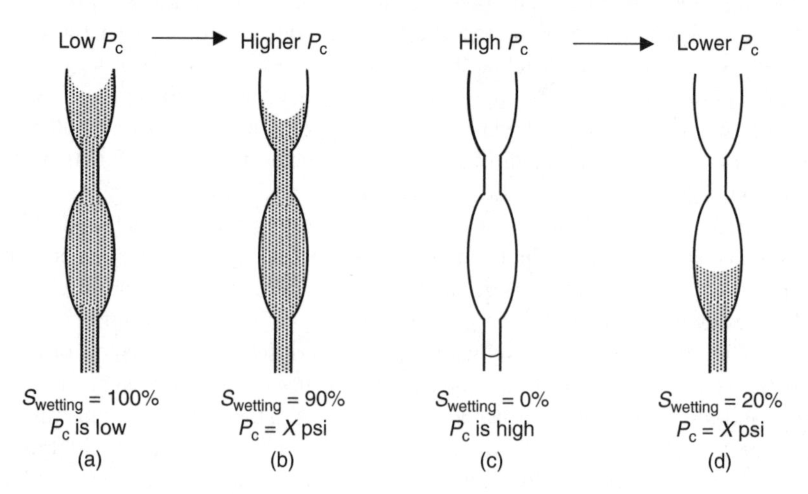

FIGURE 8.4 Schematic illustration of the relationship between capillary pressure and saturation history.

saw in the case of simple capillary tube geometries, until the adhesion force is balanced by the weight of the column of fluid. The saturation obtained by this process, as seen in Figure 8.4d, is only 20%. Therefore, comparing Figures 8.4b and d for an identical capillary pressure, 90 and 20% wetting-phase saturations are obtained. This simplified example shows that the dependence of the capillary pressure–saturation (wetting phase) relationship on the saturation process or history can be realized, that is, for a given capillary pressure, a higher value of saturation is obtained if the porous medium is being desaturated or drained than if the porous medium is being resaturated or imbibed with the wetting-phase fluid.

In summary, the capillary pressure–saturation relationship is dependent on the size and distribution of the pores, the fluids and the solids that are involved, and the history of the saturation process. Therefore, in order to use the capillary pressure data properly, it is necessary to consider these factors before the data are actually applied to reservoir engineering calculations.

8.6 LABORATORY MEASUREMENT OF CAPILLARY PRESSURE

Laboratory measurement of capillary pressure is confined to two types of capillary pressure processes: drainage and imbibition. In imbibition, the wetting phase displaces a nonwetting phase: in drainage, the nonwetting phase displaces a wetting phase in a porous medium. In all the laboratory measurement methods, wetting-phase saturation is increased in the imbibition process while in the drainage process wetting-phase saturation is decreased from a maximum value to the irreducible minimum by increasing the capillary pressure from 0 to a large positive value. The plots of capillary pressure data obtained in such a fashion are called *imbibition* and *drainage capillary pressure curves,* respectively.

It is generally agreed that water originally occupied the pore spaces of the reservoir rocks that were invaded by the migrating oil from the source rock, displacing some of the water and reducing the water saturation to some irreducible value. At the time of discovery, the pore spaces are filled with a connate water saturation and an oil saturation. All laboratory experiments are basically designed to mimic or simulate the saturation history of the reservoir. The drainage capillary pressure curve that is constructed by displacing the wetting phase by a nonwetting phase (e.g., water by oil or gas), and under the presumption that water is the wetting phase, primarily establishes the fluid saturations that are found when the reservoir is discovered. The imbibition capillary pressure curve establishes the fluid saturations in a porous medium when, for instance, oil is displaced by water in a water-wet rock.

The capillary pressure behavior studied so far was primarily based on uniform single capillaries. However, porous geological materials, e.g., reservoir rocks are far different from such simple geometries; reservoir rocks are composed of interconnected pores of various sizes. Additionally, wettability of the pore surfaces varies due to variation in the mixture of minerals in contact with the reservoir fluids. Obviously this leads to a variation of capillary pressure as a function of fluid saturation, pore geometry, and an overall mean wettability of the reservoir rock. In almost all laboratory measurement methods, capillary pressures are measured on cylindrical core plug samples of the representative formation. So, essentially, various measurement methods actually measure the average imbibition and drainage capillary pressure–saturation relationships, somewhat equivalent to measuring these relationships for a bundle of several capillaries of varying tortuosities. Generally, two types of core samples can be used to carry out capillary pressure measurements: preserved state or native state or cleaned and dried samples. For cleaned and dried samples, the drainage cycle on these samples can begin with 100% wetting-phase saturation down to the irreducible wetting-phase saturation. Although cleaning and drying may be generally acceptable, conventional cleaning and drying may not provide representative capillary pressure–saturation relationships if these processes alter the pore geometry (e.g., when the rock contains delicate clays such as illite). However, considering the strong proportionality of capillary pressure to adhesion tension or wettability, the use of appropriate core material always is an issue. Conversely, drainage capillary pressure curves that are measured usually involve the reduction of wetting phase saturation from 100% to an irreducible value. If preserved- or native-state core samples are to be used for capillary pressure measurements to maintain the original wettability, such samples would then already have an irreducible or initial water saturation. Then the core sample is normally flooded with representative formation water or brine so that a maximum in the wetting-phase saturation is obtained (under the presumption that the sample is water-wet). This saturation is then reduced in steps by injecting a nonwetting phase to mimic the original reservoir fluid distribution. Subsequently, the imbibition cycle can be completed by increasing the wetting-phase saturation. Some of the most commonly used laboratory techniques for measuring capillary pressures are addressed in the following sections, beginning with the pioneering experiments of Leverett[1] on capillary pressures of sandpacks.

8.6.1 Leverett's Capillary Pressure Experiments

The pioneering work of Leverett[1] in 1941 introduced the concept of capillary pressure measurement to the petroleum industry. Leverett conducted capillary pressure–saturation experiments on six different unconsolidated sandpacks. The sands used in his experiments were of different mesh sizes. Out of the six sands that he used, two sands contained clay-like material. In conducting his capillary pressure experiments, long tubes filled with sand were saturated with water and suspended vertically. The experiments were performed in such a manner that both imbibition as well as drainage capillary pressure curves were defined. The drainage curves were obtained by desaturating (from 100% water- or wetting-phase saturation) the water-saturated sandpack with one of its end lowered into a container having free water level. The water saturation in the tube was then determined at various positions above the free water level in the container. The imbibition curves were obtained by lowering the tubes packed with dry sand (100% air or nonwetting-phase saturation) into the water container so that water was imbibed by the sandpack due to capillary forces. Again the saturations were measured at various heights above the free water level in the container.

As shown in Figure 8.5, Leverett plotted the imbibition and drainage data obtained in this manner by expressing the capillary pressures in terms of a

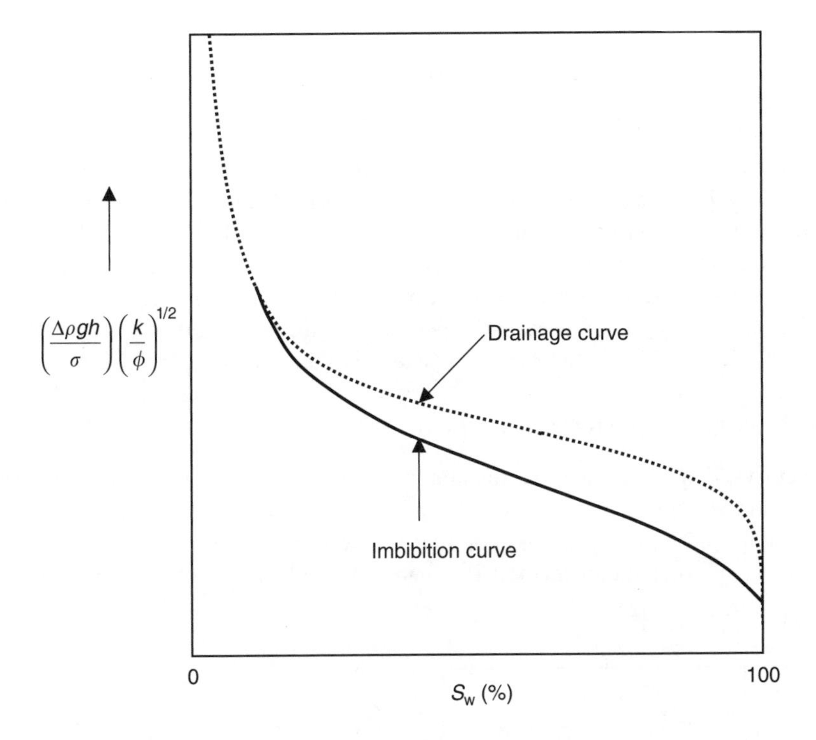

FIGURE 8.5 Schematic illustration of Leverett's[1] dimensionless function vs. water saturation data obtained from drainage and imbibition. Curves shown are correlated from the measured height saturation data.

dimensionless correlating function vs. water saturation. The dimensionless function used by Leverett is defined as

$$\text{Leverett's dimensionless function} = \left(\frac{\Delta\rho gh}{\sigma}\right)\left(\frac{k}{\phi}\right) \tag{8.19}$$

where $\Delta\rho$ is the density difference between water and air, g the gravitational constant, h the measured height during drainage and imbibition, σ the air–water surface tension, k the absolute permeability of the sandpack, and ϕ the porosity of the sandpack.

The variables in Equation 8.19 can be expressed in any consistent set of units to yield a dimensionless function. Leverett observed that the data for four of the sands fell satisfactorily near the two curves, one for imbibition and the other for drainage. From Figure 8.5, another important observation can also be made; the difference between the drainage and the imbibition curves. This difference in the curves is due to a hysterisis effect that is dependent on the saturation process.[1] The hysterisis effect in capillary pressure curves is discussed in Section 8.7.3 in greater detail. The correlating dimensionless function was proposed by Leverett so that the capillary pressure data from different sands could be expressed in a generalized form. Out of the six sands studied by Leverett, two of the sands containing clay-laden material, however, did not correlate as per the data presented for the four sands. The results in fact indicated that more water was retained by the clay-laden sands at large values of h, the height, and less at small ones, in comparison to the four clean sands. Leverett ascribed this observation to the fact that clays absorb water, that is, the water so taken up being held more tightly than the same amount of water would be held by capillarity. It should, however, be realized that even though Leverett's experiments constituted the most fundamental and original work attempted toward measuring or establishing the capillary pressure–saturation relationships, it is not feasible to determine the capillary properties of naturally occurring rock materials by a method such as Leverett's. Therefore, other means of measuring the capillary pressure have been devised and are routinely used in the petroleum industry as part of the special core analysis (SCAL). These methods are discussed in the following three sections.

8.6.2 Porous Diaphragm Method

The porous diaphragm method, illustrated in Figure 8.6, is perhaps one of the simplest methods to determine the capillary pressure–saturation relationships for core plug samples. The method was proposed by Welge and Bruce.[2] The primary component of the diaphragm method is a permeable membrane of uniform pore size

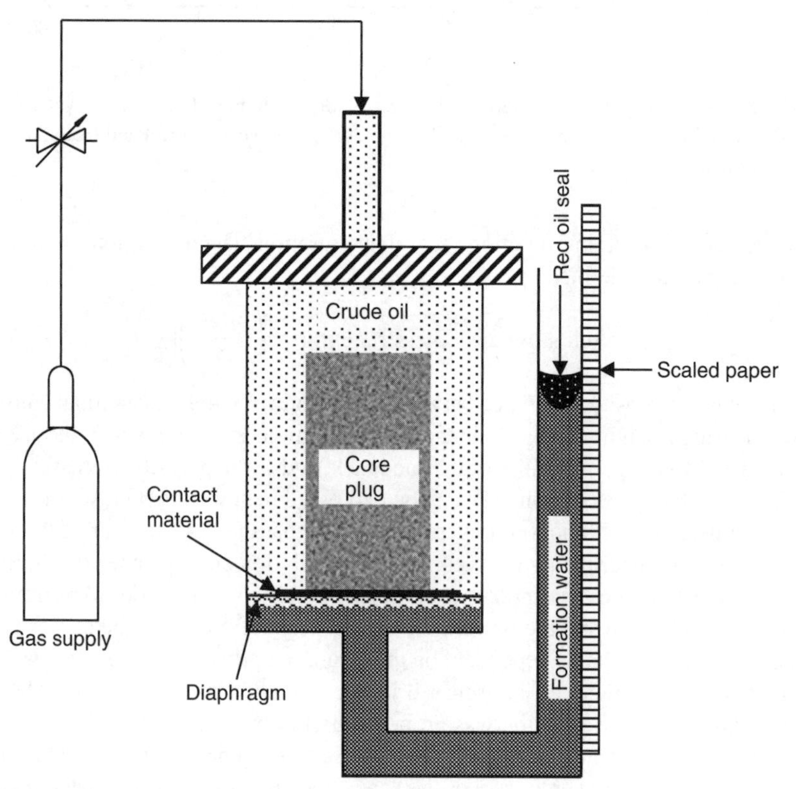

FIGURE 8.6 Schematic illustration of the experimental setup for the porous diaphragm method for capillary pressure measurement.

distribution containing pores of such size that the selected displacing fluid does not penetrate the diaphragm when the pressure applied to the displacing phase is lower than some selected maximum pressure of investigation. Materials such as fritted glass, porcelain, and others have been used successfully as diaphragms. A suitable material such as Kleenex is sandwiched between the core plug sample and the diaphragm to aid in establishing the fluid contact between the former and the latter. As shown in Figure 8.6, the core sample is 100% saturated with reconstituted formation water or brine. The test sample is subjected to displacement of water by crude oil in a stepwise fashion. The pressure is applied to the crude oil by nitrogen. The core sample is allowed to approach a state of static equilibrium at each pressure level. The saturation of the core sample at each point is calculated based on the water produced from the sample which is noted from the movement of the red oil meniscus in the tube that has a scale of squared paper attached to it, as shown in Figure 8.6. The test is terminated at a certain value of maximum pressure at which no more water is produced, also an indication that irreducible water saturation is reached. In the diaphragm method, any combination of fluid pairs can be used: gas, oil, or water. Although most determinations of capillary pressure by this method are drainage tests, imbibition curves can also be obtained by incorporating suitable modifications in the setup.

Another variant of the diaphragm method is the technique proposed by Cole,[3] developed primarily to determine the magnitude of the connate water saturation. The technique is similar to the original diaphragm method. Briefly, the procedure consists of saturating a core 100% with the reservoir water and then placing the core on a 100% water-saturated porous membrane that is permeable to water only and subjected to pressure drops imposed during the experiment. Air is admitted into the core chamber and the pressure is raised until a small amount of water is displaced through the porous, semi-permeable membrane into a graduated cylinder. The air pressure is maintained constant until no more water is displaced, verified by monitoring the water level in the graduated tube. The core sample is subsequently removed and water saturation is determined gravimetrically. The core sample is then placed back in the chamber, the pressure is increased, and the procedure is repeated until the water saturation is reduced to a minimum value. After completion of the experiment (drainage cycle), the data can be plotted as capillary pressure vs. saturation, as shown in Figure 8.7 and used to determine the connate or irreducible water saturation. The other important features of the capillary pressure curve is discussed in Section 8.7.

8.6.3 Mercury Injection Method

The mercury injection technique is one of the most commonly used methods for measuring the capillary pressure of reservoir rock samples. The method was originally proposed by Purcell[4] in 1949. The mercury injection technique was developed to accelerate the determination of the capillary pressure–saturation relationships. The essential components of the mercury injection apparatus include a mercury pump for injection, a chamber to house the core sample, pressure gauges, and arrangement for volume measurement (see Figure 8.8). A clean and dried core

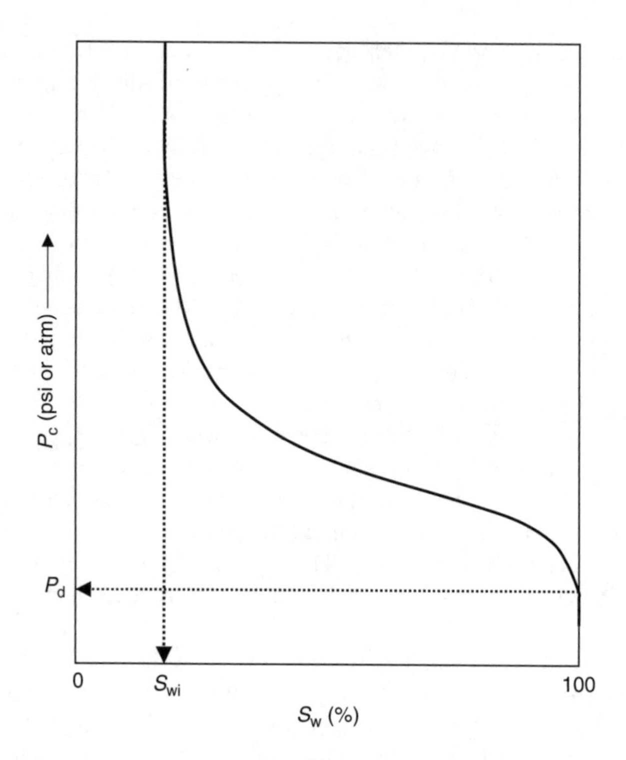

FIGURE 8.7 Schematic of a typical drainage capillary pressure curve generated from the data measured using the porous diaphragm method.

sample of known pore volume and absolute permeability is placed in the mercury chamber and evacuated. Subsequently, mercury is forced in the core sample under pressure to displace the air. The volume of mercury injected at each pressure determines the nonwetting-phase saturation because mercury is normally a nonwetting phase and air is a wetting phase. The injection procedure is continued until the core sample is filled with mercury or the injection pressure reaches some predetermined value. The pressures and saturations measured in this fashion determine the drainage capillary pressure–saturation curve (mercury displacing air). After reaching the maximum value of mercury saturation in the sample, a mercury withdrawal capillary pressure curve can be determined by decreasing the pressure in increments and recording the volume of mercury withdrawn. The process continues until a limit is approached where the mercury ceases to be withdrawn. The capillary pressure–saturation relationship determined in this manner basically constitutes the imbibition capillary pressure–saturation curve (air displacing mercury). As an example, a typical drainage and imbibition mercury-injection capillary pressure data[5] is shown in Figure 8.9.

The method discussed above obviously has several advantages that include the following: significantly reduced test time, increased range of pressure investigated, ease of implementation, and the measured data can also be used to determine pore

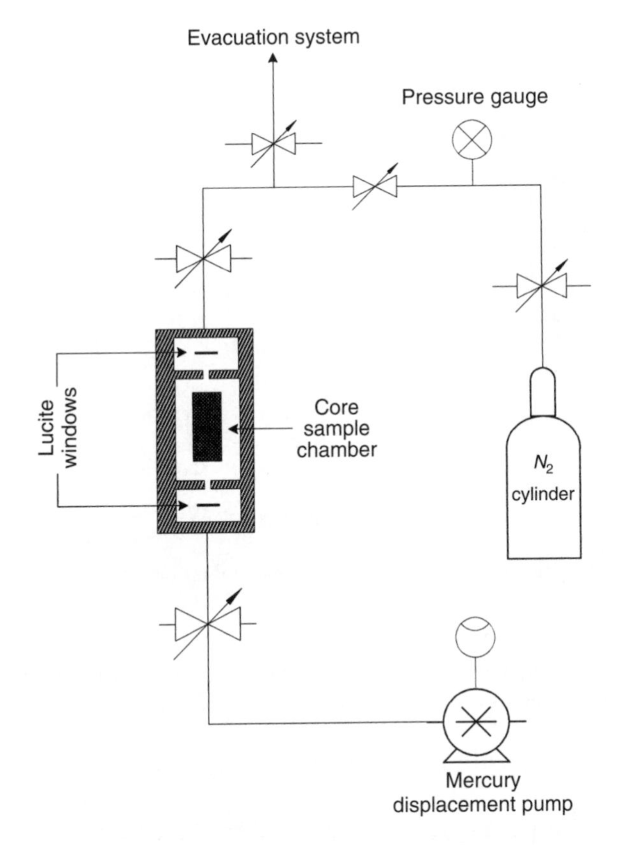

FIGURE 8.8 Schematic of a mercury injection capillary pressure measurement setup.

size distribution. The disadvantages of the method are permanent loss of the core sample and toxicity of mercury vapors, which constitutes a potential health hazard.

8.6.4 Centrifuge Method

A third commonly used method for determining the capillary pressure properties of reservoir rocks is the centrifuge method. Apparently, the first capillary pressure experiments with a centrifuge published in the petroleum literature were those of McCullough et al.,[6] followed by Hassler et al.[7] in 1944. However, the centrifuge procedure that is used today was introduced by Slobod et al.[8] The procedure described here is taken from the work of Slobod et al. The determination of the capillary pressure curve, for example, for an air–water system, begins with complete water saturation of a cleaned and dried core sample that is placed in a centrifuge tube according to an arrangement such as the one shown in Figure 7.11. The system is rotated at a number of different speeds (so that the nonwetting phase, air, displaces water, the wetting phase) selected to cover pressure differences between phases required for the particular core. At each speed, the rate of rotation is maintained constant until no

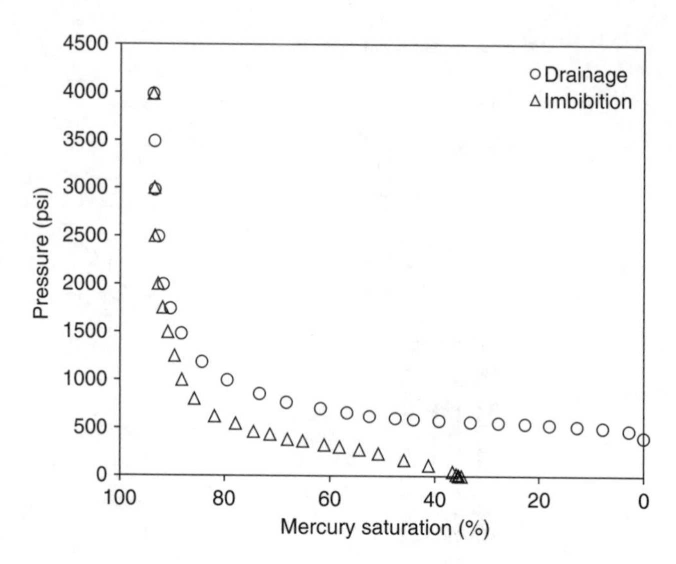

FIGURE 8.9 A typical example of drainage and imbibition mercury-injection capillary pressure data.

additional water is displaced by air. The speed of rotation is converted into capillary pressure at the inlet end of the core according to the following equation:[9]

$$(P_c)_i = (1.096\times10^{-6})\ \Delta\rho N^2(r_e - 0.5L)L \tag{8.20}$$

where $(P_c)_i$ is the capillary pressure at the inlet end of the core in gram-force/cm^2, $\Delta\rho$ the density difference in g/cm^3, N the speed of rotation in revolutions/min, r_e the outer radius of the core or the distance from the center of rotation to the end face of the core in cm, and L the length of the core in cm.

The average water saturation in the core sample is determined by subtracting the volume of displaced water from the original water content. However, the capillary pressure calculated using Equation 8.20 is the capillary pressure at the inlet end of the core, whereas the saturation measured from the amount of fluid displaced is the average saturation. Therefore, to use centrifuge-derived capillary pressure data, it must be related to the saturation at the inlet end. Tiab and Donaldson[9] have discussed the procedures for calculating the saturations at the inlet end using the approximate and theoretically exact methods. The data obtained in this manner constitutes the drainage capillary pressure curve, because air is always the nonwetting phase when air–water or air–oil pairs are considered. By following a somewhat similar procedure after incorporating alterations in the setup, imbibition capillary pressure curves can also be obtained by using the appropriate fluid pairs.

8.7 CHARACTERISTICS OF CAPILLARY PRESSURE CURVES

We will now focus on the salient features or characteristics of typical capillary pressure–saturation curves. First consider the capillary pressure data presented in Figure 8.7, measured by the diaphragm method for the air–water system. Figure 8.7 easily identifies the capillary pressure curve bound within two saturation end points, and corresponding to these two end points there are capillary pressure end points. These two end points are some of the most notable features or characteristics of the capillary pressure curves. In order to understand capillary pressure–saturation relationships, the scales of saturation and capillary pressure are examined individually in the following sections.

Another important aspect of capillary pressure curves is the noticeable difference between the imbibition and the drainage capillary pressure curves seen in Figure 8.5 based on Leverett's work on sandpacks, and Figure 8.9, for a North Sea core plug sample. This particular difference in the imbibition and drainage capillary pressure curves is due to the phenomenon of *capillary hysterisis*, which is discussed in Section 8.7.3. Another important characteristic of the capillary pressure curves to be considered is their relationship with permeability. Even though a given rock permeability does not impart any specific or particular characteristic to the capillary pressure curves, it certainly influences the location of the capillary pressure–saturation curves when these data are plotted on a single graph for rock samples of different permeabilities (see Section 8.7.4).

8.7.1 Saturation Scale

The saturation scale in Figure 8.9 includes the mercury saturation, S_{Hg}, or the air saturation, which is $(1 - S_{Hg})$, beginning with 0% mercury saturation, or 100% air saturation, which can also be construed as 100% water saturation. This is because in this case, the wetting phase (air) is being displaced by the nonwetting phase (mercury); that is, the drainage curve can also be carried out with an air–water pair and the process is designed to mimic the upward migration of hydrocarbons in a reservoir rock. Tracing the saturation–capillary pressure path backward on the curve that begins with 100% water saturation in Figure 8.5 or Figure 8.7 (0% mercury or 100% air in Figure 8.9), a minimum or irreducible saturation, S_{wi}, is reached. The process starts with 0 capillary pressure, where the water- or wetting-phase saturation is 100% and the phase is continuous. Because the saturation of the wetting phase is reduced in the drainage process, the wetting phase becomes disconnected from the bulk wetting phase. Eventually, all of the wetting phase remaining in the pore space become completely isolated, where its hydraulic conductivity is lost and it is termed *irreducible wetting-phase saturation*. The 100% water saturation basically represents the conditions that existed prior to the hydrocarbon migration, whereas finite minimum irreducible saturation represents the connate water that resulted from the migration of hydrocarbons in the reservoir rock after gravity and capillary forces equilibrated.

8.7.2 Pressure Scale

Looking at the pressure scale at 100% water saturation in Figure 8.7, we find that a finite capillary pressure is necessary to force the nonwetting phase into capillaries filled with the wetting phase. This minimum capillary pressure, the starting point of the capillary pressure curve, is known as the *displacement* or *threshold pressure*, P_d, and is sometimes also referred to as the *pore entry pressure*. The middle portion of the capillary pressure curve indicates the gradual increase in the capillary pressure, reducing the saturation of the wetting phase. The other end of the capillary pressure scale basically indicates that irrespective of the magnitude of the capillary pressure, water saturation or the wetting-phase saturation cannot be minimized further. At this end of the capillary pressure curve, all of the remaining wetting phase is discontinuous, resulting in the capillary pressure curve becoming almost vertical. Therefore, in summary, at conditions above the capillary pressure at S_{wi}, capillary forces are entirely dominant, whereas outside the capillary pressure at 100% wetting-phase saturation, the conditions are analogous to the complete dominance of gravity forces, and within these two P_c–S_w end points both gravity as well as capillary forces can be considered as being active.

8.7.3 Capillary Hysterisis

The drainage process establishes the fluid saturations that are found when the reservoir is discovered. In addition to the drainage process, the other principal flow process of interest involves the reversal of the drainage process by displacing the nonwetting phase with the wetting phase, such as the imbibition of water in Leverett's sandpacks (Figure 8.5) or the withdrawal of mercury (reduction in mercury saturation or increase in air saturation) shown in Figure 8.9. Both processes, are shown, drainage as well as imbibition, are shown in Figures 8.5 and 8.9. The two capillary pressure–saturation curves are not the same. This difference in the two curves is *capillary hysterisis*. Sometimes, the processes of saturating and desaturating a core is also called capillary hysterisis.[10]

This difference between the two capillary pressure curves is generally attributed to wettability, or specifically it is considered as closely related to the fact that advancing and receding contact angles of fluid interfaces on solids are different.[11] In some cases, in the oil–brine systems, the contact angle or wettability may change over time. Thus, if a rock sample that has been thoroughly cleaned with solvents is aged in reservoir oil for a certain period of time, it may behave as if it were oil-wet, whereas it may become water-wet if it is exposed to formation water instead.

Another mechanism that has been proposed to explain or justify capillary hysterisis is called the *ink-bottle effect*,[12] discussed in Section 8.5. This phenomenon explains why a given capillary pressure corresponds to a higher saturation on the drainage curve than on the imbibition curve. For example, in the case of data reported by Leverett on sandpacks, for a dimensionless factor value of 0.4 the difference between the drainage and imbibition saturation values is as high as 60%, whereas in the case of Figure 8.9, for a capillary pressure of 500 psi, the difference between the drainage and imbibition mercury saturation values is as high as 70%. The other way of looking at capillary hysterisis is to compare the drainage and imbibtion capillary pressures for the same fluid saturation (iso-saturation); when, as seen in Figure 8.9, compared with a mercury saturation of 68%, results in the drainage and imbibition capillary pressures of 770 and 380 psi, respectively.

8.7.4 Capillary Pressure and Permeability

Figure 8.10 shows the air–mercury capillary pressure data for various core samples having different permeability values. In this case, capillary pressure data are plotted as a function of mercury saturation for five different core samples varying in absolute permeability in a range of k_1 to k_5 mD. Figure 8.10 shows decreases in permeability have corresponding increases in the capillary pressure at a constant value of mercury saturation. In other words, when iso-saturation data are compared, the sample having a permeability of k_1 mD has the lowest capillary pressure, whereas the one with permeability of k_5 mD has the highest capillary pressure. Although, the general trend of capillary pressure–saturation curves remains unchanged, the magnitude of the capillary pressures does change when data are compared for samples of different permeabilities. Therefore, the capillary pressure–saturation–permeability relationship is a reflection of the influence of pore sizes; since the smaller diameter pores invariably have large capillary pressures and lower permeabilities.

8.8 CONVERTING LABORATORY CAPILLARY PRESSURE DATA TO RESERVOIR CONDITIONS

Capillary pressure measurements that are carried out in the laboratory by any of the methods discussed in Section 8.6 normally make use of fluid pairs that do not necessarily exist in the reservoirs from which the core samples originate. For example, the mercury injection method uses air–mercury pair, whereas in the diaphragm or the centrifuge method, quite frequently air–water or synthetic oil–water pairs are used. Although, these types of pairs are commonly used in laboratories for experimental convenience, they pose one problem: these fluids do not normally possess the same physical properties as the reservoir gas, oil, and water. Specifically, laboratory fluid pairs do not have the same surface or interfacial tension as the reservoir system and,

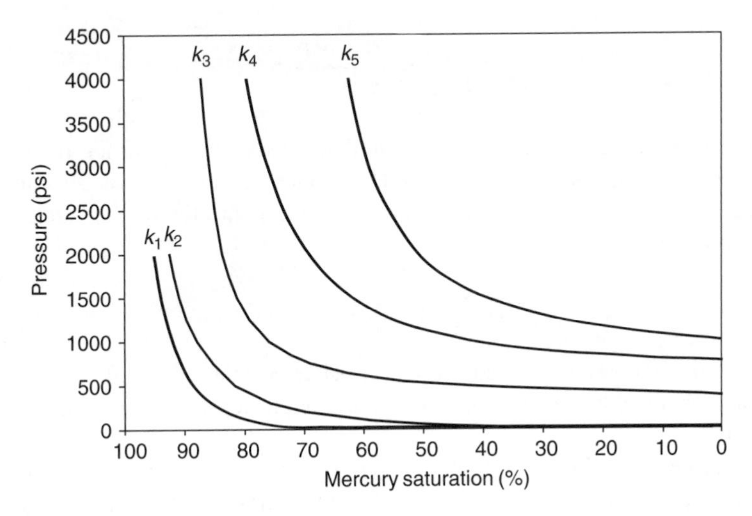

FIGURE 8.10 Variation of capillary pressure with permeability ($k_1 > k_2 > k_3 > k_4 > k_5$).

additionally, the contact angles in the laboratory and the reservoir conditions may also differ. Therefore, it becomes necessary to convert laboratory-measured capillary pressure data to reservoir condition capillary pressure data.

Essentially two techniques are available for correcting laboratory data to reservoir conditions. However, before these two techniques are addressed, first the capillary pressure data that was presented by Purcell[4] for six different sandstones must be considered. There were two different sets of capillary pressure data on the same sandstones, one measured by the mercury injection (air–mercury) and the other measured by the porous diaphragm method (air-5% by wt. sodium chloride in water). When Purcell compared the two set of capillary pressure measurements, it was found that the air–mercury capillary pressure was equivalent to approximately five times the air–water capillary pressure, which is also evident from the generalized capillary pressure equation that relates to the surface or interfacial tension, contact angle, and pore radius.

For air–mercury capillary pressure:

$$P_{\text{cam}} = \frac{2\sigma_{\text{am}}\cos\theta_{\text{am}}}{r} \tag{8.21}$$

For air–water capillary pressure:

$$P_{\text{caw}} = \frac{2\sigma_{\text{aw}}\cos\theta_{\text{aw}}}{r} \tag{8.22}$$

where P_{cam} and P_{caw} are the air–mercury and air–water capillary pressures, σ_{am} and σ_{aw} the air–mercury and air–water surface tensions; θ_{am} and θ_{aw} the contact angles for air–mercury and air–water systems, and r the pore radius.

Equating Equations 8.21 and 8.22 on the basis of equal value of r, since it is the same sample:

$$\frac{P_{\text{cam}}}{P_{\text{caw}}} = \frac{\sigma_{\text{am}}\cos\theta_{\text{am}}}{\sigma_{\text{aw}}\cos\theta_{\text{aw}}} \tag{8.23}$$

Purcell assumed a value of 480 and 70 dynes/cm as values for air–mercury and air–water surface tensions, respectively. He also assumed a value of 140° and 0° as contact angles for mercury and water against solid, respectively. Substituting these values in Equation 8.23

$$P_{\text{cam}} = 5.25\ P_{\text{caw}} \tag{8.24}$$

Amyx et al.,[13] however, pointed out that a better correlation of Purcell's data was generally obtained by neglecting the contact angle terms in Equation 8.23, i.e., $P_{\text{cam}} = 6.86\ P_{\text{caw}}$

Therefore, even though the previous conversion may not be construed as conversion of laboratory capillary pressure data to reservoir conditions, it is basically the conversion from one fluid pair to the other, that is, air–mercury to air–water, which is somewhat similar to the two techniques discussed in the following text.

Considering a specific case where the laboratory values are determined with gas and water, Equation 8.22 becomes

$$[P_{\text{cgw}}]_{\text{L}} = \frac{2\sigma_{\text{gw}}\cos\theta_{\text{gw}}}{r} \tag{8.25}$$

where $[P_{\text{cgw}}]_L$ is the laboratory-measured capillary pressure, σ_{gw} the gas–water surface tension at laboratory pressure and temperature conditions; and θ_{gw} the contact angle for the gas–water system at laboratory pressure and temperature conditions. The capillary pressure that would exist if reservoir fluids, oil, and water were used in the same pore space would be

$$[P_{\text{cow}}]_{\text{R}} = \frac{2\sigma_{\text{ow}}\cos\theta_{\text{ow}}}{r} \tag{8.26}$$

where $[P_{\text{cow}}]_{\text{R}}$ is the reservoir condition capillary pressure, σ_{ow} the oil–water interfacial tension at reservoir pressure and temperature conditions; and θ_{ow} the contact angle for the oil–water system at reservoir pressure and temperature conditions.

Comparing Equations 8.25 and 8.26 for the laboratory and reservoir conditions

$$[P_{\text{cow}}]_{\text{R}} = \frac{\sigma_{\text{ow}}\cos\theta_{\text{ow}}}{\sigma_{\text{gw}}\cos\theta_{\text{gw}}}[P_{\text{cgw}}]_{\text{L}} \tag{8.27}$$

Thus reservoir capillary pressure can be calculated on the basis of laboratory capillary pressure when the surface and interfacial tension data and contact angle data at the pertinent conditions are known. Note that the relationship expressed by Equation 8.27 assumes that the saturations as measured in the laboratory remain equal to the saturations in the reservoir. As Chapter 7 shows, surface and interfacial tension data can be measured by the pendant drop method, whereas contact angle data can be determined from sessile drop method. For surface and interfacial tension measurements, very accurate values can be obtained for both laboratory and reservoir conditions, however, the exact determination of contact angles for a representative porous medium, if not impossible, is certainly difficult. Additionally, the cosine of contact angle varies between −1 and +1, causing significant variation in the converted capillary pressure values and therefore, quite often the contact angles are neglected from Equation 8.27, which reduces it to merely a function of surface and interfacial tensions:

$$[P_{\text{cow}}]_{\text{R}} = \frac{\sigma_{\text{ow}}}{\sigma_{\text{gw}}}[P_{\text{cgw}}]_{\text{L}} \tag{8.28}$$

Ahmed,[10] however, stated that even after the laboratory capillary pressure data is converted to reservoir condition capillary pressure data, it may be necessary to make further corrections for permeability and porosity because the core sample that was used in performing the laboratory capillary pressure test may not be representative of the average reservoir permeability and porosity. Equation 8.28 is therefore, modified to

$$[P_{cow}]_R = \frac{\sigma_{ow}}{\sigma_{gw}}[P_{cgw}]_L \sqrt{\frac{\phi_R k_L}{\phi_L k_R}} \tag{8.29}$$

where ϕ_R and ϕ_L are the average reservoir porosity and core porosity, respectively and k_R and k_L the average reservoir permeability and core permeability, respectively.

8.9 AVERAGING CAPILLARY PRESSURE: THE *J* FUNCTION

The capillary pressure data measured in the laboratory are normally based on individual core plug samples, representing an extremely small part of the entire reservoir. Moreover, because of the heterogeneity of reservoir rocks, no single capillary pressure curve can be used for the entire reservoir. Therefore, it becomes necessary to combine all the capillary pressure data to classify a particular reservoir. The first attempt to develop such a relationship that could combine the capillary pressure data of varying porosity and permeability was in fact made by Leverett.[1] As seen from Equation 8.19, the capillary pressure data from sands of different porosity and permeability values could be expressed in a generalized form, that is, the data of four of the sands that Leverett tested fell satisfactorily near the two curves, one for imbibition and one for drainage. Similarly, data presented in Figure 8.10 also clearly indicate a functional relationship between capillary pressure–saturation and permeability. Therefore, it can also be stated that the capillary pressure-saturation relationships are affected by the porosity and permeability of the sample. Hence, it becomes necessary to evaluate the various sets of capillary pressure–saturation data with respect to the porosity and permeability of the core sample from which they were obtained. The approach that is commonly used in the petroleum industry is actually based on Leverett's dimensionless function, called the *J function* (sometimes referred to as *Leverett's J function*) and expressed as

$$J(S_w) = \left(\frac{P_c}{\sigma}\right)\left(\frac{k}{\phi}\right)^{1/2} \tag{8.30}$$

where P_c is the capillary pressure (Leverett's function has $P_c = \Delta\rho g h$, since Leverett measured height), σ the interfacial tension; ϕ the porosity fraction; and k the permeability.

Use of any consistent set of units would render $J(S_w)$ to be dimensionless. As an example, the J function–saturation data for North Sea reservoir rocks from the same formation are presented in Figure 8.11.

Some authors alter Equation 8.30 by including the entire adhesion tension term in the denominator to P_c in order to account for the wettability effect, as shown in Equation 8.31. Anderson,[11] however, stated that as long as all the capillary pressure measurements are made with reservoir fluids on cores with the representative reservoir

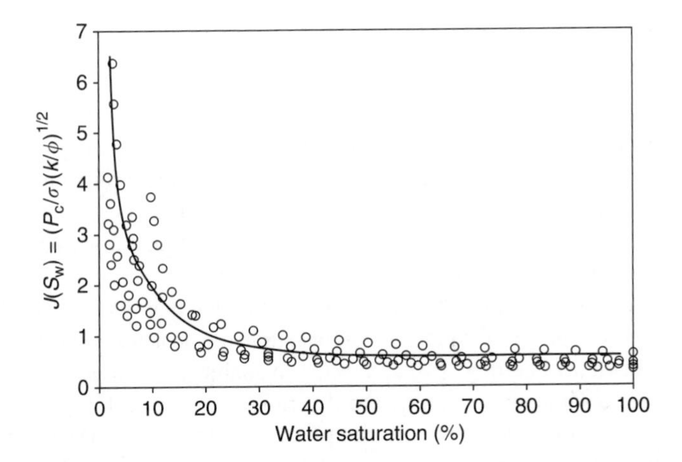

FIGURE 8.11 *J* function vs. water saturation for North Sea reservoir rock samples from the same formation.[5]

wettability, the $\cos\theta$ term merely acts as a constant multiplier without affecting the results, whereas problems would arise when different fluids are used.

$$J(S_w) = \left(\frac{P_c}{\sigma\cos\theta}\right)\left(\frac{k}{\phi}\right)^{1/2} \tag{8.31}$$

The *J* function was originally proposed as a means of converting all the capillary pressure–saturation data to a universal curve. In many cases, all of the capillary pressure data from a formation is reduced to a single composite curve when the *J* function is plotted vs. the saturation. Swanson[14] compared vacuum–mercury capillary pressure measurements with porous plate oil–brine measurements in a strongly water-wet sandstone plug. He obtained good agreement between the two sets of measurements with Equation 8.30 for the *J* function that neglects the contact angle. On the other hand, Omoregie[15] compared vacuum–mercury capillary pressure measurements with air–brine and air–oil centrifugal measurements for North Sea sandstone samples. Omoregie in fact used Equation 8.31 for the *J* function that considers the contact angle and obtained good agreement between the three sets of measurements. However, significant variations exist in the correlation of the *J* function with water saturation that differ from formation to formation. Hence, no universal correlation can be obtained. Rose and Bruce[16] have evaluated capillary pressure characteristics of several formations. A plot of the *J* function vs. the water saturation for these formations indicates a substantial scatter in the data; however, independent correlation is formulated for each material considered. Brown[17] presented data on capillary pressure for samples from the Edwards formation in the Jourdanton field and used Equation 8.31 to group the capillary pressure data. When data for all cores from the formation were plotted, a considerable dispersion of the data points was observed, although the trend of the correlation was found to be good. Brown, however, found that the *J* function plots could be improved by segregating the data on a textural basis.

8.10 CALCULATION OF PERMEABILITY FROM CAPILLARY PRESSURE

The first relationship that was developed between permeability and capillary pressure was reported by Purcell[4] in which he described the mercury injection technique for capillary pressure measurement. This was developed for capillary pressure measurement on both core samples as well as drill cuttings. Purcell justified the development of permeability and capillary pressure relationship based on the argument that experimental measurement of the former requires samples of regular shape and appreciable dimensions, the procurement of which is expensive. Therefore, Purcell's objective was to develop a P_c–k relationship that used capillary pressure data measured on drill cuttings (usually easily available as opposed to core samples) using the mercury injection technique to determine the permeability, thus eliminating the requirement of obtaining core samples for permeability measurements.

The capillary pressure–permeability relationship presented by Purcell is in fact based on the analogy between Poiseuille's equation and the generalized capillary pressure equation that is related to the adhesion tension and the pore radius. The equation presented by Purcell is developed as follows.

The rate of flow, Q, of a fluid of viscosity, μ, through a single cylindrical tube or capillary of radius, r, of length, L, is given by Poiseuille's equation:

$$Q = \frac{\pi r^4 \Delta P}{8\mu L} \tag{8.32}$$

where ΔP is the pressure drop across the tube.

Since the volume, V, of this capillary is $\pi r^2 L$, Equation 8.32 may be written as:

$$Q = \frac{V r^2 \Delta P}{8\mu L^2} \tag{8.33}$$

The capillary pressure for this single tube, is given by the capillary pressure equation that signifies the minimum pressure required to displace a wetting liquid from or inject a nonwetting liquid into a capillary of radius, r:

$$P_c = \frac{2\sigma\cos\theta}{r} \tag{8.34}$$

Substituting the value of r from Equation 8.34 into Equation 8.33:

$$Q = \frac{(\sigma\cos\theta)^2 V \Delta P}{2\mu L^2 P_c^2} \tag{8.35}$$

If the porous medium is conceived to be comprised of N capillary tubes of equal length but random radii, the total flow rate can be expressed as:

$$Q_t = \frac{(\sigma\cos\theta)^2 \Delta P}{2\mu L^2} \sum_{i=1}^{N} \frac{V_i}{P_{ci}^2} \tag{8.36}$$

On the other hand, the flow rate Q_t, through this same system of capillaries is also given by Darcy's law:

$$Q_t = \frac{kA\Delta P}{\mu L} \tag{8.37}$$

The equality of Equation 8.36 and 8.37 for Q_t now leads to a relationship between permeability, capillary pressure, and pore volume:

$$k = \frac{(\sigma\cos\theta)^2}{2AL}\sum_{i=1}^{N}\frac{V_i}{P_{ci}^2} \tag{8.38}$$

Equation 8.38 can be further simplified by expressing the volume, V_i, of each capillary as a percentage, S_i, of the total void volume, V_T, of the system:

$$S_i = \frac{V_i}{V_T} \tag{8.39}$$

Moreover, since the product of AL is the bulk volume of the system and ϕ is the porosity fraction:

$$\phi = \frac{V_T}{AL} \tag{8.40}$$

Equation 8.38 now reduces to

$$k = \frac{(\sigma\cos\theta)^2}{2}\phi\sum_{i=1}^{N}\frac{S_i}{P_{ci}^2} \tag{8.41}$$

Purcell, however, rightly noted that even though Equation 8.41 relates the permeability of a system of parallel cylindrical capillaries of equal length, but various radii, to the porosity of the system and to the capillary pressure; such a hypothetical porous medium seldom exists in naturally occurring rock formations. Therefore, to correct such a simplified approach, Purcell modified Equation 8.41 and introduced a so-called lithology factor, λ, to account for the differences between the flow in hypothetical porous media and that in the naturally occurring rocks. Finally, introducing conversion factors and expressing the summation as an integral of saturation over P_c^2 in the entire saturation range from 0 to 1, Equation 8.41 becomes:

$$k = 10.24(\sigma\cos\theta)^2\phi\lambda\int_{S=0}^{S=1}\frac{dS}{P_c^2} \tag{8.42}$$

where k is the permeability in mD, ϕ the porosity fraction, S the fraction of total pore space occupied by liquid injected or forced out of the sample, P_c the capillary pressure in psi, σ the surface or interfacial tension in dynes/cm, and θ the contact angle.

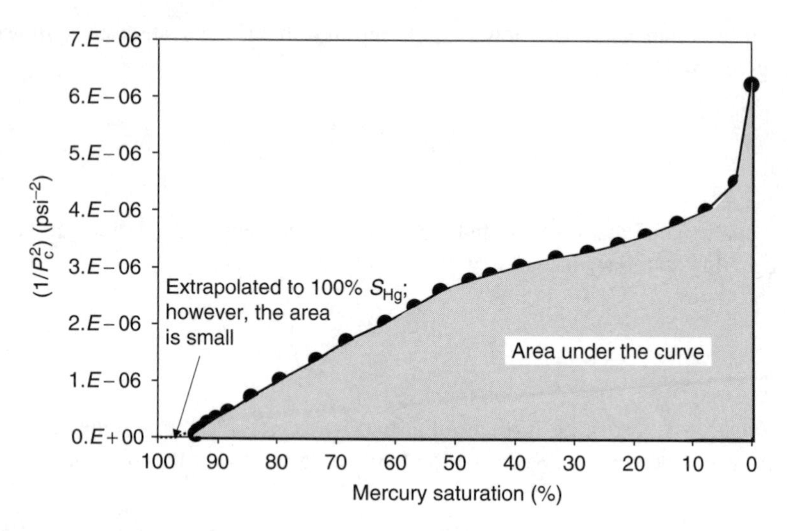

FIGURE 8.12 Plot of inverse of the square of capillary pressure vs. mercury saturation. The area under the curve is used for determination of permeability from capillary pressure data by the Purcell[4] method.

In Equation 8.42, the integral is evaluated by plotting the inverse of P_c^2 vs. saturation and determining the area under the curve, as shown in Figure 8.12.

On the basis of measured mercury injection capillary pressure data and measured permeability data of 27 different samples from Upper Wilcox and Paluxy sands, Purcell determined the value of lithology factor, λ, for each of these samples. The value of λ for the tested samples ranged from about 0.1 to 0.4, with an average value of 0.216. Purcell reported a reasonably good agreement between the calculated and measured values of permeability, when the average value of $\lambda = 0.216$ was used in Equation 8.42.

8.11 EFFECT OF WETTABILITY ON CAPILLARY PRESSURE

As seen in the capillary pressure expressions in Equations 8.1 and 8.34, capillary pressure is directly related to the wetting characteristics of the porous media. The very definition of capillary pressure is in terms of the difference in the phase pressures of nonwetting and wetting phases and the adhesion tension that is a function of the contact angle and the surface or interfacial tension. Therefore, considering the dependence of capillary pressure on wettability, a discussion on the effect of wettability on capillary pressure is warranted and is mainly based on results published in the literature.

Anderson's[11] work is probably the most exhaustive publication that deals with the effect of wettability on capillary pressure. Anderson, basically studied the effect of wettability on capillary pressure by evaluating the characteristics of capillary pressure curves, reported in the literature,[18] of different wettabilities. The oil–water capillary pressure–saturation curves for two strongly water-wet and two strongly oil-wet systems were reviewed. Specifically, the effect of wettability on capillary pressure curves was determined on the basis of external work required for oil displacing water, which is a drainage curve in a water-wet sample, and water displacing oil, which is a forced imbibition curve for a water-wet core and is described by the following two equations:

$$\Delta W_{\text{ext}} = -\phi V_{\text{b}} \int_{S_{\text{w1}}}^{S_{\text{w2}}} P_{\text{c}} \text{d}S_{\text{w}} \tag{8.43}$$

$$\Delta W_{\text{ext}} = \phi V_{\text{b}} \int_{S_{\text{o1}}}^{S_{\text{o2}}} P_{\text{c}} \text{d}S_{\text{o}} \tag{8.44}$$

where V_{b} is the bulk volume of the core, ϕ the porosity, P_{c} the capillary pressure, and S_{w} and S_{o} are the water and oil saturation, respectively.

In Equations 8.43 and 8.44, if capillary pressure is in N/m^2 and bulk volume is in m^3, ΔW_{ext} will be in Nm or Joule. The area under the drainage curves of the two water-wet samples was found to be relatively large because a great deal of work is necessary for the oil to displace water. However, when the area under the imbibition curve for these two samples is considered (Equation 8.44), it is found to be much smaller than the area under the drainage curves. Hence, more work is necessary for the oil to displace water than vice versa. This demonstrates the degree of water-wetness and the effect it has on capillary pressure. When oil is the strongly wetting fluid, the roles of oil and water are reversed from the strongly water-wet case. Again, in this case, the areas under the capillary pressure curves indicate that the work required for the nonwetting phase to displace the wetting phase is much larger compared to the reverse displacement. This indicates the degree of oil-wetness and its effect on capillary pressures.

Anderson has stated that as the oil–brine–rock system becomes neutrally wetted, the area under the drainage curve is reduced because less work is necessary for drainage as the preference of the rock surface for the wetting phase begins to diminish. Therefore, if strongly water-wet and strongly oil-wet systems tending toward neutral wettability are considered, the relative ease with which the nonwetting phase would be able to enter the increasingly smaller pores will increase.

In addition to this, Anderson also reviewed the work presented by Morrow and Mungan[19] and Morrow,[20] which examined the effect of wettability on capillary pressure using sintered porous polytetraflouroethylene cores. Air and a variety of nine organic liquids and water were used to obtain a wide range of contact angles and wettability while keeping the geometry fixed. The contact angles were measured on a flat smooth surface of the test core material, while drainage and imbibition capillary pressures were measured by the porous plate technique. The measured contact angles for the air and various liquid pairs ranged from 20° to about 100°. The measured drainage and imbibition capillary pressures were scaled by taking a ratio of P_c/σ, plotted against the liquid-phase saturations for all the fluid pairs (having different contact angles) on one single plot (one for drainage and the other for imbibition), so that the effect of wettability on capillary pressure could be investigated.

The drainage capillary pressure curves were found to be almost independent of the wettability for contact angles of 50° and less. Between the contact angle of 22° and 50°, there appears to be a slight influence of wettability on capillary pressure but much smaller than what would be predicted by an equation similar to Equation 8.23 (with surface tensions and contact angles pertinent to the fluid pairs). When the contact angles were less than 22°, they found no measurable effect of contact angle or wettability on the drainage capillary pressures. The behavior observed in the case of this study is in fact somewhat similar to the data reported by Purcell[4] on mercury injection and air–water capillary pressures of various sandstones, that is, at a particular water or mercury saturation if the mercury and air–water capillary pressures are scaled with respect to the surface tensions of 480 and 70 dynes/cm, respectively, the resulting values agree with each other quite well.

However, when the scaled imbibition curves from Morrow and Mungan[19] and Morrow[20] were evaluated in Anderson,[11] it was found that unlike the scaled drainage curves, the scaled imbibition curves were generally much more sensitive to the contact angle and hence the wettability. However, for contact angles less than 22°, the imbibition curves were found to be insensitive to contact angle. The different behavior observed in the cases of drainage and imbibition curves was attributed to the concept of advancing and receding contact angles on rough surfaces.[21]

Anderson[11] also compared the capillary pressure results measured on a single core from an East Texas Woodbine reservoir,[22] in the native state and cleaned state. Initially, the capillary pressure was measured on the core plug in its native state, exhibiting mixed wetting characteristics. The plug was then cleaned, dried, and saturated with brine, which rendered it water-wet; subsequently capillary pressure curve starting from a 100% brine saturation was measured. The behavior of this curve in its native state and cleaned state, that is, mixed-wet vs. water-wet was found to be very different. The two capillary pressure curves crossed over at a water saturation of 40%; below which the capillary pressures were lower and higher for the native state and the cleaned state, respectively, whereas the opposite was true for capillary pressures above the water saturation of 40%. This is explained by the fact that at the beginning of the capillary pressure measurement, in the mixed wettability (native state) plug, oil enters the large oil-wet pores, thus requiring a lower capillary pressure to displace water from the large pores when they are oil-wet vs. water-wet. However, subsequently, having most of the water from the larger pores already displaced, oil begins to enter the remaining smaller pores that are water-wet and filled with water, thus suddenly experiencing a higher capillary pressure that starts to rise rapidly. Where, in the case of the cleaned plug which is water-wet, the invasion of larger to smaller pores (which are all water-wet) by oil occurs in a much more gradual fashion imparting a particular type of capillary pressure behavior in comparison with the native state.

8.12 PRACTICAL APPLICATION OF CAPILLARY PRESSURE

After having discussed the various fundamental aspects of capillary pressure–saturation relationships, this chapter concludes by studying the various practical applications of capillary pressure–saturation relationships. Even though all capillary pressure curves may appear to merely contain a functional relationship between the phase pressure differences in nonwetting phase and wetting vs. saturation, within certain saturation and pressure limits; in practice, a significant amount of very useful information can be obtained from these curves. The type of information derived from capillary pressure–saturation relationships is in fact very essential to the development of reliable reservoir descriptions. Specifically, these relationships are necessary for the determination of pore size distribution, pore throat sorting, assessment of connate water saturation to calculate oil-in-place, determine the heights of fluid columns and transition zones, location of fluid contacts, and also tasks such as modeling the oil displacement either by free water imbibition or water injection. Therefore, considering this significance of capillary pressure–saturation relationships, their practical application in these areas is discussed in the following subsections.

8.12.1 Pore Size Distribution

The pore size and its relative distribution in a reservoir rock are important parameters for many fluid transport properties of porous media. Specifically, they influence fluid saturation distribution in a porous medium, porosity, permeability, and to some extent, wettability. Therefore, knowledge of pore size distribution in the reservoir rock is very important when qualitatively evaluating these rock and rock–fluid properties in relation to the pore size distribution. The technique of mercury injection is commonly used to determine the pore size distribution in a rock sample. The process of mercury injection used for measuring capillary pressure data basically probes the internal structure of the porous media. Mercury porosimetry, which is the forced intrusion of mercury into a porous medium, has been used to characterize pore-space microstructure since Washburn[23] suggested how to obtain a pore size distribution from measurement of volume injected vs. pressure applied. Mercury injection data are basically utilized to obtain a fingerprint of the reservoir rock. Considering the advantage of the mercury injection technique (very high pressure can be achieved allowing very small pores to be invaded) a large number of pressure data points in a wide range can be taken, based on which detailed characterization of the capillarity of the rocks can be accomplished.

Pore sizes in a reservoir rock can be characterized as small pores, medium pores, and large pores. Sometimes, they are also classified as nanopores, intermediate pores, and micropores. Figure 8.13 illustrates the concept of three different pore sizes based on a typical mercury injection capillary pressure curve. Figure 8.13 also shows, that the capillary pressure curve is divided into three zones, corresponding to the three different pore sizes. Therefore, the capillary pressure curve indicates the sequence in which mercury invades the pores of different sizes (i.e., large to small pores). In the initial part of the curve, mercury invades the largest pores resulting in lower capillary pressures that gradually increase as mercury starts to enter the medium pores, followed by smaller pores (see Figure 8.13). So, as the applied pressure increases, the radius of the pores that can be filled with mercury decreases and consequently the total amount (cumulative) of mercury intruded increases, or in other words, mercury (nonwetting liquid) will not penetrate the pores until sufficient

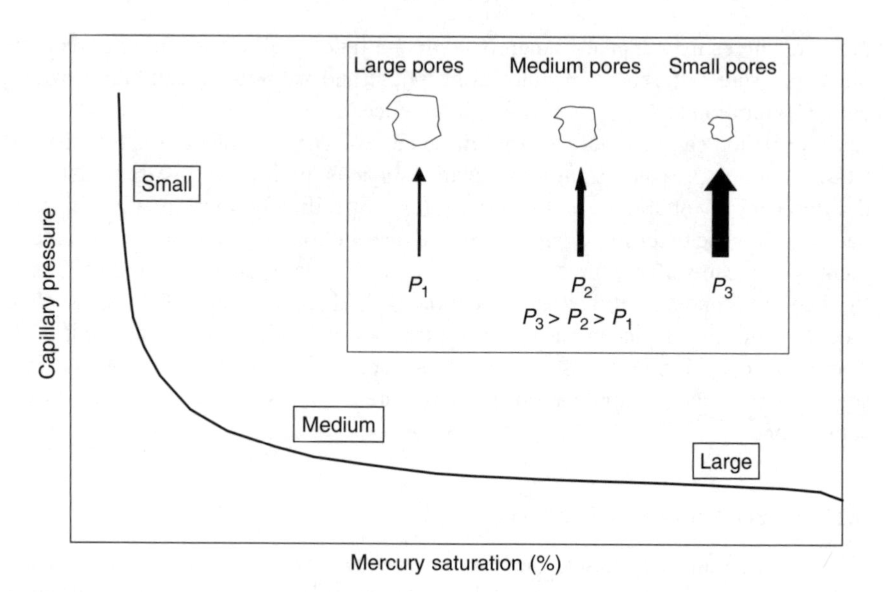

FIGURE 8.13 A schematic representation of mercury invasion into large, medium, and small pores, as part of the mercury injection process for measurement of capillary pressure.

pressure is applied to force its entrance. The determination of the pore size by mercury penetration is based on the behavior of nonwetting liquids in capillaries, shown in Equation 8.21 that can be arranged as:

$$r = \frac{2\sigma_{\mathrm{am}}\cos\theta_{\mathrm{am}}}{P_{\mathrm{cam}}} \tag{8.45}$$

where r is the pore aperture radius. Other variables and their values have been defined previously. By using a consistent set of units, the pore aperture radius can be obtained in mm, cm, or microns (10^{-6} m or μm).

It is important to notice that the variation in the mercury injected and the corresponding pressure is proportional to the number of pores with that radius in the sample. Therefore, the pressure with which mercury penetrates the sample determines the pore radius and the incremental volume introduced determines the relative number of pores with that radius in the sample. The incremental volume introduced or percent of pore space vs. pore radius give a convenient means to obtain the distribution characteristics. As an example, the pore size distribution for a sandstone and a carbonate sample is shown in Figure 8.14.

Another useful parameter that is sometimes employed to characterize the pore sizes on the basis of capillary pressure data is the pore size distribution index, λ (not to be confused with Purcell's lithology factor), proposed by Brooks and Corey.[24] They proposed the following relationship for the drainage capillary pressure based on experimental data:

$$P_{\mathrm{c}} = P_{\mathrm{ce}}(S_{\mathrm{w}}^{*})^{-1/\lambda} \tag{8.46}$$

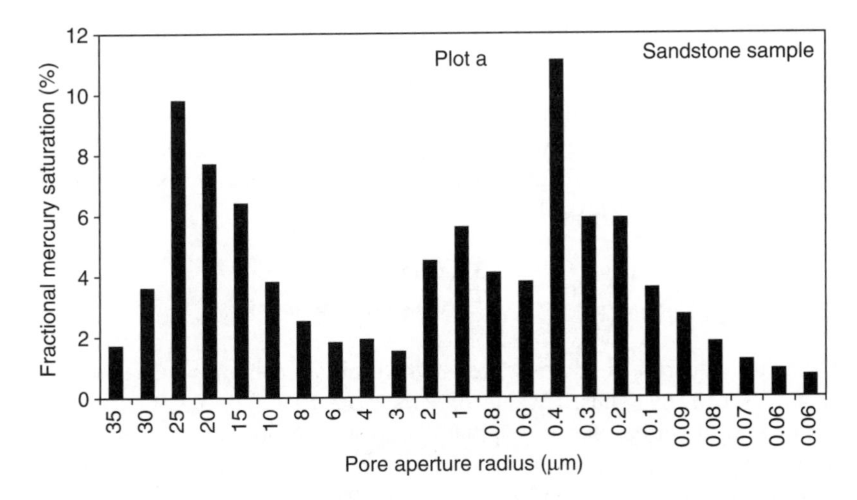

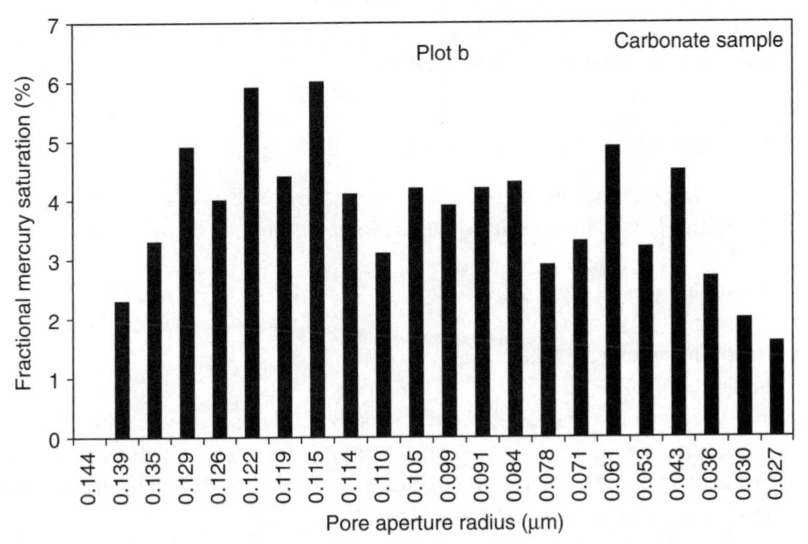

FIGURE 8.14 Pore size distribution for a sandstone (Plot a) and a carbonate (Plot b) sample plotted as pore aperture radius vs. fractional mercury saturation.

where P_c is the drainage capillary pressure, P_{ce} the capillary entry pressure or the displacement pressure, S_w^* the normalized wetting-phase saturation defined by $S_w^* = (S_w - S_{wi})/(1 - S_{wi})$, S_w the wetting-phase saturation, and S_{wi} the irreducible wetting-phase saturation

Therefore, a plot of $\ln(P_c)$ vs. $\ln(S_w^*)$ from the drainage capillary pressure data yields a straight line with a slope of $-1/\lambda$ from which the value of λ can be determined. A small value for λ indicates a wide range of pore sizes, while a large value indicates a narrow range.

8.12.2 Pore Throat Sorting

Pore throat sorting (PTS) basically provides a measure of pore geometry and the sorting of pore-throats within a rock sample.[25] PTS ranges from 1.0 (perfect sorting) to 8.0 (essentially no sorting) with the majority of rock samples falling between 1.2 and 5.0.[25] The values of PTS are computed using the following sorting coefficient equation developed by Trask:[26]

$$\mathrm{PTS} = \sqrt{\frac{\text{3rd Quartile pressure}}{\text{1st Quartile pressure}}} \tag{8.47}$$

First and third quartile pressures represent the capillary pressures at 25 and 75% saturation, respectively. Thus, a value of PTS close to 1 indicates that the porous medium is well sorted, that is, capillary pressures do not change significantly (a plateau), and increasing amounts of mercury can be intruded into the pore spaces at similar applied pressures. However, a value of PTS much greater than 1 indicates that the sample is poorly sorted, evidenced by the rapid jump in the capillary pressure curve after crossing the 25% saturation.

8.12.3 Connate Water Saturation

Essentially three methods are at the reservoir engineer's disposal for the determination of interstitial or connate or irreducible water saturation. These methods are:

1. Core analysis
2. Logging
3. Capillary pressure data

Chapter 6 discussed in great detail fluid saturation and the determination of connate water saturation from core analysis and Chapter 5 discussed the use of laboratory resistivity measurements and true resistivity from well logs to determine the water saturation.

The estimation of connate water saturation distributions in hydrocarbon reservoirs is usually based on drainage capillary pressure data, mainly because drainage curves are designed to mimic the hydrocarbon migration process in a completely water-filled reservoir rock. The connate water saturation on the drainage capillary pressure curve represents the remaining discontinuous wetting phase in the pore space; the point at which the capillary pressure curve becomes almost vertical or asymptotic. Thus, irrespective of the pressure difference between the nonwetting phase and the wetting phase, the wetting-phase saturation in the pore space remains unchanged. For instance in mercury injection capillary pressure data, the connate water saturation corresponds to the maximum mercury saturation in the pore spaces (see Figure 8.9), that is, S_{wi} or $S_{wc} = 1 - S_{Hg(maximum)}$, whereas in the case of air–water capillary pressure data, connate water saturation will be the minimum amount of water saturation that can be achieved by the displacement of air.

Considering that fluid distribution is predominantly controlled by pore sizes and its distribution; technically, any drainage curve can be utilized in determining connate water saturation because the drainage process represents the desaturation of any wetting phase by any nonwetting phase, taking place when hydrocarbons migrate in water-filled rocks. This is clearly evident from the mercury injection and air–water capillary pressure data on various sandstones and limestones that was presented by Purcell, that is, both capillary pressure curves end at almost identical maximum mercury saturation and minimum water saturation, respectively. It should, however, be noted that even though the saturation scales are analogous, the magnitude of capillary pressures in both cases is different because of varying wetting characteristics that can be reconciled by using Equation 8.23 as it is or by neglecting the contact angle terms.

Quite frequently, to increase the confidence in the connate water saturation determined from the drainage curves, it is also reconciled with values available from other sources such as core analysis or log data.

8.12.4 Zonation, Fluid Contacts, and Initial Saturation Distribution in a Reservoir

One of the most important practical applications of the concept of capillary pressure curves pertains to the zonation, fluid contacts, and initial fluid saturation distribution in a hydrocarbon reservoir prior to its exploitation. Thus, capillary pressure data can reveal a significant amount of information on the internal structure of a hydrocarbon reservoir in terms of the fluid distribution and its respective locations. However, to obtain this information from capillary pressures, $P_c - S_w$ data need to be first converted to height saturation data. This can be achieved by considering the generalized form of a combination of Equations 8.17 and 8.18:

$$h = \frac{2\sigma\cos\theta}{r\Delta\rho g} \tag{8.48}$$

By using Equation 8.34, Equation 8.48 can be expressed in terms of capillary pressure

$$h = \frac{P_c}{\Delta\rho g} \tag{8.49}$$

where h is the height above the plane of 0 capillary pressure between the nonwetting and wetting fluids, P_c the capillary pressure, $\Delta\rho$ the density difference between nonwetting and wetting phase at reservoir conditions, and g the gravitational constant.

In Equation 8.49, when a value of g is 9.81m/sec^2, P_c is in N/m^2, and the density difference is in kg/m^3

$$h\ (\text{in m}) = \frac{0.102 P_c(\text{in N/m}^2)}{\Delta\rho(\text{kg/m}^3)} \tag{8.50}$$

whereas in oil-field units, when g is 32.2 ft/sec^2, P_c is in lb$_{force}$/in^2, and the density difference is in lb$_{mass}$/ft^3:

$$h\text{ (in ft)} = \frac{P_c(\text{in lb}_{mass}/\text{in}^2) \times 32.2(\text{ft/sec}^2)}{32.2(\text{ft/sec}^2)\Delta\rho(\text{lb}_{mass}/\text{ft}^3)} = \frac{144P_c}{\Delta\rho}\ (1\text{ft} = 12\text{ in.}) \quad (8.51)$$

Equation 8.49 provides a relationship between capillary pressure and the height above the plane of 0 capillary pressure; that is, capillary pressure data are easily converted to height–saturation data based on which the zonation, fluid contacts, and fluid distribution in a hydrocarbon reservoir are determined. It should be noted, however, that prior to converting P_c–S_w data to h–S_w data, capillary pressures should be converted from laboratory conditions to representative reservoir conditions based on the methods described earlier.

In order to understand the application of height–saturation data to determine the fluid distribution, zonation, and fluid contacts in a reservoir, a drainage capillary pressure curve such as the one in Figure 8.9 must be considered, the data for which have been converted to reservoir conditions (for an oil–water system) and subsequently to height–saturation, as shown in Figure 8.15. Thus the figure essentially shows the water saturation distribution in an oil–water system. Also shown in Figure 8.15 are the following important concepts with regard to the internal structure pertaining to the fluids in the reservoir.

8.12.4.1 Free Water Level

The surface of the free water level (FWL) in Figure 8.15 is represented by the base of the height–saturation curve, that is, below this level free water such as an aquifer

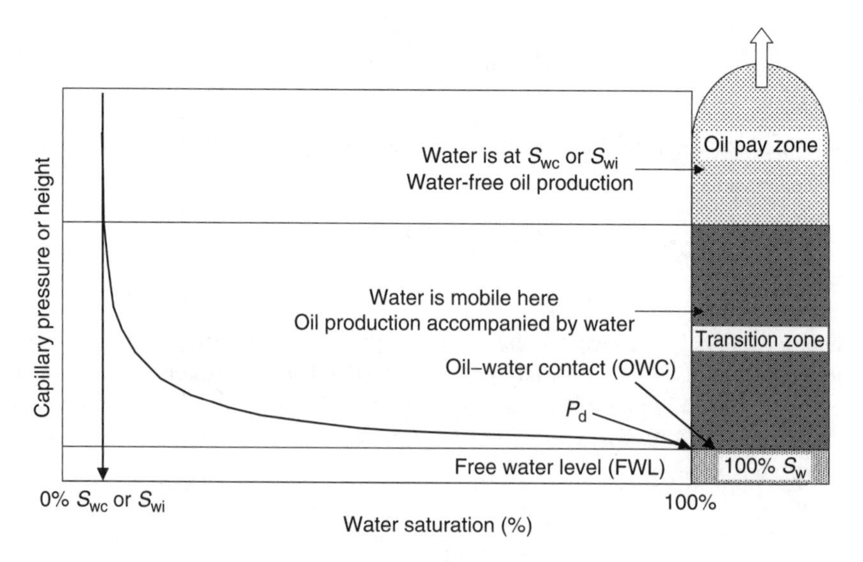

FIGURE 8.15 Profile of fluid distribution, zonation, and fluid contacts based on the capillary pressure or height vs. water saturation data.

exists in the reservoir. At the FWL, the capillary pressure is 0. Therefore, the height–saturation data presented in Figure 8.15 represent height above this free water level. This data can also be plotted as depth vs. water saturation, meaning that height above the FWL will be increasing while depth above the FWL will be decreasing. However, it should be noted that the FWL from this point extends up to a certain height and at this point water saturation is still 100%. Yet a finite capillary pressure exists that is called *capillary entry pressure* or *displacement* or *threshold pressure*. So, in this zone between the two points ($P_c = P_d$; S_w = 100%) and ($P_c = 0$; S_w = 100%) only water is produced.

8.12.4.2 Oil–Water Contact

The oil–water contact (OWC) is defined as the uppermost depth in the reservoir where a 100% water saturation exists, or in other words, the OWC and 100% water saturation point on the height or depth saturation curve is represented by the point ($P_c = P_d$; S_w = 100%). The OWC can also be mathematically expressed by the following relationships:

In terms of height above the FWL,

$$\text{OWC} = \frac{144P_d}{\Delta\rho} \tag{8.52}$$

where OWC is the oil–water contact in ft, P_d the displacement pressure in psi (from capillary pressure curve), and $\Delta\rho$ the oil and water density difference in $\text{lb}_{\text{mass}}/\text{ft}^3$.

In terms of depth above the FWL,

$$\text{OWC} = \text{FWL} - \frac{144P_d}{\Delta\rho} \tag{8.53}$$

8.12.4.3 Transition Zone

Figure 8.15 shows that water saturations are gradually changing from 100% water in the water zone to irreducible water saturation some vertical distance above the water zone or OWC. This vertical area is referred to as the *transition zone* and must exist in any reservoir where there is a bottom water table. The transition zone is therefore defined as the vertical thickness over which water saturation ranges from 100% to irreducible water saturation. So, the transition zone lies between the point at which the capillary pressure curve departs the displacement pressure and begins the asymptotic trend due to the achievement of irreducible water saturation (see Figure 8.15). Considering that water is mobile in the transition zone, any oil production from the transition zone may also be accompanied by water production. The creation or existence of the transition zone is in fact one of the major effects of capillary forces in a hydrocarbon reservoir.

It should, however, be pointed out that the thickness of the transition zone may range from few feet to several hundred feet in some reservoirs.[10] The relationship as expressed in Equation 8.48 suggests that the height above the FWL increases with decreasing density difference. This means that in a gas reservoir having a gas–water contact, the thickness of the transition zone is minimum since the density difference is large. Similarly, in the case of a light oil (smaller density) the density difference is large, resulting in a lower thickness of the transition zone. However, in the case of heavy oils (larger density), the density difference is small thus increasing the thickness of the transition one.

Equation 8.48 also shows that as the radius of the pores increases, the value of the height above the FWL decreases. Therefore, a reservoir rock system having small pore sizes has a longer transition zone than a system comprised of large pore sizes. Additionally, the more uniform the pore sizes, the flatter the transition zone of the capillary pressure curve. Similarly, since the thickness of the transition zone also varies with permeability; a high permeability reservoir rock system has shorter transition zones than low permeability reservoirs.[10]

8.12.4.4 Oil Pay Zone or Clean Oil Zone

The oil pay zone or the clean oil zone is represented by the zone above the upper demarcation line of the transition zone, as shown in Figure 8.15. Since the oil pay zone contains water at its irreducible saturation, the oil production from the clean oil zone is water-free.

8.12.4.5 Fluid Saturation in the Gas Zone

In order to calculate the fluid saturation in the gas zone, it is necessary to consider all three phases of gas, oil, and water. If all three phases are continuous:

$$P_{cgw} = P_{cgo} + P_{cow} \tag{8.54}$$

because

$$P_{cgo} = P_g - \mathrm{P_o} \tag{8.55}$$

$$P_{cow} = P_o - P_w \text{ (under the assumption that the system is water-wet)} \tag{8.56}$$

and combination of the Equations 8.55 and 8.66 results in

$$P_{cgo} + P_{cow} = P_g - P_w = P_{cgw} \tag{8.57}$$

where P_{cgw} is the capillary pressure at a given height above the free water surface, determined by using gas and water, P_{cgo} the capillary pressure at a height above the free oil surface, determined by using gas and oil, and P_{cow} the capillary pressure at a given height above the free water surface, determined by using oil and water.

If the wetting phase becomes isolated or disconnected, then the wetting-phase saturation has a minimum value, and at all heights above the point of discontinuity the wetting-phase saturation cannot be less than this minimum value.[13] The fluid saturations above the free oil surface can then be determined by the following relationships. S_w, at height, h, is calculated using oil and water as continuous phases.

The total liquid saturation, S_t at height, h, is calculated using gas and oil as the continuous phases and the height denoted by the free oil surface. Therefore,

$$S_t = 1 - S_g \tag{8.58}$$

$$S_o = S_t - S_w = 1 - S_g - S_w \tag{8.59}$$

PROBLEMS

8.1 A gas bubble is confined in a capillary tube of internal diameter 0.0007 cm. Oil and water are present on either sides of the gas bubble, thus forming an interface having $\theta_{gas\text{–}oil}$ and $\theta_{gas\text{–}water}$ of 35° and 15°, respectively. The gas–oil and gas–water surface tension values are 25 and 72 dynes/cm (mN/m), respectively. Assuming static conditions, calculate the pressure in the gas and the oil phases if the pressure in the water phase is 110 kN/m^2.

8.2 A 0.03 cm internal diameter glass capillary tube contains gas, oil, and water, thus forming two interfaces, namely; gas–oil and oil–water. The contact angles for these two interfaces are 35° and 25°, while the surface and interfacial tensions are 15 and 30 mN/m respectively. Calculate the gas–water capillary pressure.

8.3 If a generalized situation such as the one depicted in Figure 8.2, is applied to a reservoir oil and brine having a density of 0.85 g/cm^3 and 1.05 g/cm^3 respectively, what would be the height of capillary rise in a 0.0002 cm internal diameter capillary, and the corresponding capillary pressure? Additional data can be taken from problem 8.2.

8.4 The oil–water capillary pressure data was measured in the laboratory using Isopar-L and water on a reservoir rock core sample originating from a Middle Eastern field. However, the contact angle and interfacial tension for the reservoir oil and formation water in the reservoir were not the same as those measured for Isopar-L and water. Using the following data, convert the capillary pressures from laboratory conditions to reservoir conditions.

$$\theta_{\text{Isopar-L-water}} = 0°,\ \theta_{\text{reservoir oil-formation water}} = 35°,$$

$$\sigma_{\text{Isopar-L-water}} = 35\ \text{mN/m},\ \sigma_{\text{reservoir oil-formation water}} = 25\ \text{mN/m}.$$

Water Saturation (fraction)	Capillary Pressure (kN/m^2)
0.30	2000
0.35	800
0.50	600
0.65	500
0.70	50

8.5 The following capillary pressure data are measured for a core plug from the North Sea. Calculate the absolute permeability from the following sample data and capillary pressure. Lithology factor, $\lambda = 0.3$, $\theta_{air-mercury} = 140°$, $\sigma_{air-mercury} = 480$ dynes/cm, $k = 2.0$ mD (Klinkenberg corrected), and $\phi = 38\%$.

Mercury Saturation (fraction)	Capillary Pressure (psia)
0.000	234
0.041	257
0.075	263
0.121	273
0.158	281
0.208	287
0.280	301
0.326	314
0.375	324
0.415	334
0.455	346
0.500	362
0.533	379
0.582	401
0.615	425
0.689	501
0.731	576
0.779	712
0.838	1125
0.850	1312
0.869	1875
0.880	2625
0.885	3000

8.6 For the capillary pressure data in problem 8.5, calculate the pore throat sorting coefficient and the pore size distribution index. Make appropriate comments on the quality of this rock.

8.7 From the capillary pressure data presented in problem 8.5, calculate the height of oil–water contact (OWC) and thickness of the transition zone in ft, and map the reservoir in terms of fluid contacts, zones, and water saturation distribution. The free water level (FWL) is at 0 ft. Additional data include the reservoir oil formation water IFT of 25 dynes/cm and a contact angle of 30° (values are representative of reservoir conditions).

REFERENCES

1. Leverett, M.C., Capillary behavior in porous solids, *Trans. AIME*, 142, 152, 1941.
2. Welge, H.J. and Bruce, W.A., The restored state method for determination of oil in place and connate water, *Drilling Production and Practices*, American Petroleum Institute, 1947.

3. Cole, F., *Reservoir Engineering Manual*, Gulf Publishing Company, Houston, TX, 1969.
4. Purcell, W.R., Capillary pressures — their measurement using mercury and the calculation of permeability therefrom, *Trans. AIME*, 186, 39, 1949.
5. Dandekar, A.Y., Unpublished work, 1999.
6. McCullough, J.J., Albaugh, F. and Jones, P.H., Determination of the interstitial water content of oil and gas sand by laboratory tests of core samples, *Drilling Production and Practices*, American Petroleum Institute, 1944.
7. Hassler, G.L., Brunner, E. and Deahl, T.J., The role of capillarity in oil production, *Trans. AIME*, 155, 155, 1944.
8. Slobod, R.L., Chambers, A., and Prehn, W.L., Use of centrifuge for determining connate water, residual oil, and capillary pressure curves of small core samples, *Trans. AIME*, 192, 127, 1951.
9. Tiab, D. and Donaldson, E.C., *Theory and Practice of Measuring Reservoir Rock and Fluid Transport Properties*, Gulf Publishing Company, Houston, TX, 1996.
10. Ahmed, T., *Reservoir Engineering Handbook*, Butterworth-Heinemann, Woburn, MA, 2001.
11. Anderson, W.G., Wettability literature survey — part 4: effects of wettability on capillary pressure, *J. Pet. Technol*, 1283–1300, 1987.
12. McCardell, W.M., A review of the physical basis for the use of the J-function, Eighth Oil Recovery Conference, Texas Petroleum Research Committee, 1955.
13. Amyx, J.W., Bass, D.M., Jr., and Whiting, R.L., *Petroleum Reservoir Engineering,* McGraw-Hill , New York, 1960.
14. Swanson, B.F., Microporosity in reservoir rocks: its measurement and influence on electrical resistivity, *Log Analyst*, 26, 42, 1985.
15. Omoregie, Z.S., Factors affecting the equivalency of different capillary pressure measurement techniques, *SPE Formation Evaluation*, 146–155, 1988.
16. Rose, W. and Bruce, W.A., Evaluation of capillary characters in petroleum reservoir rock, *Trans. AIME*, 1949.
17. Brown, H.W., Capillary pressure investigations, *Trans. AIME*, 1951.
18. Killlins, C.R., Nielsen, R.F., and Calhoun, J.C., Capillary desaturation and imbibition in porous rocks, *Producers Monthly*, 18, 30, 1953.
19. Morrow, N.R. and Mungan, N., Wettability and capillarity in porous media, Report RR-7, Petroleum Recovery Research Institute, Calgary, 1971.
20. Morrow, N.R., Capillary pressure correlations for uniformly wetted porous media, *J. Can. Pet. Technol.*, 15, 49, 1976.
21. Morrow, N.R., The effects of surface roughness on contact angle with special reference to petroleum recovery, *J. Can. Pet. Technol.*, 14, 42, 1975.
22. Richardson, J.G., Perkins, F.M., and Osaba, J.S., Differences in behavior of fresh and aged East Texas Woodbine core, *Trans. AIME*, 204, 86, 1955.
23. Washburn, E.W., A method of determining the distribution of pore sizes in a porous material, *Proc. Nat. Acad. Sci.*, 7, 115, 1921.
24. Brooks, R.H. and Corey, A.T., Hydraulic properties of porous media. Hydrology Paper 3, Colorado State University, Fort Collins, 1964.
25. Jennings, J.B., Capillary pressure techniques: application to exploration and development geology, *AAPG Bull.*, 10, 1196, 1987.
26. Trask, P.D., *Origin and Environment of Source Sediments in Petroleum*, Gulf Publishing Company, Houston, TX, 324, 1932.

9 Relative Permeability

9.1 FUNDAMENTAL CONCEPTS OF RELATIVE PERMEABILITY

The discussion in Chapter 4 on permeability referred to absolute permeability of a porous medium when a single-phase fluid saturation was considered. Darcy's law, as originally formulated and developed, was considered to apply when the porous medium was 100% saturated with a homogeneous single-phase fluid. However, petroleum reservoirs having such simple single-phase fluid systems seldom exist; usually reservoir rock systems are saturated with at least two or more fluids such as interstitial water or formation water, gas, and oil. These multiphase fluid systems play a very important role in the reservoir flow processes when petroleum reservoirs are produced by primary recovery mechanism or immiscible displacement methods involving the injection of gas or water. It is under these circumstances that more than one fluid phase is flowing or is *mobile* through a porous medium; thus the flow of one fluid phase interferes with the other. This interference is a competition for the flow paths and must be described accurately for hydrocarbon recovery from petroleum reservoirs. Chapter 4 also outlined the feasibility of quantifying fluid flow characteristics of reservoir rocks saturated with single-fluid systems through the concept of absolute permeability derived on the basis of Darcy's law. However, the direct application of this concept on an *as-is* basis for multiphase fluid saturations is not appropriate. Therefore, in order to accurately describe the simultaneous multiphase fluid-flow characteristics in petroleum reservoirs, it becomes necessary to introduce and define the concept of *relative permeability*.

The concept of relative permeability provides a mechanism of quantifying the amount of flow for each phase in a multiphase situation. The concept of relative permeability is fundamental to the study of the simultaneous flow of immiscible fluids through porous media. In many instances, relative permeability data selected to represent the subsurface or reservoir fluid flow behavior have more effect on the ultimate answer than any other parameter used in reservoir engineering equations. Thus it is important that a reservoir engineer has a good understanding of the relative permeability behavior of a given porous medium.

Relative permeability can also be considered as a dimensionless term devised to adapt the Darcy equation to multiphase flow conditions. If a single fluid is present in a rock, its relative permeability is 1.0. The determination of relative permeability allows comparison of the different abilities of fluids to flow in the presence of each other, since the presence of more than one fluid generally inhibits flow. In the case of two mobile and immiscible fluid phases flowing through a porous medium, the flow behavior is characterized by two-phase relative permeability; when three fluid phases are flowing, the flow process is described by three-phase relative permeability.

Therefore, in a given porous medium containing any two mobile fluid phases, it is possible to define any of the following two-phase relative permeabilities:

- Gas–oil
- Gas–water
- Oil–water

However, in the case of a three-phase relative permeability, the flow is always characterized by gas–oil–water relative permeabilities.

It is these individual relative permeabilities that are required in various reservoir engineering calculations, reservoir simulations such as in assessing the nature and efficiency of displacement mechanism and ultimate recovery of hydrocarbons from petroleum reservoirs.

9.2 MATHEMATICAL EXPRESSIONS FOR RELATIVE PERMEABILITY

The gas, oil, and water relative permeabilities are normally denoted by k_{rg}, k_{ro}, and k_{rw}, respectively. A distinction is usually made between two-phase relative permeabilities and three-phase relative permeabilities, such as k_{ro} (two phase) or k_{ro} (three phase). Relative permeabilities can either be expressed as percentage or fraction and are usually expressed by the ratio of *effective permeability* to *absolute permeability*. Thus, relative permeabilities are the result of normalizing effective permeability values by absolute permeability. The relative permeability is generally expressed mathematically by the simple equation:

$$k_r = \frac{k_e}{k} \tag{9.1}$$

where k_r is the relative permeability and is dimensionless, k_e the effective permeability in mD or D, and k the base permeability of the porous medium in mD or D.

The relative permeability of a specific fluid phase such as gas, oil, or water is mathematically expressed as:

$$k_{rg} = \frac{k_{eg}}{k} \tag{9.2}$$

$$k_{ro} = \frac{k_{eo}}{k} \tag{9.3}$$

$$k_{rw} = \frac{k_{ew}}{k} \tag{9.4}$$

where k_{rg}, k_{ro}, and k_{rw} are the relative permeabilities of gas, oil, and water phases, respectively, and are dimensionless and k_{eg}, k_{eo}, and k_{ew} the effective permeability of gas, oil, and water phases, respectively.

In Equations 9.2–9.4 for relative permeability, the effective permeability is a relative measure of the conductance of the porous medium for one fluid phase when the medium is saturated with multiple fluid phases. This definition of effective permeability indicates that the medium can have a distinct and measurable conductance to each fluid phase present in the medium.

However, when a porous medium saturated with more than one fluid phase is considered (i.e., an oil–water system) it is possible that the saturations may range from 20% oil and 80% water to 80% oil and 20% water. Therefore, the question is: at what fluid saturation value is the effective permeability specified?

Unlike the absolute permeability (a single value for a rock sample or a porous medium saturated 100% with a single-phase fluid) studied in Chapter 4, many values of effective permeability now exist; one for each particular condition of fluid saturation. As a matter of fact, the effective permeability is a function of the prevailing fluid saturation along with other properties such as wetting characteristics, pore geometry, capillary characteristics, and surface forces. Therefore, it becomes necessary to specify fluid saturation when stating or reporting the effective permeability of a particular fluid phase in a given porous medium. Thus, for the effective permeability shown in Equations 9.2–9.4, the values are always specified as some numerical values at a given saturation condition. Naturally the relative permeabilities are also automatically specified as dimensionless numerical values at a given saturation condition. However, the denominator in Equations 9.1–9.4 can also be construed as effective permeability at 100% fluid saturation (porous medium fully saturated by either gas, oil, or water). It is normally assumed that effective permeability of the porous medium is the same for all fluids at 100% saturation, which is referred to as the absolute permeability or simply permeability of the porous medium. The denominator in relative permeability equations is sometimes also referred to as the base permeability.

9.3 SALIENT FEATURES OF GAS–OIL AND WATER–OIL RELATIVE PERMEABILITY CURVES

Relative permeability data are typically reported or presented in the form of relative permeability curves. Similar to the capillary pressure–saturation curves, the relative permeability–saturation curves or plots also represent the relative permeability values in a certain fluid phase saturation that ranges typically between the irreducible wetting-phase saturation and in almost all cases the residual oil saturation or the corresponding wetting-phase saturation $1 - S_{or}$. As mentioned earlier, if the relative permeability–saturation data are for the gas–oil, gas–water, or oil–water, collectively, the data are referred to as two-phase relative permeability curves. This section studies their various important characteristics with the help of typical gas–oil and oil–water relative permeability curves.

To understand the characteristic features of the relative permeability curves, typical gas–oil and oil–water relative permeability data, shown in Figures 9.1 and 9.2, are considered. It should, however, be noted that these curves do not represent the relative permeability data of any particular porous medium but are representative of typical relative permeability data and are used merely to study the various important features of such data. Considering that relative permeability data signify relative conductive capacity of a porous medium, relative permeability data presented in Figures 9.1 and 9.2 are in almost all cases obtained by conducting laboratory-scale displacement tests (or fluid flow experiments) on reservoir rock core plug samples. These laboratory measurement procedures and techniques are discussed in Section 9.4.

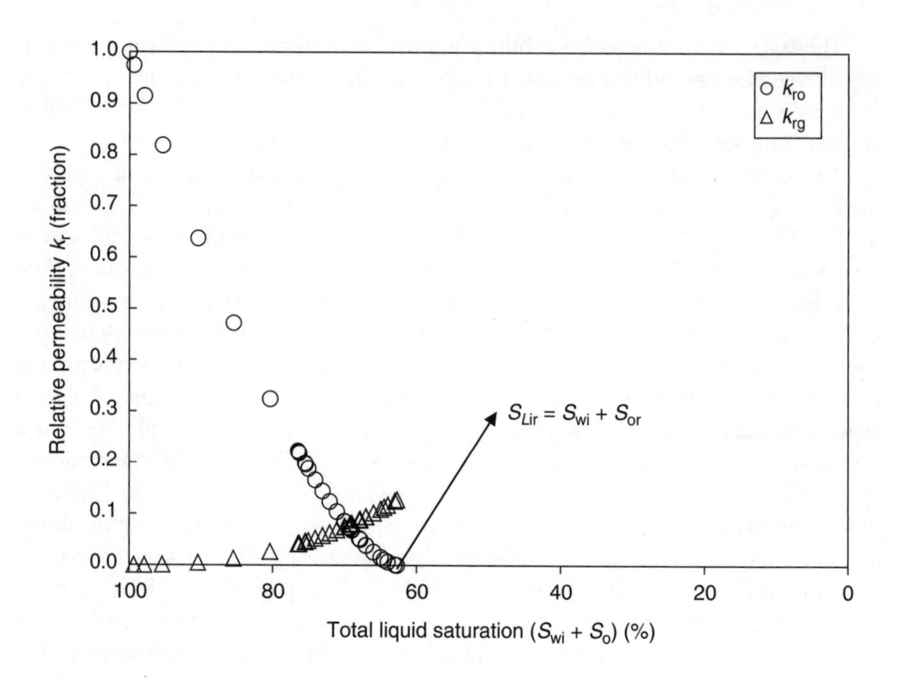

FIGURE 9.1 A typical gas–oil relative permeability curve. Saturation scale is composed of irreducible water saturation.

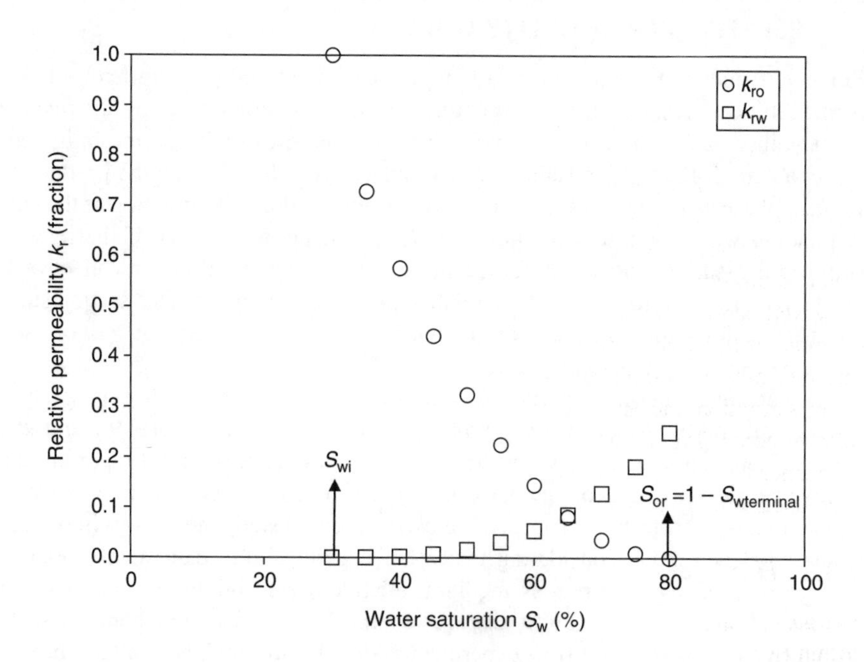

FIGURE 9.2 A typical oil–water relative permeability curve.

Relative permeability curves presented in Figures 9.1 and 9.2 consist of the elements described in the following sections.

9.3.1 The End-Point Fluid Saturations

As mentioned earlier, relative permeability data are usually plotted as relative permeability–saturation curves. The saturation on the x axis typically ranges from the irreducible wetting-phase saturation to the residual oil saturation. For gas–oil relative permeability, the starting saturation is referred to as total liquid-phase saturation that consists of the irreducible water saturation, S_{wi} with the remainder as oil saturation, totaling to 100% liquid saturation. As gas displacement is carried out, the total liquid-phase saturation begins to reduce; the irreducible water saturation remaining constant, while the oil saturation reduces. The gas and oil relative permeability curves end at residual oil saturation, basically consisting of a summation of S_{wi} and S_{or}, or normally referred to as S_{Lir}. The liquid-phase saturation changes that take place in gas–oil relative permeability curves are summarized in Figure 9.3.

For oil–water relative permeability curves, wetting phase saturation on the x axis (water in most cases) begins with the irreducible water saturation, S_{wi} (the remainder liquid being the initial oil saturation). As water is injected into the core sample, it displaces the oil, increasing the saturation of the former and decreasing the saturation of the latter, respectively. The desaturation of oil continues until the residual oil saturation is achieved, also expressed as $1 - S_{wterminal}$. So, basically, the oil and water relative permeability curves begin at S_{wi} and terminate at S_{or} (see Figure 9.4).

9.3.2 The Base Permeabilities

Relative permeabilities can be expressed as any specified base permeability, that is, the effective permeability divided by the absolute air (usually Klinkenberg corrected) or liquid permeability or usually the effective oil permeability at irreducible water saturation (used in Figures 9.1 and 9.2). If the effective permeability at the irreducible water saturation is used, then the oil effective permeability at irreducible water saturation is always employed for calculating the relative permeability. However, careful understanding of which base permeability is used is needed. For example, if the oil effective permeability at the oil saturation of 50% or water saturation of 50% is 100 mD and if the base permeability is 110 and 120 mD at 100% water saturation and an irreducible water saturation of 20%, respectively, then the relative permeability could be either 100/110 or 100/120, which are obviously not the same. However, the important issue is to realize which base permeability has been used in determining the relative permeability.

9.3.3 End-Point Permeabilities and Relative Permeability Curves

The gas–oil as well as oil–water relative permeability curves are basically confined within two end points. These end points are referred to as end-point relative permeabilities and are defined as follows.

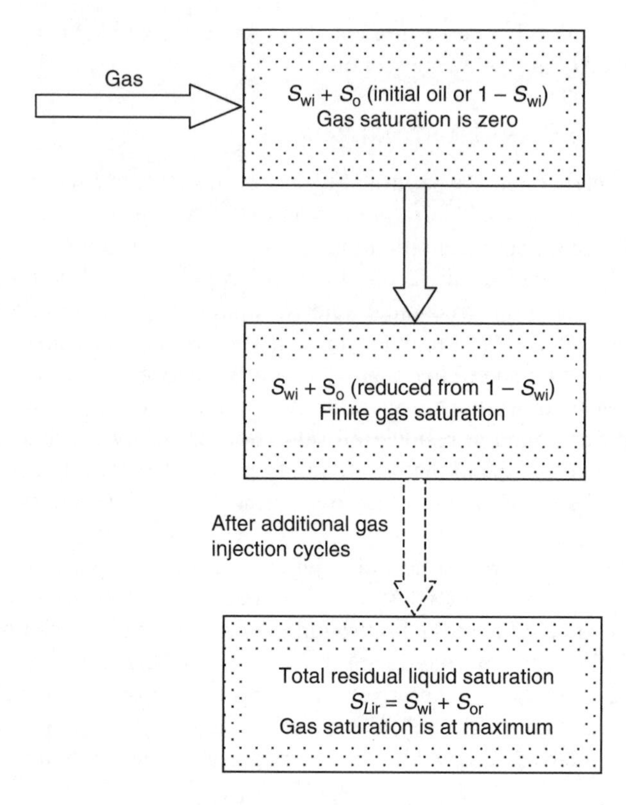

FIGURE 9.3 Saturation changes taking place in a gas–oil relative permeability curve.

9.3.3.1 Gas–Oil Relative Permeability Curves

At 100% liquid saturation, the oil relative permeability, k_{ro}, always equals 1 if the base permeability used is the effective permeability to oil at S_{wi}, or $k_{ro} = (k_{eo}$ @ S_{wi}/k_{eo}@$S_{wi}) = 1$. At this point, irrespective of the base permeability used, k_{rg} is always equal to 0 because gas is immobile.

At the residual liquid-phase saturation, the oil relative permeability, k_{ro}, always equals 0 because the liquid phase(s) is immobile ($S_{wi} + S_{or}$). However, at this saturation, the gas relative permeability is at its maximum value because gas is the only phase that is mobile. These relationships or boundary conditions can also be expressed as:

$$\text{At } S_L = 100\%,\ k_{ro} = 1;\ k_{rg} = 0 \text{ and } S_L = S_{Lir},\ k_{ro} = 0;\ k_{rg} = \text{maximum}$$

where S_L is the total liquid saturation (oil + water), initially 100% when the test begins, and S_{Lir} the residual liquid saturation, which is the sum of S_{wi} and S_{or}.

9.3.3.2 Oil–Water Relative Permeability Curves

At the initial or irreducible water saturation, the oil relative permeability, k_{ro}, always equals 1 if the base permeability used is the effective permeability to oil at S_{wi}, or

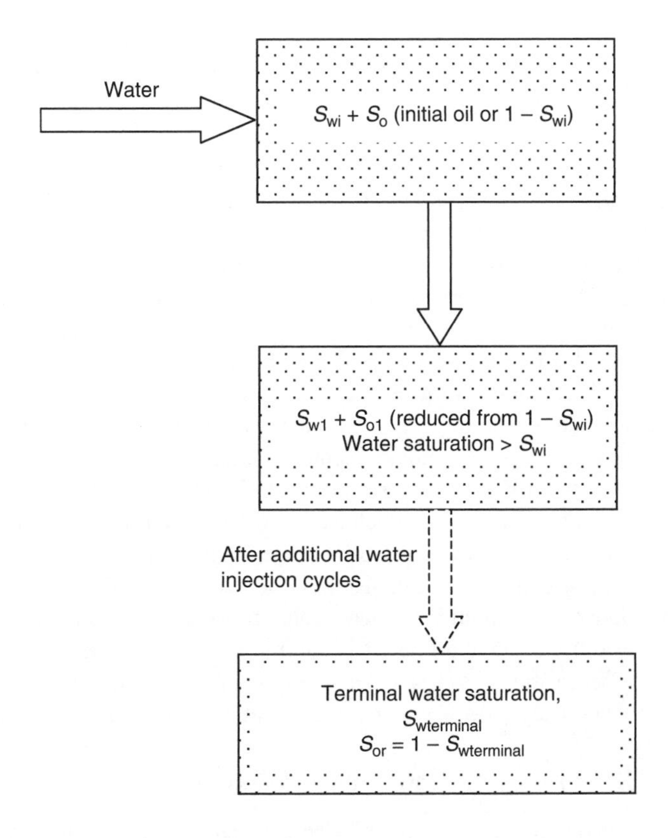

FIGURE 9.4 Saturation changes taking place in an oil–water relative permeability curve.

$k_{ro} = k_{eo}@S_{wi}/k_{eo}@S_{wi} = 1$. At this point irrespective of the base permeability used, k_{rw} is always equal to 0 because water is immobile.

At residual oil saturation, the oil relative permeability, k_{ro}, always equals 0 because the oil phase is immobile (S_{or}). However, at this saturation water relative permeability is at its maximum value because water is the only phase that is mobile.

These relationships or boundary conditions can also be expressed as:

$$\text{At } S_{wi},\ k_{ro} = 1;\ k_{rw} = 0 \text{ and } S_w = 1 - S_{or},\ k_{ro} = 0;\ k_{rw} = \text{maximum}$$

where S_w is the water saturation, which is at S_{wi} when the test begins, S_{or} the residual oil saturation equal to $1 - S_{wterminal}$, and $S_{wterminal}$ the terminal water saturation.

Figures 9.1 and 9.2 show individual gas, oil (from a gas–oil pair), and oil (from a oil–water pair) or water relative permeability curves. These individual relative permeability curves begin and end according to the end-point relationships described here. Two of the points (or the *y*-axis values) on these relative permeability curves are, however, always fixed, that is, oil relative permeability at irreducible water saturation is always 1 (under the assumption that the base permeability is effective permeability to oil at S_{wi}, which happens to be true in many cases), whereas the conjugate (either gas or water) phase relative permeability values are always 0. Other sections of individual phase relative permeabilities are, however, dependent on one important variable, the residual liquid saturation (in case of gas–oil) or just residual oil saturation (in case of oil–water). If the difference between the irreducible water saturation and the terminal water saturation $(1 - S_{or})$ is large, the relative permeability curves have fairly wide ranges whereas the opposite is true, that is, relative permeability curves occur in a rather narrow range if these saturation differences are not significant.

For the magnitude and curvature of the individual phase relative permeabilities, the relative permeability of the gas and the oil phase begins to rapidly increase and decrease, respectively, as soon as the total liquid-phase saturation starts to reduce as gas displacement progresses. A similar observation can also be made regarding oil and water relative permeabilities, that is, they decrease and increase, respectively, as water saturation rises. The other notable feature includes the attainment of very high (approaching nearly 100% or 1) relative permeability of a nonwetting phase such as gas at nonwetting-phase saturation much less than 100% or 1, typically happening at the residual liquid saturation for a gas–oil relative permeability. However, if oil is the wetting phase, a similar behavior can also be observed for the oil–water case, that is, high water relative permeability at water saturations much less than 100% or 1. Relative permeability of oil at initial water saturation obtains a value of 1 (by definition if the base permeability is $k_{eo}@S_{wi}$) and yet the oil saturation is not 100% even if the porous media do contain some water phase that is however, immobile.

9.3.4 The Direction of the Relative Permeability Curves

The direction of a relative permeability curve with regard to saturation history is another very important characteristic, that is, whether the curve is produced by a drainage process or imbibition process, under the assumption that wettability of the rock sample is known *a priori*. The drainage relative permeability curve applies to the process in which the wetting phase is, or has been, decreasing in magnitude; the imbibition process assumes the wetting-phase saturation is, or has been, increasing in magnitude.

Taking Figure 9.1 into consideration, a decrease in the total liquid-phase saturation is evident when moving along the x axis starting with 100% liquid saturation. Note that, in the case of gas–oil relative permeability data, the process is always drainage as gas is always the nonwetting phase in comparison to the oil and water phases. So, basically, all gas–oil or gas–water relative permeability curves are drainage relative permeability curves.

If Figure 9.2 is now considered and if water is the wetting phase and relative permeabilities are measured by the displacement of oil by water, the curves are called *imbibition curves*. If oil displaces water and relative permeabilities are measured from this displacement, the curves are called *drainage curves*. The converse is true if instead of water, oil is the wetting phase.

9.4 LABORATORY MEASUREMENT OF RELATIVE PERMEABILITY

Since relative permeability data basically signify the relative conductive capacity or flow behavior of a porous medium when it is saturated with more that one fluid phase, the most obvious laboratory measurement technique from which relative permeability data can be determined is the flow experiment. Laboratory measurement techniques for obtaining the two-phase relative permeability data based on the flow experiments are fairly well established. Essentially two different types of flow experiments can be conducted in reservoir rock samples from which relative permeability data are determined. These methods are called *steady state* (SS) and *unsteady state* (USS). In addition to the SS and USS methods, the centrifuge technique has also been used to determine relative permeability data. However, the flow mechanism in centrifuge experiments is very different from that in reservoirs, except those going through gravity drainage. Therefore use of this technique for determining relative permeabilities for reservoir analysis has never gained popularity, although this method reduces the time needed to conduct experiments, especially on tight cores. Therefore, the centrifuge technique is not described here.

The procedure for determining gas–oil and oil–water relative permeabilities from the SS and USS displacement tests require very comprehensive testing programs composed of various steps beginning with the initial preparation of the reservoir rock sample and ending with the determination of final fluid saturations, as discussed in Section 9.4.1. Additionally, considering that reservoir rock samples, fluid samples, and pressure and temperature conditions are the key elements of a successful relative permeability testing program, these issues do merit a discussion, provided in Sections 9.4.2 and 9.4.3, respectively. Section 9.4.1 discusses the establishment of initial water saturation, and with the determination of base permeability, these constitute two important steps of relative permeability determination that are discussed in Sections 9.4.4 and 9.4.5, respectively. The SS and USS displacement tests are normally carried out in displacement apparatus or relative permeability rigs. These rigs are basically set up by integrating various individual components, such as displacement pumps, core holder, high-pressure tubing, a set of valves, and a variety of electronic controls. A detailed description of a typical displacement apparatus is provided in Section 9.4.6. The generic procedure for conducting the SS and USS displacement tests and determining gas–oil and oil–water relative permeability data derived from these tests is described in Sections 9.4.7 and 9.4.8, respectively.

Apart from the SS and USS methods, the centrifuge technique is also utilized for the measurement of relative permeability data. However, considering the immense popularity of the steady-state and the unsteady-state techniques and their routine use, concentration is focused on these two methods. In addition to these laboratory methods, other indirect approaches also used for the determination of relative permeability data are covered in Section 9.5.

9.4.1 Flowchart for Relative Permeability Measurements

Measurements for relative permeability on core plug samples usually include both gas–oil and oil–water data. A whole core sample recovered from a particular formation can typically yield several small core plug samples. To obtain a spread for relative permeability values, several core plug samples are usually tested in the laboratory because single one-off core plug sample measurements are usually not sufficient to adequately describe flow behavior that is representative of the reservoir. Dimensions of a typical core plug sample usually constitute a small fraction of the reservoir which supports use of several core plug samples instead of one. Therefore, it is always desirable to conduct relative permeability measurements on a number of core plug samples ranging from anywhere between 50 to 100. To accomplish relative permeability testing for a large number of samples, a comprehensive special core analysis (SCAL) program (since relative permeability testing is part of SCAL) is usually designed.

This SCAL program contains a number of experimental steps, summarized by the flowchart shown in Figure 9.5, as an example. Note that the matrix presented in Figure 9.5 serves as a general guideline that can be altered depending on the actual requirements of a particular study. The prominent steps of the testing program include preparation of the core plug samples to obtain the initial conditions (i.e., the establishment of irreducible water saturation) and the actual displacement experiments to obtain the data that is subsequently used for determining the gas–oil and oil–water relative permeabilities. Almost always, gas–oil relative permeability measurements precede the oil–water measurements. The final or terminal step involves the determination of final fluid saturations in the core plug samples, accomplished by the Dean–Stark extraction method. The terminal measured fluid saturations can be used to back-calculate (based

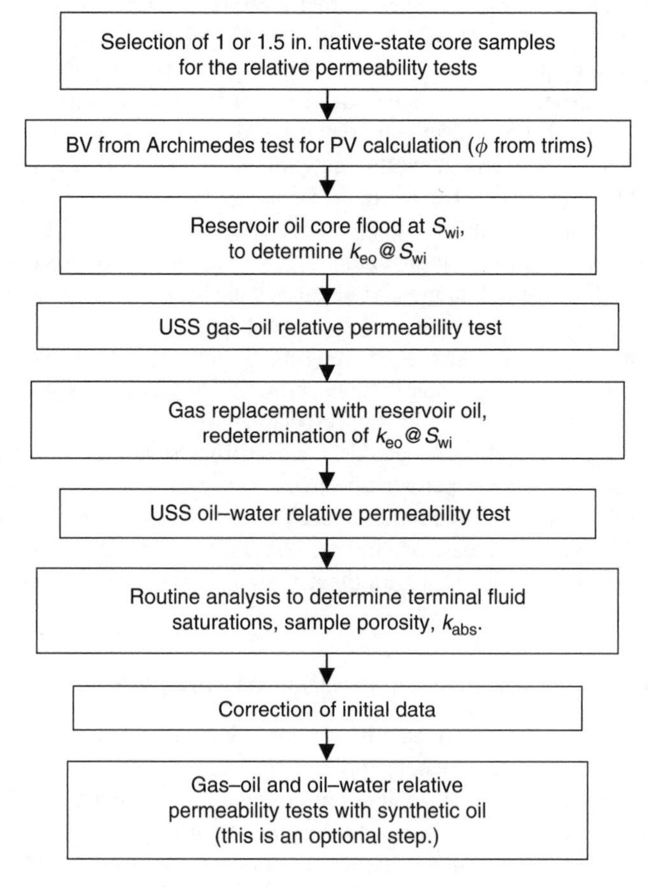

FIGURE 9.5 Relative permeability testing matrix.

on the record of fluid volume or mass balances in the intermediate steps) the saturations that existed in the sample prior to beginning the SCAL program for relative permeability testing. This in fact constitutes a very crucial step when dealing with preserved core-plug samples as initial saturations are not known (or an *indirect* value may be available from log data, or plug-end trim data that need to be reconfirmed or corrected). However, this is also an important step when dealing with cleaned core plugs because the Dean–Stark determined values can also be used in reconfirming the initial fluid saturations in the core plug sample (obtained from fluid volumes or mass balances). In addition to these prominent steps, the program also includes several intermediate steps that are necessary to complete the relative permeability SCAL program.

9.4.2 Core Plug Samples Used in Relative Permeability Measurements

Considering that relative permeability measurements in the laboratory are carried out on reservoir rock core samples; similar to some of the earlier-discussed reservoir rock properties, the use of particular core plug samples for carrying out such measurements always becomes an important issue. Specifically, considering relative permeability data are affected by factors such as wettability, the use of core plug samples having representative reservoir wettability is certainly very important and cannot be overlooked.

Therefore, the question always arises: what type of rock samples should be used for carrying out the measurements, cleaned or preserved? Thus, considering the importance of wettability on relative permeability, preserved samples are always preferred assuming that at least the probability of such samples maintaining the original reservoir wettability is definitely much higher in comparison to cleaned core plugs where cleaning fluids may have a tendency to alter wettability by removing the natural coating on grain surfaces. However, other general recommendations such as the use of core plugs from the central noninvaded section of the whole core, in view of drilling mud filtrate invasion, also apply when preserved core plug samples are considered.

One practical problem with the use of preserved core samples is that the initial saturations are unknown, especially the irreducible water saturation that actually is the starting point of all relative permeability measurements. However, this problem can be overcome by using the core plug-end trim data and considering that as representative of the entire core plug. This assumption can also be verified by some of the methods discussed in Chapter 6 on fluid saturations. Additionally, at the termination of all relative permeability testing, the core plug sample can be subjected to the Dean–Stark extraction process from which initial water saturation data can be back-calculated based on the consideration of mass or volume balance.

If for some reason preserved samples are not available, and cleaned and dried samples are to be used, then the samples should be cleaned preferably with nonreactive type of cleaning agent so that they do not alter wetting characteristics. If the samples are cleaned, they should be aged after attaining the initial conditions (e.g., after achieving the initial water saturation). This step, as mentioned earlier in Chapter 7 is called *the restoration of the reservoir rock wettability*. Although, questions are always raised with regard to the usefulness or validity of this process, it nevertheless is the only option available in the absence of native core material.

One significant advantage of using cleaned core plug samples is that all the fluid saturations are known right from the beginning stage because testing begins with the establishment of irreducible water saturation.

9.4.3 Displacement Fluids and Test Conditions

It is theoretically possible to use a wide variety of displacement fluids, including air, humidified nitrogen, synthetic oils (e.g., ISOPAR-L), model oil (e.g., *n*-decane), flashed or degassed reservoir oil, live reservoir oil, pure water, formation brine, or reconstituted or laboratory-prepared brine (based on known compositions). The choice of displacement fluids is, however, often dictated by factors such as availability of physical samples of reservoir oil and brine, special issues related to the handling of reservoir fluids (e.g., preserving the solution gas in live oils), impact of using displacement fluids that are not native to the formation, and effects related to the phase behavior of live oils (e.g., asphaltene or wax deposition). Any decision with respect to the use of a particular set of fluids is made based on careful evaluation of the various factors mentioned.

However, some of the most commonly used sets of fluids include humidified nitrogen-synthetic oil/degassed crude oil for gas–oil relative permeability measurements and laboratory-reconstituted brine–synthetic oil/degassed reservoir oil is used for oil–water relative permeability tests. Quite frequently, both gas–oil as well as oil–water relative permeability measurements are also repeated using two different pairs of fluids to evaluate the effect of factors such as viscosity, surface tension properties, and wettability (see flowchart in Figure 9.5).

The other important issue with regard to displacement testing addresses the selection of pressure and temperature conditions. Test conditions that can be used are outlined in the following text.

9.4.3.1 Room Condition Tests

In room condition tests, the displacement experiments are carried out at ambient temperature and at whatever pressures that result from the injection of gas and water. Gas injection tests are usually carried out at constant differential pressures, whereas water injection is carried out at either constant flow rate or constant differential pressure, the former being much more common that the latter. Although, these test conditions are fairly easy to control in the laboratory, the tests have a major disadvantage because these are not representative of reservoir conditions; the same being true for reservoir fluids as these are degassed or flashed fluids; reservoir fluids are at much higher pressures and temperatures and also contain dissolved gases.

9.4.3.2 Partial Reservoir Condition Tests

The only difference between partial reservoir condition tests and room condition tests is the use of reservoir temperatures instead of ambient temperatures. Other parameters are identical to room condition tests.

9.4.3.3 Reservoir Condition Tests

These displacement tests are carried out at reservoir conditions (pressure and temperature) and are by far the most desired among the three options; primarily for two reasons:

1. The reservoir oil used is usually the *live* reservoir condition crude oil, which is representative of the formation.
2. Since reservoir oil is used, the pressure and temperature conditions are automatically controlled at values above the saturation pressure of the oil. This way the rock pore space also continues to be in contact with the native oil thus significantly reducing the chances of wettability alteration.

As far as the gas phase is concerned, humidified nitrogen can be used as it usually develops miscibility with reservoir oil at very high pressures, thus precluding the possibility of miscibility development. Similarly, the water phase used is usually the actual formation brine. Even though the reservoir conditions tests are by far the most representative tests as they are carried out at reservoir conditions with live reservoir fluids, the precise control of the experimental conditions is nevertheless a very challenging and difficult task indeed.

9.4.4 Establishment of Initial Water Saturation

The minimum water saturation from capillary pressure data or core analysis data (Dean–Stark or retort) is considered as an estimate of water saturation in the oil column at the time of discovery. Typically, this is the same saturation that is used for initializing the relative permeability measurements. This starting point is probably the most significant parameter when dealing with the relative permeability measurements because the scale, qualitative, and quantitative characteristics of the relative permeability curves are dictated by the value of S_{wi}. When native-state or preserved core plug samples are used, it is assumed that these samples already have established irreducible water saturation. However, when cleaned core samples are used, the value of irreducible water saturation needs to be established in the sample before relative permeability testing, as described in Figure 9.5, can begin. The following sections describe the procedures of establishing the initial water saturation from a practical point of view when preserved or cleaned core samples are used.

9.4.4.1 Preserved Core Plug Samples

When preserved core plugs are used, it is normally assumed that they already contain initial or irreducible water saturation. This is true under the presumption that the drilling fluid invasion has not taken place and no water has been expelled from the core when coring took place, or the core plug sample has been taken from a central noninvaded section of the whole core. Additionally, considering that preserved core plug samples are used in order to maintain reservoir wettability so that the relative permeability data measured are representative of reservoir wettability, the initial water saturation determined from the plug-end trim is assumed to represent the initial water saturation in the entire core plug sample. Therefore, from this point onwards the relative permeability measurements by some of the techniques described later can begin. Even though a significant advantage is gained based on the fact that the original reservoir wettability is maintained, some degree of uncertainty is introduced because the relative permeability testing begins with an *assumed* value of the initial water saturation from the end trim. However, the actual initial water saturation that existed in the core plug sample can always be back-calculated on the basis of mass or volume balances, after termination of the entire relative permeability testing.

9.4.4.2 Cleaned Core Plug Samples

For cleaned core plug samples, the establishment of initial water saturation begins with saturating the core plug with water. The condition at which the water saturation is carried out depends on the choice of test conditions: room, partial, or complete reservoir condition tests using reconstituted or actual formation brine. After the core plug is fully saturated with water, the absolute permeability of the sample is determined using Darcy's law. In a subsequent step, the hydrocarbon phase (degassed oil or live reservoir oil) is injected in the core plug sample saturated with water. The process is terminated when no more water is produced. Subsequently, based on the total amount of water produced and the pore volume of the core plug, the initial or irreducible water saturation S_{wi} is determined. This value of S_{wi} is then compared with other sources, such as the log data; if the two values are similar, the test stops at this point. However, if the S_{wi} from the log data is lower than the S_{wi} from the coreflood, an oilflood with a very viscous oil is normally carried out in order to squeeze some more water from the core plug thus resulting in a lowered irreducible water saturation. It is possible to obtain lower values of S_{wi} by flooding the core plug with a very viscous oil because high viscosity results in much higher pressure drops eventually removing additional water from the sample. The high viscosity oil in the core plug is then replaced with reservoir oil and the core plug is then subjected to aging in reservoir oil with the objective of restoring the original reservoir wettability. Another point noteworthy here is that the initial water saturation that was established in the reservoir due to the hydrocarbon migration process differs from the process in which this value is established under laboratory conditions, the former being the result of a combination of gravity–capillary forces and the latter being viscous dominated. Therefore, an additional experimental step such as one using a very high viscous oil becomes necessary in order to achieve similar values of S_{wi}.

9.4.5 Determination of Base Permeability

All tests for measuring the gas–oil or oil–water relative permeabilities basically begin with the measurement of base permeability that is, either the absolute permeability or the effective permeability to oil at S_{wi}. The determination of base permeability is the starting point of any relative permeability measurement. As described in Equation 9.1, the relative permeability of a particular phase is defined as the ratio of its effective permeability to base permeability. As far as the base permeability is concerned, either the absolute permeability or effective permeability to oil at irreducible water saturation can be used. If absolute permeability is used, the value obtained from the routine core analysis for the preserved-state core plugs, after termination of the relative permeability testing is employed, whereas for cleaned core plugs, the value determined during the establishment of initial water saturation is employed (see Section 9.4.4.2).

However, when the effective permeability to oil at irreducible water saturation is used as the base permeability, the procedure for preserved-state core plugs involves evacuation and an oilflood. This oilflood is normally preceded by evacuation of the core plug in order to remove trapped gas (solution gas released during the pressure reduction process). The evacuation process is carried out in a large airtight container (similar to a desiccator) with provision for evacuation that houses a glass beaker or something similar filled with reservoir oil in which the core plug sample is submerged. The evacuation process continues for several hours and results in the replacement of trapped gas by the reservoir oil in the core plug sample. Next, the core plug is subjected to an oilflood, usually at constant flow rate, and based on the sample dimensions, flow rate used, oil viscosity, and observed steady pressure drop, the effective permeability to oil at S_{wi} is calculated using direct application of Darcy's law. The Darcy equation can be directly applied because water phase is immobile and only oil is flowing, essentially a single-phase flow. As an example, the inlet pressure development and the calculation of effective permeability to oil at S_{wi} for a chalk core plug sample is shown in Figure 9.6.

For the determination of effective permeability to oil at S_{wi} for cleaned core plug sample, the value is determined as part of the procedure used for establishment of initial water saturation. The core plug is first fully saturated with water followed by an oil flood down to irreducible water saturation that continues until a steady pressure

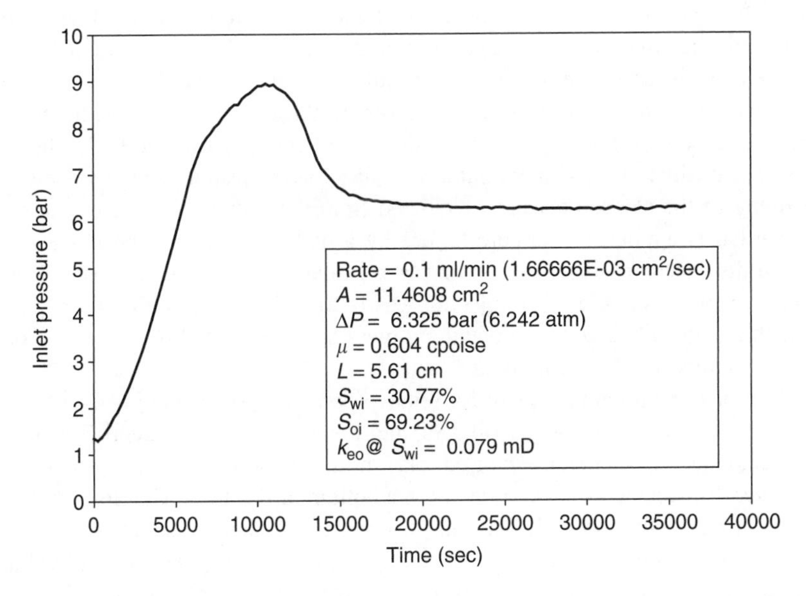

FIGURE 9.6 Inlet pressure development for an oilflood in a core sample containing S_{wi}. The effective permeability value is calculated from the experimental data and Darcy's law.

drop is obtained from which the effective permeability is determined using direct application of Darcy's law.

As discussed in Section 9.3.3, the effective permeability to oil determined in this fashion is one of the end-point effective permeability. The other two end-point effective permeabilities are that of gas (for a gas–oil curve) and water (for a oil–water curve) at residual oil saturations of S_{org} and S_{orw}, respectively, which are determined in a manner similar to the base permeability measurement, the only difference being the immobile oil phase saturation instead of immobile water phase saturation. These measurements are carried out in the terminal steps of relative permeability testing.

9.4.6 Displacement Apparatus for Relative Permeability

Figure 4.7 shows the schematic of a displacement apparatus for absolute permeability, which is in fact a typical setup also employed when conducting relative permeability tests. Although, the apparatus shown is geared for conducting the unsteady-state tests, by incorporating a few modifications, the same setup can also be used for conducting the steady-state experiments.

These relative permeability rigs are usually designed to handle high pressures and high temperatures that may be up to 10,000 psi and 250°C, so that displacement tests at reservoir conditions can also be carried out. The primary components of a displacement apparatus include a Hassler core holder, floating piston storage vessels for reservoir oil and brine, produced fluids collector, displacement pump, hydraulic pump (for confining pressure), back-pressure regulator, differential pressure transducer, source of gas supply, and gasmeter (for recording gas production). All these components are interconnected by high-pressure tubing that also includes a set of valves at strategic locations in the entire lineup for pertinent fluid flow control. Almost all the components, except the pumps, are placed in a climatic air bath to maintain a constant test temperature, normally the reservoir temperature.

The displacement of test fluids (oil or water), through a core sample mounted in a Hassler-type core holder is accomplished by injecting a suitable hydraulic fluid through a positive displacement pump into the floating piston storage vessels. The confining or the sleeve pressure, calculated from the initial reservoir pressure and overburden is applied to the core holder by a hydraulic pump. The gas used for gas–oil displacement experiments for gas–oil relative permeability can be directly supplied from gas bottles. The injection gas is sometimes humidified by passing it through a packed bed column containing glass beads saturated with water in order to avoid outdrying of the core plug.

The most common procedure for relative permeability testing involves the displacement of oil by gas (for gas–oil data) and water (for oil–water data). The gas–oil displacements are generally carried out in a top–down manner, whereas the oil–water displacements are carried out in a bottom–up manner. The flow of various fluids during gas and water displacement experiments is achieved through a setup of different flow loops and two- and three-way high-pressure valves. Liquid volumes (oil and water) produced as a function of time, from the core plug sample during the tests are normally measured by a produced fluid separator equipped with optical or acoustic sensors. The production of gas is, however, measured by a gasmeter. Simultaneous production data measured in this manner along with observed pressure drops, flow rates, and fluid properties are employed in the calculation of the individual gas–oil and oil–water relative permeabilities.

9.4.7 Steady-State Technique

The measurement of relative permeabilities to oil and water in unconsolidated sands using the steady-state method were first reported in 1939 by Leverett.[1] The steady-state method for a two-fluid system (gas–oil or oil–water) basically involves injecting two phases at a certain volumetric ratio until stabilization of both the pressure drop across the core and the effluent volumetric ratios. The saturations of the two fluids in the core are then determined, typically, by weighing the core or by performing a mass-balance calculation for each phase. Individual relative permeability data are calculated from the direct application of Darcy' law.

The various experimental steps in a steady-state process can be illustrated by the sequence of events as shown in Figure 9.7 (for oil–water) and summarized by the following points:

1. The process starts with complete water saturation of the core sample in the case of a clean core, followed by the oilflood down to irreducible water saturation, and the determination of the effective permeability to oil at S_{wi}. However, in the case of a preserved core sample, the process starts with the determination of the effective permeability to oil at S_{wi} (i.e., $k_{eo}@S_{wi}$)
2. In the subsequent steps, the objective is to increase the water-phase saturation steadily so that a number of data points on the relative permeability curve can be obtained. The two fluids, oil and water, are simultaneously injected into the core sample through a mixer head at a certain volumetric flow rate ratio, and the volume of fluids produced and pressure drop is recorded. Initially, the ratio of water: oil is small as the water saturation in the core is gradually increased from S_{wi}. The simultaneous injection of oil and water is continued until the injection ratio is equal to the production ratio, a condition at which the system is considered to be in steady state and the existing saturations are considered stable. The phase saturations and the individual oil- and water-phase effective permeabilities at the specific saturations are calculated as per the following procedure.

Saturation Calculation

Total mass of rock plus fluids is

$$M_{rf} = M_r + S_{w1}PV\rho_w + (1 - S_{w1})PV\rho_o \quad (9.5)$$

Rearrangement of Equation 9.5 gives

$$S_{w1} = \frac{M_{rf} - M_r - PV\rho_o}{PV(\rho_w - \rho_o)} \quad (9.6)$$

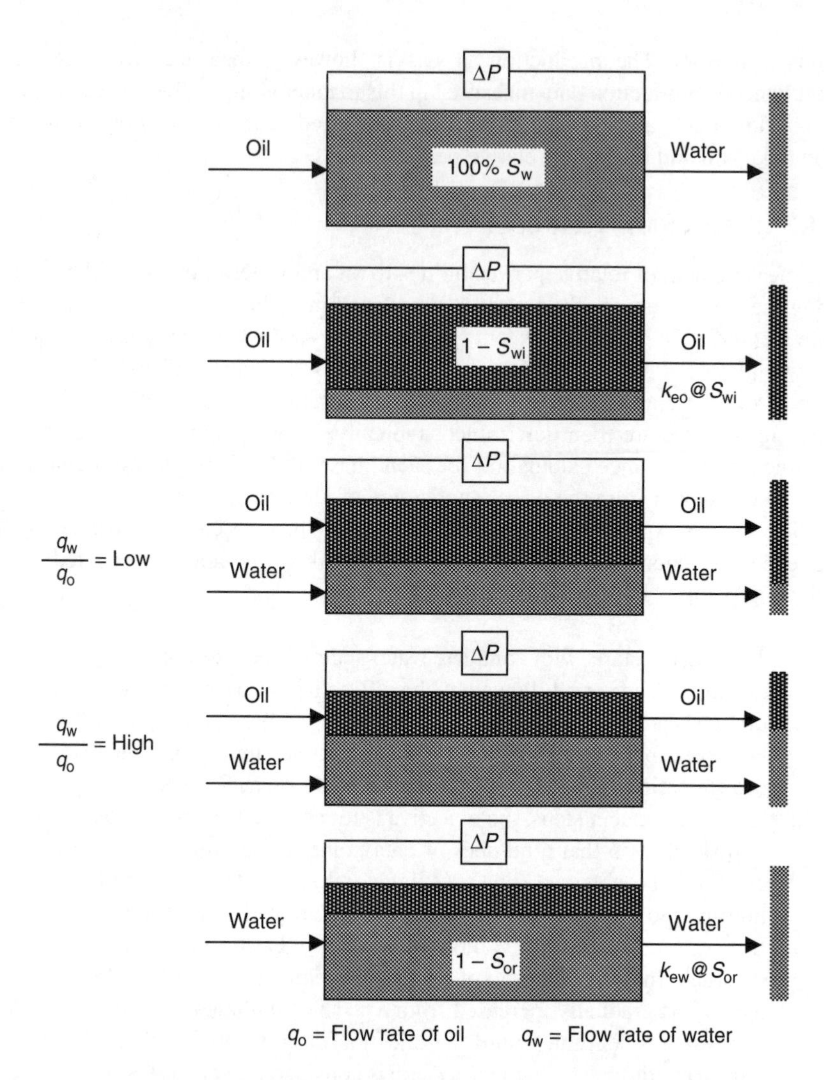

FIGURE 9.7 Schematic representation of the sequence of various events that take place during the steady-state displacement test for determination of relative permeabilities.

where S_{w1} is the new water saturation (at the first injection ratio) which is greater than S_{wi}, oil saturation is $(1 - S_{w1})$, M_{rf} the mass of rock plus fluids, M_r the mass of dry rock, PV the pore volume, ρ_o the oil density, and ρ_w the water density.

Effective Permeability Calculation

$$k_{eo} = \frac{q_o \mu_o L}{A \Delta P_o} \text{ at } S_{w1} \tag{9.7}$$

$$k_{ew} = \frac{q_w \mu_w L}{A \Delta P_w} \text{ at } S_{w1} \tag{9.8}$$

where k_{eo} and k_{ew} are the effective permeability of oil and water, respectively, in mD or D, q_o and q_w the oil and water flow rates, respectively, μ_o and μ_w the oil and water viscosities at test conditions, respectively, ΔP_o and ΔP_w the oil- and water-phase pressure drops, considered as equal under the assumption of 0 capillary pressure, respectively, and A and L the cross-sectional area and the length of the core sample, respectively.

3. The volumetric flow rate ratio at which the oil and water phases are simultaneously injected into the core sample is then gradually increased so that the oil phase is replaced by the water phase, so that the water-phase saturation increases (see Figure 9.7). The process is terminated when steady state is reached and is then followed by an even higher water oil ratio of injection. The saturation and the individual phase permeabilities are calculated according to the procedure described earlier.
4. In the final step, only water is injected down to residual oil saturation. Based on the flow rate of water injection and the observed steady pressure drop, the effective permeability to water at residual oil saturation or $k_{ew}@S_{or}$ is calculated by directly applying Darcy's law. This point constitutes the other end point of the oil–water relative permeability curve.

In a nutshell, the data obtained from the steady-state process can be summarized as follows:

First End Point from Step 1

k_{ro} @ S_{wi} = 1 (if $k_{eo}@S_{wi}$ is the base permeability) and k_{rw} @ S_{wi} = 0 (only oil is flowing).

Second Point on the Relative Permeability Curve from Step 2

$$k_{ro}@S_{w1} = \frac{k_{eo}@S_{w1}}{k_{eo}@S_{wi}}$$

and

$$k_{rw}@S_{w1} = \frac{k_{ew}@S_{w1}}{k_{eo}@S_{wi}}$$

Additional points on the relative permeability curves are obtained as water saturation increases, S_{w2}, S_{w3}, and so on, and the calculations are as shown above.

Last End Point from Step 4

$$k_{ro}\ @\ S_{or} = 0 \text{ and } k_{rw}@S_{or} = \frac{k_{ew}@S_{or}}{k_{eo}@S_{wi}}$$

Finally, these data can be plotted as water saturation vs. the oil and water relative permeabilities. Calculations similar to these for gas–oil relative permeability can also be carried out for gas displacement.

Even though the calculation of relative permeabilities is relatively simple, the entire process can be time consuming, as the time required for achieving a steady state may be rather inordinate. Additionally, if mass balance is used for the determination of saturations, the procedure involves repeated removing and mounting of the core sample after every step that can be quite cumbersome and prone to uncertainties, because fluid loss and damage to the core during the disassembly and reassembly process can cause errors in the measured saturations and resulting relative permeabilities.

9.4.8 Unsteady-State Technique

The unsteady-state method is primarily based on the interpretation of an immiscible displacement process. For a two-phase system; basically a core that is either in the native/preserved state or restored state after cleaning and aging, at the saturation conditions that exist in the reservoir, is flooded with one of the displacing phases. Typically, the flood phase is gas (for gas–oil relative permeability) or water (for oil–water relative permeability) since in the reservoir one or the other of these phases displaces oil.

The various experimental steps in the unsteady-state process can be illustrated by the sequence of events in Figure 9.8 (for oil–water). Similar to the steady-state process, the first step is the determination of the effective permeability to oil at irreducible water saturation, $k_{eo}@S_{wi}$, already described. Subsequently, the injection of water at a constant flow rate is initiated; the accompanying pressure drop and the volume of oil produced is recorded as a function of time. In this manner, water saturation in the core sample increases from the irreducible value of S_{wi}. As the injection of water progresses, additional oil and some water is also produced, eventually leading to a plateau of cumulative oil production vs. time. After the production of oil stops, only water is produced from the outlet end of the core sample. As an example, the oil and water production data as a function of time for an unsteady-state coreflood on a chalk core plug sample are shown in Figure 9.9.

The injection of water is continued further at the residual oil saturation, enabling calculation of the effective permeability of water at this particular saturation condition, $k_{ew}@S_{or}$, also the other end point of the relative permeability curve. This value is expressed as the end-point relative permeability to water by dividing $k_{ew}@S_{or}$ by $k_{eo}@S_{wi}$ (assuming this is base permeability).

For the determination of relative permeability data from the unsteady-state experiments, two different methods can be used for the calculation of water saturation: the alternate method[2] and the Johnson–Bossler–Naumann[3] method also known as the JBN method. Buckley and Leverett[4] developed the equations governing the displacement of one fluid by another in a porous medium. They assumed linear, incompressible flow, and negligible capillary forces. Ten years later, Welge[5] presented a method based on the Buckley–Leverett theory to calculate the saturations and the ratio of relative permeabilities of the displacing phase and the displaced phase. The JBN method that is commonly used to date is actually based on Welge's solution of the flow equation, developed for the first time to calculate the individual phase relative permeabilities from displacement data. The development of these various theories are presented in Section 9.4.8.1. The calculation procedure for using the alternate method is presented in Section 9.4.8.2.

The alternate method yields the relative permeabilities as a function of the average water saturation in the core, simplifying calculations because it is only necessary to apply Darcy's law to the displacement process and subsequently plot the relative

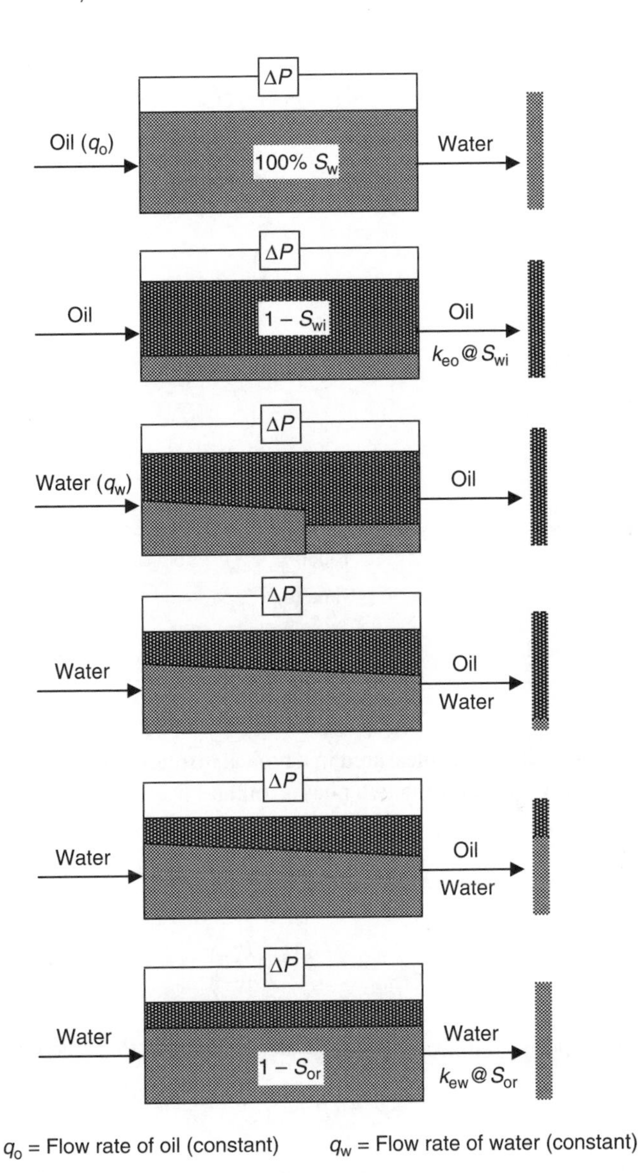

FIGURE 9.8 Schematic representation of the sequence of various events that take place during the unsteady-state displacement test for determination of relative permeabilities.

permeabilities as a function of the average water saturation. The JBN method calculates relative permeabilities as a function of effluent (outlet) fluid saturation. The JBN method, however, requires that the capillary end effects should be minimized by conducting the displacement at sufficiently high flow rates.

9.4.8.1 Buckley–Leverett to Welge to Johnson–Bossler–Naumann

Before describing the Buckley–Leverett theory of immiscible displacement in one dimension, the fractional flow equation that relates the ratio of relative permeabilities

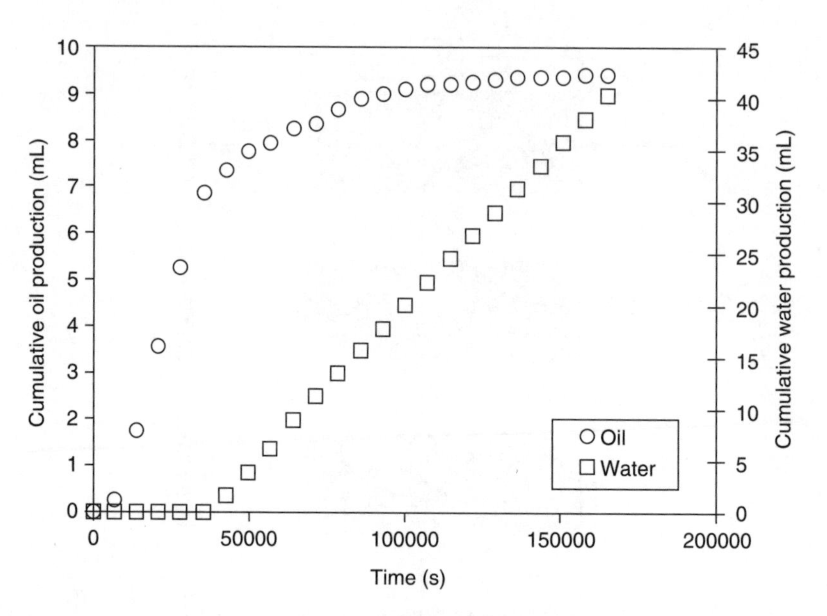

FIGURE 9.9 Oil and water production data collected during an unsteady-state relative permeability test.

to the experimental data that is measured in a typical displacement experiment must be derived. From Darcy's law, for the two phases, oil and water:

$$q_o = -\frac{k_o A}{\mu_o}\left(\frac{\partial P_o}{\partial x}\right) \quad (9.9)$$

$$q_w = -\frac{k_w A}{\mu_w}\left(\frac{\partial P_w}{\partial x}\right) \quad (9.10)$$

The capillary pressure in the system, if water is assumed to be the wetting phase is

$$P_c = P_o - P_w \quad (9.11)$$

Since oil and water are considered to be incompressible; the continuity equation applies to each phase

$$\frac{\partial q_o}{\partial x} = -\phi A \frac{\partial S_o}{\partial t} \quad (9.12)$$

$$\frac{\partial q_w}{\partial x} = -\phi A \frac{\partial S_w}{\partial t} \quad (9.13)$$

Additionally,

$$S_o + S_w = 1.0 \quad (9.14)$$

A combination of Equations 9.12–9.14 yields –

$$\frac{\partial}{\partial x}(q_o + q_w) = 0 \tag{9.15}$$

which means that the total flow rate $q_t = q_o + q_w$ is constant. Now, if Equations 9.9–9.11 are combined to eliminate P_o and P_w

$$q_o = -\frac{k_o A}{\mu_o}\left[-\frac{q_w \mu_w}{k_w A} + \frac{\partial P_c}{\partial x}\right] \tag{9.16}$$

The production data also defines the fractional flow, f_w, of the flowing stream

$$f_w = \frac{q_w}{q_o + q_w} \tag{9.17}$$

By neglecting the effect of capillary pressure gradient in Equation 9.16,

$$q_o = \frac{k_o A}{\mu_o}\left[\frac{q_w \mu_w}{k_w A}\right] \tag{9.18}$$

and substituting the value of q_o from Equation 9.18 in Equation 9.17

$$f_w = \frac{q_w}{\dfrac{k_o A}{\mu_o}\left[\dfrac{q_w \mu_w}{k_w A}\right] + q_w} \tag{9.19}$$

or

$$f_w = \frac{1}{1 + \left[\dfrac{k_o \mu_w}{k_w \mu_o}\right]} \tag{9.20}$$

If k_o and k_w in Equation 9.20 are expressed in terms of the product of relative permeability and the base permeability

$$f_w = \frac{1}{1 + \left[\dfrac{k_{ro} \mu_w}{k_{rw} \mu_o}\right]} \tag{9.21}$$

Equation 9.20 or 9.21 is the fractional flow equation for the displacement of oil by water for a horizontal displacement and neglecting the capillary pressure gradient. For a typical set of oil–water relative permeabilities the fractional flow usually has the shape indicated in Figure 9.10, with saturation limit S_{wi} and $1 - S_{or}$, between which the fractional flow increases from 0 to unity.

9.4.8.1.1 Buckley–Leverett theory

In 1942, Buckley and Leverett[4] presented the basic equation for describing immiscible displacement in one dimension. For water displacing oil, the equation determines

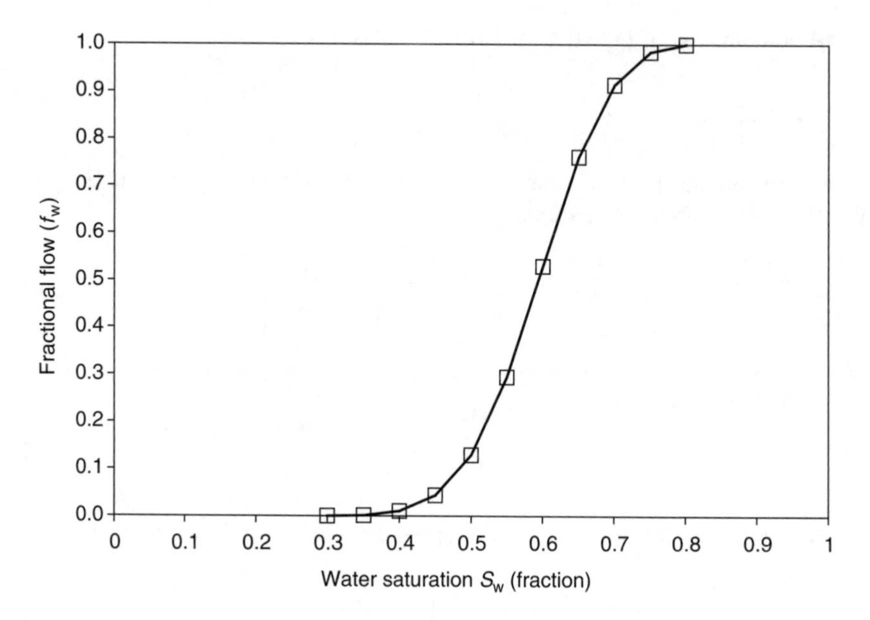

FIGURE 9.10 A typical fraction flow curve generated from oil–water relative permeabilities and viscosities.

the velocity of a plane of constant water saturation traveling through a linear system. The Buckley–Leverett model discussed in this section is based on the following assumptions:

- One-dimensional immiscible flow occurs for two incompressible fluids
- No mass transfer between fluids
- Flow is horizontal
- Diffuse flow, that is, displacement, occurs at very high injection rates so that the effects of capillary and gravity forces are negligible
- Viscosity is constant
- Homogeneous porous medium rock; ϕ and k are constant
- Water is injected at $x = 0$ (inlet face) at constant rate q_w.

The conservation of mass of water flowing through the volume element $A\phi\,dx$, as shown in Figure 9.11, may be expressed as:
Mass flow rate in – Mass flow rate out = Rate of increase of mass in the volume element

$$(q_w\rho_w)_x - (q_w\rho_w)_{x+\Delta x} = A\phi\,dx\frac{\partial}{\partial t}(\rho_w S_w) \tag{9.22}$$

or

$$(q_w\rho_w)_x - \left[(q_w\rho_w)_x + \frac{\partial}{\partial x}(q_w\rho_w)dx\right] = A\phi\,dx\frac{\partial}{\partial t}(\rho_w S_w) \tag{9.23}$$

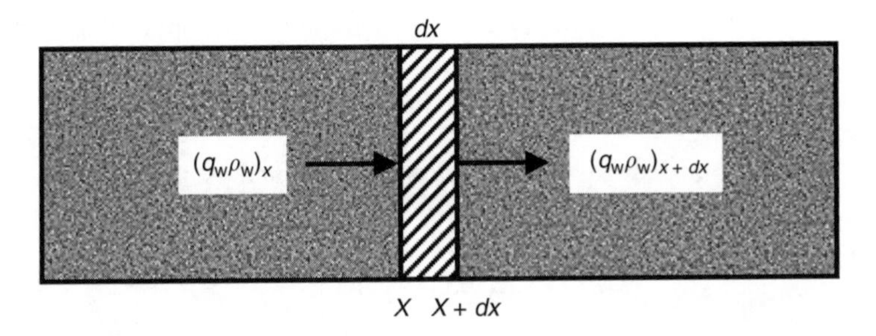

FIGURE 9.11 Schematic of a homogenous porous medium of porosity ϕ, intrinsic permeability k, and area A of which the volume element $A\phi$ dx is considered for the application of conservation of mass equation for the development of the Buckley–Leverett theory.

which reduces to

$$\frac{\partial}{\partial x}(q_w\rho_w) = -A\phi\frac{\partial}{\partial x}(\rho_w S_w) \tag{9.24}$$

where, S_w is the water saturation and ρ_w is the water density.

For the incompressible displacement, water density, ρ_w is assumed constant

$$\left.\frac{\partial q_w}{\partial_x}\right|_t = -A\phi\left.\frac{\partial S_w}{\partial t}\right|_x \tag{9.25}$$

The right-hand side of Equation 9.25 can be expressed as

$$\left(\frac{\partial S_w}{\partial t}\right)_x = -\left(\frac{\partial S_w}{\partial x}\right)_t\left(\frac{dx}{dt}\right)_{S_w} \tag{9.26}$$

substituting the Equation 9.26 into Equation 9.25

$$\left(\frac{\partial q_w}{\partial S_w}\right)_t\left(\frac{\partial S_w}{\partial x}\right)_t = -A\phi\left(-\frac{\partial S_w}{\partial x}\right)_t\left(\frac{dx}{dt}\right)_{S_w} \tag{9.27}$$

$$\left(\frac{\partial q_w}{\partial S_w}\right)_t = A\phi\left(\frac{dx}{dt}\right)_{S_w} \tag{9.28}$$

Now, the velocity of a plane of constant saturation, V_{Sw} can be expressed as

$$V_{Sw} = \left(\frac{dx}{dt}\right)_{Sw} = \frac{1}{A\phi}\left(\frac{\partial q_w}{\partial S_w}\right)_t \tag{9.29}$$

As seen in Equation 9.15, the total flow rate, q_t is constant, and from Equation 9.17, q_w can be expressed as $q_w = f_w q_t$, that allows Equation 9.29 to be written as

$$\left(\frac{dx}{dt}\right)_{Sw} = \frac{q_t}{A\phi}\left[\left(\frac{df_w}{dS_w}\right)_t\right]_{Sw} \tag{9.30}$$

Equation 9.30 is the Buckley–Leverett equation which implies that, for a constant rate of water injection q_t, the velocity of a plane of constant water saturation, V_{Sw} is directly proportional to the derivative of the fractional flow equation evaluated at that saturation.

The Buckley–Leverett equation can now be integrated as

$$\int_{o}^{x_{S_w}} dx = \left(\frac{df_w}{dS_w}\right)_{S_w} \frac{1}{A\phi}\int_{o}^{t} q_t dt \tag{9.31}$$

$$X_{S_w} = \frac{W_i}{A\phi}\left(\frac{df_w}{dS_w}\right)_{S_w} \tag{9.32}$$

In Equation 9.32, $W_i = q_t t$, is the cumulative water injected at time, t, and it is assumed, as an initial condition, that $W_i = 0$ when $t = 0$. Therefore, at a given time after the start of injection (W_i constant) the position of different water saturation planes can be plotted, using Equation 9.32, merely by determining the slope of the fractional flow curve for the particular value of each saturation.

By dividing Equation 9.32 by L, the total length of the porous medium (a core sample) –

$$\frac{X_{S_w}}{L} = Q_{wi}\left(\frac{df_w}{dS_w}\right)_{S_w} \tag{9.33}$$

where the left-hand side of the equation represents the normalized position ranging from 0 to 1, and the ratio $Q_{wi} = W_i/LA\phi$ is the pore volume of water injected ($LA\phi$ is the total pore volume).

A typical plot of df_w/dS_w vs. water saturation is shown in Figure 9.12. However, when the value of X_{S_w} or X_{S_w}/L is calculated, and plotted as a function of water saturation, a curve as in Figure 9.13 results. Clearly, the plot of saturations is showing an impossible physical situation, since two saturations can be found at each position. Buckley–Leverett solution to this problem is to modify the plot by defining a saturation discontinuity or a shock as indicated by the vertical line in Figure 9.13. The position of the shock is chosen such that the two areas enclosed by the shock and the dashed line are equal. As per Equation 9.33, the saturation profiles for arbitrary amounts of pore volumes of water injected; Q_{wi} can be obtained by *compressing* or *stretching* the profile.

9.4.8.1.2 Welge's extension solution

A more elegant method of achieving the same results as in the previous section was presented by Welge[5] in 1952. In order to understand Welge's extension solution, consider the water saturation distribution shown in Figure 9.14, as a function of distance or position for a fixed time, t, prior to the breakthrough. At this particular instance, the maximum water saturation, $S_w = 1 - S_{or}$, has moved a distance x_1, its velocity being proportional to the slope of the fractional flow curve, while the flood front saturation

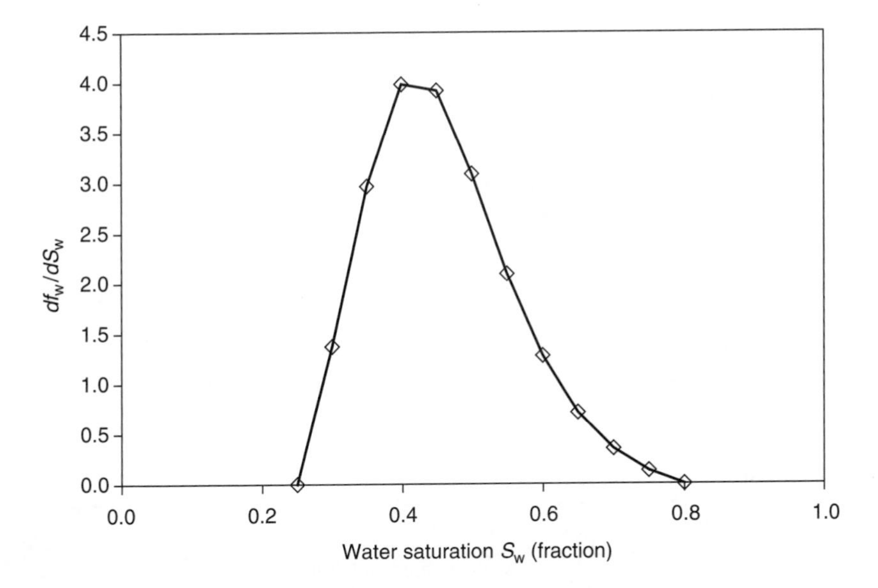

FIGURE 9.12 A typical plot of df_w/dS_w vs. water saturation, S_w, leading to a physically impossible situation.

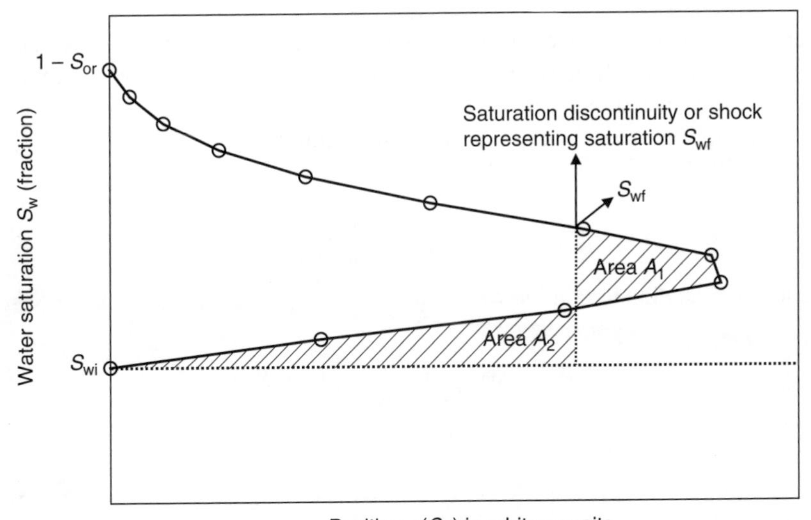

FIGURE 9.13 Buckley–Leverett solution to the physically impossible situation: A saturation discontinuity or shock is introduced so that the two areas A_1 and A_2 are equal.

S_{wf} is located at position x_2 measured from the injection point. Based on this consideration, the following material balance can be written as

$$W_i = A\phi x_2(\overline{S}_w - S_{wi}) \tag{9.34}$$

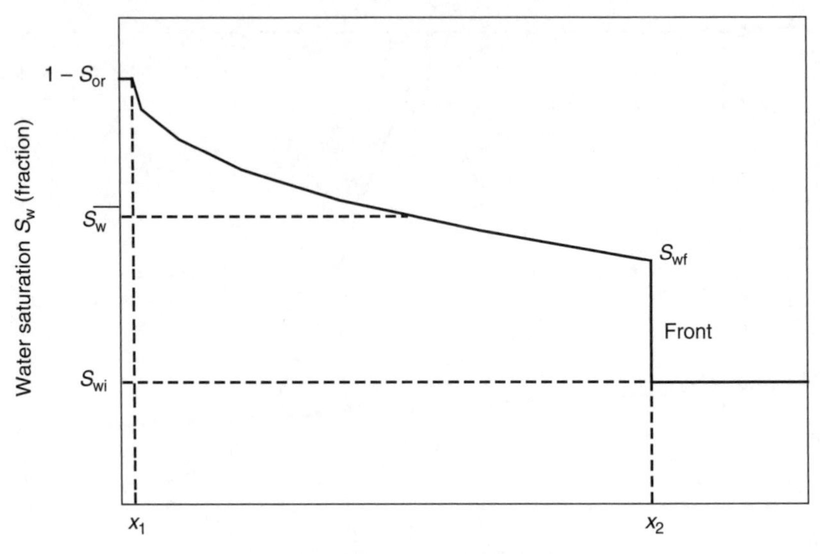

FIGURE 9.14 Water saturation distribution as a function of distance or position for a fixed time, t, prior to the breakthrough and used for Welge's extension solution.

where W_i is the cumulative water injected at fixed time t (which is less than the break-through time), A the cross-sectional area, ϕ the porosity of the medium, x_2 the position or distance shown in Figure 9.14, $\bar{S}_w$ the average water saturation behind the front, and S_{wi} the irreducible or connate water saturation.

From Buckley–Leverett theory

$$X_2 = \frac{W_i}{A\phi}\left(\frac{df_w}{dS_w}\right)_{S_{wf}} \tag{9.35}$$

that allows Equation 9.34 to be written as

$$W_i = A\phi\,\frac{W_i}{A\phi}\left(\frac{df_w}{dS_w}\right)_{S_{wf}}(\bar{S}_w - S_{wi}) \tag{9.36}$$

or

$$\bar{S}_w = S_{wi} + \frac{1}{\left(\dfrac{df_w}{dS_w}\right)_{S_{wf}}} \tag{9.37}$$

However, the value of S_{wf} is unknown and hence $(df_w/dS_w)_{S_{wf}}$ is also unknown.

The average water saturation behind the shock front can also be determined by integrating the saturation profile from the inlet to the position of the front x_f or x_2 (see Figure 9.14)

$$\bar{S}_w = \frac{\int_0^{x_2} S_w dx}{\int_0^{x_2} dx} \tag{9.38}$$

The numerator in Equation 9.38 can be integrated by parts

$$\bar{S}_{w} = \frac{(S_{wf}x)\big|_{0}^{x_2} - \int_{1-S_{or}}^{S_{wf}} x \mathrm{d}S_{w}}{x_2} \tag{9.39}$$

$$\bar{S}_{w} = \frac{S_{wf}x_2 + \int_{S_{wf}}^{1-S_{or}} x \mathrm{d}S_{w}}{x_2} \tag{9.40}$$

$$\bar{S}_{w} = S_{wf} + \frac{1}{x_2}\int_{S_{wf}}^{1-S_{or}} x \mathrm{d}S_{w} \tag{9.41}$$

Using Equation 9.35 and $x = W_i/A\phi(df_w/dS_w)$ from Buckley–Leverett theory, the previous equation becomes –

$$\bar{S}_{w} = S_{wf} + \frac{\frac{W_i}{A\phi}\int_{S_{wf}}^{1-S_{or}}\left[\frac{\mathrm{d}f_w}{\mathrm{d}S_w}\right]\mathrm{d}S_w}{\frac{W_i}{A\phi}\left[\frac{\mathrm{d}f_w}{\mathrm{d}S_w}\right]_{S_{wf}}} \tag{9.42}$$

$$\bar{S}_{w} - S_{wf} = \frac{\int_{S_{wf}}^{1-S_{or}} \mathrm{d}f_w}{\left[\frac{\mathrm{d}f_w}{\mathrm{d}S_w}\right]_{S_{wf}}} \tag{9.43}$$

$$\bar{S}_{w} - S_{wf} = \frac{(f_w)_{1-S_{or}} - (f_w)_{S_{wf}}}{\left[\frac{\mathrm{d}f_w}{\mathrm{d}S_w}\right]_{S_{wf}}} \tag{9.44}$$

However, $(f_w)_{1-S_{or}}$ becomes 1 because at this saturation condition only water is flowing, whereas $(f_w)_{S_{wf}}$ is the fractional flow at S_{wf} (f_{wf}) or the shock front, reducing Equation 9.44 to

$$\bar{S}_{w} - S_{wf} = \frac{1 - f_{wf}}{\left[\frac{\mathrm{d}f_w}{\mathrm{d}S_w}\right]_{S_{wf}}} \tag{9.45}$$

or

$$\left[\frac{\mathrm{d}f_w}{\mathrm{d}S_w}\right]_{S_{wf}} = \frac{1 - f_{wf}}{\bar{S}_w - S_{wf}} = \frac{1}{\bar{S}_w - S_{wi}} \tag{9.46}$$

The significance of Equation 9.46 can be graphically illustrated in Figure 9.15, which is basically the Welge construction for determining the water saturation and the fractional flow at the shock. To satisfy Equation 9.46, the tangent to the fractional flow curve, from the point $S_w = S_{wi}$; $f_w = 0$, must have a point of tangency with coordinates $S_w = S_{wf}$; $f_w = f_{wf}$, and the extrapolated tangent must intersect the line $f_w = 1$.

However at breakthrough, the shock front arrives at $x = L$ and the water saturation at the outlet equals S_{wf}. Therefore, in order to obtain the water saturation at

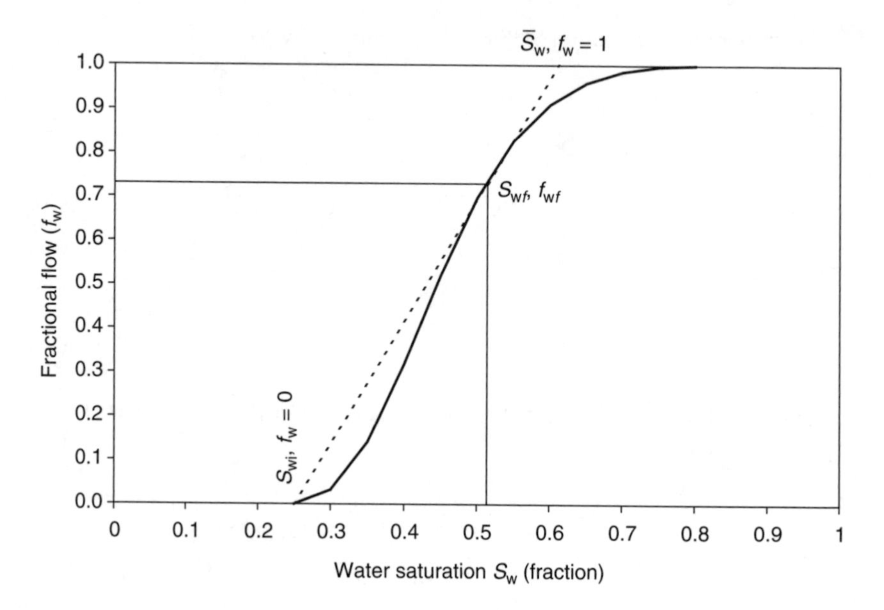

FIGURE 9.15 Tangent to the fractional flow curve, Welge's extension solution.

the outlet after breakthrough, tangents can be constructed to the fractional flow curve for water saturations greater than S_{wf}. The water saturation and the water fractional flow at each point of tangency correspond to the water saturation S_{wL} and the fractional flow of water f_{wL} at the outlet ($x = L$). The average water saturation in the medium is obtained by extrapolating each of these tangents to intersect the saturation scale at $f_w = 1$. This relationship can be mathematically expressed by an equation similar to Equation 9.46

$$\bar{S}_w = S_{wL} + \frac{1 - f_{wL}}{\left[\dfrac{df_w}{dS_w}\right]_{S_{wL}}} \tag{9.47}$$

It can be easily seen that Equation 9.47 can also be derived by integrating the saturation profile over the length of the porous medium

$$\bar{S}_w = \frac{1}{L}\int_0^L S_w dx \tag{9.48}$$

From Equation 9.33, at $x = L$

$$Q_{wi} = \frac{1}{\left(\dfrac{df_w}{dS_w}\right)_{S_{wL}}} \tag{9.49}$$

where Q_{wi} is the pore volumes of water injected, where $1 - f_{wL} = f_{oL}$ is the fractional flow of oil at the outlet end that can be obtained by differentiation of the cumulative volume of oil produced (Q_{op}) with respect to the water injected, that, $f_{oL} = dQ_{op}/dQ_{wi}$. Therefore, Equation 9.47 can be expressed in terms of the outlet water saturation as

$$S_{wL} = \bar{S}_w - Q_{wi}\frac{dQ_{op}}{dQ_{wi}} \tag{9.50}$$

When unsteady-state relative permeability tests are carried out, the experimental data recorded include, fluid production, and injection data on the basis of which the average water saturation is calculated

$$\bar{S}_w = S_{wi} + \frac{\text{cumulative volume of oil produced}}{\text{pore volume of the sample}} = S_{wi} + Q_{op} \tag{9.51}$$

that leads to the determination of water saturation at the outlet end of the porous medium from Equation 9.50.

For the Welge method, the relative permeability ratio can also be calculated on the basis of viscosity data and the value of f_{wL} or f_{oL} –

$$f_{wL} = -\frac{1}{1 + \left[\dfrac{k_{roL}\mu_w}{k_{rwL}\mu_o}\right]} \quad \text{or} \quad f_{oL} = \frac{1}{1 + \left[\dfrac{k_{rwL}\mu_o}{k_{roL}\mu_w}\right]} \tag{9.52}$$

where rearrangement gives

$$\frac{k_{rwL}}{k_{roL}} = \frac{\mu_w}{\mu_o}\frac{(1 - f_{oL})}{f_{oL}} \tag{9.53}$$

Thus, in summary the water saturation vs. relative permeability ratio plots can be constructed using the Welge approach when fluid production and injection data are available from unsteady-state displacement experiments.

9.4.8.1.3 Johnson–Bossler–Naumann method

The method developed by Johnson et al.[3] is probably the most commonly used data reduction method for determining the relative permeability relationships for each of the flowing phases. The application of JBN method requires data such as the pore volumes of fluids injected and produced, pressure drop across the sample, and fluid property data such as viscosities.

Johnson et al. basically adopted Welge's approach toward the determination of the average and the outlet water saturation (assuming water is the displacing phase) and Equations 9.50 and 9.51 are used. The major contribution of Johnson et al. was the methodology they proposed for the determination of individual-phase relative permeabilities. As described in Equation 9.54, the procedure begins with the integration of the pressure gradient across the core sample

$$\Delta P = \int_o^L \frac{\partial P}{\partial x}dx \tag{9.54}$$

The pressure gradient $\partial P/\partial x$ can also be expressed by Darcy's law as

$$\frac{\partial P}{\partial x} = -\frac{\mu_o u f_o}{k k_{ro}} \tag{9.55}$$

where k is the base permeability (absolute, or effective permeability to oil at initial conditions of S_{wi}), f_o the fractional flow of oil, k_{ro} the relative permeability to oil, μ_o the viscosity of oil, and u the average velocity of approach, Q/A.

Using the Buckley–Leverett theory

$$dx = dx_{sw} = LQ_{wi} df'_w = L\frac{df'_w}{f'_{wL}} \tag{9.56}$$

$$\Delta P = -\frac{\mu_o u L}{k f'_{wL}} \int_o^{f'_{wL}} \frac{f_o}{k_{ro}} df'_w \tag{9.57}$$

$$\int_o^{f'_{wL}} \frac{f_o}{k_{ro}} df'_w = -\frac{k f'_{wL} \Delta P}{\mu_o u L} \tag{9.58}$$

However, before the start of water injection when the core sample is at its initial saturation conditions, using Darcy's law,

$$\left(\frac{u}{\Delta P}\right)_i = -\frac{k k_{ro,max}}{\mu_o L} \tag{9.59}$$

$$\int_o^{f'_{wL}} \frac{f_o}{k_{ro}} df'_w = -\frac{f'_{wL}}{k_{ro,max}} \frac{\left(\frac{u}{\Delta P}\right)_i}{\left(\frac{u}{\Delta P}\right)} \tag{9.60}$$

where $k_{ro,max}$ for the initial conditions will be 1 if the base permeability is effective permeability to oil at S_{wi}.

Rapoport and Leas[6] defined a new quantity in terms of intake capacity $(u/\Delta P)$ called relative injectivity,denoted by I_r and mathematically expressed as

$$I_r = \frac{\left(\frac{u}{\Delta P}\right)}{\left(\frac{u}{\Delta P}\right)_i} \tag{9.61}$$

Therefore in Equation 9.61, $(u/\Delta P)$ may be considered intake capacity at any given displacement or flood stage, whereas $(u/\Delta P)_i$ represents intake capacity at the very initiation of the flood, at the moment when practically only oil is flowing through the system in the presence of immobile water.

By differentiating Equation 9.60 with respect to f'_{wL} and using the definition of I_r –

$$\frac{f_{oL}}{k_{roL}} = \frac{d(f'_{wL}/I_r)}{d(f'_{wL})} \tag{9.62}$$

With the use of Equation 9.49, a practical form of Equation 9.62 can be obtained for the determination of relative permeability to oil:

$$k_{roL} = f_{oL}\frac{d\left[\dfrac{1}{Q_{wi}}\right]}{d\left[\dfrac{1}{Q_{wi}I_r}\right]} \tag{9.63}$$

The relative permeability to water can be calculated from Equation 9.53 to

$$k_{rwL} = k_{roL}\frac{\mu_w(1-f_{oL})}{\mu_o f_{oL}} \tag{9.64}$$

In summary, for a given fluid production, injection, pressure drop and fluid property data, first the outlet water saturation is calculated from Equation 9.50, whereas k_{roL} and k_{rwL} are calculated from Equations 9.63 and 9.64 and are the relative permeabilities of the oil and water phase, respectively, at that particular outlet water saturation, S_{wL}. This calculation procedure is repeated for all the collected data, on the basis of which the entire saturation-relative permeability curve spanning the two end points of $S_w = S_{wi}$ and $S_w = 1 - S_{or}$, for a given system is established.

Another variant of the JBN method is a method proposed by Jones and Roszelle[7] (not discussed here) that also combines the Welge analysis and the differentiation of pressure drop and flow rate information for the determination of relative permeabilities.

9.4.8.2 Relative Permeabilities from the Alternate Method

The alternate method is fairly simple to apply because it primarily involves the calculation of average water saturation from Equation 9.51, whereas the corresponding effective and relative permeabilities are calculated as follows (shown for oil–water relative permeability):

Effective Permeabilities

$$k_{eo} = \frac{q_o\mu_o L}{A\Delta P} \text{ at } \bar{S}_w \tag{9.65}$$

$$k_{ew} = \frac{q_w\mu_w L}{A\Delta P} \text{ at } \bar{S}_w \tag{9.66}$$

The pressure drop, ΔP, is the corresponding pressure drop for that particular time instance at which the average water saturation is determined. The values of q_o and q_w represent the volumetric flow rates of oil and water, respectively; the former is calculated from the slope of the volume of oil produced vs. time curve at that particular time, whereas the latter is usually a fixed value at which water is injected into the core sample. In this manner, values of average water saturations and corresponding effective permeabilities of all the other steps are determined up to the terminal water saturation, the point at which the residual oil saturation and the end-point effective permeabilities are obtained (oil is 0 and water is maximum). Based on the calculated effective permeabilities, the relative permeabilities of the oil and water phase are also determined by using either the absolute permeability or the effective permeability to oil at S_{wi}, as base permeability.

9.4.9 Capillary End Effect

In both the steady-state and unsteady-state method flow tests for the determination of relative permeabilities; one of the major problems encountered is the so-called *capillary end effect*. The capillary end effect arises from the saturation discontinuity existing at the outlet face of the porous medium (core sample) when mounted for a flow test. The fluids flowing through the sample are discharged into a region that is void of the porous medium, resulting in all the fluids being at the same pressure, that is, capillary pressure is 0. However, immediately within the pore spaces of the rock at the outflow face, the capillary pressure is not 0 and capillary pressure conditions require that the saturation of the wetting phase approach 100%.

This phenomenon results in the establishment of a saturation gradient in the flow system. Specifically, the result of the strong variation in capillary pressure at the outlet, and the corresponding change in saturation is the capillary end effect. The other practical consequence of the capillary end effect is the retardation of wetting phase breakthrough in a displacement process, due to the excessive buildup of the wetting-phase saturation at the outflow face, a delay in the wetting phase breakthrough occurs that can introduce errors in the calculated relative permeabilities.

The theoretical saturation gradient for a linear system can also be developed on the basis of Darcy's law and the fundamental equation of capillary pressure is shown in the following equations.[8]

For the wetting and nonwetting phases

$$-\,\mathrm{d}P_{\mathrm{w}} = \frac{q_{\mathrm{w}}\mu_{\mathrm{w}}\mathrm{d}L}{k_{\mathrm{w}}A} \tag{9.67}$$

$$-\,\mathrm{d}P_{\mathrm{nw}} = \frac{q_{\mathrm{nw}}\mu_{\mathrm{nw}}\mathrm{d}L}{k_{\mathrm{nw}}A} \tag{9.68}$$

For the capillary pressure term

$$P_{\mathrm{c}} = P_{\mathrm{nw}} - P_{\mathrm{w}} \tag{9.69}$$

or

$$\mathrm{d}P_{\mathrm{c}} = \mathrm{d}P_{\mathrm{nw}} - \mathrm{d}P_{\mathrm{w}} \tag{9.70}$$

By combining Equations 9.67–9.70

$$\frac{dP_c}{dL} = \frac{1}{A}\left[\frac{q_w\mu_w}{k_w} - \frac{q_{nw}\mu_{nw}}{k_{nw}}\right] \quad (9.71)$$

where dP_c is the capillary pressure gradient within the core of length, L, A the cross-sectional area of the core, q_w and q_{nw} are the volumetric flow rates of the wetting and the nonwetting phases, respectively, μ_w and μ_{nw} the viscosities of the wetting and the nonwetting phases, respectively, and k_w and k_{nw} the permeabilities of the wetting and the nonwetting phases, respectively.

Additionally, considering that capillary pressure is a function of saturation, the saturation can also be expressed as a function of the core length

$$\frac{dP_c}{dL} = \frac{dP_c}{dS_w}\frac{dS_w}{dL} \quad (9.72)$$

where S_w is the wetting-phase saturation.

Equation 9.72 then becomes –

$$\frac{dS_w}{dL} = \frac{1}{A}\left[\frac{q_w\mu_w}{k_w} - \frac{q_{nw}\mu_{nw}}{k_{nw}}\right]\frac{1}{dP_c/dS_w} \quad (9.73)$$

where dS_w/dL is the change in wetting-phase saturation with length.

The saturation gradient dS_w/dL within a flow system can be determined by graphical integration of Equation 9.73, using capillary pressure and relative permeability data.

In order to determine the magnitude of the capillary end effect, Richardson et al.[8] studied the saturation gradients in a long core apparatus. The core sample was 30.7 cm long with a diameter of 6.85 cm and a porosity of 17.7%. The test apparatus was designed to determine the pressure in each of the flowing phases at different positions along the core. The gas–oil relative permeability relationships were determined for various flow rates and pressure gradients across the core. Additionally, the capillary pressure characteristics of the system were also measured.

Richardson et al. compared the experimentally measured saturation gradient with the theoretical saturation gradient (determined from Equation 9.73) as a function of the core length for a gas–oil system at two different (high and low) sets of gas (nonwetting phase) and oil (wetting phase) flow rates. A good correspondence between the experimentally and theoretically determined saturation gradients and the reduction of the end effect in the case of high flow rates was observed for data presented by Richardson et al. The capillary end effect can be qualitatively illustrated by the plot in Figure 9.16, which shows the wetting-phase saturation as a function of the distance from the outlet end of the core sample.

These capillary end effects may be overcome by using a high rate of flow and high pressure differential. Alternatively, the end effects can also be minimized by preparing each end of the sample with porous disks. The capillary end effect observed in laboratory flow tests for relative permeability measurements is purely a laboratory artifact and is nonexistent in the case of reservoir flow process.

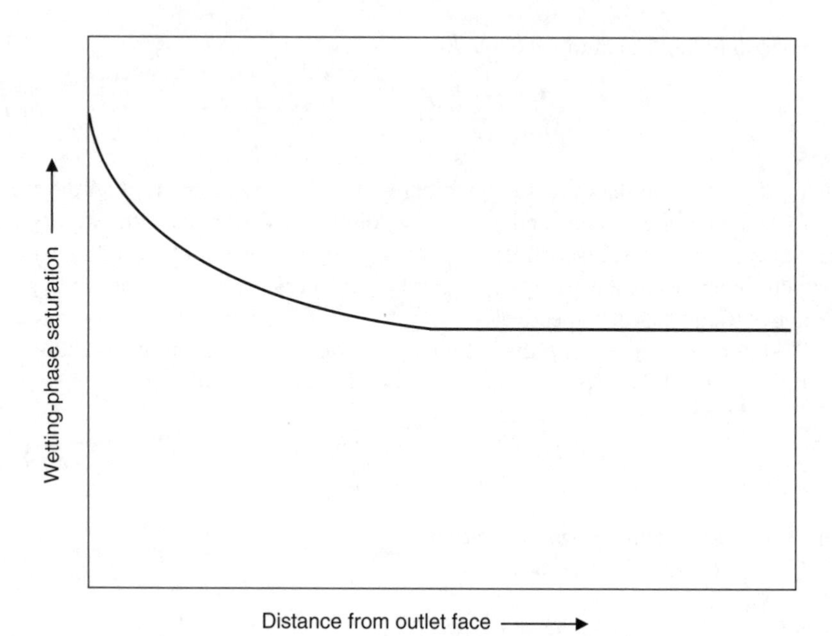

FIGURE 9.16 Schematic representation of capillary end effect.

9.5 DETERMINATION OF RELATIVE PERMEABILITY FROM CAPILLARY PRESSURE DATA

The determination of relative permeabilities from capillary pressure data is actually based on the relationship between absolute permeability and capillary pressure, developed by Purcell[9] (see Chapter 8 on capillary pressure) and expressed as

$$k = 10.24(\sigma \cos \theta)^2 \phi \lambda \int_{S=0}^{S=1} \frac{\mathrm{d}S}{P_c^2} \tag{9.74}$$

Equation 9.74 can be readily adapted to the computation of both the nonwetting phase and the wetting phase relative permeabilities defined as the ratio of the effective permeability at a given saturation to the intrinsic permeability of the medium. By generalizing Equation 9.74 and considering the capillary pressure data for displacement of the wetting phase, the following equations can be written for the wetting and nonwetting phase, respectively,

$$k_w = 10.24(\sigma \cos \theta)^2 \phi \lambda \int_{S=0}^{S=S_w} \frac{\mathrm{d}S}{P_c^2} \tag{9.75}$$

and

$$k_{nw} = 10.24(\sigma \cos \theta)^2 \phi \lambda \int_{S=S_w}^{S=1} \frac{\mathrm{d}S}{P_c^2} \tag{9.76}$$

where k_w and k_{nw} are the effective permeabilities to the wetting and nonwetting phases, respectively.

Since Equation 9.74 represents the absolute permeability calculated from the capillary pressure over the entire saturation range from 0 to 1, Equations 9.75 and 9.76 can be divided by Equation 9.74 so that the relative permeabilities of the wetting and nonwetting phases are defined as

$$k_{rw} = \frac{k_w}{k} = \frac{\int_{S=0}^{S=S_w} \frac{dS}{P_c^2}}{\int_{S=0}^{S=1} \frac{dS}{P_c^2}} \tag{9.77}$$

$$k_{rnw} = \frac{k_{nw}}{k} = \frac{\int_{S=S_w}^{S=1} \frac{dS}{P_c^2}}{\int_{S=0}^{S=1} \frac{dS}{P_c^2}} \tag{9.78}$$

where k_{rw} and k_{rnw} are the relative permeabilities of the wetting and nonwetting phases, respectively.

In defining Equations 9.77 and 9.78, the interfacial tension (σ), contact angle (θ), lithology factor (λ), and the porosity (ϕ) are assumed to be same for a given rock–fluid system. Fatt and Dykstra,[10] following the basic method of Purcell for calculating the absolute permeability from capillary pressure data, developed an expression for relative permeability by considering the lithology factor λ as a function of saturation. The lithology factor is essentially the correction factor for deviation of the path length from length of the porous medium. Fatt and Dykstra further assumed that the deviation of the path length was a function of the radius of the conducting pores so that

$$\lambda = \frac{a}{r^b} \tag{9.79}$$

where r is the radius of the pore, and a and b are constants for the material. Fatt and Dykstra assumed a value of $b = ½$, which reduces Equations 9.77 and 9.78 (after using Equation 9.79 and a value of $b = ½$) to

$$k_{rw} = \frac{\int_{S=0}^{S=S_w} \frac{dS}{P_c^3}}{\int_{S=0}^{S=1} \frac{dS}{P_c^3}} \tag{9.80}$$

$$k_{rnw} = \frac{\int_{S=S_w}^{S=1} \frac{dS}{P_c^3}}{\int_{S=0}^{S=1} \frac{dS}{P_c^3}} \tag{9.81}$$

In using Equations 9.77 and 9.78 or 9.80 and 9.81 for the determination of relative permeabilities from capillary pressure data, the denominators are evaluated over the entire saturation range resulting in one value, whereas the numerators are evaluated as follows. For example, if P_c data are available as a function of water saturation ranging from 45 to 100 %, then the numerator for wetting-phase relative permeability equation is evaluated for 0 to 45%, 0 to 50%, 0 to 75%, and 0 to 100%, whereas the numerator for nonwetting-phase relative permeability equation is evaluated from 45 to 100%, 50 to 100%, and 75 to 100%.

9.6 FACTORS AFFECTING RELATIVE PERMEABILITY MEASUREMENTS

The concept of relative permeability is simple; however, the measurement and interpretation of relative permeability vs. saturation curves is not. Even though relative permeability is strictly a function of fluid saturation, evidence exists that relative permeability may be a function of many more parameters than just fluid saturation. We know this mainly because fluid saturation or the distribution of fluid itself in a porous media is primarily a function of variables such as wettability, obviously resulting in relative permeability being affected by various factors. Bennion et al.[11] address steady-state relative permeability measurement of bitumen–water systems and provide the most comprehensive list of factors that affect relative permeability–saturation relationships. Relative permeability can be affected by many physical parameters including:

- Fluid saturations
- Saturation history (hysterisis effects)
- Magnitude of initial-phase saturations, especially the value of S_{wi}
- Wettability
- Effect of rock pore structure
- Overburden stress
- Clay and fines content
- Temperature
- Interfacial tension and viscosity
- Displacement rates

Refer to Bennion et al.[11] for the complete list of publications that discuss these various factors affecting relative permeability functions.

A quick glance at the various factors clearly indicates that most of these are connected to the type of laboratory procedures used, and more specifically, types of fluids and rock samples and the conditions used for performing laboratory relative permeability tests. Therefore, the general consensus among researchers is that in order to obtain the most representative relative permeability data, experimental conditions during the test must be duplicated as closely as possible to reservoir conditions. This involves the use of well-preserved or restored-state reservoir core material, the use of real reservoir fluids (brines and oils) in the tests, and the operations at complete reservoir conditions of pressure, temperature, and appropriate net confining stress.

Although a very detailed discussion of these various factors affecting relative permeability is not presented here, a review of most of these factors, based on material from the literature, is presented in this section. All these factors affect relative permeability in one way or another; however, the factor that has the most dominant effect on relative permeability relationships is wettability of the given fluid–rock system. Hence, considering the significance of wettability, its effect on relative permeability relationships is discussed in a more detailed manner than some of the other factors.

9.6.1 Effect of Fluid Saturation, History of Saturation, and Initial Water Saturation

Relative permeabilities are primarily the functions of fluid saturations, that is, they are directly proportional to fluid saturations. As saturation of a particular phase increases, its relative permeability increases (see Figures 9.1 and 9.2). However, a major difference affecting fluid saturation and relative permeability is the effect of saturation history on relative permeability. Like the capillary pressure–saturation relationship, the relative permeability–saturation relationship also exhibits hysterisis effect and relative permeability curves show hysterisis between drainage processes (wetting phase decreasing) and imbibition processes (wetting phase increasing). If the relative permeability data are obtained by increasing the saturation of the wetting phase, the process is termed *imbibition* or *resaturation*, whereas the process is termed *drainage* if the data are obtained by decreasing the saturation of the wetting phase.

It is generally agreed that pore spaces of reservoir rocks were originally filled with water, and subsequently, hydrocarbons moved into the reservoir displacing some of the water and reducing the water to some minimum saturation. When discovered, reservoir pore spaces are filled with connate water saturation and hydrocarbon saturation. If gas is injected in the reservoir and if the hydrocarbon phase is oil, then the process is essentially a drainage process as gas (nonwetting) displaces the oil (wetting).

The same history must be duplicated in the laboratory to eliminate the effects of hysterisis. The laboratory procedure is to first prepare the core sample by achieving the irreducible water saturation by a drainage process, oil displacing water. Subsequently, the gas injection is carried out to displace the oil, which is also a drainage process where the wetting phase saturation is continuously decreased.

For oil–water relative permeability measurements, the initial preparation of the core sample is carried out in exactly the same manner as that for the gas–oil relative permeability tests. However, the displacement of oil by water (assumed as the wetting phase) is essentially an imbibition process because now the wetting phase saturation is continuously increased by injecting water. This type of relative permeability data is intended for application to water drive or waterflooding calculations.

The differences between the drainage and imbibition processes of measuring the relative permeability data can be illustrated by the relative permeability curves shown in Figure 9.17. Note that the imbibition process causes the nonwetting phase (oil) to lose its mobility at higher values of its saturation than does the drainage process. However, the drainage process causes the wetting phase to lose its mobility at higher values of its saturation than the imbibition process.

A brief qualitative discussion of the effect that water saturation has on gas–oil relative permeability measurements was presented by Owens et al. in 1956.[12] Although no data were shown, the authors stated that the presence of water saturation has no

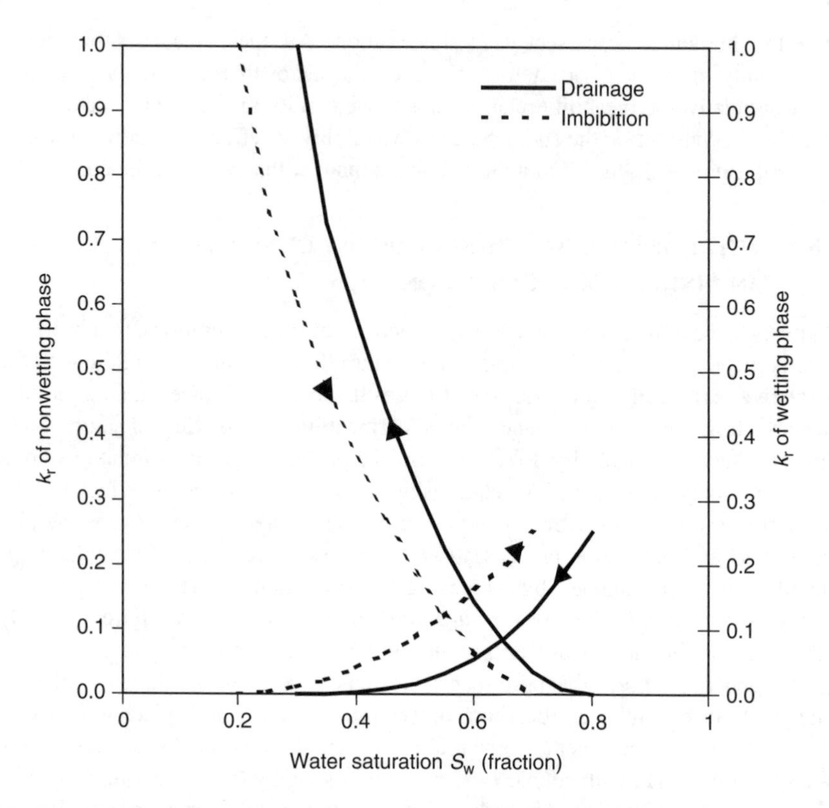

FIGURE 9.17 The differences between drainage and imbibition relative permeability curves.

effect on gas–oil relative permeability if the water is immobile. In 1990, Narahara et al.[13] reported their results on the study of the effect of connate water on gas–oil relative permeabilities for water-wet and mixed-wet Berea sandstone. The connate water saturation was varied from 0 to about 27% for both water-wet and mixed-wet core samples. The gas–oil relative permeability data were measured by the unsteady-state method and the centrifuge technique. Based on their study, Narahara et al. concluded that connate water had no effect on gas–oil primary drainage relative permeabilities for both wetting systems provided the water phase was immobile during the flow test, the gas–oil relative permeabilities were expressed as functions of total liquid saturation, and the effective permeability to oil at connate water was used as the base permeability for relative permeability. Tang and Firoozabadi[14] also studied the effect of connate water saturation on gas and oil relative permeabilities of untreated (water-wet) and treated (intermediate gas-wet) Berea sandstone cores. In the case of untreated samples, they observed a significant reduction in the gas relative permeability when connate water saturation increased from 0 to 11%. However, the opposite was observed for treated samples: gas relative permeability was unchanged and the oil relative permeability reduced substantially when the connate water saturation was increased from 0 to 7.5%. Based on their own results and discrepancies with observations of Narahara et al., Tang and Firoozabadi stated that the effect of initial water saturation on gas and oil relative permeability needed further investigation.

Caudle et al.[15] found that relative permeability curves measured on water-wet sandstone were dependent on the initial water saturation when addressing the effect of initial water saturation on oil–water relative permeabilities. Decreasing the initial water saturation changed the location and shape of the curves. Craig[16] has also stated that the initial water saturation strongly influences relative permeability curves in strongly water-wet rocks, however, S_{wi} has little effect on curves measured on oil-wet rocks as long as the value is less than 20%.

9.6.2 Effect of Wettability on Relative Permeability

Wettability affects relative permeability because it controls the location, flow, and relative distribution of fluids in the pore space. A 1987 review by Anderson[17] is the most comprehensive literature review dealing with the effect of wettability on relative permeability. Anderson[17] discusses the effect of wettability on relative permeability for all types of wettability systems: strongly wetted, uniformly wetted, fractional wetted, and mixed wettability systems.

In evaluating the effect of wettability on oil–water relative permeability data for strongly water-wet and oil-wet systems, Anderson stated that the differences in relative permeabilities measured in strongly water-wet and oil-wet systems are caused by the differences in fluid distributions. For example, in a strongly water-wet system, at S_{wi}, the water is located in the small pores where it has very little effect on the flow of oil and the oil effective permeability is relatively high often approaching the absolute permeability. In contrast, the effective water permeability at S_{or} is very low because some of the residual oil is trapped as ganglia in the center of the larger pores, where it is effective in lowering the water permeability. Therefore, water permeability at S_{or} is much less than the oil permeability at S_{wi}. However, a strongly oil-wet system reverses the positions of the two phases. The oil permeability at S_{wi} is relatively low because the residual water or connate water blocks the flow of oil, whereas the water permeability at S_{or} is high because the residual oil is located in the small pores and as a film on the surface where it has little effect on the flow of water. As a result, the ratio of the two permeabilities can approach 1 or even greater. Donaldson and Thomas,[18] Owens and Archer,[19] and Morrow et al.[20] have presented oil–water relative permeability data for different rock–fluid systems where the wettability of the system was varied from strongly water-wet to strongly oil-wet, such that relative permeability data were reported for contact angles of 0°, 180°, and at contact angle values between the two wettability extremes. All these data sets indicate that at any given saturation, water relative permeability increases while the oil relative permeability decreases as the wettability of the system shifts from strongly water-wet to strongly oil-wet.

Additionally, Craig[16] presented several rules of thumb indicating the differences in the relative permeability characteristics of strongly water-wet and strongly oil-wet cores, as shown in Table 9.1. Although the relative permeability data of many strongly water-wet and strongly oil-wet systems indicate agreement with Craig's rules of thumb (see Anderson[17] for a complete list of references); exceptions are to be considered due to the dependence of relative permeability on initial water saturation and pore geometry.

In addition to the strongly water-wet and strongly oil-wet systems, Anderson also reviewed the relative permeability data presented by Fatt and Klikoff[21] and

TABLE 9.1
Craig's[16] Rules of Thumb Relating Wettability and Relative Permeability

Characteristics	Water-Wet	Oil-Wet
Initial water saturation, S_{wi}	Greater than 20–25%	Generally less than 15%
S_w at which $k_{ro} = k_{rw}$	Greater than 50%	Less than 50%
k_{rw} at $1 - S_{or}$; based on $k_{eo}@S_{wi}$ as base permeability	Generally less than 30%	Greater than 50%

Richardson et al.[22] on fractional and mixed wettability systems, respectively. Fatt and Klikoff measured the relative permeability ratio (k_{rw}/k_{ro}) in fractionally wetted sandpacks that were formed by mixing treated and untreated sand grains together. Data were presented as the ratio of k_{rw}/k_{ro} as a function of water saturation for weight fraction of oilwet sand grains of 1 (strongly oil-wet), 0.75, 0.5, 0.25, and 0 (strongly water-wet). For a given water saturation, the ratio of k_{rw}/k_{ro} decreased as the weight fraction of oil-wet sand grains decreased from 1 to 0. Richardson et al. presented oil–water relative permeability data on native state East Texas Woodbine cores. These cores were later shown by Salathiel[23] to have mixed wettability in their native state. The oil–water relative permeability for these cores were presented for four different conditions: first in their native state (mixed wet), after which oil-flooding was carried out followed by waterflooding in additional cycles, and in a final step in which the cores were cleaned and dried and then the relative permeability ratio was measured by waterflooding. The behavior of the relative permeability ratio as the core was cleaned and rendered water-wet contrasts with the behavior observed for the uniformly wetted and fractionally wetted systems, that is, the relative permeability ratio at a given water saturation was higher for strongly water-wet system and the more oil-wet curves were to the right of the strongly water-wet curve.

In discussing the issues related to the effect of wettability on relative permeability, Anderson also highlighted the importance of using preserved or native-state core material in the determination of relative permeability data. His argument is based on the fact that original reservoir wettability is preserved only in the native-state samples or alternatively if cleaned core material is used, an attempt has been made to restore the original wettability by following certain procedures that are in practice and considered to restore original wetting conditions (e.g., see Chapter 7).

9.6.3 Effect of Rock Pore Structure

Morgan and Gordon[24] studied the influence of pore geometry on oil–water relative permeability data with a review primarily based on the relative permeability data of sandstones obtained from a commercial laboratory and their own analysis of photomicrographs of thin sections from the ends of the core plugs used for the relative permeability tests. In general, rocks with large pores tend to have low irreducible water saturations thus resulting in a relatively large amount of pore space available for the flow of fluids. This condition allows higher end-point permeabilities and a larger saturation change to occur during two-phase flow, that is, wider relative permeability curves spanning a broad water saturation range. On the other hand, rocks with small pores tend to have higher irreducible water saturations that leave little room for the flow of fluids. Consequently, the end-point permeabilities are of a lower magnitude and the saturation change is small during two-phase flow, that is, relative permeability functions are defined over a narrow water saturation range. Morgan and Gordon also stated that postdepositional alterations can form more than one type of reservoir rock from a single original rock type, and depending on the type of these alterations, pertinent changes in the relative permeability characteristics occur.

9.6.4 Effect of Overburden Stress (Confining Stress)

In routine and special core analysis, properties such as porosity, absolute permeability, and relative permeability are quite frequently measured on rock samples that are not under net overburden pressure. However, under actual reservoir conditions, the rocks experience a net overburden pressure equal to the gross overburden from the reservoir depth less the pressure of the fluids (or pore pressure) in the pores of the rock. If these measurements are carried out at 0 net overburden or at nonrepresentative values, systematic error will be introduced into reservoir engineering calculation such as well productivity, reserves, and simulation. Therefore, it is important to evaluate the effect of net overburden on properties such as relative permeability.

Although, the effect of overburden on absolute permeability has been studied by a number of investigators, relatively few attempts have been made on studying the effect of overburden on relative permeability. Ali et al.,[25] however, performed a systematic study of the effect of net overburden pressure on porosity and absolute and relative permeabilities. Their evaluation was based on dynamic displacement experiments on small consolidated core samples under net overburden pressures up to 6000 psi. The following points summarize the results obtained by Ali et al. regarding the effect of overburden on relative permeability.

Irreducible water saturation and residual oil saturation increase when the net over burden pressure is increased from 1000 to 6000 psi. The S_{wi} and S_{or} increase with increasing net overburden, however, the increase appears to be marginal, around 3% in both cases. Ali et al. attributed this observation to increased capillarity due to the pore space compressibility.

The oil and water end-point relative permeability shows a decrease as the net overburden is increased from 1000 to 6000 psi. However, this decrease also appears to be rather small.

Relative permeability data revealed a pronounced reduction in k_{ro} with increase in net overburden pressure compared to the negligible effect on k_{rw}. Ali et al. explained this phenomenon on the basis of sand grains coming closer together with increasing overburden causing a general shift in the pore throat diameter distribution toward smaller values. For a given value of S_w, this leads to redistribution of the wetting phase (water) to occupy more pore throats. While this does not cause any significant change in k_{rw}, it leads to more blockage of oil flow and hence reduces k_{ro}.

9.6.5 Effect of Clay Content and Movement of Fines

Amaefule et al.[26] conducted laboratory studies to elucidate the role of formation damage processes (clay swelling and fines movement) in the determination of relative permeability data. They observed anomalous trends in laboratory-derived oil–water relative permeability data in rock samples containing mobile fines and water-sensitive clays. Under such conditions, oil–water relative permeability data tend to exhibit nonmonotonic trends with saturation, slightly S-shaped water relative permeability with a "bend-over" at high water saturations and a rebound in water relative permeability at residual oil saturation with reversal in the flow direction.

The observed characteristics are, however, indicative of adverse physicochemical interactions between the flowing phases and rock. While high flow velocity of the displacing phase is preferred for overcoming the capillary end effect, this may have an added disadvantage if the flow velocity exceeds the critical velocity for mobilization of fines. On the other hand if the injected brine is incompatible with the clay or is not in ionic equilibrium with the rock, clay swelling occurs. Both flow velocity (higher than critical velocity for fines mobilization) and clay swelling can affect the saturation–relative permeability relationships.

9.6.6 Effect of Temperature

Akin et al.[27] addressed the effect of temperature on heavy oil–water relative permeabilities by presenting a detailed review of the various experimental investigations of temperature effects on relative permeability. The literature reviewed by Akin et al. included oil–water relative permeability measured in the temperature range from room conditions to as high as 500°F and for a variety of rock-fluid systems. While some of these studies do show a temperature effect on end-point saturations or oil and water relative permeabilities; generally, no conclusive agreement seems to be made on the effect of temperature on oil–water relative permeabilities. Akin et al. attributed the inconsistency or divergence of experimental data on temperature-relative permeability studies to factors such as errors in saturation measurements, errors caused by neglect of capillary end effects, wettability variations with differing oils and brines, assumptions made to relate experimental procedures and calculations, and the inadequacy of mathematical models used to represent multiphase flow conditions.

9.6.7 Effect of Interfacial Tension, Viscosity, and Flow Velocity

Many studies have focused on evaluating the effect of interfacial tension, viscosity, and flow velocity on relative permeabilities via a dimensionless number that relates viscous forces and capillary forces. This dimensionless number is called the *capillary number* denoted by N_c and is generally defined as

$$N_c = \frac{\mu v}{\sigma \phi} \tag{9.82}$$

where μ is the fluid viscosity, v the flow velocity, σ the interfacial tension, and ϕ the porosity in fraction.

The viscous forces are defined by the fluid viscosity, flow velocity, and flow path length. Capillary forces are defined through the surface or interfacial tension. Any consistent set of units can be used in Equation 9.82 to determine the value of the capillary number.

The discussion presented in this section with regard to the effect of capillary number or its constituents on relative permeability functions is primarily based on the data presented by Blom et al.,[28] Fulcher et al.,[29] and Lefebvre.[30] Most of these studies focused on the effect of capillary number and its constituents on relative permeability functions.

Blom et al.[28] studied the relative permeability at near-critical conditions (characterized by very low interfacial tension) using a glassbead pack as porous media and a fluid system of methanol (wetting phase) and normal hexane (nonwetting phase). They studied the effect of flow velocity, interfacial tension, and capillary number on relative permeabilities. The effect of flow velocity on relative permeabilities was evaluated by keeping constant IFT values at 0.29 and 0.06 mN/m, respectively. At an IFT of 0.29 mN/m, a slight enhancement in the relative permeability to the nonwetting phase and no change in the wetting-phase relative permeability is observed when flow velocity is increased by a factor of 3.5 from 14 and 49 m/d. However, at the low IFT value of 0.06 mN/m, the relative permeability to both phases is increased by increasing the flow velocity by a factor of 2.5 from 12 to 30 m/d.

The effect of IFT on relative permeability was evaluated by varying the IFT from 0.29 to 0.01 mN/m at a flow velocity of around 14m/d. The relative permeability curves showed a clear dependence on IFT. The relative permeability to the nonwetting phase increased gradually when the IFT decreased by a factor of 30. At 0.01 mN/m, the nonwetting-phase relative permeability approaches a unit-slope line for which nonwetting relative permeability is simply equal to the nonwetting-phase saturation, somewhat similar to the relative permeability behavior observed in the case of near-miscible conditions (i.e., X-shaped curves). The wetting-phase relative permeability is, however, not affected until the IFT value is decreased below 0.06 mN/m.

Blom et al. also evaluated the effect of capillary number on relative permeability data. They, however, used a different definition of capillary number: the viscous gradient in the numerator was replaced by a product of absolute permeability and pressure difference across the sample. Their calculations revealed that relative permeability is low and displays a pronounced curve at low values of the capillary number; whereas it increases and straightens out at a higher capillary number, that is, lower capillary activity at high value of N_c, resulting in X-shape curves typical of near-miscible conditions.

Fulcher et al.[29] evaluated the effects of the capillary number (as defined by Equation 9.82) and its constituents on a series of relative permeabilities determined by the steady-state technique for Berea sandstone employing 2% calcium chloride as the aqueous phase and a synthetic oil as the oleic phase. The IFT and viscosity effects were studied by using isopropyl alcohol and glycerin, respectively. The initial variable altered in this study was the flow velocity varied from 4.9 m/d (minimum rate to avoid capillary end effects) to 24.4 m/d, for which little or insignificant change occurred in the relative permeability curves. The IFT effects were evaluated by maintaining a constant wetting-phase (water) viscosity and a constant flow velocity of 12 m/d while varying the IFT from 37.9 to 0.0389 mN/m. Similar to the results of Blom et al., a significant increase in the relative permeabilities was observed at the lowest attainable IFT value of 0.0389 mN/m, at which the two curves started to approach linearity, a behavior akin to the near-miscible condition of X-shaped curves.

The wetting-phase viscosity effects on relative permeability were studied by maintaining a constant IFT of around 30 mN/m and a constant flow velocity of 12 m/d, while varying the viscosity from approximately 1 to 1000 cP. As the wetting-phase viscosity increased its relative permeability values also increased and tended toward linearity. However, nonwetting-phase values decreased in approximately the same order of magnitude as wetting-phase values increased.

In order to evaluate the effect of capillary number on relative permeabilities; Fulcher et al. plotted capillary numbers as a function of the oil and water relative permeabilities for an iso-saturation of $S_w = 50\%$. As the capillary number increased, the ability of the wetting phase to flow (k_{rw}) also increased. However, for the nonwetting phase, the trend appeared to be unclear. Fulcher et al. concluded that both relative permeabilities were found to be functions of IFT and the viscosity variables individually rather than the capillary number.

Lefebvre[30] also performed a systematic study to mainly evaluate the effect of capillary number on oil–water relative permeability characteristics. However, the sample he used were sintered artificial-porous materials such as Teflon, stainless steel, and alumina, with fluid systems comprising of a wide variety of liquids primarily used to impart particular wetting, interfacial, or viscous properties for the displacement experiments. The capillary number used by Lefebvre is, however, the inverse of what has been defined in Equation 9.82, (IFT in the numerator and the denominator is the product of flow velocity and the fluid viscosity). For the Teflon (oil-wet) core, the oil–water relative permeability data were measured at three different capillary numbers ranging from about 10^3 to 10^6. These capillary numbers were obtained by varying all three constituents of the capillary number: flow velocity, IFT, and the displacing fluid viscosity. Also, the displacing fluid and the displaced fluid viscosity were approximately the same. The data for Teflon clearly show the influence of capillary numbers on both the wetting- and nonwetting-phase relative permeabilities. An appreciable effect of capillary number on relative permeability seems to exist; both the wetting- and nonwetting-phase relative permeabilities significantly increase with decreasing capillary number (or increasing capillary number as per Equation 9.82). Based on the results obtained, Lefebvre stated that relative permeability measurements must be made under conditions similar to those found in a reservoir, especially with respect to fluid properties such as interfacial tension and viscosity.

9.7 PECULIARITIES OF RELATIVE PERMEABILITY DATA

In the unsteady-state method, a displacing phase such as gas or water is injected into a core sample containing the irreducible water saturation (the balance being oil), at a steady rate and the relative permeability is calculated from the pressure drop and produced fluids using the JBN method. However, in some instances it is not possible to obtain a complete relative permeability curve because of the so-called *piston-like displacement*. In a piston-like displacement, the period of simultaneous two-phase production is completely absent. In other words, the oil production completely ceases after the displacing fluid breakthrough occurs. The concept of piston-like displacement is illustrated by the oil and water production profile for a waterflood carried out in a North Sea chalk core sample[31] shown in Figure 9.18.

However, the very basis of methods such as the JBN is the calculation of relative permeability data from the simultaneous two-phase (displaced and the displacing phase) production data, the absence of which due to piston-like displacement allows the calculation of only end-point relative permeabilities at S_{wi} and S_{or}, respectively (see Figure 9.19). If low viscosity oils are used in a water-wet core, the likelihood of obtaining a piston-like displacement is much greater. Therefore, viscous oils are normally used in a water-wet core to prolong the period of two-phase production because the flow before breakthrough gives no information about the relative permeability.

Mohanty and Miller[32] attributed the piston-like displacement to the capillary end effect. Before breakthrough, water saturation becomes greater than S_{wi} at the core outlet, but no water is produced as long as $P_c > 0$ (excessive build-up of wetting-phase saturation). When the water saturation is sufficiently high for $P_c = 0$, water production begins while oil stops flowing and oil saturation reaches its residual value, S_{or}. As a result the flood front inside the core is dispersed, but the effluent profile has the appearance of a piston-like displacement. The JBN method is then rendered ineffective to determine the relative permeability curves because most of the flood front is disguised by the end effect. Archer and Wong[33] actually proposed the use of a reservoir simulator to interpret relative permeability characteristics from laboratory waterflood history for the cases of piston-like displacement where the JBN method is inapplicable.

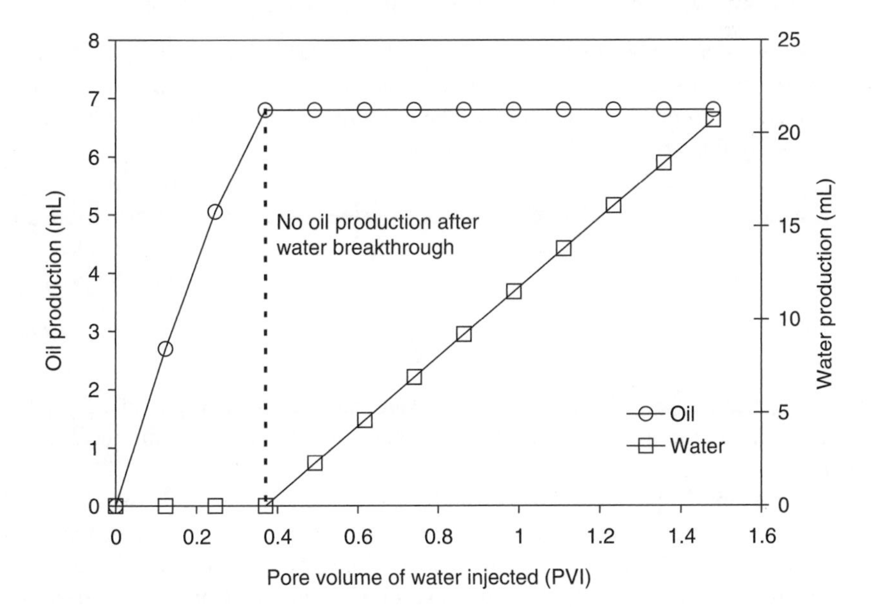

FIGURE 9.18 Oil and water production profile from a waterflood showing that oil production ceases after water breakthrough, indicating a piston-like displacement.

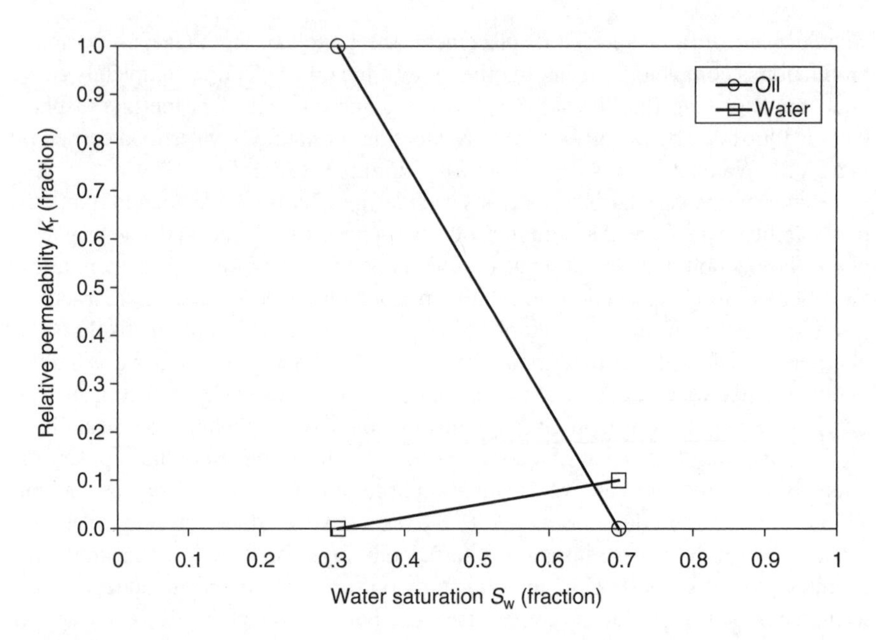

FIGURE 9.19 Oil and water relative permeabilities for a piston-like displacement, resulting in only the end-point values.

9.8 ASSESSING THE VALIDITY OF RELATIVE PERMEABILITY DATA AND DETERMINATION OF COREY EXPONENTS

Work done by a number of researchers has shown that valid relative permeability data often yield a straight line on a log–log plot when the relative permeability data are plotted vs. normalized saturations. The normalized saturations for an oil–water system are defined by the following equations:

$$S_{on} = \frac{1 - S_w - S_{or}}{1 - S_{wi} - S_{or}} \tag{9.83}$$

$$S_{wn} = \frac{S_w - S_{wi}}{1 - S_{wi} - S_{or}} \tag{9.84}$$

where S_{on} and S_{wn} are normalized oil and water saturations, respectively, S_w is the given water saturation, S_{wi} the irreducible water saturation, and S_{or} the residual oil saturation.

As an example, plots of S_{on} vs. k_{ro} and S_{wn} vs. k_{rw} are shown in Figures 9.20 and 9.21, respectively. As seen in these figures, both plots result in a straight line with slopes of 2. These slopes are referred to as the Corey[34] exponents for oil and water, and are denoted by N_o and N_w. Although the original work by Corey was carried out on gas–oil systems, the concepts have been found to apply also to oil–water systems.

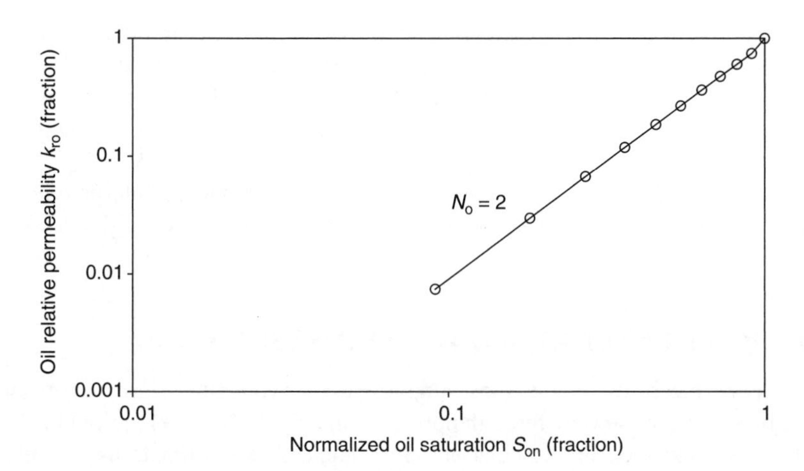

FIGURE 9.20 Log–log plot of normalized oil saturation, S_{so}, vs. oil relative permeability, k_{ro}, resulting in a straight line with Corey exponent $N_o = 2$.

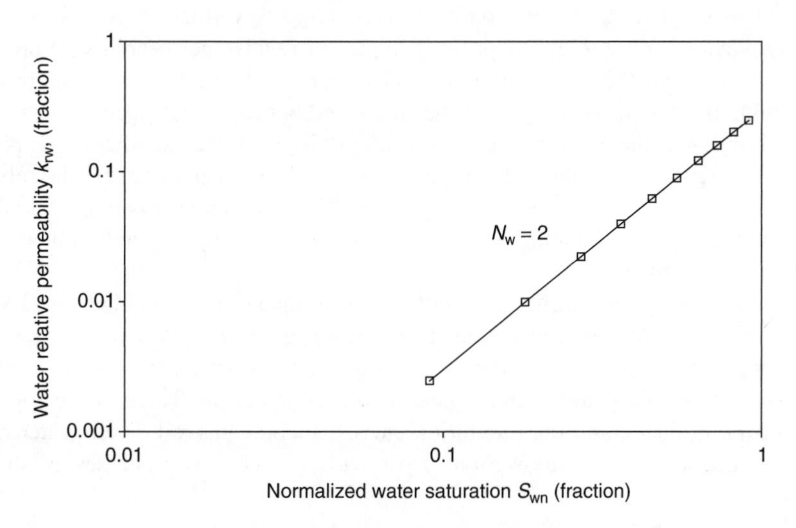

FIGURE 9.21 Log–log plot of normalized water saturation, S_{wn}, vs. water relative permeability, k_{rw}, resulting in a straight line with Corey exponent $N_w = 2$.

The behavior seen in Figures 9.20 and 9.21 therefore allows expression of the oil and water relative permeability data according to the following equations:

$$k_{ro} = k_{ro}\text{ (end point at } S_{wi}\text{) } [S_{on}]^{N_o} = k_{ro}\text{ (end point at } S_{wi}\text{)}\left[\frac{1 - S_w - S_{or}}{1 - S_{wi} - S_{or}}\right]^{N_o} \quad (9.85)$$

$$k_{rw} = k_{rw}\text{ (end point at } S_{or}\text{) } [S_{wn}]^{N_w} = k_{rw}\text{ (end point at } S_{or}\text{)}\left[\frac{S_w - S_{wi}}{1 - S_{wi} - S_{or}}\right]^{N_w} \quad (9.86)$$

If the base permeability used is the effective permeability to oil at S_{wi}, then k_{ro} at S_{wi} is unity and Equation 9.85 reduces to a simple expression relating the oil relative permeability and the normalized oil saturation raised to the power of Corey exponent for oil. The consistency of Equations 9.85 and 9.86 can be readily realized; at $S_w = S_{wi}$, $k_{ro} = k_{ro}@S_{wi}$ and $k_{rw} = 0$; at $S_w = 1 - S_{or}$, $k_{ro} = 0$ and $k_{rw} = k_{rw}@S_{or}$. Equations 9.85 and 9.86 are also valuable in interpolating and extrapolating the relative permeability curves and also in assessing the validity of the laboratory data.

9.9 SIGNIFICANCE OF RELATIVE PERMEABILITY DATA

Relative permeability is the most important reservoir fluid-flow parameter in describing multiphase flow through porous media. Such data are required in almost all the flow and recovery calculations. In a nutshell, the ultimate use of relative permeability models is to help design, optimize, and analyze oil displacement processes. Although relative permeability is just one part of the overall hydrocarbon recovery scenario, reservoir characterization, gravitational effects, hydrocarbon fluid phase behavior, and mass transfer processes all interact to determine the amount of oil that can be economically recovered. Nevertheless, relative permeability plays a central role. The primary impact of relative permeability on process design is through the fluid mobilities and fractional flows. Total fluid mobilities determine the resistance to flow of the fluids and hence impact injectivity and the overall timing of the process and the severity of viscous fingering (displacing phase such as water fingering through a high viscosity oil) or channeling and the robustness of a process to heterogeneities in general. The fractional flows impact producing water–oil ratio, producing gas–oil ratio, breakthrough timing, and ultimate and incremental recoveries.

For relative permeability data, either the drainage or the imbibition relative permeability curves are considered. In the drainage process a nonwetting phase such as gas displaces oil or water in a porous medium, whereas in the imbibition process the wetting phase such as water displaces gas or oil. The specific application of the drainage and the imbibition curves, such as gas–oil or oil–water relative permeabilities, for the various hydrocarbon recovery processes is shown below.

Drainage curves are important for the following:

1. Solution gas drive (primary recovery, oil generally wetting relative to gas)
2. Gravity drainage (gas displaces drained oil)
3. Gas injection processes

Imbibition curves are important for the following:

1. Waterflood calculations
2. Water influx (encroachment of water from an aquifer)
3. Oil displacing gas (oil moving into a gas cap)

9.9.1 Example of Practical Application of Relative Permeability Data

The mathematical formulation for multiphase flow in petroleum reservoirs consists of the fluid flow equations that are written for either all the individual fluid components or the fluid phases in the reservoir. These flow equations are obtained by a combination of the principle of mass conservation, Darcy's law, and an equation of state (to represent the fluid-phase behavior). It is in these reservoir fluid flow equations that relative permeability directly enters through Darcy's law. In order to demonstrate how relative permeability data are used in these flow equations, the development of basic fluid flow equations for a two-phase oil–water flow model is described in this section.

Consider a cubic element of a porous medium (shown in Figure 9.22) having porosity ϕ and bulk volume $\Delta x \Delta y \Delta z$, through which flow of oil and water is taking place in all three directions: x, y, and z. For example, the oil phase enters the cubic element at Darcy velocities of V_{ox}, V_{oy}, and V_{oz}, and exits at velocities $V_{ox+\Delta x}$, $V_{oy+\Delta y}$ and $V_{oz+\Delta z}$, respectively. The mass conservation equation applied to the oil phase is written as:

$$\begin{bmatrix}\text{Mass of oil entering}\\ \text{the cubic element in}\\ \text{time increment } \Delta t\end{bmatrix} - \begin{bmatrix}\text{Mass of oil exiting}\\ \text{the cubic element in}\\ \text{time increment } \Delta t\end{bmatrix} = \begin{bmatrix}\text{Mass of oil accumulated in}\\ \text{the cubic element in time}\\ \text{increment } \Delta t\end{bmatrix} \tag{9.87}$$

or

$$[\rho_o V_{ox} \Delta y \Delta z \Delta t + \rho_o V_{oy} \Delta x \Delta z \Delta t + \rho_o V_{oz} \Delta x \Delta y \Delta t] - [\rho_o V_{ox+\Delta x} \Delta y \Delta z \Delta t + \rho_o V_{oy+\Delta y} \Delta x \Delta z \Delta t + \rho_o V_{oz+\Delta z} \Delta x \Delta y \Delta t] = [\rho_o(\phi \Delta x \Delta y \Delta z) S_o]_{t+\Delta t} - [\rho_o(\phi \Delta x \Delta y \Delta z) S_o]_t \tag{9.88}$$

where ρ_o is the density of oil and S_o the oil saturation.

The division of Equation 9.88 by a product of $\Delta x \Delta y \Delta z \Delta t$ gives

$$\left[\frac{\rho_o V_{ox} - \rho_o V_{ox+\Delta x}}{\Delta x}\right] + \left[\frac{\rho_o V_{ox} - \rho_o V_{oy+\Delta y}}{\Delta y}\right] + \left[\frac{\rho_o V_{oz} - \rho_o V_{oz+\Delta z}}{\Delta z}\right] = \left[\frac{(\rho_o \phi S_o)_{t+\Delta t} - (\rho_o \phi S_o)_t}{\Delta t}\right] \tag{9.89}$$

From first principles,

$$\frac{\partial f}{\partial x}(x) = \lim_{\Delta x \to o} \frac{f(x + \Delta x) - f(x)}{\Delta x} \tag{9.90}$$

Therefore, taking limits, Δx, Δy, Δz, and $\Delta t \to 0$, yields

$$-\left[\frac{\partial \rho_o V_{ox}}{\partial x} + \frac{\partial \rho_o V_{oy}}{\partial y} + \frac{\partial \rho_o V_{oz}}{\partial z}\right] = \frac{\partial}{\partial t}(\rho_o \phi S_o) \tag{9.91}$$

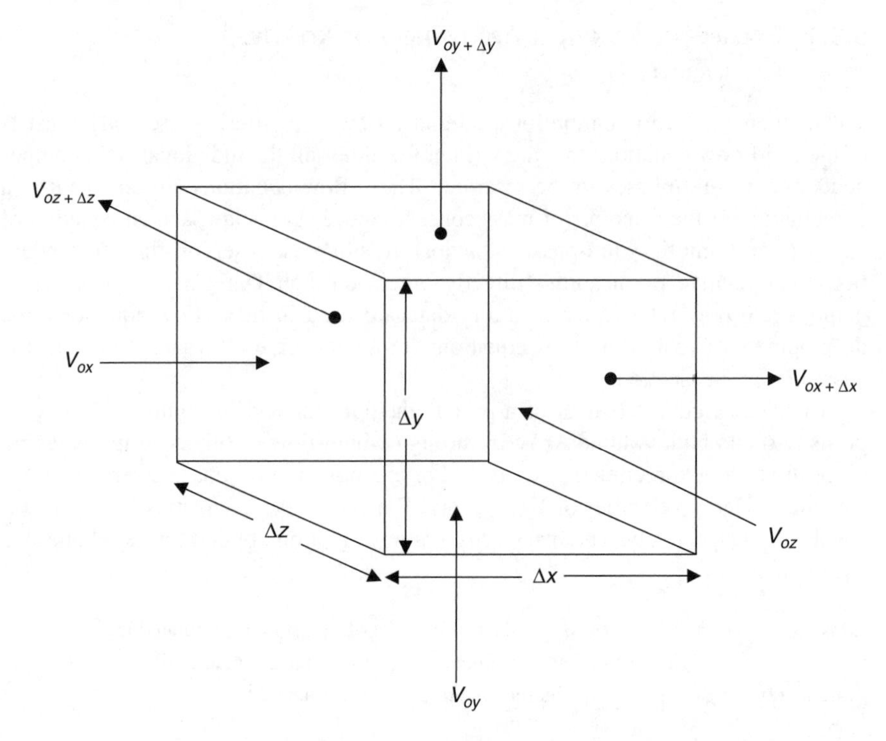

FIGURE 9.22 Schematic representation of flow of oil through a cubic element of a porous medium of porosity ϕ and bulk volume $\Delta x \Delta y \Delta z$ used in the derivation shown in Section 9.9.1.

In Equation 9.91, V_{ox}, V_{oy}, and V_{oz} are the Darcy velocities expressed as

$$V_{ox} = -\frac{kk_{ro}}{\mu_o}\frac{\partial P_o}{\partial x} \tag{9.92}$$

$$V_{oy} = -\frac{kk_{ro}}{\mu_o}\frac{\partial P_o}{\partial y} \tag{9.93}$$

$$V_{oz} = -\frac{kk_{ro}}{\mu_o}\frac{\partial P_o}{\partial z} \tag{9.94}$$

where k and k_{ro} [$f(S_w)$] are the base permeability and the relative permeability of oil, respectively, μ_o the viscosity of oil, and ∂P_o the pressure gradient in the oil.

Similar equations can also be developed for the water phase and also extended for the three-phase flow of gas, oil, and water. This derivation is in fact similar to the Buckley–Leverett theory applied for simultaneous flow of oil and water taking place in all three directions.

The other equations that are also considered include the saturation and the capillary pressure equations

$$S_o + S_w = 1 \tag{9.95}$$

and

$$P_{cow} = P_o - P_w = f(S_w) \tag{9.96}$$

Equations 9.91–9.96 demonstrate the use of not only relative permeability data but also capillary pressure as function of saturations and are in fact the backbone of a reservoir simulator. These equations are normally solved for obtaining the solution of pressure and saturation as a function of position and time.

9.10 THREE-PHASE RELATIVE PERMEABILITY

The flow of the three phases of gas, oil, and water occurs in a variety of circumstances in petroleum reservoirs. Specifically, it occurs in processes such as primary production below bubble point pressure in reservoirs with water drive, in gas or water alternating gas (WAG) injection, and in thermal oil recovery. The recovery processes such as WAG, frequently require consideration of coexisting gas, oil, and water phases and the impact of saturation cycles as water and gas slugs move through the reservoir. In the case of immiscible hydrocarbon gas WAG flooding, these considerations include the displacement of oil by simultaneous gas–water flow. Therefore, to understand the simultaneous fluid movement of the three phases, estimates of three-phase relative permeabilities are necessary.

While two-phase relative permeabilities such as gas–oil, gas–water, and oil–water are often time consuming to obtain, only two principal displacement paths exist: the saturation of one phase may either increase or decrease. However, in the experimental determination of three-phase relative permeability, in addition to the measurement of saturations, pressure drops, and fluxes in three flowing phases, an infinite number of different displacement paths exist because any three-phase displacement involves the variation of two independent saturations. Therefore, the measurement of three-phase relative permeability poses a particular challenge.

Considering the complexity involved in the direct experimental determination of three-phase relative permeability data, methods that allow these data to be calculated from the relatively easily measured two-phase gas–oil and oil–water data must be used. This is normally accomplished by applying empirical models to predict the three-phase relative permeability, k_{ro} as a function of the oil relative permeability in the presence of water only, k_{row}, and the oil relative permeability in the presence of gas (and normally in the presence of irreducible or connate water), k_{rog}.

9.10.1 REPRESENTATION OF THREE-PHASE RELATIVE PERMEABILITY DATA

Three-phase relative permeability data are usually plotted on ternary diagrams (see Figure 9.23) to illustrate the changes in the relative permeability values when three phases are flowing simultaneously. It should, however, be noted that Figure 9.23 merely shows how three-phase relative permeability data are presented and does not represent any specific three-phase relative permeability values. The three corners of the ternary diagram represent 100% gas, 100% oil, and 100% water saturations, whereas opposite ends (sides of the triangle) of these corners represent 0% saturation

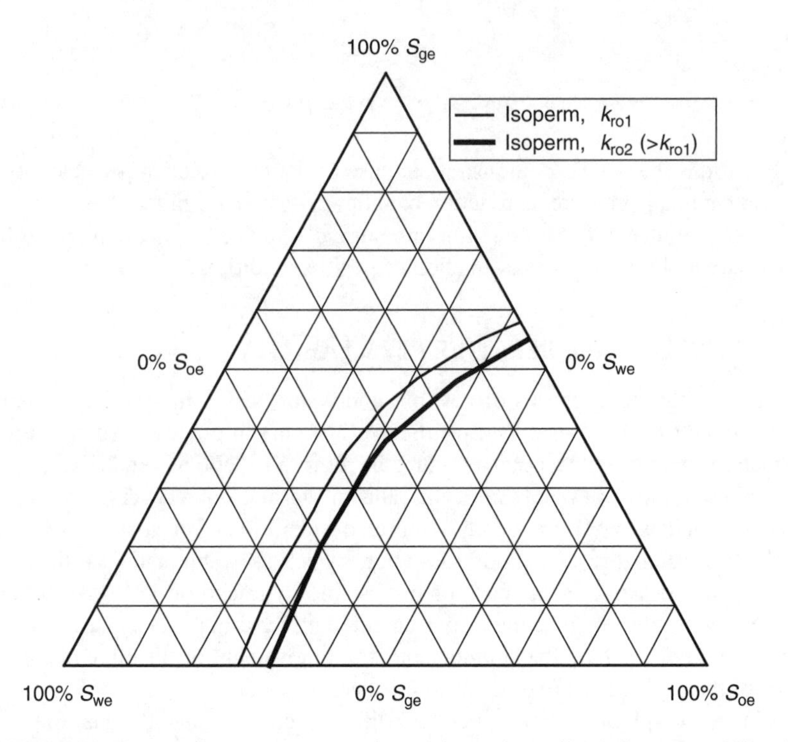

FIGURE 9.23 Ternary diagram representation of three-phase oil relative permeability (k_{ro}). The two k_{ro} curves shown in the diagram do not represent any particular three-phase relative permeability data but are given to basically illustrate the concept of three-phase relative permeability.

of that particular phase. In Figure 9.23, the three-phase oil relative permeability data are shown as curves of constant percentage relative permeability or isoperms showing dependence on the saturation values of all the three phases in the porous medium. Therefore, data inside the triangle represent the three-phase relative permeability values as a function of the three-phase saturations, whereas values lying on the sides of the triangle represent the two-phase saturation relative permeability data, that is, gas–oil and oil–water.

In a three-phase system, it is found that the relative permeability of water (assuming that water is the wetting phase) depends only on the saturation of water since water can only flow through the smallest interconnected pores that are present in the rock. Similarly, gas (least wetting phase) relative permeability depends only on the saturation of gas, or in other words, gas relative permeability should be dependent on the total liquid saturation and independent of how much of that total is composed of either phases (oil or water). However, the dependence of oil relative permeability on the saturations of the other phases can be established by the following reasoning. The oil phase has a greater tendency than the gas phase to wet the solid. Additionally, the IFT between oil and water is less than the ST between gas and water. Oil occupies pores of the rock that are dimensionally between those occupied by the water and the gas. Therefore, in general, the relative permeability of each phase (gas, oil, and water) in a three-phase system is essentially represented by the following functions of saturations:

$$k_{rg} = f(S_g) \tag{9.97}$$

$$k_{ro} = f(S_g, S_w) \tag{9.98}$$

$$k_{rw} = f(S_w) \tag{9.99}$$

For illustration purposes, as seen in Figure 9.23, for an oil saturation of 30% and gas and water saturations of 46 and 24%, respectively, the relative permeability to oil is given by k_{ro1}, whereas for the same oil saturation of 30% and gas and water saturations of 30 and 40%, respectively, it is noted that the relative permeability to oil is given by k_{ro2}, that is higher than k_{ro1}. These observations indicate that by changing the gas and water saturations, the flow characteristics of the oil are changed.

9.10.2 Empirical Models for Three-Phase Relative Permeability

The practical approach employed by empirical models for estimating the three-phase relative permeability data is based on the two independent sets of two-phase relative permeability data.

For a gas–oil system,

$$k_{rog} = f(S_g) \tag{9.100}$$

$$k_{rg} = f(S_g) \tag{9.101}$$

For an oil–water system,

$$k_{row} = f(S_w) \tag{9.102}$$

$$k_{rw} = f(S_w) \tag{9.103}$$

where k_{rog} and k_{row} are defined as the relative permeability to oil in the gas–oil and the oil–water system, respectively, whereas the symbol k_{ro} is reserved for the oil relative permeability in the three-phase system.

Stone[35,36] proposed two empirical models of three-phase relative permeability that are widely used in the petroleum industry. These models have become a benchmark against which experimental measurements, and other models, are compared. The Stone I model is described in the following text.

In 1970, Stone[35] developed a probability model to estimate the three-phase relative permeability data from laboratory-measured two-phase data. In the Stone I model, the underlying assumption is that the gas-phase relative permeability is the same function of gas saturation in the three-phase system as in the two-phase gas–oil system.

Similarly, the water-phase relative permeability is the same function of water saturation in the three-phase system as in the two-phase oil–water system. Stone suggested that a nonzero residual oil saturation, called the *minimum oil saturation*, S_{om} exists when oil is displaced simultaneously by gas and water. This minimum oil saturation is, however, different than the residual oil saturation S_{org} and S_{orw} in the gas–oil and the oil–water system, respectively. Stone introduced the following normalized saturations for the three-phase system

$$S_{ge} = \frac{S_g}{1 - S_{wi} - S_{om}} \tag{9.104}$$

$$S_{oe} = \frac{S_o - S_{om}}{1 - S_{wi} - S_{om}} \tag{9.105}$$

$$S_{we} = \frac{S_w - S_{wi}}{1 - S_{wi} - S_{om}} \tag{9.106}$$

Equations 9.104 to 9.106 are consistent because their summation also yields 1. The oil relative permeability in a three-phase system is defined as

$$k_{ro} = S_{oe}\beta_g\beta_w \tag{9.107}$$

where β_g and β_w are determined from

$$\beta_g = \frac{k_{rog}}{1 - S_{ge}} \tag{9.108}$$

$$\beta_w = \frac{k_{row}}{1 - S_{we}} \tag{9.109}$$

where S_{om} is minimum oil saturation, k_{rog} the oil relative permeability as determined from the gas–oil two-phase relative permeability at S_g, and k_{row} the oil relative permeability as determined from the oil–water two-phase relative permeability at S_w.

Considering the difficulty in selecting the minimum oil saturation, Fayers and Mathew[37] suggested an expression for determining S_{om}:

$$S_{om} = \alpha S_{orw} + (1 - \alpha)S_{org} \tag{9.110}$$

where

$$\alpha = 1 - \frac{S_g}{1 - S_{wi} - S_{org}} \tag{9.111}$$

and S_{org} and S_{orw} are the residual oil saturations in the gas–oil and oil–water relative permeability systems, respectively.

However, the version of the Stone I model normally used in reservoir simulators was proposed by Aziz and Settari[38] and includes a normalization to ensure that the model reduces smoothly to the two-phase data when $S_g = 0$ or $S_w = S_{wi}$:

$$k_{ro} = \frac{S_{oe}}{(1 - S_{ge})(1 - S_{we})}\left[\frac{k_{rog}k_{row}}{k_{ro}@S_{wi}}\right] \tag{9.112}$$

In Equation 9.112, $k_{ro}@S_{wi}$ is the oil relative permeability for an oil–water displacement measured at irreducible water saturation with no gas present.

PROBLEMS

9.1 Water injection is carried out in a 50 ft thick, 300 ft wide and 400 ft long gridblock (much like a rectangular core; called gridblock in reservoir simulation terminology), which has a porosity of 33% and absolute permeability of 500 mD. The injection of water is carried out at a residual oil saturation of 20%. The pressure drop across the gridblock is 5 atm. The relative permeability of water (k_{rw}) at residual oil saturation of 20% or water saturation of 80% is 0.67. Water viscosity is 1.0 cP. Calculate the water flowrate in barrels/day.

9.2 A preserved core plug of 34.63% porosity and a bulk volume of 51.05 cm^3 was used to carry out gas–oil and water–oil displacement experiments for determination of relative permeability. Due to the preserved nature of the plug, the initial saturation of oil and water in the plug (prior to carrying out any tests) were unknown. Moreover, considering the heterogeneity of the core plug, saturations measured on the plug trim were considered to be unreliable. The testing program on the core plug was carried out in the following manner and sequence:
First, a gasflood was carried out that resulted in an oil production of 6.9 cm^3 and water production of 1.2 cm^3.
Second, the core plug, after completion of the gas flood, was resaturated with crude oil to replace the gas. The plug also took additional oil in place of the produced water. Third, the core plug was subjected to a waterflood that resulted in an oil production of 6.8 cm^3. Finally, the Dean–Stark extraction was performed on the core plug right after the termination of the waterflood. This resulted in the saturations as $S_o = 40\%$ and $S_w = 60\%$. Determine the initial oil and water saturations that existed in the preserved core plug prior to carrying out any of the displacement tests.

9.3 An oil-water steady-state displacement experiment was carried out on a 5.0 cm long and 3.0 cm diameter sandstone core plug. The porosity of the plug is 25% and the grain density is 2.65 g/cm^3. Oil and water densities are 0.85 g/cm^3 and 1.05 g/cm^3, while viscosities are 2.0 cP and 1.0 cP respectively. The differential pressure for the test is 12.94 psi. Other data are provided in the following table. Calculate and plot the oil–water relative permeability data.

q_o (cm^3/min)	q_w (cm^3/min)	Wet Weight of Core Plug (g)
21.20	0.00	77.7415
18.00	0.50	78.0947
10.60	1.40	78.4479
3.00	2.90	78.8011
0.40	6.50	79.1543
0.00	24.30	79.5075

9.4 For laboratory waterflood data, the relative permeability of oil is measured and the fractional flow data as a function of saturation is also available. The S_{wi} and S_{or} values are 16.7% and 79.2%, respectively. The base permeability is k_{oil} at S_{wi}. The value of k_{rw} at S_{or} is 0.145. Construct the oil–water relative permeability, and relative permeability ratio plots as a function of water saturation. The oil and water viscosities are 1.81 and 0.42 cP, respectively.

S_w (%)	f	k_{ro}
16.7	0.0000	1.0000
40.1	0.5837	0.2090
44.8	0.6889	0.1580
53.9	0.8392	0.0834
58.0	0.8889	0.0582
63.7	0.9421	0.0310
67.8	0.9694	0.0166
76.5	0.9990	0.0006
77.5	0.9997	0.0002
78.8	0.9998	0.0001
79.2	1.0000	0.0000

9.5 The oil–water relative permeability data for a reservoir condition coreflood are given in the following table. A separate centrifuge test on the same core sample resulted in a residual oil saturation of 14% (which is believed to be the true value) and the end-point relative permeability to water as $k_{rw} = 0.9$. Extend/extrapolate (not by hand) the oil–water relative permeability curve to the centrifuge test residual oil saturation so that the relative permeability data can be used in a reservoir simulation study.

S_w, (fraction)	k_{ro}	k_{rw}
0.075	1.000	0.000
0.233	0.288	0.097
0.251	0.251	0.106
0.276	0.203	0.118
0.301	0.166	0.129
0.327	0.132	0.140
0.344	0.110	0.152
0.368	0.091	0.171
0.387	0.079	0.185
0.407	0.062	0.194
0.425	0.051	0.201
0.447	0.040	0.231
0.468	0.031	0.253
0.486	0.022	0.274
0.505	0.015	0.290
0.521	0.009	0.313
0.543	0.005	0.336
0.556	0.002	0.373
0.560	0.001	0.389

9.6 The oil and water relative permeabilities for a chalk core plug are expressed by the following equations:

$$k_{rw} = 0.52\,(S_w - 0.25)^3$$

$$k_{ro} = 3.62\,(0.75 - S_w)^3$$

Determine the values of irreducible water saturation, residual oil saturation and the end-point relative permeabilities to oil and water.

9.7 Determine the Corey exponents for the following oil–water relative permeability data:

$$k_{eo} \text{ at } S_{wi} = 0.204 \text{ mD}$$

$$k_{ew} \text{ at } S_{or} = 0.128 \text{ mD}$$

S_w(%)	k_{ro}	k_{rw}
14.3	1.0000	0.000
63.3	0.1200	0.342
66.0	0.0513	0.376
68.1	0.0161	0.398
68.7	0.0113	0.400
69.1	0.0094	0.402
72.9	0.0025	0.436
74.8	0.0011	0.464
76.4	0.0004	0.503
76.9	0.0002	0.530
77.8	0.0000	0.529
79.4	0.0000	0.627

9.8 For the oil–water capillary pressure data given in the following table, calculate the oil and water relative permeabilities based on the equations that are an extension of Purcell's method which defines the absolute permeability and capillary pressure relationship of a porous medium.

S_w(%)	P_c(psi)
100	325
90	410
80	440
70	480
60	530
50	580
45	640
40	710
35	800
30	940
25	1160
20	1500

9.9 The relative permeability data for an oil–water system is characterized by Corey exponents of 2 for both oil and water. The initial water saturation and the residual oil saturation for this system is 0.25 and 0.20, whereas the end-point relative permeabilities of the oil and water at these saturations are 0.9 and 0.3, respectively. The oil–water viscosity ratio for this system is 10. Calculate and plot the water saturation profiles vs. the normalized position for 0.1, 0.2, 0.4, 0.8, 7.63 and breakthrough (BT) pore volumes (PV) of water injected using a combination of Buckley–Leverett theory and Welge's extension solution.

9.10 Calculate the oil–water relative permeabilities by the JBN method using the following data collected during a laboratory unsteady-state displacement experiment.

Q_{wi}(PV of water injected)	Q_{op}(PV of oil produced)	ΔP(psi)
0.42	0.371	425.70
0.50	0.388	396.00
0.62	0.405	376.20
0.75	0.422	346.50
0.95	0.441	316.80
1.19	0.458	306.90
1.52	0.477	287.10
1.97	0.493	277.20
2.58	0.510	267.30
3.45	0.525	257.40
4.68	0.540	247.50
6.51	0.555	237.60
9.32	0.570	227.70
13.86	0.585	217.80
21.78	0.600	207.90
37.02	0.615	198.00
71.43	0.630	188.10
174.00	0.645	183.15

Additional data for this experiment include: $S_{wi} = 0.20$, $S_{or} = 0.15$, $(\mu_o/\mu_w) = 7$, $k_{ro}@S_{wi} = 1.0$, and $k_{rw}@S_{or} = 0.35$. Stabilized pressure drop for base permeability measurement is 350 psi.

REFERENCES

1. Leverett, M.C., Flow of oil-water mixtures through unconsolidated sands, *Trans. AIME*, 132, 381–401, 1939.
2. Loomis, A.G. and Crowell, D.C., Relative permeability studies: gas-oil and water-oil systems, *Bulletin* 599, U.S.B.M., 1–30, 1962.
3. Johnson, E.F., Bossler, D.P., and Naumann, V.O., Calculation of relative permeability from displacement experiments, *Trans. AIME*, 216, 370–372, 1959.
4. Buckley, S.E. and Leverett, M.C., Mechanism of fluid flow in sands, *Trans. AIME*, 146, 107–116, 1942.

5. Welge, H.J., A simplified method for computing oil recovery by gas or water drive, *Trans. AIME*, 195, 91–98, 1952.
6. Rapoport, L.A. and Leas, W.J., Properties of linear waterfloods, *Trans. AIME*, 198, 139–148, 1953.
7. Jones, S.C. and Roszelle, W.O., Graphical techniques for determining relative permeability from displacement experiments, *J. Pet. Technol.*, 807–817, 1978.
8. Richardson, J.G., Kerver, J.K., Hafford, J.A., and Osoba, J.S., Laboratory determination of relative permeability, *Trans. AIME*, 195, 187–196, 1952.
9. Purcell, W.R., Capillary pressures – their measurement using mercury and the calculation of permeability therefrom, *Trans. AIME*, 186, 39, 1949.
10. Fatt, I. and Dykstra, H., Relative permeability studies, *Trans. AIME*, 192, 249–256, 1951.
11. Bennion, D.B., Sarioglu, G., Chan, M.Y.S., Courtnage, D., and Wansleeben, J., Steady-state bitumen-water relative permeability measurements at elevated temperatures in unconsolidated porous media, Society of Petroleum Engineers (SPE) paper number 25803.
12. Owens, M.U., Parrish, O.R., and Lamoreaux, U.E., An evaluation of a gas drive method for determining relative permeability relationships, *Trans. AIME*, 207, 275–280, 1956.
13. Narahara, G.M., Pozzi, A.L., Jr., and Blackshear, T.H., Jr., Effect of connate water on gas/oil relative permeabilities for water-wet and mixed-wet Berea rock, Society of Petroleum Engineers (SPE) paper number 20503.
14. Tang, G. and Firoozabadi, A, Relative permeability modification in gas-liquid systems through wettability alteration to intermediate gas-wetting, *SPE Reservoir Evaluation & Engineering*, 427–436, 2002.
15. Caudle, B.H., Slobod, R.L., and Brownscombe, E.R., Further developments in the laboratory determination of relative permeability, *Trans. AIME*, 192, 145–150, 1951.
16. Craig, F.F., *The Reservoir Engineering Aspects of Waterflooding,* Monograph Series, SPE, Richardson, TX, 1971.
17. Anderson, W.G., Wettability literature survey – Part 5: the effects of wettability on relative permeability, *J. Pet. Technol.*, 1453–1468, 1987.
18. Donaldson, E.C. and Thomas, R.D., Microscopic observations of oil displacement in water-wet and oil-wet systems, Society of Petroleum Engineers (SPE) paper number 3555.
19. Owens, W.W. and Archer, D.L., The effect of rock wettability on oil-water relative permeability relationships, *J. Pet. Technol.*, 873–878, 1971.
20. Morrow, N.R., Cram, P.J., and McCaffery, F.G., Displacement studies in Dolomite with wettability control by Octanoic acid, *Soc. Pet. Eng. J.,* 221–232, 1973.
21. Fatt, I. and Klikoff, W.A., Effect of fractional wettability on multiphase flow through porous media, *Trans. AIME*, 216, 426–432, 1959.
22. Richardson, J.G., Perkins, F.M., and Osoba, J.S., Differences in the behavior of fresh and aged East Texas Woodbine cores, *Trans. AIME*, 204, 86–91, 1955.
23. Salathiel, R.A., Oil recovery by surface film drainage in mixed-wettability rocks, *J. Pet. Technol.*, 1216–1224, 1973.
24. Morgan, J.T. and Gordon, D.T., Influence of pore geometry on water-oil relative permeability, Society of Petroleum Engineers (SPE) paper number 2588.
25. Ali, H.S., Al-Marhoun, M.A., Abu-Khamsin, S.A., and Celik, M.S., The effect of overburden pressure on relative permeability, Society of Petroleum Engineers (SPE) paper number 15730.

26. Amaefule, J.O., Ajufo, A., Peterson, E., and Durst, K., Understanding formation damage processes: an essential ingredient for improved measurement and interpretation of relative permeability data, Society of Petroleum Engineers (SPE) paper number 16232.
27. Akin, S., Castanier, L.M., and Brigham, W.E., Effect of temperature on heavy oil/water relative permeabilities, Society of Petroleum Engineers (SPE) paper number 54120.
28. Blom, S.M.P., Hagoort, J., and Soetekouw, D.P.N., Relative permeability at near-critical conditions, *Soc. Pet. Eng. J.,* 5, 172–181, 2000.
29. Fulcher, R.A., Jr., Ertekin, T., and Stahl, C.D., Effect of capillary number and its constituents on two-phase relative permeability curves, *J. Pet. Technol.,* 249–260, 1985.
30. Lefebvre du Prey, E.J., Factors affecting liquid-liquid relative permeabilities of a consolidated porous medium, *Soc. Pet. Eng. J.,* 39–47, 1973.
31. Dandekar, A.Y., Unpublished data, 1999.
32. Mohanty, K.K. and Miller, A.E., Factors influencing unsteady relative permeability of a mixed-wet reservoir rock, *SPE Formation Evaluation*, 349–358, 1991.
33. Archer, J.S. and Wong, S.W., Use of a reservoir simulator to interpret laboratory waterflood data, Society of Petroleum Engineers (SPE) paper number 3551.
34. Corey, A.T., The interrelation between gas and oil relative permeabilities, *Producers Monthly*, 19, 38–41, 1954.
35. Stone, H.L., Probability model for estimating three-phase relative permeability, *J. Pet. Technol.,* 22, 214–218, 1970.
36. Stone, H.L., Estimation of three-phase relative permeability and residual oil data, *J. Can. Pet. Technol.,* 12, 53–61, 1973.
37. Fayers, F.J . and Matthew, J.D., Evaluation of normalized Stone's methods for estimating three phase relative permeabilities, *Soc. Pet. Eng. J.,* 224–232, 1984.
38. Aziz, K. and Settari, T., *Petroleum Reservoir Simulation*, Applied Science Publishers, London, 1979.

10 Introduction to Petroleum Reservoir Fluids

10.1 INTRODUCTION

Petroleum reservoir fluids, or *reservoir fluids*, from a petroleum engineering perspective, broadly refer to the hydrocarbon phase and the water phase that exist in deep formations or *petroleum reservoirs.* Simply put, fluids that exist in petroleum reservoirs are petroleum reservoir fluids. Water is also present in petroleum reservoirs but its influence on properties of the hydrocarbon phase is a minor consideration.

From a practical perspective, the term petroleum reservoir fluids generally refers to the hydrocarbon phase in a petroleum reservoir. This phase or "petroleum" is a mixture of different types of naturally occurring hydrocarbon constituents, that is, molecules containing carbon and hydrogen existing in either the gaseous, liquid, or in rare cases, the solid state, depending upon prevailing pressure and temperature conditions in the reservoirs. Therefore, virtually all hydrocarbons or petroleum are produced from the reservoirs in gaseous or liquid forms and are broadly referred to as *natural gas* or *crude oil*, or sometimes as *reservoir gases* or *reservoir oils*. These reservoirs are then specifically characterized as gas reservoirs or oil reservoirs, depending upon the state of the hydrocarbons they produce (i.e., gas or liquid).

Generally, when the hydrocarbon phase is composed of smaller molecules, the resulting mixture is usually a typical naturally occurring hydrocarbon gas, or vice versa, that is when larger molecules compose the hydrocarbon phase, the resulting mixture is a typical naturally occurring crude oil. Thus, the physical state of petroleum reservoir fluids is gaseous, liquid, or solid, whereas their chemical makeup is principally hydrocarbons; compounds of carbon and hydrogen.

10.2 CHEMISTRY OF PETROLEUM

Although all petroleum reservoir fluids are constituted primarily of carbon and hydrogen, the molecular constitution or chemical composition differs widely. However, this wide variation is predominantly found in crude oils instead of natural gases because crude oil is made up of larger molecules, whereas natural gas is made up of smaller molecules. As the molecular size becomes larger, the chemistry of petroleum becomes even more complicated, resulting in different types of classifications as per the structure of the larger molecules found in the overall mixture. Therefore, since the petroleum engineer spends his or her professional life working with these reservoir fluids composed of mixtures of various hydrocarbon constituents of different types, it is important to understand the chemistry of petroleum, that is, the different types of compounds that make up these mixtures.

The branch of chemistry that deals with petroleum is called *organic chemistry* because it involves the chemistry of compounds of carbon and is connected to living organisms that make up the two major sources of organic material, petroleum and coal, from which organic compounds are obtained. Both these compounds are products of decayed plants and animals and are prominently considered in the organic theory of the formation of petroleum. However, organic chemistry nomenclature is complicated because of the various organic structures found in naturally occurring hydrocarbon fluids. Organic chemicals and the various organic structures studied as part of organic chemistry are usually named on the basis of a system developed by the International Union of Pure and Applied Chemistry (IUPAC). According to this system, hydrocarbons are divided into two main classes: *aliphatics* and *aromatics*. Aliphatics are subdivided into families or series called alkanes, alkenes, alkynes, and cycloaliphatics.

In addition to these hydrocarbon classes, petroleum reservoir fluids also contain some nonhydrocarbon components described in Section 10.2.6.

10.2.1 Alkanes

This particular series of hydrocarbons has a general formula, C_nH_{2n+2}, where n denotes the number of carbon atoms. The alkanes are named through the combination of a prefix (signifying the number of carbon atoms) and a suffix that ends with "-ane." Compounds belonging to this series are sometimes called *saturated hydrocarbons* because the carbon atoms are attached to as many hydrogen atoms as possible, that is, the carbon atoms are saturated with hydrogen. However, they are commonly known as *paraffins* in the petroleum industry. Examples of well-known alkanes or paraffins are given in Table 10.1. The structural formulas for methane (smallest hydrocarbon found in reservoir fluids) and propane are shown in Figure 10.1.

However, as the carbon number increases, carbon atoms may be connected in continuous chains or as branches with more than two carbon atoms linked together. Figure 10.2 shows this connection for pentane having carbon number five. Continuous chain hydrocarbons are known as *normal hydrocarbons* and a prefix "*normal*," or simply "*n*-," is attached to the name. The prefix "*iso*-" is used for substances with two methyl groups (CH_3) attached to carbon atoms at the end of an otherwise straight chain (see Figure 10.2). The prefix "*neo*-" denotes three methyl groups attached to carbon atoms at the end of a chain (similar to the case of *iso*-). For example, pentane has three different structures, *normal*-pentane, *iso*-pentane, and *neo*-pentane. Obviously, the number of possible structures or configurations increases with the number of carbon atoms in the molecule. For example, decane (carbon number 10) has 75 such configurations. These configurations are called *structural isomers* or, simply, isomers, meaning they have the same molecular formula, C_4H_{10} or C_5H_{12}, but different arrangements of atoms. Therefore, even though the molecular weight remains the same, certain physical properties are different. For example, *n*-butane has a boiling point of 31.1°F, but *iso*-butane boils at 10.9°F. At normal pressure and temperature conditions, when carbon numbers range from 1 to

TABLE 10.1
Common Normal Alkanes

Name	Carbon Number	Formula	Molecular Weight[a] (g/g mol)
Methane	1	CH_4	16.043
Ethane	2	C_2H_6	30.070
Propane	3	C_3H_8	44.097
n-Butane	4	*n*-C_4H_{10}	58.124
n-Pentane	5	*n*-C_5H_{12}	72.150
n-Hexane	6	*n*-C_6H_{14}	86.178
n-Heptane	7	*n*-C_7H_{16}	100.205
n-Octane	8	*n*-C_8H_{18}	114.232
n-Nonane	9	*n*-C_9H_{20}	128.259
n-Decane	10	*n*-$C_{10}H_{22}$	142.286

[a] Molecular weight can be in any units such as g/g mol, lbm/lbm mol, or *kg*/*kg* mol.

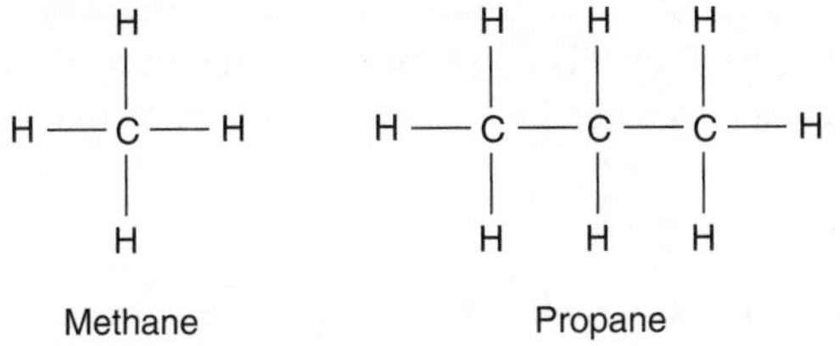

Methane Propane

FIGURE 10.1 Structural formulas for methane and propane.

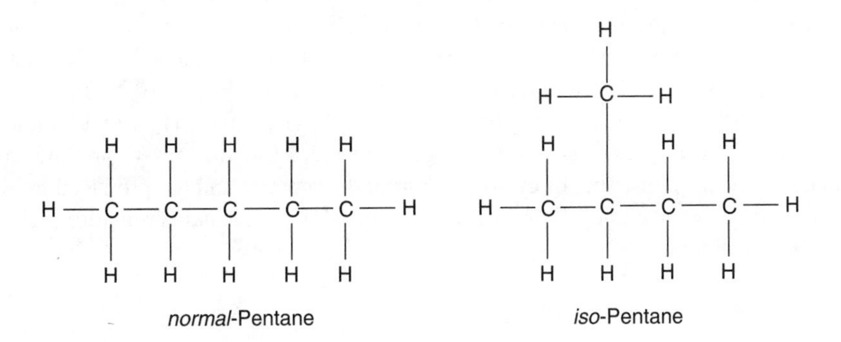

FIGURE 10.2 Structural formulas for isomers of pentane.

4, the resulting compounds are gases; while carbon numbers that range from 5 to 17 are mainly liquids. Carbon numbers higher than 17 have a solid appearance.

10.2.2 Alkenes

This family of hydrocarbons is also called *olefins* and is commonly known by that name in the petroleum and petrochemical industries. The general formula for olefins is C_nH_{2n}. Common examples of alkenes or olefins include ethylene and propylene. The presence of olefins in naturally occurring hydrocarbons is usually a rare occurrence. On the other hand, olefins are actually produced in petrochemical complexes based on the feedstock, usually composed of light alkanes.

10.2.3 Alkynes

The general formula for alkynes is C_nH_{2n-2}. Common examples of alkynes include substances such as acetylene and propyne. Similar to alkenes, alkynes are also rarely found in naturally occurring hydrocarbons and are usually produced in petrochemical plants that use lighter hydrocarbons as feedstock.

10.2.4 Cycloaliphatics

In many hydrocarbon compounds, carbon atoms are arranged in rings instead of chains, as seen in the case of normal alkanes. These types of compounds are called *cyclic compounds* and include cycloalkanes and cycloalkenes. Cycloalkanes or cycloparaffins are commonly known in the petroleum industry as *naphthenes*. The general formula is given by C_nH_{2n}. Common examples of naphthenes include cyclopentane and cyclohexane and have structures as shown in Figure 10.3.

10.2.5 Aromatics

This class of hydrocarbons is also called *arenes*. The name *aromatics* comes from the fact that many compounds belonging to this class have very pleasant odors. However, these compounds are generally very toxic and some are carcinogenic. Benzene, having the chemical formula C_6H_6, is one of the most commonly known aromatic compounds, characterized by the six carbon atoms arranged in a hexagonal ring structure, known as the *benzene ring*. The six hydrogen atoms are associated with each carbon, as shown in the benzene ring structure in Figure 10.4. Other commonly found aromatics include toluene (C_7H_8) and xylene(s) (C_8H_{10}). In addition to these simple aromatics, many of the larger molecules found in reservoir fluids are condensed rings consisting of cycloparaffins and aromatics and are produced by the joining of rings and chains. Examples of condensed ring and chain aromatics include tetralin and lindane.

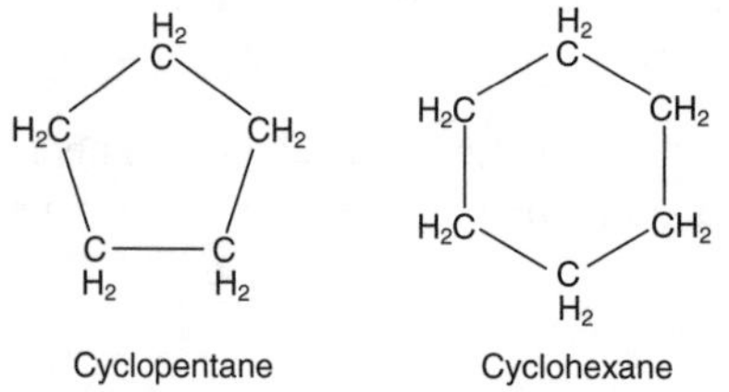

Cyclopentane Cyclohexane

FIGURE 10.3 Structural formulas for cyclopentane and cyclohexane.

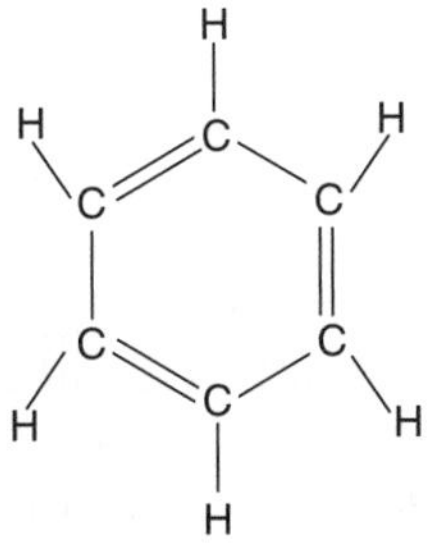

FIGURE 10.4 Structural formula for benzene.

10.2.6 Nonhydrocarbons in Reservoir Fluids

In addition to the hydrocarbons just discussed, reservoir gases and oils also contain certain nonhydrocarbon components. Commonly found nonhydrocarbon components include nitrogen (N_2), carbon dioxide (CO_2), and hydrogen sulfide (H_2S). Reservoir gases or oils that contain hydrogen sulfide are called *sour gases* or *sour crudes*. Those devoid of H_2S are normally termed *sweet gases* or *sweet oils*.

10.3 THE SOLID COMPONENTS OF PETROLEUM

All activities in the oil and gas industry, upstream (production), midstream (transportation), and downstream (refining), are frequently affected by the solid components of petroleum. The solid components of petroleum are sometimes referred to as unique phases and include gas hydrates, waxes, asphaltenes, and diamondoids. These so-called solid components or compounds that form the unique phases are generally the result of particular pressure, temperature, and compositional or chemical changes occurring in reservoir gases and oils. The appearance of these solid components of petroleum in the reservoir pore space, pipelines, and refining operations can severely restrict the flow of fluids, which is somewhat akin to the arterial blockage that restricts the flow of blood in human bodies. Therefore, solid components are usually considered a nuisance in almost all activities of the oil and gas industry. A brief discussion of these four unique phases is given in the following sections.

10.3.1 Gas Hydrates

These are defined as solid, semistable compounds that are basically light hydrocarbon molecules (e.g., methane or ethane) occupying the geometric lattices of water molecules and having snow-like appearances. For example, natural gas hydrates are usually formed whenever natural gas, water, and appropriate temperature and pressure conditions exist. Gas hydrates are generally considered a nuisance because they plug flow lines. However, in recent years, the energy potential of geologic gas hydrate formations, known to exist in the arctic regions and the Gulf of Mexico, has received considerable attention as a future energy resource.

10.3.2 Waxes

Waxes can be defined as solids deposited from crude oils when cooled below a certain temperature. The deposited solids are called waxes and are generally composed of heavy paraffins (e.g., *n*-hexadecane and higher). The wax deposition tendency of a reservoir oil is characterized by wax appearance temperature (WAT); that is, the lowest temperature at which wax appears when the sample is cooled down at a certain rate. The appearance of wax can create major problems by plugging flow lines and process equipment and is primarily a surface problem rather than a reservoir problem.

10.3.3 Asphaltenes

Asphaltenes are perhaps one of the most widely investigated solid components or fractions of petroleum. From a chemistry perspective, asphaltenes can be characterized as organic materials consisting of aromatic and naphthenic ring compounds containing nitrogen, sulfur, and oxygen molecules. The asphaltene fraction of a crude oil is defined as the organic part of the oil that is not soluble in normal alkane solvents such as pentane or heptane. Asphaltenes generally exist as a colloidal suspension stabilized by resin molecules (aromatic ring systems) in oil. The stability of asphaltic dispersions depends on the ratio of resin to asphaltene molecules. Pure asphaltenes are solid, dry, black powders and are nonvolatile, while pure resins are heavy liquids or sticky solids and are as volatile as hydrocarbons of the same size. As soon as asphaltene stability is disturbed by pressure, temperature, or compositional changes in the crude oil, asphaltenes precipitate and can cause major blockage problems in the production, transportation, and refining operations of the petroleum industry.

10.3.4 Diamondoids

Diamondoids are saturated, polycyclic organic compounds. The name *diamondoid* originates from their structure, which is similar to a diamond lattice. Examples of diamondoids include adamantane and diamantane, which are pure components. Therefore, one major difference between asphaltenes and diamondoids is that asphaltenes are chemically much more complex than diamondoids. Similar to the formation of gas hydrates, deposition of waxes, and asphaltenes, the deposition of diamondoids from petroleum streams is associated with changes in pressure, temperature, and composition of reservoir fluids. These changes normally affect the solubility properties of diamondoids in gas and liquid phases, resulting in diamondoid phase segregation that may lead to their deposition. However, the deposition of diamondoids is not as common as asphaltenes or waxes as is evident from the scarcity of literature dealing with diamondoids.

10.4 CLASSIFICATION OF RESERVOIR GASES AND OILS

Reservoir gases and oils can generally be classified according to their chemical characteristics or constituents and their physical properties. Reservoir gases are mostly composed of hydrocarbons of the alkane or paraffin series, containing primarily methane or ethane in the range of 80 to 90% by volume. The remainder of these gases consists of propane and other heavier hydrocarbons and impurities. Therefore, considering the limited range of components, compositional analysis of reservoir gases is readily obtained by techniques such as gas chromatography and low-temperature distillation. For physical classification of reservoir gases, gas gravity is frequently used as a characterization parameter. Gas gravity is defined as the ratio of gas density and air density at the same pressure and temperature conditions.

10.4.1 Chemical Classification of Reservoir Oils or Crude Oils

The chemical classification of reservoir oils is not as simple as gases because the size range of molecules can be huge. For instance, the smallest molecule is methane, CH_4, with a molecular weight of 16 (12 carbon atoms and 4 hydrogen atoms), while the largest molecules can have molecular weights as high as 10,000. Therefore, instead of identifying each and every component in reservoir oils, which is practically impossible anyway, the average chemical analysis of oils includes paraffins-isoparaffins-aromatics-naphthenes-olefins (PIANO) or PONA analysis, and in the latter case all paraffins are combined while other groups remain as they are. However, considering the rarity of olefins in reservoir oils and the lumping of all the paraffins, chemical analysis is simplified or reduced to only the determination of paraffins, naphthenes, and aromatics, known as PNA analysis.

Since the smaller molecules (carbon number 6 and less) are predominantly paraffins, classification is usually based on analysis of the oils after most of the light molecules are removed. This type of analysis gives the basic chemical nature of a given reservoir oil, which is of great significance not only from the production perspective, but also in terms of refining aspects because it directly relates to the type of end (or refined) products produced from a particular crude oil. Based on this type of analysis, reservoir oils dominated by a particular group are called paraffinic, naphthenic, or aromatic.

10.4.2 Physical Classification of Crude Oils

In certain cases, if a chemical classification such as PNA analysis is not available, reservoir oils are classified according to various physical properties that include specific gravity, color, sulfur content, refractive index, odor, and viscosity. Among these various physical properties, the most important is specific gravity. In fact, the commercial value of a crude oil, to a great extent, is determined by its specific gravity. When crude oils are sold to refiners, they are traded as per their specific gravities.

The specific gravity of a crude oil (or any liquid) is defined as the ratio of the density of the oil and the density of water at specified pressure and temperature conditions, mathematically expressed as

$$\gamma_o = \frac{\rho_o}{\rho_w} \tag{10.1}$$

where γ_o is the specific gravity of oil, which is dimensionless, ρ_o the density of oil at a specified pressure and temperature, and ρ_w the density of water at a specified pressure and temperature.

As long as consistent units for ρ_o and ρ_w are used, Equation 10.1 always yields the same specific gravity.

Since the density of oil is a function of pressure and temperature, it is necessary to designate standard conditions for reporting specific gravity. The petroleum industry has adopted as standard the atmospheric pressure and temperature of 60°F or 15.55°C. Therefore, when the specific gravity is given as γ_o (60°/60°), it means that the density of both oil and water are taken at 60°F and atmospheric pressure.

The petroleum industry also uses another gravity scale known as API (American Petroleum Institute) gravity, defined as

$$°API = \frac{141.5}{\gamma_o} - 131.5 \quad (10.2)$$

where °API is the the API gravity and γ_o the specific gravity at 60°/60°.

As seen in Equation 10.2, the API gravity of water is 10, whereas numbers greater than 10 are for materials having specific gravities less than 1, normally the case with crude oils.

Based on the API gravities of crude oils, a gross classification that is used primarily from a refining point of view includes fairly loose terms such as light oil (high API), medium oil, heavy oil, and extra heavy oil (low API). Although this classification serves the basic purpose of categorizing reservoir oils from a commercial perspective, it is certainly inadequate from a phase behavior or reservoir engineering point of view. Therefore, a more rigorous classification is used here and is described in the following section.

10.5 FIVE RESERVOIR FLUIDS

Although the classification of petroleum reservoir fluids as natural gas or crude oil (and subsequent classifications described in Section 10.4.2) is satisfactory from a very broad perspective, it is certainly insufficient for proper characterization of various properties and phase behaviors from a reservoir engineering standpoint. Therefore, petroleum reservoir fluids are classified into the following types[1]:

1. Black oils
2. Volatile oils
3. Gas condensates
4. Wet gases
5. Dry gases

Each of these are classified based on various properties and the phase behavior that each exhibits at different pressure and temperature conditions, resulting in different approaches used by reservoir engineers and production engineers. The very basic characteristics used to identify these fluids as black oils, volatile oils, gas condensates, wet gases, or dry gases include properties such as API gravity, viscosity, color of the liquid hydrocarbons, and chemical composition. This type of classification is also considered as part of field identification and is presented in Table 10.2.

TABLE 10.2
Basic Characteristics of the Five Reservoir Fluids

Reservoir Fluid	API Gravity deg(°)	Viscosity (cP)	Color of Stock Tank Liquid[a]
Black oils	15–40	2 to 3–100 and up	Dark, often black
Volatile oils	45–55	0.25–2 to 3	Brown, orange or green
Gas condensates	Greater than 50	In the range of 0.25	Light colored or water white
Wet gases	Greater than 60	In the range of 0.25	Water white
Dry gases	No liquid is formed, hence the name "dry"	0.02–0.05	–

[a] After the reservoir fluids are produced, they are processed in surface facilities to reduce the pressure and temperature and separate the vapor phase from the liquid phase. In the final stage, the liquid phase ends up in what is called a stock tank that is normally operated at atmospheric pressure. The liquid in this stock tank is referred to as stock tank liquid.

However, a more rigorous classification based on the phase behavior of these five reservoir fluids is presented in Chapter 12.

10.6 FORMATION WATERS

Even though the term petroleum reservoir fluids primarily refers to the naturally occurring hydrocarbons, gas or oil production from petroleum reservoirs is frequently accompanied by water. Water has been addressed earlier from a saturation standpoint, that is, how much of the pore space is occupied by water. However, this part of the book deals with the properties of that particular water phase that occupies a given pore space in a reservoir rock. Water in the reservoir pore space is normally referred to as reservoir water, formation water, oilfield water, or simply brine since most formation waters are mixtures of water and sodium chloride (NaCl). The other constituents of formation waters include potassium chloride, calcium chloride, magnesium chloride, and strontium chloride. Formation waters are primarily characterized by their salinities (total concentration of dissolved solids), composition, and type of solutes, density, and viscosity. However, similar to reservoir gases and oils, characteristics of formation waters vary from formation to formation and bear no relationship to seawater, either in concentration of solids or in the distribution of the various constituents present. Generally formation waters contain much higher concentrations of solids than seawater. A detailed discussion of formation water properties is presented in Chapter 17.

REFERENCE

1. McCain, W.D, Jr., *The Properties of Petroleum Fluids*, PennWell Publishing Co., Tulsa, OK, 1990.

11 Introduction to Phase Behavior

11.1 INTRODUCTION

Chapter 10 explains that petroleum reservoir fluids are mixtures of naturally occurring hydrocarbons that may exist in gaseous, liquid, or in some rare cases, solid states, depending upon the prevalent pressure and temperature conditions in the petroleum reservoirs. Similarly, when these hydrocarbon fluids are produced from the reservoirs and are captured at the surface, they may be produced either in gaseous or liquid states or in both states. This particular state of the hydrocarbon mixture both in the reservoir and on the surface (i.e., gas or liquid) is primarily dictated or is a result of the pertinent pressure and temperature conditions that exist in these locations. However, as stated in Chapter 10, these petroleum reservoir fluids are generally complex mixtures of a number of hydrocarbon (paraffins, naphthenes, and aromatics) and even some nonhydrocarbon components. Therefore, the state of reservoir fluids as gas, liquid, or solid is actually controlled by not only the prevailing pressure and temperature conditions but also by the chemistry and composition (e.g., how much of a particular component is present in a hydrocarbon mixture and what percentage of a mixture is composed of methane or ethane) of a given system. Hence, the domain in which petroleum reservoir fluids behave in a certain fashion is defined by *pressure*, *temperature*, *chemistry*, and *composition*. Or, in other words, the state of a system is fully defined when pressure, temperature, composition, and chemistry are specified.

During the producing life of a particular petroleum reservoir, most of these variables continuously change and as a result, a variation occurs in the state of their existence, the amount in which they are present, and their physical properties, which is broadly characterized as *phase behavior*. Knowledge of phase behavior and properties of reservoir fluids is of great significance because it enables the petroleum engineer to evaluate recovery in terms of standard volumes of gas (volume of gas at standard conditions of atmospheric pressure and 60°F) and stock tank barrels of liquid (crude oil) that may be obtained upon production to the surface of a unit volume of reservoir fluid.

Therefore, the primary purpose of this chapter is to introduce the fundamental concepts of phase behavior. However, before studying the phase behavior of petroleum reservoir fluids (i.e., reservoir gases and oils), it is first necessary to understand some simple systems, such as pure components. Moreover, since reservoir fluids are basically mixtures of various pure components, the study begins with the phase behavior of pure components that are subsequently extended to simple two-component and multicomponent well-defined model systems. However, based on our understanding of phase behavior of these simple systems presented here, phase behavior of the five reservoir fluids will be explored in Chapter 12.

11.2 DEFINITION OF TERMS USED IN PHASE BEHAVIOR

Before studying the important phase behavior concepts, one needs to understand the definitions of the various terms that are used when describing the phase behavior of a given system. The following text highlights the definition of the terms and some other important distinctions.

11.2.1 PHASE

The term *phase* defines any homogenous and physically distinct part of a system that is separated from other parts of the system by definite bounding surfaces. The most common examples are water vapor, liquid water, and ice. Definite boundaries between water vapor–liquid water, water vapor–ice, and liquid water–ice exist; thus it can be said that the gas, liquid, and solid phases make up a three-phase system. A distinction is sometimes made by some authors between gas and vapor since vapor is defined as any substance in the gaseous state, usually a liquid under ordinary ambient conditions. However, in dealing with hydrocarbon fluids, it is convenient to think of gas and vapor as being synonymous and thus these words are used interchangeably throughout this text.

11.2.2 PRESSURE, TEMPERATURE, AND INTERMOLECULAR FORCES

The manner in which hydrocarbons behave in a certain fashion when pressure and temperature are altered can be readily explained by considering the behavior of individual molecules. In fact, pressure, temperature, and intermolecular forces are a reflection of the number of molecules and their kinetic energy. For example, pressure is a reflection of the number of times the molecules of a gas strike the wall of its container. Obviously if the molecules are forced together by either adding more gas in the same container volume or reducing the volume of the container, the pressure increases. However, temperature is simply a physical measure of the average kinetic energy of the molecules of the material; or, in other words, temperature reflects the quantity of energy of motion of the molecules of the material. Intermolecular forces are forces of attraction and repulsion between molecules that change as the distance between the molecules change.

11.2.3 EQUILIBRIUM

A condition at which a material appears to be at rest, that is, not changing in volume or changing phases is called equilibrium.

11.2.4 COMPONENT AND COMPOSITION

In phase behavior terminology, a component is defined as an entity, constituent, a given compound, or a substance. For example, methane, ethane, carbon dioxide, or nitrogen are called components. Composition or concentration broadly defines the amount in which a particular component is present in a given mixture. For example, if half of a mixture of methane, ethane, and propane is composed of methane, then the composition of methane is 50% or 0.5 (expressed as a fraction). Therefore, when the composition of a hydrocarbon mixture, such as reservoir gas or oil, is specified, the relative amounts or proportion (either on mass, volume, or molar basis) in which the various components present in the overall mixture are indicated. Composition of reservoir fluids is normally given in terms of mol% (normalized to 100%) or mole fraction (normalized to 1). The mole fraction (or mol% when multiplied by 100) is simply the ratio of moles of a component in the mixture and the total moles. For example, in a mixture consisting of components A and B, the mole fraction of A is expressed in moles of A/(moles of A + moles of B). The concept of composition can be expressed mathematically by this simple summation:

$$\sum_{i=1}^{n} Z_i = 1 \tag{11.1}$$

where Z_i is the composition of the ith component in a mixture consisting of n number of components.

11.2.5 Distinction between Gases and Liquids

The basic difference between gases and liquids is primarily related to the distance between molecules. In gases, the molecules are relatively far apart, while in liquids the molecules are fairly close together. Therefore, gases are compressible, while liquids are incompressible since a repelling force between molecules causes the liquid to resist further compression. Additionally, when gases are placed in containers, they fill the containers completely and do not exhibit free surfaces. Liquids take the shape of the containers but exhibit free surfaces.

11.2.6 Types of Physical Properties

The two types of physical properties are termed *intensive* or *extensive*. Intensive properties are those that do not depend on the amount of material present, such as density and viscosity. Extensive property values are determined by the total quantity of matter present, such as volume and mass.

11.2.7 Phase Rule

The *phase rule* describes the possible number of degrees of freedom in a closed system at equilibrium in terms of the number of separate phases and the number of chemical constituents in the system. The phase rule was deduced from thermodynamic principles expressed by Gibbs[1] in the 1870s and is expressed mathematically as

$$F = C - P + 2 \tag{11.2}$$

where F is the degrees of freedom, C the number of components, and P the number of phases.

The degrees of freedom F are the number of independent intensive variables that need to be specified in value to fully determine the state of the system. Typically, the variables that need to be fixed so that the conditions of a system or a component at equilibrium may be completely defined are pressure, temperature, and composition.

A system is called invariant when $F = 0$, univariant when $F = 1$, bivariant when $F = 2$, and trivariant when $F = 3$. For example, when ice, liquid water, and water vapor are present in equilibrium, the system is invariant as $C = 1$ and $P = 3$, resulting in a value of $F = 0$. The phase rule is again applied during the discussion of phase behavior of pure components and simple binary systems.

11.3 PHASE BEHAVIOR OF A PURE COMPONENT

As mentioned earlier, systems consisting of a single, pure component will first be considered here. These systems behave differently from binary, ternary, or multicomponent mixture systems that are made up of two or more components. In particular, phase behavior is defined as the conditions of pressure and temperature at which different phases exist or coexist. Existing phases are identified or characterized by their specific volumes or densities. After examining the phase behavior of single-component systems, the phase behavior of systems or mixtures that contain two or more components is discussed.

11.3.1 PHASE DIAGRAM OF A PURE COMPONENT

The phase behavior of a pure component is characterized by what is known as a *phase diagram*. A phase diagram is basically a pressure–temperature plot that describes the conditions under which the various phases of a component are present. Phase diagrams are sometimes also called pressure–temperature plots. Figure 11.1 shows the phase diagram of a single-component system. For a single-component system, phase behavior is primarily defined by two lines: the vapor pressure curve and the melting point curve. Two points on the vapor pressure curve called the critical point and the triple point also define phase behavior and are identified as CP and TP, respectively, in Figure 11.1.

11.3.1.1 Vapor Pressure Curve

The vapor pressure curve joined by TP–CP separates the pressure–temperature (PT) conditions at which the component is a liquid from the conditions at which it is a gas. The PT coordinates which lie above the vapor pressure (VP) curve indicate conditions at which the component is in a liquid phase. The PT coordinates that lie below the VP curve represent conditions at which the component is in a gas phase. However, PT points that lie exactly on the VP curve indicate conditions at which both gas and liquid phases coexist in equilibrium.

Figure 11.1 also shows paths of a vertical (ABC) and a horizontal (DEF) line that cross the vapor pressure curve. The vertical line represents the reduction (moving from A to C) or increase (moving from C to A) in pressure at constant temperature or isothermal conditions. If the expansion (A to B) at isothermal conditions is considered,

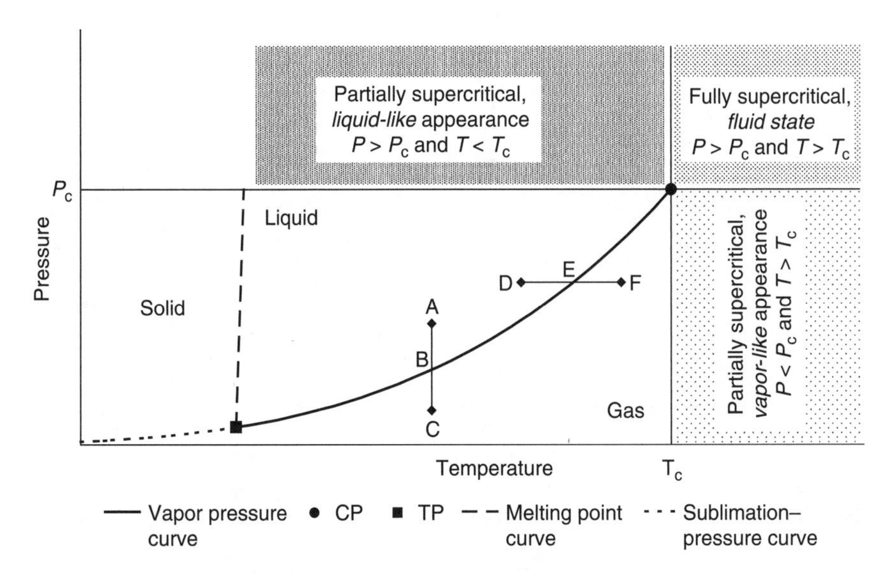

FIGURE 11.1 Phase diagram of a pure component.

the point at which the first few molecules leave the liquid phase and form a small bubble of gas is called the *bubble-point pressure* (point B). However, precisely the opposite is observed if we consider the compression (C to B) at isothermal conditions, so the point at which a small drop of liquid is formed is called the *dew-point pressure* (also point B). Therefore, for a pure component the bubble-point and dew-point pressures are equal to the vapor pressure at a given temperature of interest.

Similar to the isothermal expansion or compression, the behavior of a pure component at constant pressure or isobaric conditions can also be studied by varying the temperature. For line DEF, if temperature is increased from D to E, point E now represents the *bubble-point temperature;* if temperature is decreased from F to E, point E now signifies the *dew-point temperature.*

11.3.1.2 Critical Point

The upper limit of the VP curve is called the *critical point.* The pressure and temperature conditions represented by this point are called *critical pressure*, P_c, and *critical temperature*, T_c, respectively. The critical point represents the maximum pressure and temperature at which a pure component can form coexisting phases. For a pure component, the critical temperature may also be defined as the temperature above which the gas cannot be liquefied, regardless of the applied pressure. Another classical definition of critical point is the state of pressure and temperature at which the intensive properties of the gas and the liquid phases are continuously identical. It should, however, be noted here that these definitions are strictly valid for pure-component systems and are not applicable for those containing more than one component (mixtures). Critical pressure and critical temperature data of some of the common normal alkanes are shown in Figure 11.2.

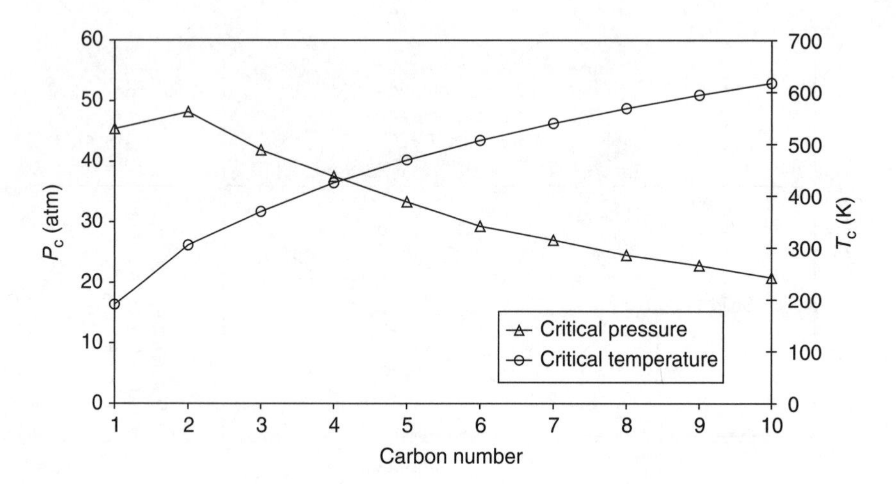

FIGURE 11.2 Critical pressures and critical temperatures of common normal alkanes.

11.3.1.3 Triple Point

The point TP on the vapor pressure curve is called the *triple point* and represents the pressure and temperature conditions at which all three phases (gas, liquid, and solid) of a component coexist under equilibrium conditions. For example, the triple point of water is 0.01°C. Recalling the phase rule, it can be stated that every single-component system is invariant at triple point, that is, the fixed unique point.

11.3.1.4 Melting Point Curve

The melting point curve (solid–liquid equilibrium curve) is the almost vertical line above the triple point that separates the solid conditions from liquid conditions. Similar to the VP curve, *PT* conditions that fall exactly on this line indicate a two-phase system, that is, the coexistence of liquid and solid.

11.3.1.5 Sublimation–Pressure Curve

The sublimation-pressure curve represents solid–vapor equilibrium. The VP curve below the triple point divides the conditions at which the component is solid from those at which it is gas. Dry ice or solid carbon dioxide is a good example of this region of the phase diagram.

11.3.1.6 Conditions Outside the P_c–T_c Boundary

In Figure 11.1, the pressure–temperature conditions outside the P_c–T_c boundary are shown by three different types of shaded regions. The conditions at which both pressure and temperature are greater than P_c and T_c is called a *completely supercritical region*, where a distinction between gas and liquid phase cannot be made and the component is said to be in the fluid state. However, the fluid phase assumes the properties of a vapor-like or a liquid-like phase depending on the proximity to the critical point.

A component is said to be *partially supercritical* and demonstrate *liquid-like behavior* if conditions exist where only the pressure is greater than the critical pressure while the temperature is less than its critical temperature. Similarly, conditions where only the temperature is greater than the critical temperature and the pressure is less than its critical pressure, the component is partially supercritical and demonstrates a gas or a *vapor-like* behavior. The particular distinction between partial and complete supercriticality has never been reported before.

11.3.2 Pressure–Volume Diagram

Another convenient means of illustrating the phase behavior of a pure-component system is through the pressure–specific volume (*PV*) relationships. The *PV* relationship for *n*-butane is shown in Figure 11.3. The two-phase region shown in Figure 11.3 is clearly illustrated. A solid line and a dashed line show the bubble-point and dew-point curves, respectively. The area within these two curves indicates the conditions at which both gas and the liquid phases coexist. This area is often called the *saturation envelope*. Point CP is the critical point of *n*-butane, having a critical pressure of 551.1 psia and critical temperature of 305.69°F. The specific volume corresponding to the critical pressure is the critical volume, V_c, of *n*-butane, having a value of 0.0775 ft^3/lb. As seen in Figure 11.3, the bubble-point and dew-point curves coincide or meet at the critical point, where the distinction between the gas and the liquid phase is lost and the phase has a fixed specific volume. Similar to a characteristic value of P_c and T_c, every pure component also has a fixed critical volume V_c, that is basically the specific volume at the critical conditions.

Figure 11.3 also shows the *PV* relationships for other isotherms or constant temperatures. Considering the *PV* curve at 400°F, an isothermal expansion of *n*-butane at temperature above its critical temperature of 305.69°F does not result in any phase change. The component smoothly transitions from a partially supercritical vapor-like state ($T > T_c$ and $P < P_c$) to a fully supercritical fluid state after crossing a pressure of

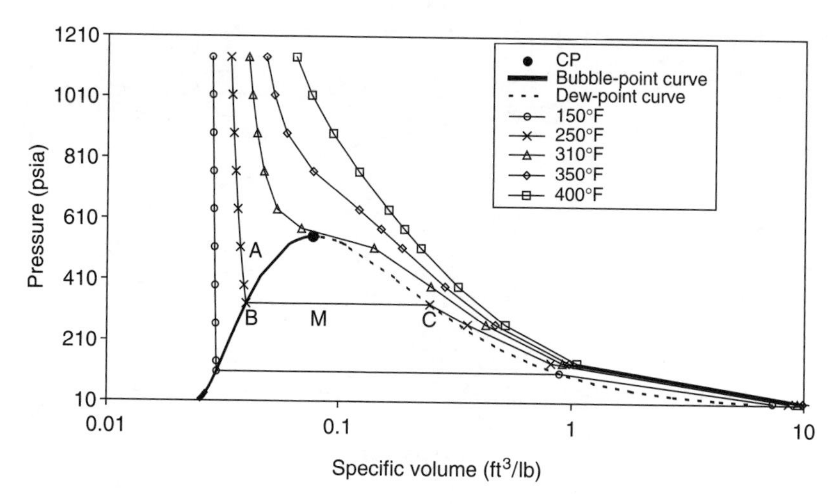

FIGURE 11.3 Phase behavior of *n*-butane described by a pressure–volume diagram.

551.1 psia, $T > T_c$ as well as $P > P_c$, without any abrupt changes in the specific volumes. However, at temperatures close to the critical temperature (PV curve at 310°F), the transition from partially supercritical vapor-like state to fully supercritical (but a liquid-like fluid state) is rather abrupt or discontinuous as seen by the specific volume changes. This behavior is attributed to the proximity of this isotherm to the critical temperature of *n*-butane. Now consider the compressed liquid state at point A for an isotherm of 250°F, that is below the critical temperature of *n*-butane. The reduction of fluid pressure causes a small increase in the volume due to the relatively incompressible nature of the fluid until the vapor pressure is reached at point B, where the first bubble evolves (bubble point). The continued expansion of the system results in changing the liquid into a vapor phase (point C). However, for a pure component, the pressure remains constant (as evidenced by the horizontal line BC) and equal to the vapor pressure, a consequence of the phase rule, until the last drop of the liquid vaporizes at point C (dew point). The fluid existing at point M forms two equilibrated phases with the vapor/liquid molar ratio equal to BM/MC. This type of pressure–volume behavior described for *n*-butane is a common feature of all pure substances.

11.3.3 Density–Temperature Behavior of a Pure Component

The relationship between the equilibrium vapor and liquid densities as a function of temperature for *n*-butane is shown in Figure 11.4. The curves basically indicate the densities of the vapor and the liquid phases that coexist in the two-phase region, that is, along the vapor pressure curve. These densities are sometimes called *saturated densities*. Figure 11.4 shows two densities approaching each other as temperature increases, eventually becoming equal at the critical point. At this point, the phases become indistinguishable.

Figure 11.4 also shows the average densities of the vapor and the liquid phases as a function of temperature, indicating a straight line that passes through the critical point. This particular property is known as the *law of rectilinear diameters*.

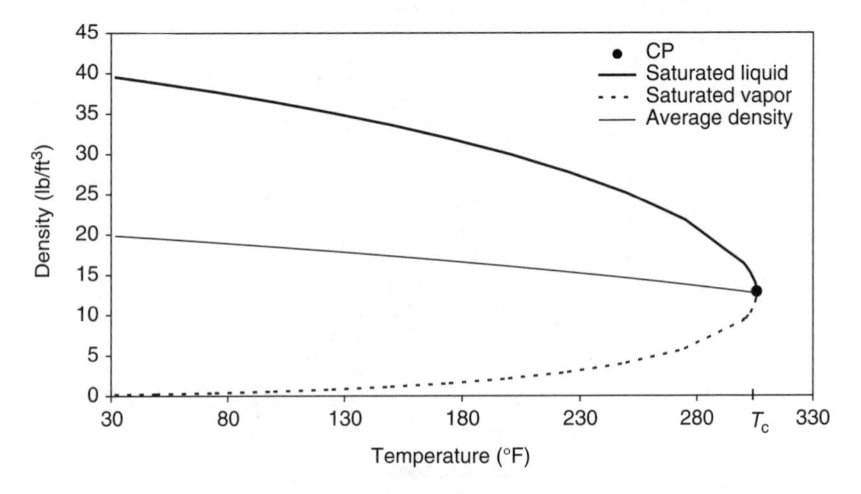

FIGURE 11.4 Saturated liquid and vapor densities of *n*-butane.

11.3.4 Determination of Vapor Pressure

Vapor pressure data of pure components can be readily obtained from laboratory measurements or by using correlations. A simple laboratory measurement setup includes a PVT (pressure–volume–temperature) cell made of suitable material such as stainless steel or titanium that has a mechanism for reducing or increasing cell volume (i.e., pressure), and a climatic air bath in which a constant test temperature can be achieved. Such a PVT cell normally has a small window through which visual information about the change of phase can be obtained. The entire vapor pressure curve can be constructed by varying the pressure by mercury injection/withdrawal or via a mechanically driven piston and observing the phase behavior at different isotherms.

When laboratory determination of vapor pressures is not available, various correlations can be used to determine vapor pressures. A commonly used vapor pressure correlation was formulated by Lee and Kesler[2] based on the concept of the *principle of corresponding states*, which basically states that substances behave similarly when they are scaled according to their critical points or compared on a scale of reduced pressure and temperature. In the Lee–Kesler correlation, the reduced vapor pressure $P_{vr} = (P_v/P_c)$ is expressed as a function of the reduced temperature $T_r = (T/T_c)$ in the following generalized form:

$$P_{vr} = f(T_r) \tag{11.3}$$

However, if the principle of corresponding states were exact, the vapor pressure curves for all the components, plotted in the reduced form, should result in one composite curve; that in practice does not occur due to differences in molecular structures of various substances. Therefore, to account for the deviation from the corresponding states principle, a third parameter called the *acentric factor* ω is introduced in the generalized relationship expressed by Equation 11.3:

$$P_{vr} = \exp(A + \omega B) \tag{11.4}$$

where A and B are expressed by the following equations:

$$A = 5.92714 - \frac{6.09648}{T_r} - 1.28866\ln(T_r) + 0.16934(T_r)^6 \tag{11.5}$$

$$B = 15.2518 - \frac{15.6875}{T_r} - 13.4721\ln(T_r) + 0.4357(T_r)^6 \tag{11.6}$$

In Equation 11.4, since the product of $A + \omega B$ is dimensionless, vapor pressure P_v takes pertinent units that are used for critical pressure, P_c. The consistency of Equation 11.4 is evident because when $T = T_c$ and $T_r = 1$, the condition at which both A and B are 0 eventually results in the equality of P_v and P_c, that is, at critical temperature vapor pressure is equal to critical pressure. Thus, Equation 11.4 offers a convenient way to estimate the vapor pressure of pure components at various temperatures if the critical pressure, critical temperature, and the acentric factor are known. Reid et al.[3] have listed values of the critical constants and ω for a number of

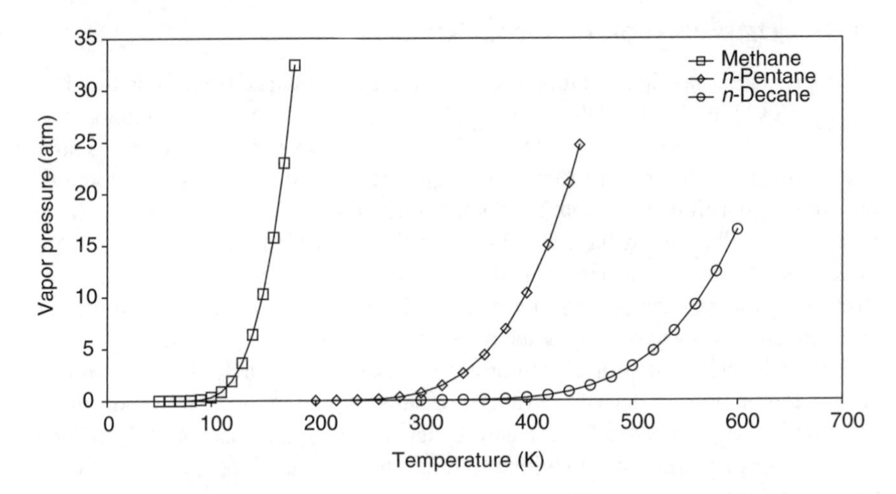

FIGURE 11.5 Vapor pressures of methane, *n*-pentane, and *n*-decane calculated by the Lee–Kesler[2] correlation.

hydrocarbon and nonhydrocarbon components in their book on properties of gases and liquids. The vapor pressure data calculated from Equation 11.4 for selected normal alkanes at various temperatures are shown in Figure 11.5.

11.4 PHASE BEHAVIOR OF TWO-COMPONENT OR BINARY SYSTEMS

For single-component systems, it can be noted from Gibb's phase rule that pressure and temperature are the only two independent variables or degrees of freedom that need to be fixed when determining the state of the component. For example, how many degrees of freedom are required for a pure component so that it exists in single phase? Since both C and P are 1, Equation 11.2 results in a value of $F = 2$, that is, pressure and temperature need to be specified in order to satisfy this condition.

However, when a second component is added to a single component, the resulting mixture is a two-component or a binary system. Here the phase behavior becomes more complex and elaborate and is the result of the introduction of another variable or a degree of freedom, composition. For example, the same question can be posed for a binary system: how many degrees of freedom are required for a two-component mixture so that it exists in single phase? Since $C = 2$ and $P = 1$, Equation 11.2 results in a value of $F = 3$, meaning pressure, temperature, and composition or concentration of at least one of the components needs to be fixed. Therefore, when comparing the phase behavior of a pure component and a binary system, several differences are clearly evident.

Although, petroleum engineers do not normally encounter the simple two-component or binary system, it certainly serves as a good forerunner to the study of the phase behavior of multicomponent systems of the five reservoir fluids. Therefore, it is instructive to first observe the differences between the phase behavior of single-component and binary systems and then extend comparison of those differences to multicomponent mixtures. The approach followed in studying binary systems is almost identical to that taken for examining pure-component systems, that is, begin with the phase diagram and follow with a discussion of various terms and important concepts.

11.4.1 Phase Diagram of a Binary System

Similar to the single-component systems, the phase behavior of a binary system is also described by a phase diagram. However, the most significant difference between the phase behavior of a pure component and a binary system is the difference in the characteristics of the phase diagram itself. Instead of a single vapor pressure curve that represents the two-phase region for a pure component, a broad region is shown in which the two phases coexist in equilibrium. This broad region is commonly referred to as the *phase envelope, saturation envelope,* or simply, the *two-phase region.* A typical phase envelope for a binary system having a fixed overall composition is shown in Figure 11.6. The various important features of the phase envelope of a binary system include critical point, bubble point, dew point, bubble-point and dew-point curves, cricondenbar, cricondentherm, retrograde dew point, and condensation and behavior of a mixture in the two-phase region. All these features are described in detail in the following sections.

11.4.1.1 Critical Point

The definition of critical point applied to pure components does not apply to binary systems. In a binary system, gas and liquid can coexist at pressures and temperatures above the critical point (see Figure 11.6). The phase envelope also exists at pressures higher than the critical pressure and temperatures higher than the critical temperatures.

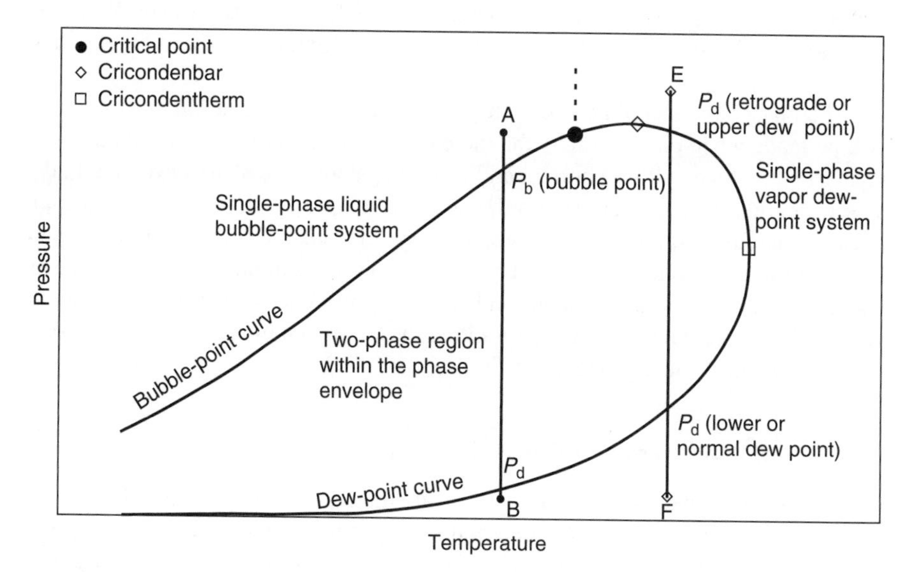

FIGURE 11.6 Typical phase behavior of a binary system.

If a vertical line is drawn beginning with the critical point, as shown in Figure 11.6, at *PT* conditions (outside the phase envelope) that lie on the left-hand side of this line, the binary mixture is in single-phase liquid. At those points that are on the right-hand side of this line, the system is in single-phase vapor. These differences are readily noticeable when the mixture is in single-phase conditions and is in proximity to the critical point. However, as the system moves away from the critical conditions, the transition from single-phase vapor to single-phase liquid is relatively smooth, and without any abrupt changes in the density values. A more rigorous definition of the critical point is the point at which all differences between the vapor and liquid phases vanish and the phases become indistinguishable.

11.4.1.2 Bubble Point and Dew Point

Consider the expansion or pressure depletion of this binary system at a constant temperature shown by line AB in Figure 11.6. According to the definition of critical point, the system is in single-phase liquid at point A. As pressure depletion continues, liquid expands until the pressure reaches a point that meets the phase envelope or the boundary of the two-phase region where a small amount of vapor is formed, primarily composed of the lightest component in the system. This point is called bubble point. Since this expansion is isothermal, the pressure at which the first gas bubble is formed is the bubble-point pressure, denoted by P_b. As pressure depletion continues below the bubble-point pressure, the system passes through the two-phase region and additional gas appears. Finally, the pressure depletion meets the other boundary of the two-phase region where a small amount of liquid remains. This particular meeting point is defined as the dew point. Again, since this is an isothermal expansion process, the dew point encountered is called dew-point pressure, denoted by P_d. Note that the dew-point pressure encountered in this particular case is the normal dew point which is different from the retrograde dew point which will be discussed later. When the expansion of the system reaches point B, the entire mixture turns into single-phase vapor.

The observations similar to the pressure expansion at constant temperature can also be made when one considers the increase in temperature at a constant pressure or an isobar, which will be a horizontal line cutting across the phase envelope. In this case when the line meets or crosses the phase envelope, the first point it encounters is called as bubble-point temperature, whereas the second point is the dew-point temperature. However, from a practical standpoint, considering the fact that reservoir temperatures remain constant and only pressure changes; instead of bubble-point and dew-point temperatures, it is either the bubble-point or dew-point pressure, which is of significance in reservoir fluids.

11.4.1.3 Bubble-Point and Dew-Point Curves

The bubble-point and the dew-point curves are the outermost boundary of the phase envelope lying on the left-hand side and the right-hand side of the critical point, respectively, as shown in Figure 11.6. The bubble-point and the dew-point curves meet at the critical point.

11.4.1.4 Cricondenbar and Cricondentherm

Cricondenbar and cricondentherm are defined as the highest pressure and highest temperature, respectively, on the phase envelope. Both conditions are shown in Figure 11.6.

11.4.1.5 Retrograde Dew Point and Condensation

Similar to the pressure depletion along line AB on the left-hand side of the critical point in Figure 11.6, if the expansion is now carried out on the right-hand side of the critical point as shown by line EF, as pressure is decreased from point E (where the mixture is single-phase vapor), the dew-point curve is encountered and a small amount of liquid appears, composed of the heaviest component in the system. The pressure at this particular point is called the retrograde dew-point pressure, denoted by P_d. This is exactly the reverse of the behavior expected, hence the name "retrograde." Because in a single-component system, a decrease in pressure causes a change of phase from liquid to gas, or vice versa, when pressure is increased. As the pressure decline continues, pressures fall within the two-phase region and more liquid appears. The liquid that appears during this pressure decline is termed *retrograde condensate*. Eventually the retrograde liquid that has formed begins to revaporize as pressure falls to even lower values and finally meets the lowest point on the dew-point curve, called *normal* or *lower dew-point pressure*. Subsequently, as the pressure depletion continues, the system turns into single-phase vapor at point F. The retrograde dew-point pressure and the normal dew-point pressure are sometimes referred to as the *upper* and the *lower dew point*, respectively.

Figure 11.6 also shows that retrograde dew point and condensation occur between the critical temperature and the cricondentherm. A similar retrograde phenomenon is observed when the phase envelope is approached isobarically, that is, temperature is changed at constant pressure between critical pressure and the pressure corresponding to the cricondentherm. Again, considering that reservoir temperature is usually constant, retrograde dew point and condensation are of great significance with respect to gas condensate reservoirs. This concept is discussed in Chapter 12.

11.4.1.6 Behavior of a Mixture in the Two-Phase Region

The behavior of a binary system in the two-phase region can be studied by observing the changes that take place in composition and density of the equilibrated vapor and liquid phase as pressures fall below dew point or bubble point. Since equilibrium vapor and liquid phase compositions have the most profound effect on phase densities, the changes that take place in the composition of the equilibrated vapor and liquid phases are evaluated first. For this purpose a binary system having an overall single-phase composition of 70 mol% methane and 30 mol% *n*-butane, respectively, is selected.

Figure 11.7 plots mole fractions of methane and *n*-butane in the equilibrated vapor and liquid phases for this binary system at 150°F (at this isotherm, the system exhibits a dew-point behavior) at various pressures below the retrograde dew point of 1848 psia. As seen in this plot, the mole fraction of methane in the equilibrium vapor phase increases while that of *n*-butane decreases at pressures below the dew point up to

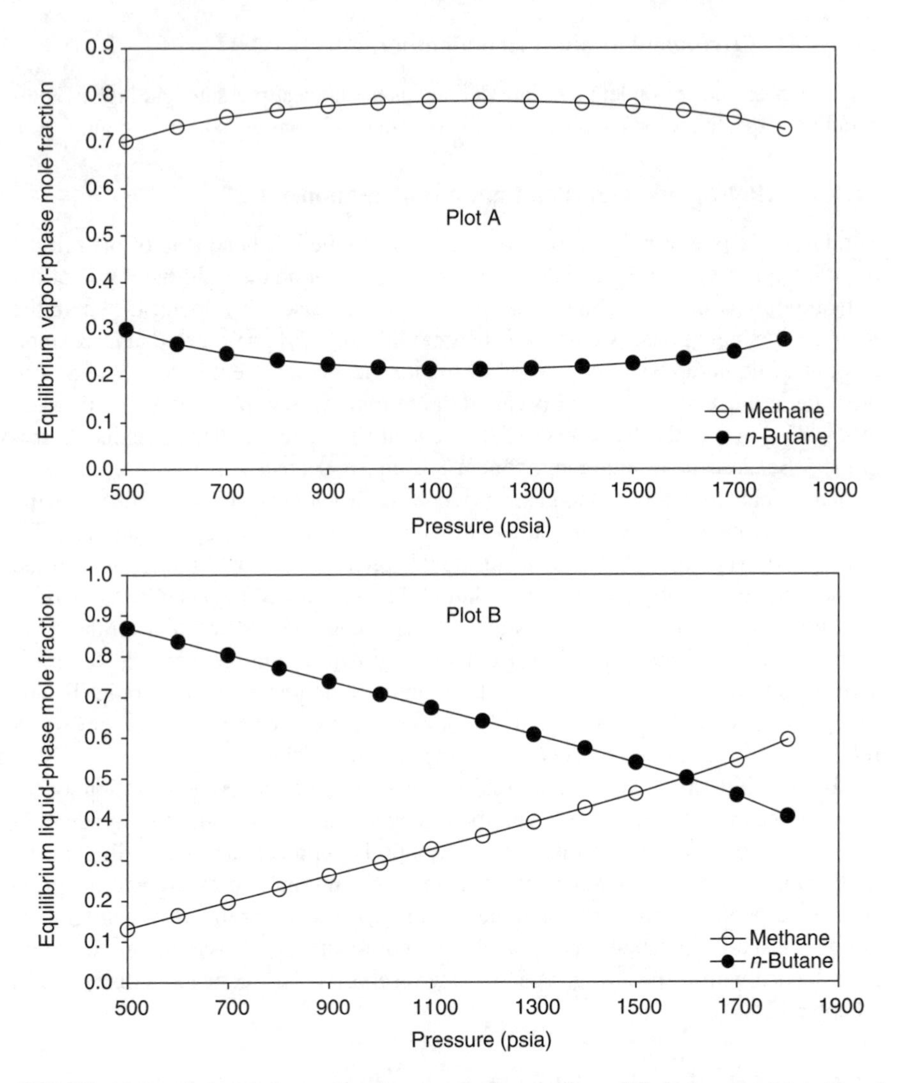

FIGURE 11.7 Mole fractions of methane and *n*-butane in equilibrium vapor (plot A) and liquid phases (plot B) at 150°F (dew-point system).

1200 psia. However, as pressures fall below 1200 psia, the mole fractions of methane and *n*-butane in the vapor phase begin to decrease and increase, respectively. This particular behavior is observed because up to 1200 psia, the vapor phase loses *n*-butane, however, below 1200 psia revaporization of *n*-butane begins, and at a value of 500 psia the entire system almost turns into a single-phase mixture having the original composition of 70 mol% methane and 30 mol% *n*-butane. However, when considering the mole fractions of methane and *n*-butane in the equilibrium liquid phase, a decrease in methane and increase in *n*-butane are evident, at all pressures below the dew point. This particular trend is seen because as pressure falls, initially more and more of *n*-butane appears in the liquid phase thus increasing its mole fraction. As pressure continues to

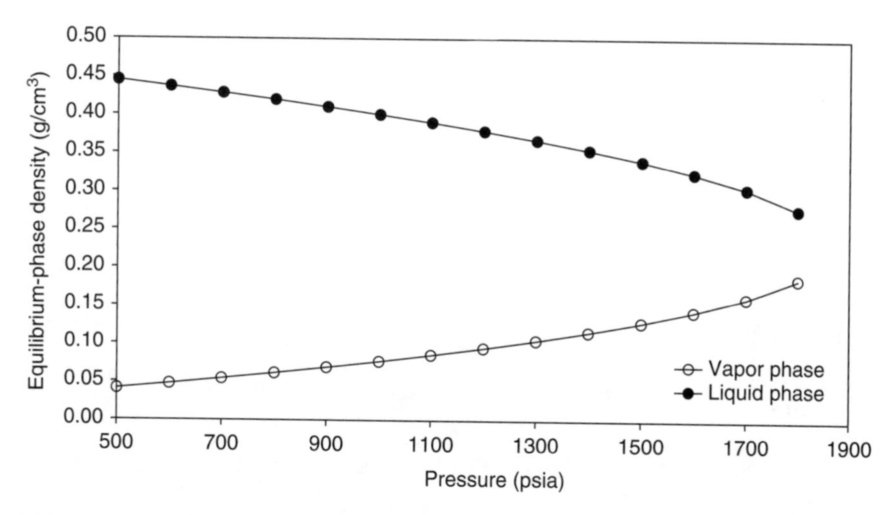

FIGURE 11.8 Densities of the equilibrium vapor and liquid phases for the methane and *n*-butane binary system at 150°F (dew-point system).

decline, revaporization of *n*-butane as well as methane (more dominant) begins and, as a result, the overall effect is an increase and decrease in their respective mole fractions.

Now the effect of compositions and pressures on the densities of the equilibrium vapor and liquid phases at various pressures below the dew point is considered. Figure 11.8 shows the density data of this binary system at various pressures and at an isotherm of 150°F. As seen in Figure 11.8, the density of the vapor phase decreases as pressure decreases, while the liquid-phase increases. Ordinarily, when pressure decreases, both densities should decrease, which is true for the vapor phase; pressure and composition generally favor this trend. As seen in Figure 11.7, up to a pressure of 1200 psia, the mole fraction of methane in the vapor phase increases, making it lighter and compounded by the falling pressure, thus results in reduction of density. However, at pressures below 1200 psia, even though a small decrease in the mole fraction of methane in the vapor phase is evidenced, the reduction in pressure tends to mask the compositional effect, causing the continuation in the reduction of density. In the liquid phase, compositional and pressure effects act against each other: pressure is reducing, but the phase is becoming increasingly heavier (see Figure 11.7). The reduction in pressure should cause a reduction in the density whereas compositionally, since the phase is getting heavier, the density should increase. Therefore Figure 11.8 data clearly show that the compositional effect is much more dominant than the pressure effect.

A behavior similar to what has been described for a dew-point case for this binary mixture can also be observed when the system exhibits a bubble-point behavior at a temperature of 100°F. The mixture is in single-phase liquid at pressures outside the phase boundary at 100°F. As pressure is reduced isothermally, a bubble point of 1991 psia is obtained. If the compositional and density data are plotted as a function of pressures below the bubble point (see Figures 11.9 and 11.10), the behavior is similar to that seen in Figures 11.7 and 11.8, respectively. The liquid phase becomes

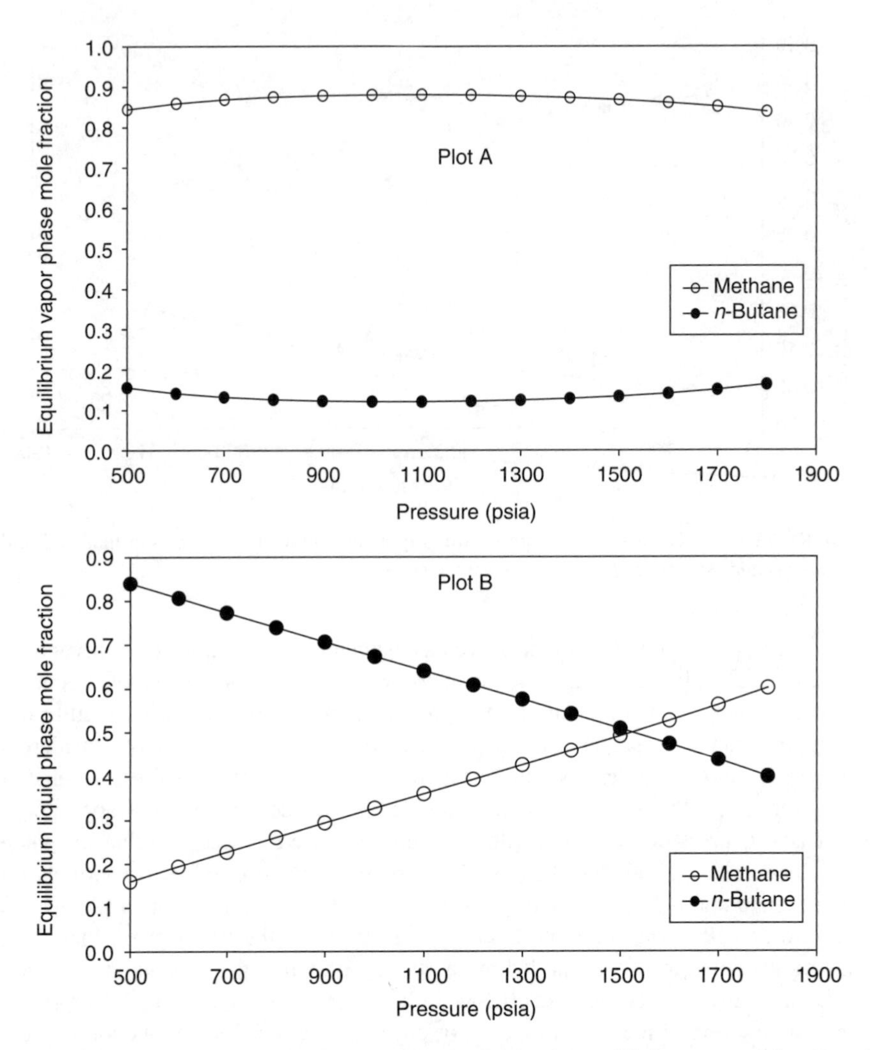

FIGURE 11.9 Mole fractions of methane and *n*-butane in equilibrium vapor (plot A) and liquid (plot B) phases at 100°F (bubble-point system).

increasingly heavier because it mainly loses its lighter component, methane, and also a small fraction of *n*-butane, as pressure declines. Initially the vapor phase is primarily composed of methane; however, after a pressure of about 1100 psia, the pressure is so low that even *n*-butane begins to vaporize and appears in the vapor phase, thus increasing its composition and decreasing the composition of methane. All these effects transpire into compositional and density data shown in Figures 11.9 and 11.10, respectively.

11.4.2 Effect of Changing the System Composition

As mentioned earlier, the phase envelope of a given mixture is determined by its overall composition. For example, the phase envelope shown in Figure 11.6 is valid for a fixed overall composition. Therefore, as soon as the overall composition is

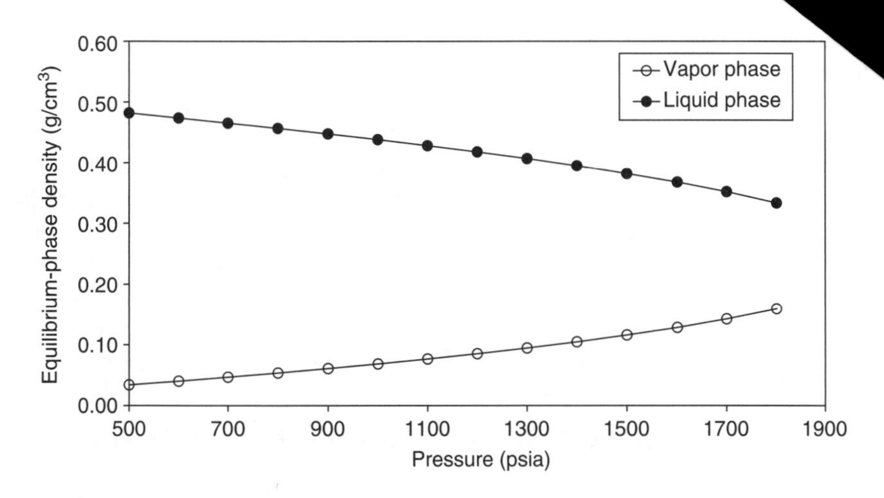

FIGURE 11.10 Densities of the equilibrium vapor and liquid phases for the methane and n-butane binary system at 100°F (bubble-point system).

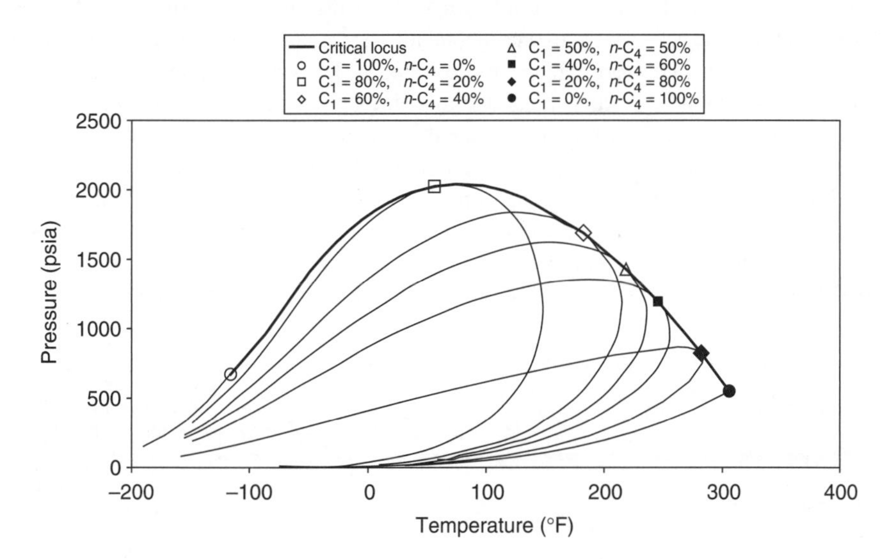

FIGURE 11.11 Phase behavior of methane and n-butane binary mixtures of varying overall compositions.

changed, the characteristics of the phase envelope also change. Or, in other words, for each possible overall composition a distinct phase envelope exists. This can be readily illustrated by plotting the *PT* conditions for phase envelopes of a number of possible overall compositions for any given binary system.

Figure 11.11 shows phase envelopes for five mixtures, each representing a different but fixed overall composition of the methane and n-butane binary system. Also shown in Figure 11.11 are the vapor pressure curves for methane and n-butane. All phase envelopes are bounded by the vapor pressure curves of the two pure ne lying toward the extreme left (methane) while the other lying xtreme right (n-butane). Also, it can be clearly observed from Figure 11.11 verall composition changes, all characteristics (size, location, cricondenn, cricondenbar, and critical point) of the phase envelopes on the PT plot ange. The mixture containing the lowest fraction of methane lies to the far right of the overall PT plot, whereas the mixture having the highest fraction of methane lies to the far left. The higher the amount of n-butane, the greater the slant toward right.

The critical temperature of different mixtures lies between the critical temperatures of the two pure components. The critical pressure, however, exceeds the values of both components as pure in most cases. The solid line shown in Figure 11.11 is the locus of critical points of mixtures of methane and n-butane.

11.5 PHASE BEHAVIOR OF MULTICOMPONENT MIXTURES

As outlined in Section 11.4, it is possible to show the phase envelopes encompassed by the two single-component vapor pressure curves, for various overall compositions, in the case of a binary system. However, as soon as a third or a fourth component is added, the number of possible overall compositions also increases, and instead of the vapor pressure curves of only two components, the remaining components also need to be considered. Therefore, phase envelopes of systems comprising more than two components cannot be readily illustrated in a simple manner. Hence, the phase envelope of a multicomponent mixture is usually shown for a fixed overall composition for a particular system that consists of n number of given components.

Figures 11.12 shows the phase envelope of a ternary system consisting of 70 mol% methane, 20 mol% n-butane, and 10 mol% n-decane. Figure 11.13 shows the phase envelope of a seven-component system consisting of normal alkanes, methane through n-hexane, and n-hexadecane for a fixed overall composition.

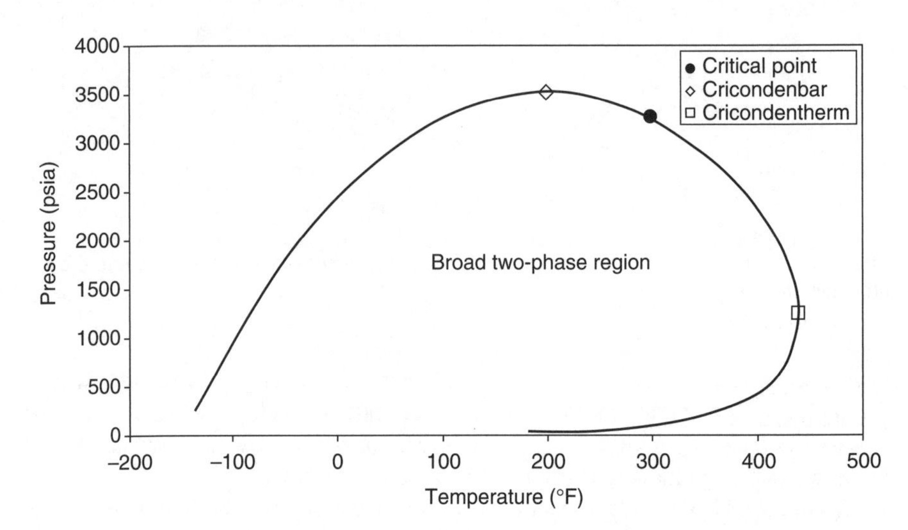

FIGURE 11.12 Phase behavior of a ternary system consisting of 70 mol% methane, 20 mol% n-butane, and 10 mol% n-decane.

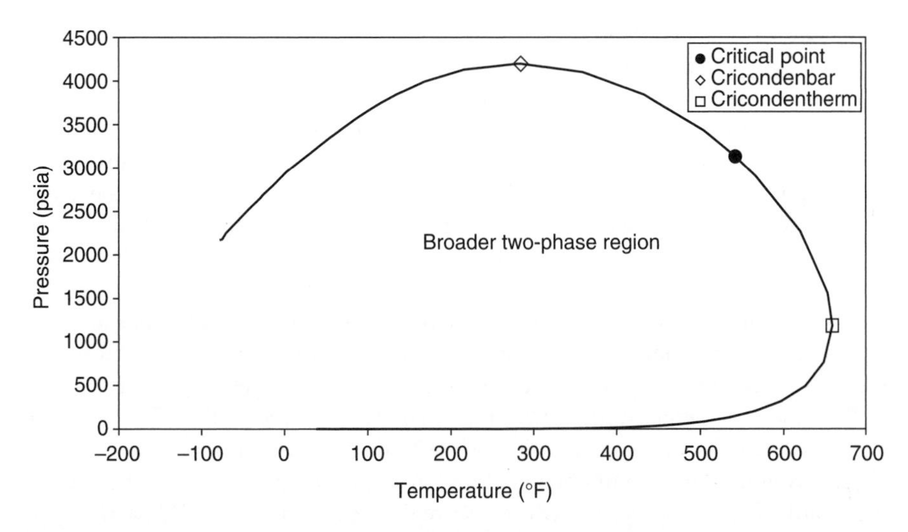

FIGURE 11.13 Phase behavior of a seven-component normal alkane system of methane through *n*-hexane and *n*-hexadecane.

As seen from these two phase envelopes, almost all the characteristic features of the phase envelope that were seen in the case of a two-component system are retained.

In general, the phase behavior of multicomponent hydrocarbon system in the liquid vapor region is quite similar to that of the binary system. However, as the system becomes more complex with a greater number of different components, the pressure and temperature ranges in which two phases exist increase significantly, or in other words, the separation between the bubble-point and dew-point curves becomes greater. The definitions of critical point, cricondentherm, cricondenbar, and so on, remain the same as that for a binary system, however, their magnitudes change with the number of components, their chemistry, and composition. The phase behavior of a typical multicomponent system shown in Figure 11.13, describes the behavior of reservoir fluids in most cases. Chapter 12 makes sense of the wide variety of shapes and sizes of phase envelopes for a variety of naturally occurring hydrocarbon mixtures, that is, the five reservoir fluids.

11.6 CONSTRUCTION OF PHASE ENVELOPES

The construction of phase envelopes basically involves the determination of bubble points and dew points of a given system at various isotherms that are plotted on a *PT* diagram. This is normally accomplished by laboratory measurements or by use of various prediction methods. The laboratory determination and the prediction methods are discussed in somewhat more detail in Chapters 15 and 16. However, for the sake of completeness a brief discussion is provided here.

A very basic laboratory determination of bubble-point and dew-point pressures involves the use of PVT cells. These PVT cells are generally capable of handling high pressures and high temperatures and are equipped with a mechanism of varying the pressures (by mercury injection/withdrawal or a mechanically driven piston) and temperatures (via a climatic air bath). The visual information for noting the formation of a new phase (a vapor or a liquid) is achieved through a special glass window. The fluid sample of a fixed overall composition is directly prepared in the PVT cell or is loaded from a separate vessel. After a homogenous single-phase sample is achieved, pressure depletion is carried out at a constant test temperature and the bubble point or dew point is determined by continuously monitoring the phase changes through the window manually or via a video recording mechanism. The sample is then taken back to single-phase conditions, a new isotherm is selected, and the procedure is repeated.

In many instances, however, it may not be possible for various reasons to obtain laboratory measurements of bubble-point or dew-point pressures. In such cases, if the overall composition of the fluid system is available, then equations of state (EOS) models are employed (see Chapter 16 on vapor liquid equilibria). Numerous EOS models exist in the literature that are frequently used to not only construct phase envelopes but also to obtain data, such as the compositions and densities of the equilibrium phases in the two-phase region. For example, all data shown in Figures 11.7–11.13 are obtained by EOS models. Several commercial and in-house PVT simulators have the capability of performing a variety of phase behavior calculations. Danesh[4] and Pedersen et al.[5] provide a detailed discussion of various EOS models.

PROBLEMS

11.1 The following diagram shows vapor pressures of components 1 and 2.
- (a) Identify the heaviest and the lightest component and justify your answer.
- (b) What is the state of component 1 at 300 psia and 0°F?
- (c) What is the state of component 2 at 200 psia and 110°F?
- (d) Graphically identify conditions (b) and (c).

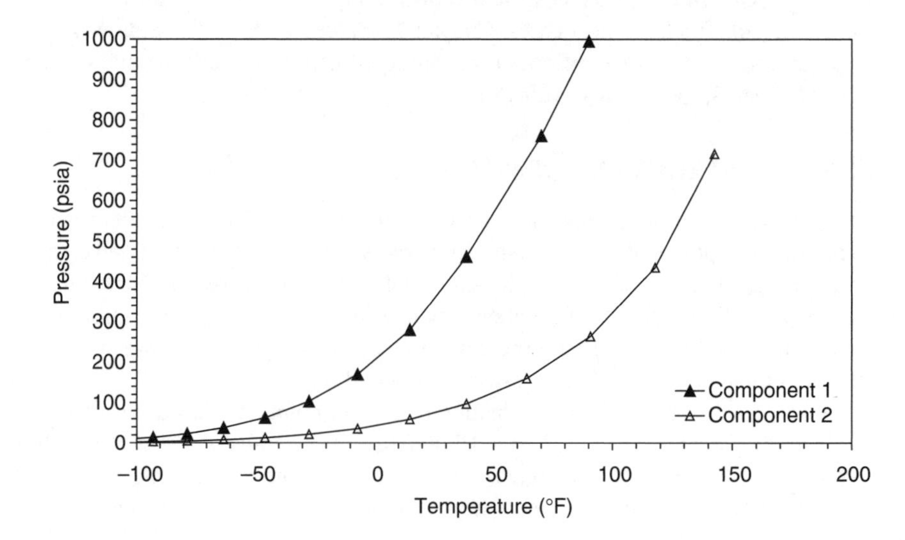

11.2 A PVT cell contains propane at 100.0°F and 1000 psia. What is the state of propane in the cell?

11.3 Pure carbon dioxide is held in a PVT cell at 50°F. Both gas and liquid are present so what is the pressure in the cylinder?

11.4 Plot the vapor pressures (on one graph) of nitrogen, carbon dioxide, and hydrogen sulfide on the reduced scales of P/P_c and T/T_c.

11.5 At what point on the *PT* diagram is a pure-component system invariant?

11.6 For a seven-component hydrocarbon system, determine the number of degrees of freedom that must be specified for the system to exist in a single-phase region.

11.7 The phase envelope of a natural gas mixture is described by the following diagram. Describe the phase behavior of this mixture along the pressure decline path at the reservoir temperature of 110°F.

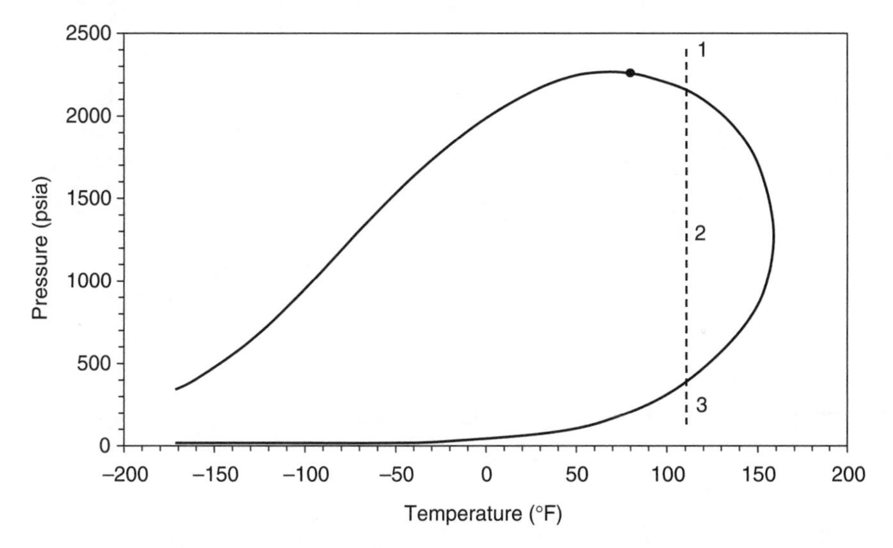

11.8 For the phase envelope shown in problem 11.7, determine T_c, P_c, cricondentherm, and cricondenbar. Also determine the bubble-point and dew-point pressures of this mixture at 50°F.

11.9 A ternary mixture of methane, carbon dioxide and *n*-butane having a fixed overall composition enters a two-phase region at a certain pressure and temperature. The composition of methane and carbon dioxide in the equilibrium vapor phase and liquid phase is measured as 85%, 12% and 15%, 30%, respectively. Determine the composition of *n*-butane in the equilibrium vapor and the liquid phase.

REFERENCES

1. Gibbs, J.W., *Elementary Principles in Statistical Mechanics; The Rational Foundation of Thermodynamics,* Scribner, New York, 1902.

2. Lee, B.I. and Kesler, M.G., A generalized thermodynamics correlation based on three-parameter corresponding states, *AIChE J.*, 4, 510, 1975.
3. Reid, R.C., Prausnitz, J.M., and Poling, B.E., *The Properties of Gases and Liquids,* McGraw-Hill, New York, 1987.
4. Danesh, A., *PVT and Phase Behavior of Petroleum Reservoir Fluids*, Elsevier Amsterdam, 1998.
5. Pedersen, K.S., Fredenslund, A., and Thomassen, P., *Properties of Oils and Natural Gases*, Gulf Publishing Company, Houston, 1989.

12 Phase Behavior Of Petroleum Reservoir Fluids

12.1 INTRODUCTION

Chapter 11 presented the phase behavior of synthetic or model binary and multicomponent systems. Although these systems were mixtures of simple well-defined hydrocarbon components, various important observations regarding phase behavior could be made. The two most important observations were: As the overall composition of the systems changed and additional components were added to the system, the phase envelopes changed. However, this particular change in phase envelopes was mainly the increase or decrease in the magnitudes of the cricondenbar, cricondentherm, critical point, and the size of the two-phase region. With such a simple model system, if a wide variation in their phase behavior exists, it can be readily realized that this variation becomes much more pronounced and elaborate when phase behavior for real reservoir fluids is considered. This happens mainly for two reasons: Numerous components make up these petroleum reservoir fluids, and as seen in Chapter 10, diverse chemical species are found in them. Therefore, phase envelopes of petroleum reservoir fluids are primarily determined by the types and quantities or by the chemistry and the overall composition of a particular mixture.

The main purpose of this chapter is to describe the phase behavior of the five reservoir fluids, (black oils, volatile oils, gas condensates, wet gases, and dry gases) on the basis of their phase envelopes. However, in addition to their phase envelopes, the various properties that are usually employed in order to distinguish or identify a particular fluid type are studied.

12.2 PREAMBLE TO THE PHASE BEHAVIOR OF PETROLEUM RESERVOIR FLUIDS

Before studying the phase envelopes of the five reservoir fluids, we must first consider Figure 11.11 because it shows the phase envelopes of a binary system of methane and *n*-butane having fixed overall compositions. As seen in this figure, as the quantity of methane in the mixture decreases, the phase envelope slants toward the right, indicating that the two-phase region existed at relatively higher temperatures on the pressure–temperature diagram. Therefore, the phase envelope of the mixture having the highest methane composition is located at lower temperatures, while the phase envelope having the highest composition of *n*-butane exists at much higher temperatures. The other readily noticeable feature of Figure 11.11 is the shift or the change in the location of the critical point as the overall composition of the system changes. As methane composition increases, the critical point trends toward the critical point of pure methane, also observed in the case of the mixture that contains the maximum fraction of *n*-butane, that is, the mixture critical point approaching the critical point of *n*-butane. However, as soon as a third even heavier component, *n*-decane, is added to a mixture of methane and *n*-butane, the size of the phase envelope increases substantially, thus covering wider ranges of pressure and temperature (see Figure 11.12). Also, the critical point shifts greatly toward the right on the phase envelope.

In general, phase behavior of petroleum reservoir fluids is in fact quite similar to what was described for simple model systems. Reservoir gases have relatively small phase envelopes and reservoir oils have relatively large phase envelopes. For reservoir gases,methane is the most dominant component of the system, resulting in narrower phase envelopes and the critical point appearing far down the left slope of the phase envelope (closer to the critical point of methane). However, reservoir oils, in addition to methane, also contain a wide range of intermediate and very large molecules, usually grouped as a plus fraction, resulting in much larger phase envelopes covering a wide range of pressure–temperature conditions and having very high critical points.

However, as seen in the phase envelope of the ternary system of methane, *n*-butane, and *n*-decane (Figure 11.12), the critical point in fact appears to the right of the cricondenbar. This happens because the component distribution is not continuous; the mixture does not contain any components between methane and *n*-butane or *n*-butane and *n*-decane. However, naturally occurring hydrocarbon mixtures are made up of a large number of components with a generally continuous distribution. The location of the critical point, right or left of the cricondenbar on the phase envelope, is a function of component distribution and the quantities in which these are present in petroleum reservoir fluids. Therefore, the critical point may or may not appear on the right-hand side of the cricondenbar. However, McCain[1] has pointed out that if the reservoir oils are deficient in intermediate components (often found in South Louisiana), the critical point appears to the right of the cricondenbar.

12.3 A BRIEF DESCRIPTION OF THE PLUS FRACTION

The plus fraction is discussed in somewhat more detail in Chapter 14; however, the effects of the magnitude of these plus fractions on the phase behavior of petroleum reservoir fluids can be quite significant. Therefore, a very brief discussion of plus fractions is provided in this section.

Petroleum reservoir fluids are generally composed of numerous components belonging to diverse chemical species. Therefore, the identification of every individual component in a given petroleum reservoir fluid is almost impossible. However, most of the lighter and intermediate components (typically the three nonhydrocarbons, if they are present, methane, and ethane through hexane) are usually clearly identified, whereas the heavy unidentified components are usually grouped as a plus fraction. This particular plus fraction is denoted by C_{7+}. In many cases the plus fraction is characterized further and instead of lumping everything into C_{7+}, the plus fraction is extended to a carbon number of 20 or 30 (e.g., C_{20+} or C_{30+}). The magnitude (composition or mol%) of the plus fraction (C_{7+}) in a reservoir fluid is used as one of indicators of fluid type.

12.4 CLASSIFICATION AND IDENTIFICATION OF FLUID TYPE

Naturally occurring reservoir fluids are generally classified into five different fluid types: *black oil*, *volatile oil*, *gas condensate*, *wet gas*, and *dry gas*.[1] Instead of generalizing the reservoir fluids as merely reservoir gases and reservoir oils, this particular type of detailed classification is very important since the fluid type is the deciding factor in many of the decisions that concern field development plan or reservoir management. Various issues such as fluid sampling, design of surface facilities, prediction of hydrocarbon reserves, and strategy for production, i.e., primary recovery or enhanced oil recovery (EOR) are all dependent on the type of reservoir fluid.

Identification of fluid type can be confirmed only by laboratory analysis, primarily including the phase behavior of reservoir fluids. In general, reservoir fluids are classified based on the location of the point representing the initial reservoir pressure and temperature with respect to the phase envelope of a given fluid. However, data available from production information, such as the initial producing gas-to-oil ratio (GOR), gravity, and color of the stock tank oil also serve to some extent as indicators of fluid type. In some cases it is quite possible that the production information may have an overlap because of which the fluid type cannot be identified. Therefore, in such cases, the reservoir fluid must be observed in the laboratory to identify its type. Although fluid classification is primarily based on the phase behavior of reservoir fluids, this ensuing discussion also provides the range of field indicators for each fluid type.

12.5 BLACK OILS

Black oils are sometimes also referred to as ordinary oils because they are the most common type of oil reserves. These types of oils are generally composed of more than 20% heptanes plus fraction indicating a large quantity of heavy hydrocarbon components. Therefore, their phase envelopes are the widest of all types of reservoir fluids, covering a wide temperature range. Due to a significant amount of heptanes plus fraction, the critical point is generally found high up the slope of the phase envelope. A typical black-oil phase envelope is shown in Figure 12.1. The curves within the phase envelope represent constant liquid volume, measured as percentage of total volume. These curves are called *iso-vols* or *quality lines*.

The pressure and temperature conditions in the reservoir and the separator are also shown in Figure 12.1. The vertical line ABC represents the pressure reduction in the reservoir at reservoir temperature. Reservoir pressures anywhere along line AB indicate that the oil is a single-phase liquid or is *undersaturated*, meaning that the oil is capable of dissolving more gas if more gas was present. As soon as the reservoir pressure reaches point B, the oil is at its bubble-point pressure and is said to be saturated from that point onward at all pressures below the bubble point.

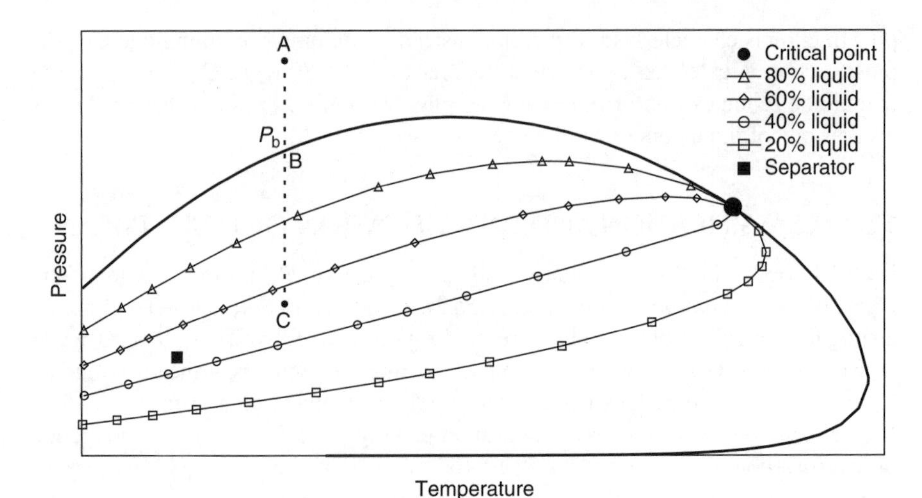

FIGURE 12.1 Phase envelope of a typical black oil.

Due to the high critical temperatures, the reservoir conditions are relatively far away from the critical temperature, which results in fairly low bubble-point pressures. A continued reduction in the pressure anywhere along line BC results in the release of more gas to form a free gas phase in the reservoir. At each pressure below the bubble-point pressure, the volume of gas on a percentage basis equals 100% minus the percentage of liquid.

Figure 12.1 shows that separator conditions are within the phase envelope in the two-phase region, lying on relatively high quality lines and indicating that a large amount of liquid arrives at the surface. The oil undergoes relatively less shrinkage as pressure and temperature conditions change, as the oil is produced. Therefore, black oils are sometimes also called *low-shrinkage oils*.

The initial producing GORs are usually within 250 to 1750 scf/STB, remaining constant when reservoir pressures are above bubble-point pressures. However, GORs may decrease initially when reservoir pressures fall below the bubble points because evolved gas remains immobile at very low saturations. As gas saturation exceeds critical gas saturation, gas also begins to flow, thus resulting in increases in GORs. The initial formation volume factor, which signifies the volume of reservoir oil required to produce a barrel of stock tank oil, is usually 2.0 res. bbl/STB or less. The stock tank liquid is very dark in color, indicating the presence of heavy hydrocarbon components. The stock tank oil gravity is usually lower than 45° API. The variation in the stock tank oil gravity is relatively small during the producing life of the reservoir. Laboratory analysis and field identification parameters of black oil are shown in Table 12.1.

12.6 VOLATILE OILS

In comparison to black oils, *volatile oils* contain relatively fewer heavy hydrocarbon components and more intermediate components (ethane through hexanes). Therefore,

TABLE 12.1
Classification of Petroleum Reservoir Fluids Based on Field Data and Laboratory Analysis

Reservoir Fluid	Field Data			Laboratory Analysis			
	Initial Producing GOR (scf/STB)	Initial API Gravity of Liquid	Color of Stock Tank Liquid	Mol% of C_{7+}	Phase Change in Reservoir	Formation Volume Factor (res. bbl/STB)	Reservoir temperature
Black oil	250–1750	< 45.0	Dark	> 20.0	Bubble point	< 2.0	$< T_c$
Volatile oil	1750–3200	> 40.0	Colored	12.5–20.0	Bubble point	> 2.0	$< T_c$
Gas condensate	> 3200	40.0–60.0	Lightly colored	< 12.5	Dew point	–	$> T_c$
Wet gas	> 50,000	Upto 70.0	Water-white	May be present in trace amounts	No phase change	–	>Cricondentherm
Dry gas	–	–	–	–	No phase change	–	>Cricondentherm

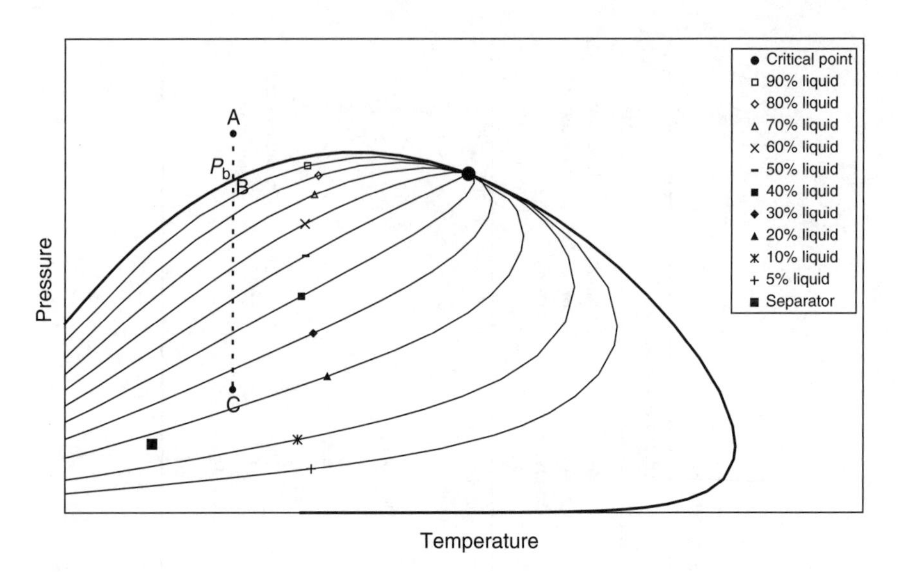

FIGURE 12.2 Phase envelope of a typical volatile oil.

the temperature range covered by the phase envelope of volatile oils is somewhat smaller compared to black oils. Also, considering the smaller percentage of heavy molecules and more intermediates, the critical point is much lower than that of black oils and lies in close proximity to the reservoir temperature, resulting in volatile oils also referred to as *near-critical oils*. The phase envelope of a typical volatile oil is shown in Figure 12.2.

The vertical line ABC shows the path taken by the isothermal reduction in pressure during production. Volatile oils generally have high bubble-point pressures. As seen in Figure 12.2, since the iso-vols are tighter and closer near the bubble-point curve; several quality lines are crossed by the pressure reduction path BC, also indicating that a small reduction in pressure causes the vaporization of a significant fraction of the oil; hence these oils are known as volatile oils. A volatile oil may become as much as 50% gas in the reservoir at only a few hundred psi below the bubble point. Due to these phase-behavior characteristics of volatile oils, the separator conditions typically lie on low quality lines. Therefore, the oil undergoes relatively high shrinkage as pressure and temperature conditions change as the oil is produced, causing volatile oils to be categorized as *high-shrinkage oils*.

The initial producing GORs of volatile oils typically range between 1750 and 3200 scf/STB, which remains constant when the reservoir pressure is above the bubble-point pressure. However, the GOR increases as production proceeds and reservoir pressure falls below the bubble-point pressure of the oil. Again, considering the fact that volatile oils are high-shrinkage oils, the initial formation volume factor is usually greater than 2.0 res. bbl/STB. The heptanes plus mole fraction in volatile oils usually ranges between 12.5 and 20.0%. The actual color of the stock tank liquid, to a great extent, depends on the compositional characteristics of the oil. However, stock tank liquids of brown, orange, or sometimes green color are common. The stock-tank oil gravity is usually higher than 40° API, and increases during production below the bubble-point pressure, particularly at high producing GORs, because significant liquid production is due to the condensation of the rich associated gases. The characteristic laboratory and field data of volatile oils are shown in Table 12.1. It should be noted here, however, that since volatile oils are near-critical oils; if the heptanes plus fraction is relatively low, the oils may exhibit a dew-point behavior (reservoir temperature higher than the critical temperature). In such cases, the stock tank oil color and its gravity may not serve as indicators for classifying volatile oils.

12.7 GAS CONDENSATES

Gas condensates are also known as *retrograde gases*. These names are derived from the fact that the phase behavior of these types of reservoir fluids is characterized by retro grade dew point and retrograde condensation. In comparison to black oils and volatile oils, gas condensate fluids contain fewer heavy hydrocarbon components in relatively less quantities and are usually much richer in the intermediate components. This results in a somewhat smaller phase envelope with a critical point moving further down the slope on the left-hand side of the phase envelope. Figure 12.3 shows a typical gas condensate phase envelope with the reservoir temperature lying between the critical temperature and the cricondentherm. At point A, the gas condensate is initially in single-phase vapor. However, as reservoir pressure decreases, the expanding fluid exhibits a retrograde dew point at point B. As pressure decline continues, liquid condenses from the gas due to retrograde condensation to form a free liquid or condensate in the reservoir. The separator conditions also lie within the phase envelope because further condensation from the produced gas occurs due to cooling. A recombination of

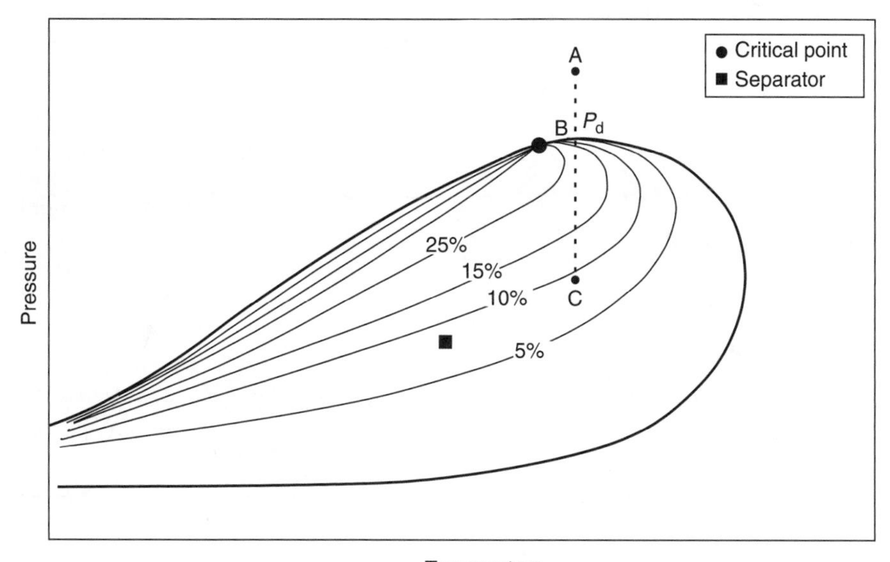

FIGURE 12.3 Phase envelope of a typical gas condensate.

the produced gas and the condensate at the surface represents the reservoir gas but not the total reservoir fluid because retrograde liquid is precipitated in the reservoir. It is commonly assumed that the condensate formed in the reservoir remains immobile due to its low saturation. However, experimental investigations by Danesh et al.[2] in glass micromodels and long cores to determine the critical condensate saturation revealed that the condensate can flow even at very low saturations. Ahmed[3] stated that near the well bore where the pressure drop is high, enough condensate might accumulate to give two-phase flow of gas and retrograde liquid.

Another major distinguishing feature between black oils or volatile oils (bubble-point systems) and gas condensates (dew-point systems) can be readily noticed by comparing the path of the vertical line BC in Figures 12.1 or 12.2 and 12.3. For bubble-point systems, the decline in pressure simply causes a reduction in the percentage of liquid, that is , the path of line BC crosses the continuously decreasing quality lines, or line BC crosses each iso-vol only once. However, in gas condensate fluids, the path of line BC initially crosses a quality line of low liquid percentage at pressures just below the dew point. As additional liquid appears due to retrograde condensation, quality lines of higher liquid percentage are now crossed. A further reduction in pressure in fact results in line BC crossing the same iso-vols for the second time, at some low pressures where the condensate begins to revaporize. This particular behavior is one of the most common features of gas condensate fluids and is generally characterized by a liquid drop-out curve, as shown in Figure 12.4. It should be noted here, however, that by the time the pressure falls below the dew point, the original phase envelope is no longer valid since the overall composition of the system changes during the production period. Therefore, special laboratory tests that simulate reservoir conditions are necessary and are described in a later part of this book (Chapter 15).

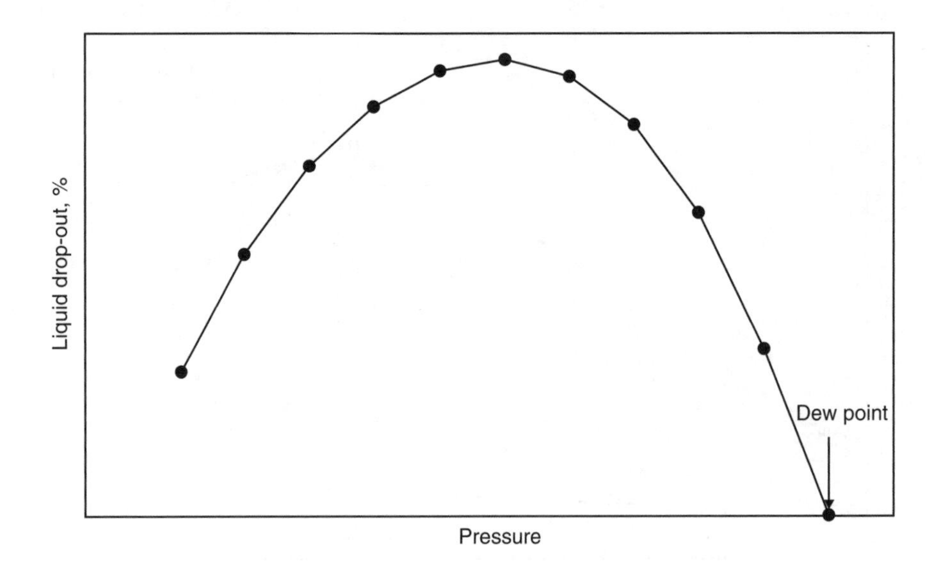

FIGURE 12.4 Liquid drop-out behavior of a gas condensate.

The GORs of gas condensate fluids are usually greater than 3200 scf/STB. The producing GOR remains constant at pressures above the dew point, while an increase is observed at pressures below the dew point. The upper limit of GOR is not well defined because values of 150,000 scf/STB have been observed.[4] However, GORs which are that high indicate that the phase envelope is relatively much smaller, having cricondentherms close to reservoir temperatures and resulting in the precipitation of very little retrograde liquid. Such types of reservoir fluids are also characterized as *lean gas condensates*. McCain in fact stated that as a practical matter, when producing GOR is above 50,000 scf/STB, the reservoir fluid can be treated as a wet gas (defined in the next section).

In general, the phase behavior of gas condensate fluids is rather sensitive to the concentration of the heptanes plus fraction, practically controlling the retrograde dew point and the subsequent condensation and its characteristics. The heptanes plus fraction in gas condensate fluids is generally less than 12.5 mol%. The stock tank liquid can be water-white or lightly colored having gravities in the range of 40 to 60°API. The API gravity of the condensate increases as pressure falls below the dew point because the condensate produced on the surface is much lighter and the fluid has already lost its retrograde liquid, which has already precipitated in the reservoir. The laboratory and field indicators of gas condensate fluids are summarized in Table 12.1.

12.8 WET GASES

The word *wet* does not mean that the reservoir fluid is wet with water but refers to the hydrocarbon liquid that condenses at surface conditions. A *wet gas* is primarily composed of predominantly smaller molecules, making its phase envelope much smaller, and it is located entirely over a temperature range below that of the reservoir. In other words, the reservoir temperature is greater than the cricondentherm. A wet gas therefore exists solely as a gas in the reservoir throughout the reduction in the reservoir pressure during depletion, line AB on the phase envelope, as shown in Figure 12.5. Since the pressure reduction path does not enter the phase envelope or the two-phase region, a wet gas does not drop out condensate in the reservoir during depletion. However, as seen in Figure 12.5, the separator conditions lie within the phase envelope, causing some liquid to be formed at the surface. A recombination of this liquid and the surface gas represents the gas in the reservoir.

Because no condensate is formed in the reservoir, the overall composition of the gas in the reservoir remains unchanged throughout the entire life of the reservoir. Therefore, the producing GORs,generally greater than 50,000 scf/STB, remain constant during the reservoir production life. The condensate produced by a wet gas on the surface is usually water-white, having a fairly high API gravity which remains constant because the overall composition of the gas in the reservoir is always the same during the life of the reservoir. The typical laboratory and field values for various wet-gas properties are provided in Table 12.1.

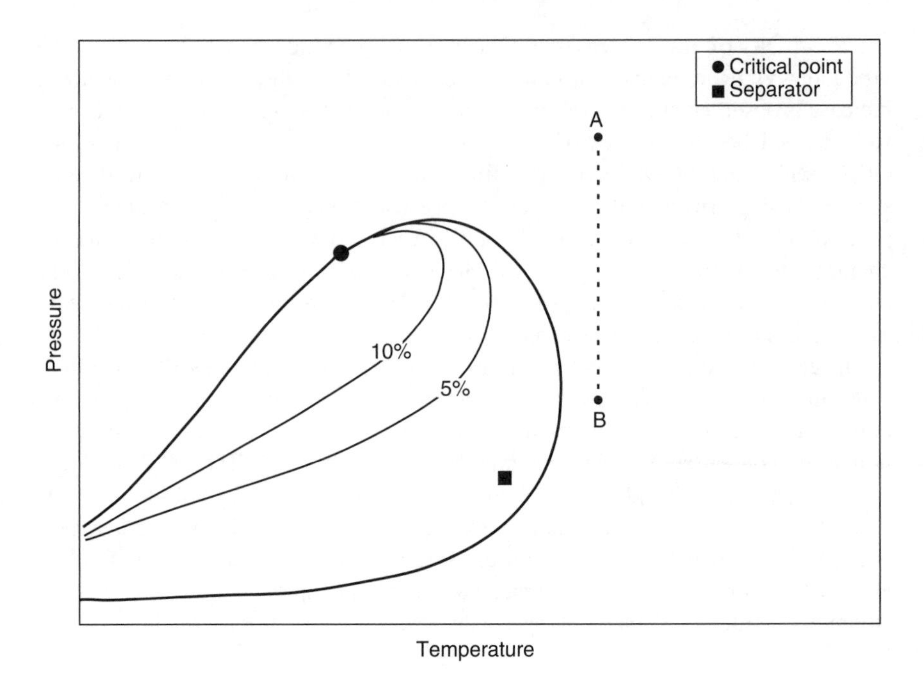

FIGUIRE 12.5 Phase envelope of a typical wet gas.

12.9 DRY GASES

The word *dry* in dry gases is used from a compositional standpoint, that is, the gas is primarily composed of methane and some intermediate components and is incapable of producing condensate even at the surface. Usually some liquid water is condensed at the surface. As shown in Figure 12.6, due to the compositional characteristics of dry gases, the phase envelope shrinks even further in comparison to wet gases and shifts toward very low temperatures. This results in both the production path in the reservoir and the separator conditions lying outside the entire phase envelope. Therefore, dry gas remains single phase in the reservoir and on the surface. A dry gas reservoir is sometimes simply referred to as a gas reservoir. The associated laboratory and field characteristics of dry gases are given in Table 12.1.

12.10 BEHAVIOR OF PETROLEUM RESERVOIR FLUIDS IN THE TWO-PHASE REGION

The phase behavior of petroleum reservoir fluids in the two-phase region can also be studied in a manner similar to behaviors studied in Chapter 11 for a well-defined binary system. Figure 12.7 shows the mole fractions of methane and C_{7+} fraction in the equilibrium vapor and liquid phases at pressures below the bubble point of 3222 psia at 260.3°F for a black-oil system. The equilibrium vapor phase and liquid phase density data are provided in Figure 12.8. A similar type of data (i.e., equilibrium phase compositions and densities for a gas condensate system having a dew

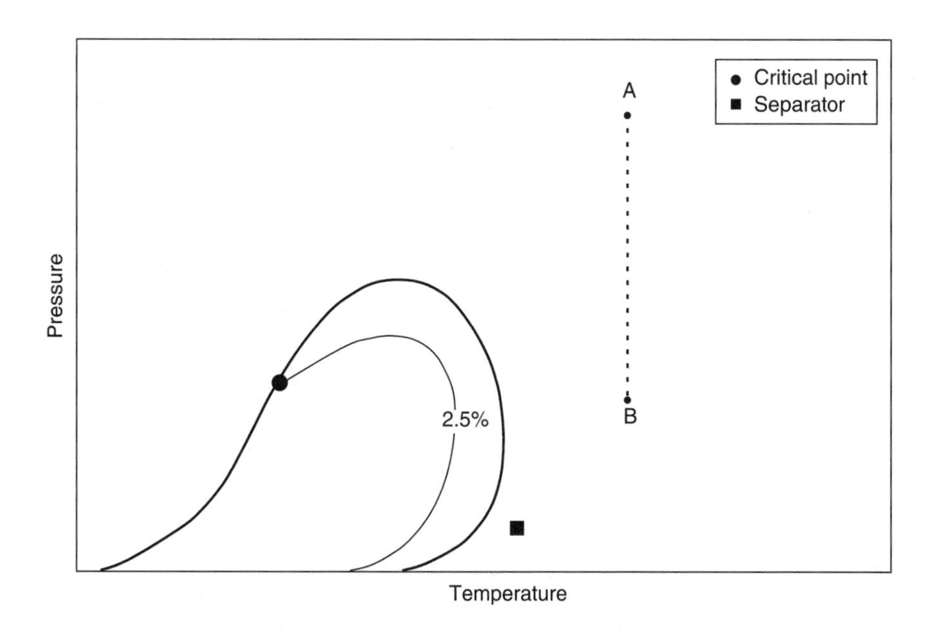

FIGURE 12.6 Phase envelope of a typical dry gas.

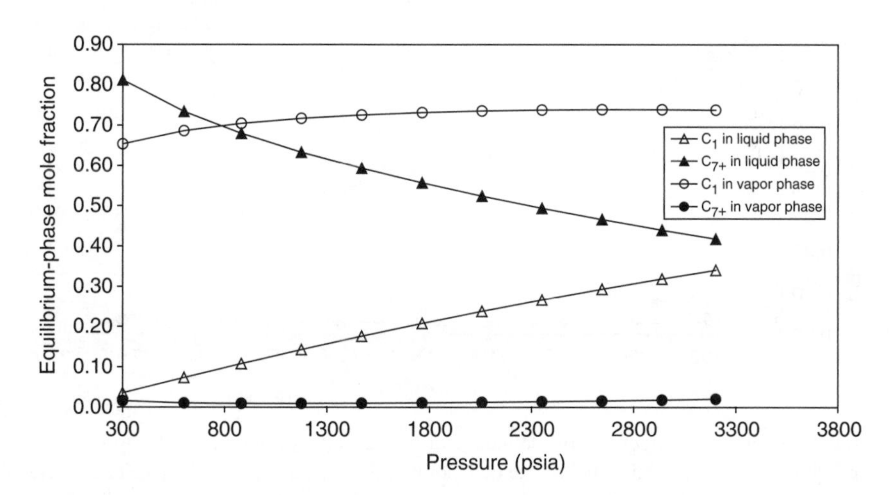

FIGURE 12.7 Mole fractions of methane and heptanes plus fraction in equilibrium vapor and liquid phases at pressures below the bubble point of 3222 psia at 260.3°F for a black-oil system.

point of 2303 psia at 290°F) is shown in Figure 12.9 and Figure 12.10, respectively. Despite the fact that these reservoir fluids are composed of several different components, the compositional data are shown only for methane and the C_{7+} fraction because these components primarily control the major aspects of phase behavior. (Note that the entire plus fraction is treated as one component.)

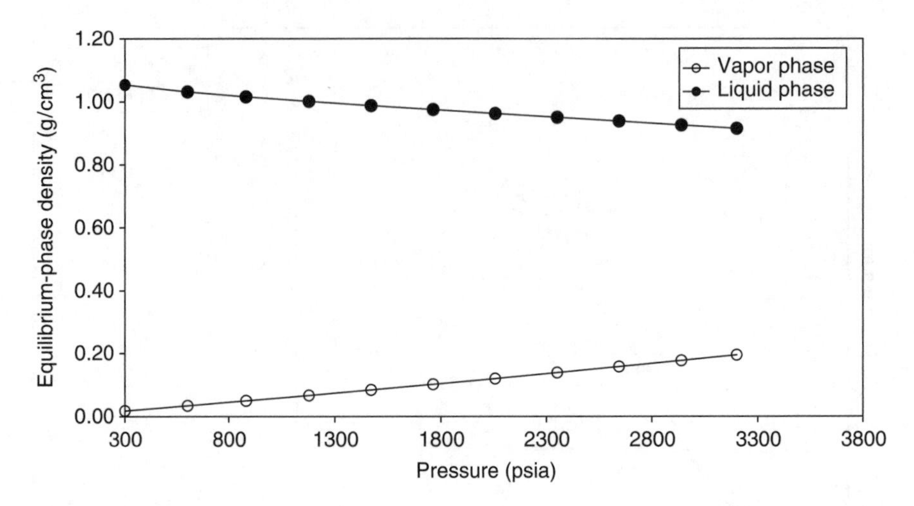

FIGURE 12.8 Densities of the equilibrium vapor and liquid phases at pressures below the bubble point of 3222 psia at 260.3°F for a black-oil system.

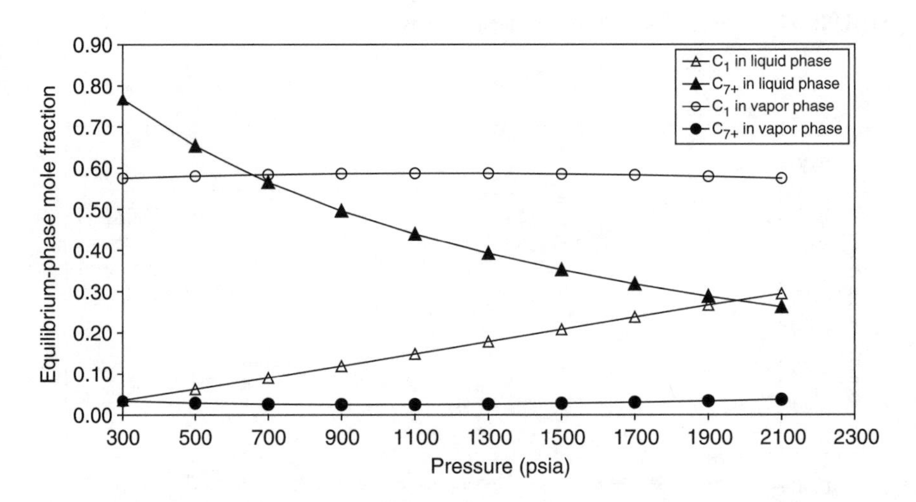

FIGURE 12.9 Mole fractions of methane and heptanes plus fraction in equilibrium vapor and liquid phases at pressures below the dew point of 2303 psia at 290°F for a gas condensate system.

As seen in Figures 12.7–12.10, the compositional and density characteristics of the equilibrium phases for both systems are similar. In fact, the behavior observed here for both the black-oil and the gas condensate system in the two-phase region is also qualitatively similar to the simple two-component system discussed in Chapter 11. These particular changes that occur in the characteristics of the equilibrium vapor and liquid phases at pressures below the bubble and dew points, occur precisely for the same reasons described in Chapter 11. Although, other components, such as the intermediates,

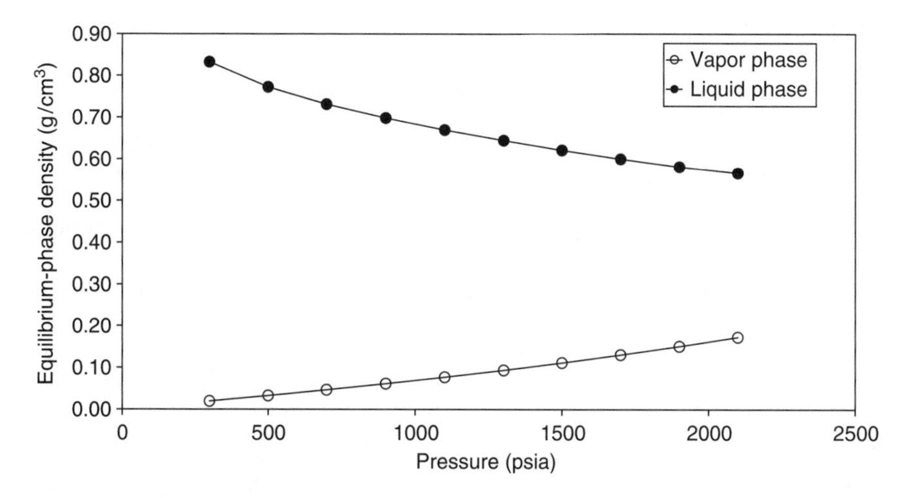

FIGURE 12.10 Densities of the equilibrium vapor and liquid phases at pressure below the dew point of 2303 psia at 290°F for a gas condensate system.

play a role as far as compositions and densities in the two-phase region are concerned, the majority of the influence comes from the lightest (methane) and the heaviest (C_{7+}) components.

PROBLEMS

12.1 The average gas–oil ratio for a fluid produced from a Middle Eastern reservoir is 300 scf/STB. The gravity of the stock tank oil is 26°API. What type of reservoir fluid is in this reservoir?

12.2 The initial reservoir pressure and temperature in a North Sea reservoir is 5000 psia and 260°F. The PVT analysis indicated the bubble-point pressure of the oil at 3500 psia. Is the reservoir fluid saturated or undersaturated? How do you know?

12.3 Compositional analysis of a reservoir fluid from a field in India reported a C_{7+} of 15.0 mol%, while the PVT analysis of this fluid indicated a formation volume factor of 2.5 res. bbl/STB. What type of reservoir fluid exists in this field?

12.4 Field data of a reservoir in the U.K. Continental Shelf initially reported a constant GOR of 5000 scf/STB and production of a light colored 45° API liquid. As production continued, the GOR and the API gravity of the liquid started to increase. What type of reservoir fluid exists in this reservoir and why?

12.5 Producing GOR from a field in the Norwegian Continental Shelf, monitored over a period of 7 years, was found to be 55,000 scf/STB. The gravity of the liquid that was produced on the surface was also found to be constant at 65°API. Classify this reservoir fluid.

REFERENCES

1. McCain, W.D., Jr., *The Properties of Petroleum Fluids*, PennWell Publishing Co., Tulsa, OK, 1990.
2. Danesh, A., Henderson, G.D., and Peden, J.M., Experimental investigation of critical condensate saturation and its dependence on interstitial water saturation in water-wet rocks, *Soci. Pet. Eng. Reservoir Evaluation Eng. J. (SPEREE)*, 336–342, 1991.
3. Ahmed, A., *Reservoir Engineering Handbook*, Butterworth-Heinemann, Woburn, MA, 2001.
4. McCain, W.D., Jr. and Bridges, B., Volatile oils and retrograde gases – what's the difference?, *Pet. Eng. Int.*, 45–46, 1994.

13 Sampling of Petroleum Reservoir Fluids

13.1 INTRODUCTION

Chapter 12 addressed the phase behavior of petroleum reservoir fluids by stating that the determination of fluid type is one of the most important aspects of field development planning or reservoir management. The fluid type in a hydrocarbon accumulation is classified primarily on the basis of laboratory analysis of phase behavior of a given fluid. In addition to the determination of fluid type, studies of pressure–volume–temperature (PVT) behavior of a given fluid also provide vital data for many reservoir engineering and production applications. However, this particular laboratory analysis is entirely dependent on the physical sample of petroleum reservoir fluid(s) from a given hydrocarbon accumulation. Therefore, it is necessary to obtain the physical sample of a reservoir fluid on which laboratory studies can be conducted, its fluid type confirmed, and PVT data obtained for overall reservoir management.

The process of obtaining a physical reservoir fluid sample from a given formation is called *sampling* and is probably the most important aspect of PVT, phase behavior, and reservoir fluid property studies. Reservoir fluid samples have the biggest influence on the quality of the measured laboratory data because if the samples are not representative, all measurements on them are invalid. It should be noted, however, that even though PVT and phase behavior measurements on the collected fluid may be accurate (under the assumption that the studies were conducted by following appropriate laboratory procedures), the results may not be representative of the actual reservoir fluid that exists in the formation. Additionally, sampling operations are also continuously under pressure from factors such as cost control and operational limitations due to which sampling can perhaps take place only once and there may not be a second chance of sampling if errors were made in the first instance. Therefore, given the objective of obtaining a valid reservoir fluid sample and various constraints and limitations, the task of sampling is indeed a very challenging operation of oil field activities.

Considering the challenges involved in obtaining representative reservoir fluid samples, Montel[1] asserts that fluid sampling is a *weak link* in what he calls a *fluid chain*. The fluid chain is basically a sequence of steps resulting in overall characterization of the reservoir fluid, which is applied to various processes such as reservoir engineering, surface processes, and geology. The main operations of the fluid chain are summarized in Figure 13.1.

Obtaining a valid representative reservoir fluid sample is fundamentally important; however, this fact may be overlooked during preparation of the laboratory study and totally ignored when results are interpreted. Therefore, the primary objective of this chapter is to introduce methods of sampling that are employed to obtain valid

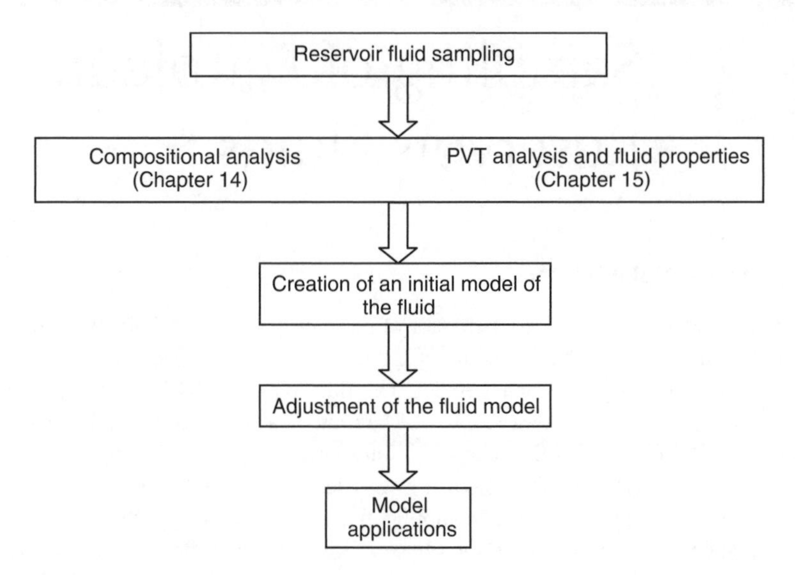

FIGURE 13.1 The fluid chain.

petroleum reservoir fluid samples for various laboratory studies. In addition to the sampling methods, various important issues, such as the preparation and conditioning of a well for obtaining valid samples, evaluating the representativity of collected samples, and proper handling of collected samples are discussed.

13.2 PRACTICAL CONSIDERATIONS OF FLUID SAMPLING

Sampling of reservoir fluids may occur at various stages of exploration, appraisal, or production. However, for proper identification of the fluid type and the performance of a proper PVT study, it is usually essential that samples be collected immediately after hydrocarbon discovery. Early sample collection is needed because as soon as reservoir pressure falls below the saturation pressure (bubble or dew point), the reservoir fluid forms two phases of gas and liquid. The mole ratio of the two phases that flow into the well generally is not equal to that formed in the reservoir. Therefore, the collection of a representative sample becomes a highly demanding task, and in many cases, an impossible task if sampling is delayed considerably.

Having established that the activities surrounding the sampling process are very important elements for a successful reservoir fluid analysis, equally important are well conditioning and use of the proper sampling method for the type of fluid. Therefore this discussion first focuses on well-conditioning procedures followed by the three basic methods of sample collection. Detailed sampling procedures are discussed in pertinent literature elsewhere.[2–5]

13.2.1 Well Conditioning

Well conditioning is an integral part of the sampling procedure and is extremely important for obtaining representative samples. Nevertheless it is often neglected or completely ignored. First it should be ensured that representative fluids are flowing out of the formation, by properly conditioning the well before sampling. The most important aspect of well conditioning is flowing the well at a reduced flow rate. This lowers the pressure draw-down, which can cause two-phase flow in the well bore, particularly if the fluid in the formation in saturated.

The problem of pressure draw-down in case of saturated gas and oil reservoirs is, however, unavoidable. The purpose of well conditioning, then, is to reduce the pressure draw-down by reducing the flow rate at the *lowest possible stable rate* at which the fluid entering the well bore more closely approximates the reservoir fluid. Although this is applicable to oil reservoirs, the method is not suitable for gas-condensate reservoirs because pressure buildup due to raised pressure may vaporize the retrograde liquid (much of this liquid phase is immobile) in the reservoir into the gas phase to form an even richer fluid than the original. For undersaturated oil reservoirs, the lowering of pressure draw-down raises the oil pressure, possibly above its original bubble point. As an example, the well conditioning of an undersaturated reservoir is represented in Figure 13.2.

Although these are general guidelines for conditioning the well for fluid sampling, the remainder of the conditioning process is usually dictated by the sampling method used and also the type of fluid, if known. El-Banbi and McCain[6] and McCain and Alexander[7] discussed specific sampling procedures for volatile oil and gas-condensate wells.

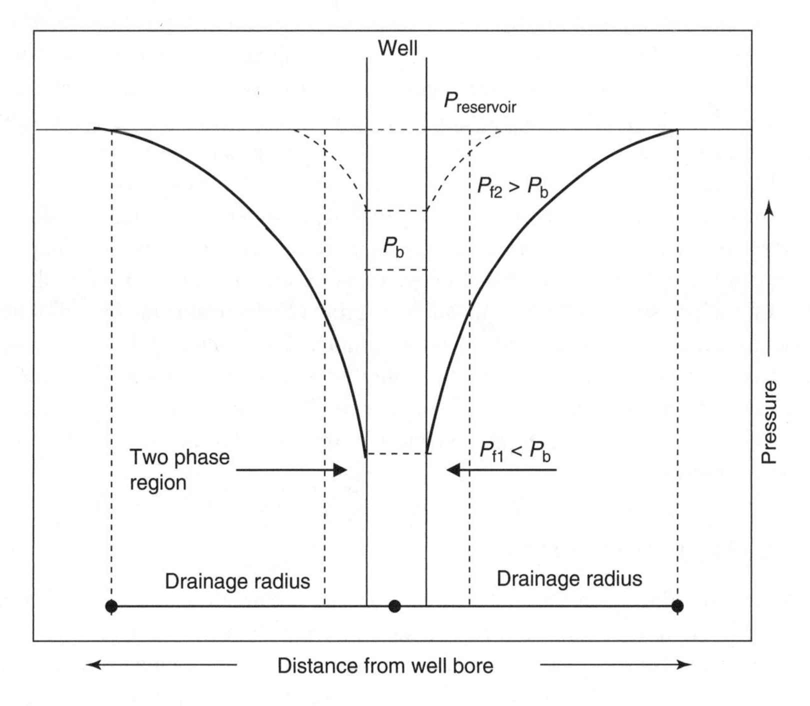

FIGURE 13.2 Well conditioning of an undersaturated reservoir. The flowing bottomhole pressure is raised from P_{f1} to P_{f2}, which is greater than the saturation pressure P_b.

13.3 METHODS OF FLUID SAMPLING

Sampling of reservoir fluids can be accomplished by three methods:

1. Subsurface (bottomhole) sampling
2. Wellhead sampling
3. Surface (separator) sampling

As the names suggest, samples of reservoir fluids are collected at these particular locations.

13.3.1 SUBSURFACE SAMPLING

Reservoir fluid samples collected downhole are called *subsurface,* or more commonly, *bottomhole samples* and are designed to draw in a representative sample of the reservoir fluid *in situ*. The essential element of these samplers is basically a floating piston-type device that is also equipped with a pressure compensation mechanism, usually a nitrogen gas charge. The gas charge maintains pressure on the collected sample so that it remains in single phase on its trip to the surface from the reservoir. The sampler is normally lowered into the well on a slick line or a conducting wireline device. Typical bottomhole samplers are capable of handling pressures up to 15,000 psi and temperatures of 350°F.

The well conditioning for bottomhole sampling involves a period of reduced flow rate that generally lasts for 1 to 4 days, depending upon formation and fluid characteristics and the drainage area affected. Following this period, the well is shut-in and allowed to reach static pressure. Again, the shut-in period generally lasts from 1 day up to 1 week or more, based primarily upon formation properties. The shut-in period forces gas into solution in the oil, thus raising the saturation pressure.

Although, the method of subsurface sampling is the most desirable, it does have some limitations. The method is generally not recommended for depleted gas-condensate reservoirs. The other problem is related to water that frequently remains at the bottom of the hole. In case of saturated oils (or the ones with a gas cap), a static pressure gradient is run and interpreted to determine the height or depth of the gas–oil contact (GOC) and oil–water contact (OWC) in the tubing. During the collection of a bottomhole sample, the sampler is lowered between the GOC and OWC. A non representative sample will be collected if the sampler is too close to GOC, while formation water is collected if the sampler is lowered below the oil–water interface.

13.3.2 WELLHEAD SAMPLING

Wellhead sampling is only possible for fluids that are single phase under wellhead conditions. Therefore, samples are obtained directly from the wellhead when it is known that the flowing conditions at wellhead are within the single-phase region. For this purpose, the wellhead flowing pressure must be sufficiently above the saturation pressure of the fluid at wellhead temperature. Hence, some information on the phase envelope (saturation curve) of the fluid must be available in advance. Under suitable conditions, wellhead sampling is the most reliable, efficient, and cost-effective sampling method available.

13.3.3 Surface Sampling

The surface or separator sampling method is by far the most commonly employed method for collecting reservoir fluid samples. This method has wider applications than either the subsurface or wellhead sampling because separator sampling is applicable for collecting samples of all types of reservoir fluids. Surface sampling is generally recommended for sampling a gas-condensate reservoir unless wellhead conditions are optimal. Separator sampling has also been used successfully for oil reservoirs. Moreover, separator samples also offer an added advantage because they can be recombined in varying proportions to achieve a desired bubble-point pressure.

This particular method of sampling also requires that the reduced gas and liquid flow rates must be monitored continually during the period of stable flow at reduced flow rate. A minimum test of 24 h is recommended, but much longer period may be needed if the pressure draw-down in the reservoir has been high. At the conclusion of the test period with a stabilized producing gas/liquid ratio, the well is properly conditioned and ready for the collection of separator samples.

The principle of separator sampling is quite simple and basically consists of drawing a given number of companion gas and liquid samples from the test separator. These individual samples are then physically recombined in accordance with the producing GOR to create the reservoir fluid or mixed in a certain ratio that yields a reservoir fluid having a particular saturation pressure. Therefore, accurate measurement of gas and liquid flow rates is one of the primary requirements for successful separator sampling. Along with the gas and liquid flow rates, information on pressure and temperature conditions of the separator is also necessary because this is used in the recombination process and also in evaluating the validity or representativity of the collected sample (see Section 13.4). All the necessary sampling information is normally recorded on a sampling sheet, as shown in Table 13.1.

Another important consideration while sampling from separators is that bottom sampling of gas lines and top sampling of liquid lines should be avoided. Any phase change in the gas and the liquid sample results in carry over of liquid in the separator gas and trapping gas in the separator liquid sample, which introduces uncertainties in the recombination process. A configuration for obtaining representative separator samples is shown in Figure 13.3. The vessel (usually about 5 L) used for collecting gas samples is fully evacuated and connected by a very short (low-volume) line. Sampling of liquids is best carried out by the controlled displacement of a liquid, either directly or via a floating piston. Typical liquid sample vessels have a capacity of ~600 cm^3.

TABLE 13.1
Sampling Sheet for Petroleum Reservoir Fluid Samples

Company:________	Sampling date:________
Well name:________	Sample type:________
Geographic location:________	

FORMATION DATA

Formation name :		
Date first well completed:		
Original reservoir pressure:	psi at	ft
Original produced GOR:	scf/bbl:	
Production rate:	bbl/day	
Separator pressure and temperature:	psi	°F
Oil gravity at 60°F:	°API	

WELL DATA

Total depth:	ft	
Last reservoir pressure:	psi at	ft
Date completed:		
Producing interval:	-	ft
Reservoir temperature:	°F	
Normal production rate:	bbl/day	
GOR:	scf/bbl	
Separator pressure and temperature:	psi	°F
Standard (base) pressure:	psi	
Tubing and casing dimensions:	in.	
Tubing and casing depth:	ft	

SAMPLING DATA

Date:		
Reservoir pressure and temperature:	psi	°F
Status of well:	Shut-in time	h
Separator pressure and temperature:	psi	°F
Flowing bottomhole pressure:	psi	
Separator gas cylinder number and size:	L	
Separator liquid cylinder number and size:	cm^3	
Bottomhole sampler number and size:	cm^3	

COMMENTS

13.4 EVALUATING THE REPRESENTATIVITY OF FLUID SAMPLES: QUALITY CHECKS

Once the reservoir fluid samples are collected for example bottomhole or separator samples, quality checks are performed to evaluate the validity of the collected samples. The most common check on bottomhole samples is the measurement of bubble-point pressure at surface temperature or reservoir temperature. If the bubble-point pressure

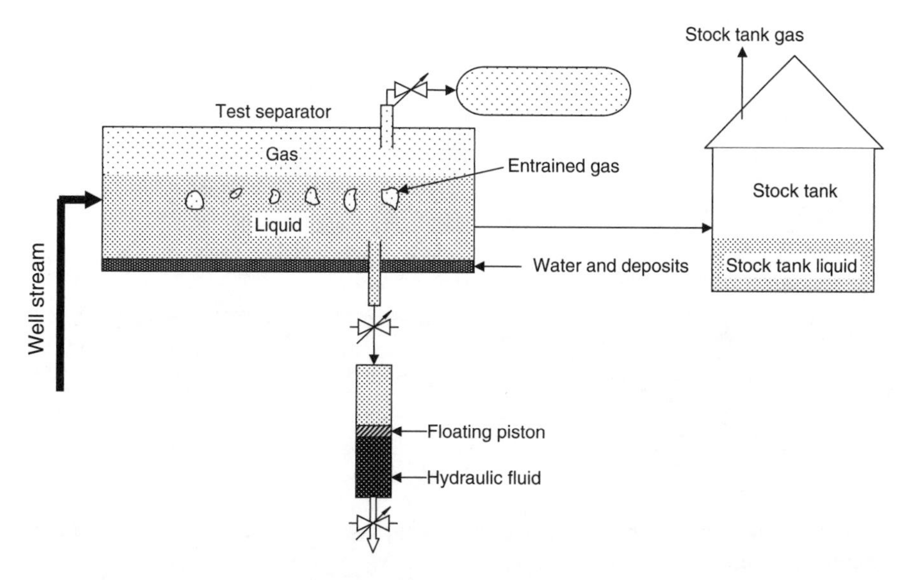

FIGURE 13.3 Configuration for obtaining representative separator gas and liquid samples.

at surface temperature exceeds the sampling pressure, then this is an indication that the sampling device either leaked oil or collected free gas. In the case of saturated reservoirs (gas cap and oil column in equilibrium), if proper well-conditioning procedures have been followed, the bubble-point pressure of the bottomhole sample normally corresponds to the existing reservoir pressure. However, in the case of undersaturated reservoirs, the bubble-point pressure of the bottomhole sample is less than the existing reservoir pressure.

For separator gas samples, the most common validity check is the opening pressure of the separator gas bottle. Representative separator gas samples at separator temperature read an opening pressure equal to the separator pressure. The other logical validity check is the measurement of dew point and bubble point of the separator gas and the liquid sample at separator temperature. For valid samples, the dew and bubble points should be the same and equal the separator pressure at separator temperature because both phases are in equilibrium at the separator conditions. The relationship between the phase envelopes of separator gas and liquid phase is illustrated in Figure 13.4. This type of relationship is similar to the equilibrium that exists between the fluid from the gas cap and the oil column in a saturated reservoir, that is, the dew point of the gas cap fluid and the bubble point of the oil column fluid at reservoir temperature equal the existing reservoir pressure.

13.5 FACTORS AFFECTING SAMPLE REPRESENTATIVITY

The presence of heavy organic solids such as aphaltenes or waxes in petroleum reservoir fluids can also have a significant impact on the characteristics of the sampled fluids. The solubility of these organic solids is generally dependent on prevalent pressure

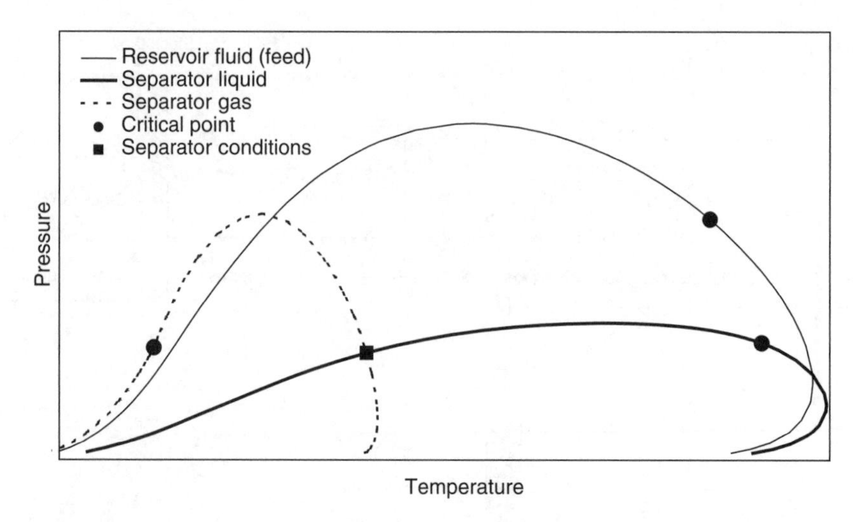

FIGURE 13.4 Relationship between the phase envelopes of separator gas, liquid, and reservoir fluid (feed).

and temperature conditions and the overall composition of the reservoir fluid. The path taken by reservoir fluids from the reservoir to the surface during the production process is accompanied by changes that take place in both pressure and temperature. These pressure and temperature changes can influence the solubility behavior of asphaltenes and waxes, causing precipitation at certain points in the installation, such as the tubing or separator. Therefore, a risk always exists that they may not be sampled quantitatively. In such a case, obviously the recombination of separator gas and liquid samples (with solids already precipitated elsewhere) does not result in a representative reservoir fluid sample because the original fluid has already lost some of its components. Unfortunately, the effect of these solid organic constituents on the phase behavior of the reconstituted reservoir fluid can be quite considerable. Therefore, extreme caution is needed when sampling such fluids.

The other factor that can affect sample representativity of an otherwise valid sample is the mishandling of fluid samples in laboratories. The possibility of mishandling occurring in a laboratory is greatest if subsamples are drawn under incorrect pressure and temperature conditions between different stages or tests of laboratory analysis. For instance, when a separator gas sample is transferred to a PVT cell at room temperature, it is quite possible that the liquid (condensate) accumulated (due to low room temperature compared to separator temperature) may remain at the bottom of the sample bottle and only the lean gas is transferred. This lean gas when recombined with the separator liquid in fact results in a nonrepresentative reservoir fluid sample. This mishandling obviously results in completely altering the overall composition of the separator gas and the sample is not useful for carrying out any meaningful laboratory analysis. In precisely the same manner, separator liquid samples can also be affected if the reservoir fluid also contains solid organic constituents. Especially, waxes may precipitate and deposit along the walls of the sample cylinder, resulting in the transfer (if carried out at room temperature) of only the clear liquid. Again, this alters the overall composition of the separator liquid and the remainder of the separator liquid in the sample bottle is rich in solid constituents, while the liquid transferred into a PVT cell is much lighter, resulting in a nonrepresentative recombination. Therefore, as a precautionary measure, separator sample bottles are always conditioned by heating to, or slightly above, the separator temperature prior to using them in any recombination process.

PROBLEMS

13.1 A North Sea gas-condensate reservoir was sampled from surface separators. The collected samples were separator gas and condensate. At the time of sampling, reservoir pressure was still above the dew-point pressure. If the separator gas and liquid samples are recombined in the producing GOR, what fluid would the recombined sample represent?

13.2 A North Slope oil reservoir was sampled from surface separators. Both the separator gas and the separator liquid were tested for validity and found to be valid and representative. Next, the samples were recombined in the producing GOR in a PVT cell. The separator liquid and the separator gas samples were transferred to the PVT cell at separator and ambient temperature, respectively. Does the recombined sample represent the reservoir fluid? If yes/no, why?

13.3 Separator gas and liquid samples are drawn from a test separator from an Alaska North Slope oil field. The separator operating conditions are 110°F and 500 psia. A validity check was carried out to test the representativity of the collected samples. The validity check revealed a bubble-point pressure of 400 psia at 110°F for the separator liquid sample. Is the separator liquid sample representative? If it is not, why? Give appropriate reasoning for your answer.

13.4 A bottomhole sample was collected from a well in an oil field in China. The sampled formation is saturated with a gas cap and an oil column in equilibrium. The validity check of the collected bottomhole sample revealed a bubble-point pressure of 4500 psia at the reservoir temperature of 212°F. The corresponding reservoir pressure at the time of sampling was recorded at 4505 psia. Is the collected bottomhole sample valid?

REFERENCES

1. Montel, F., Lecture notes, IVC-SEP Ph.D. Summer School, Lyngby, Denmark, 1995.
2. Reudelhuber, F.O., Sampling procedure for oil reservoir fluids, *J. Pet. Technol.*, 15–18, 1957.
3. Reudelhuber, F.O., Separator sampling of gas-condensate reservoirs, *Oil Gas J.*, 138–140, 1954.
4. Reudelhuber, F.O., Better separator sampling of crude-oil reservoirs, *Oil Gas J.*, 81–183, 1954.

5. American Petroleum Institute, API recommended practice for sampling petroleum reservoir fluids, *API,* 44, Dallas, TX, 1966.
6. El-Banbi, A.H. and McCain, W.D., Jr., Sampling volatile oil wells, Society of Petroleum Engineers (SPE) paper number 67232.
7. McCain, W.D. Jr., and Alexander, R.A., Sampling gas-condensate wells, *Soc. Pet. Eng. Reservoir Eng. J.*, 358–362, 1992.

14 Compositional Analysis of Petroleum Reservoir Fluids

14.1 INTRODUCTION

The phase behavior and properties of all petroleum reservoir fluids are uniquely determined by four primary variables: pressure, temperature, chemistry, and the overall *composition*. In other words, the state of a reservoir fluid system is fully defined when these four variables are specified. Also, Chapter 13 addresses the compositional analysis of reservoir fluids as an important component of the fluid chain in which one of the objectives is to create an initial compositional model of the fluid used in a variety of applications for hydrocarbon recovery. This particular compositional model is basically a description of the presence and concentration (composition) of various components in a given reservoir fluid and is handled by equations-of-state (EOS) models to simulate the phase behavior and physical properties required in various hydrocarbon recovery processes. Therefore, detailed and accurate compositional data, usually in terms of mol%, are a prerequisite for generating reliable EOS-based predictions of phase behavior and physical properties of the reservoir fluids over a wide range of conditions. These predictions are then compared with experimentally determined phase behavior and physical properties; and if required, the EOS models are tuned or calibrated and become an integral part of compositional reservoir simulators.

It should be mentioned here that phase behavior and physical properties could be measured in a laboratory on samples that are physically recombined (separator samples) or are available in the original state (bottomhole samples) without knowing the composition of reservoir fluid sample. However, such measurements do not have any meaning unless the overall composition of the reservoir fluid is known in order to evaluate the results and use them in EOS-based predictions. Therefore, considering the importance of compositional data, the compositional analysis of petroleum reservoir fluids is a vital component of any laboratory analysis. Therefore, the primary objective of this chapter is to introduce the methods employed for compositional analysis of petroleum reservoir fluids. Although, compositional data are measured under a variety of conditions, such as equilibrium vapor and liquid phases in the two-phase region, this discussion mainly focuses on the determination of the overall original composition of reservoir fluids.

14.2 STRATEGY OF COMPOSITIONAL ANALYSIS

Unfortunately, no standard methods exist in the petroleum industry for accomplishing the measurement of reservoir fluid compositions. This happens to be the case, perhaps, because petroleum reservoir fluids are composed of thousands of different components covering a wide range of boiling points and molecular weights. Obviously, the identification and quantification of each component is an almost impossible task. Therefore, different analytical techniques are employed for obtaining the molecular composition. However, this diversification inevitably leads to differences in the reported compositions, which can significantly affect the quality of EOS-based predictions.

This section focuses on the so-called "industry accepted" methods for determination of reservoir fluid compositions. The analytical protocol generally accepted today includes the compositional description of reservoir fluids by analyzing the gas and liquid samples separately and mathematically recombining their compositions in the correct ratio to obtain the composition of the original single-phase reservoir fluid. In addition to this conventional method, other advanced techniques, such as the direct determination of composition of the live reservoir fluid, have also been used successfully. The manner in which these different techniques are applied to reservoir fluid samples are described in the following subsections.

14.2.1 Surface Samples of Separator Gas and Liquid

Separator samples of the gas and liquid phases are analyzed for their compositions separately by a technique called gas chromatography. The individual samples are analyzed in their pressurized states and are subsequently recombined on the basis of producing gas–oil ratio (GOR) to determine the live reservoir fluid composition.

14.2.2 Blow-Down Method

The blow-down method is applied to bottomhole samples or physically recombined separator gas and liquid samples. A large volume of the live recombined sample is flashed at atmospheric pressure to form generally two stabilized phases of gas and liquid. These two phases are then individually analyzed by gas chromatography or a combination of gas chromatography and true boiling-point distillation. The analyzed compositions are then mathematically recombined using the ratio of the separated phases to determine the live fluid composition.

This technique can give reliable results for large samples of high-pressure liquids (black oils or volatile oils) where the error involved in measurement of the two-phase ratio is relatively small. However, for fluids such as gas condensates or wet gases, where the condensate volume formed by blow down is low, the error involved in measurement of the two-phase ratio is relatively high and the technique is unreliable.[1,2]

14.2.3 Direct Determination of Composition

The direct determination technique offers an effective alternative to the blow-down method where errors can accrue, especially in cases of fluids such as gas condensates or wet gases. In this technique, the physically recombined separator gas and liquid samples or the bottomhole sample are injected under pressure directly onto a gas chromatograph and the total sample is analyzed for composition. A small sample, of the order of microliters, of the live fluid is pinched by an auxiliary fluid (solvent) at the test pressure. The flow of the solvent directs the slim slug of the sample into a high pressure valve that is then exposed to the flow of a hot carrier gas, which injects the live fluid into the gas chromatograph.[3] In addition to the determination of original single-phase live reservoir fluid composition, the direct sampling technique has also been successfully applied to the compositional analysis of equilibrated gas and liquid phases (e.g., the two-phase region within the phase envelope) in various laboratory tests.[4,5]

14.3 CHARACTERISTICS OF RESERVOIR FLUID COMPOSITION

Before the actual techniques of gas chromatography and true boiling-point distillation used to identify and quantify the various components present in a reservoir fluid are discussed, an understanding is needed of the manner in which various constituents of a reservoir fluid are classified. Table 14.1 shows the single-phase compositional data of a black oil. As seen in this table, the entire compositional spectrum

TABLE 14.1
An Example of Black Oil Composition, Showing Well-Defined Components, Pseudo Fractions, and the Plus Fraction (Residue)

	Component	Composition (mol%)
Well defined or discrete components	Nitrogen	0.81
	Hydrogen sulfide	0.60
	Carbon dioxide	1.47
	Methane	45.93
	Ethane	7.32
	Propane	6.42
	i-Butane	1.42
	n-Butane	3.87
	i-Pentane	1.68
	n-Pentane	2.05
Pseudo fractions or SCN groups	C_6	2.93
	C_7	2.30
	C_8	2.21
	C_9	1.66
	C_{10}	1.97
	C_{11}	1.61
	C_{12}	1.39
	C_{13}	1.36
	C_{14}	1.28
	C_{15}	1.22
	C_{16}	1.09
	C_{17}	1.04
	C_{18}	0.98
	C_{19}	0.77
Plus fraction or residue	C_{20+}	6.63
		100.00

Discrete components making up a C_6 fraction

1. 2,2-DM-C_4
2. CY-C_5
3. 2,3-DM-C_4
4. 2-M-C_5
5. 3-M-C_5
6. nC_6

is divided into three different sections, consisting of well-defined components, pseudo fractions or components, and the plus fraction.

14.3.1 Well-Defined Components

Well-defined components consist of nonhydrocarbon components such as nitrogen, carbon dioxide, hydrogen sulfide, and hydrocarbon components methane through normal pentane. Methane, ethane, and propane exhibit unique molecular structures, whereas butane can exist as two isomers (*normal* and *iso*) and pentane as three isomers (*normal*, *iso*, and *neo*). The reason these are called well-defined components is because they are chemically and physically well characterized, that is, their chemical structures, molecular weights, boiling points, critical constants, densities, acentric factors, and other physical properties are well known.

14.3.2 Pseudo Fractions

As the carbon number increases (hexanes and heavier), the number of isomers rises exponentially. Additionally, components belonging to other homologous series, such as naphthenes and aromatics, also begin to appear with increasing carbon numbers. The identification of each and every one of these components is thus a rather cumbersome task. Therefore, these components are expressed as a group that consists of a number of components that have boiling points in a certain range. These groups are commonly denoted as *single carbon number (SCN) components* or *pseudo fractions*. These components are called pseudo fractions because they represent not just one component but a group of components that boil in a particular range. As an example, the components that make up a C_6 fraction are shown in Table 14.1. The chemical and physical properties of these pseudo fractions are not well defined as they are in the case of pure well defined components. Therefore, these pseudo components are generally characterized by properties, such as average boiling points, average molecular weights, and average densities, or specific gravities, that are also employed to estimate their critical constants and acentric factors. These estimation methods, also known as fluid characterizations, are discussed in Section 14.6. Sometimes, the pseudo fractions are also analyzed further to determine their paraffin–naphthene –aromatic (PNA) distribution.[6]

14.3.3 Plus Fraction

Heavier components that are not identified by pseudo fractions are referred to as *residue* or *plus fractions*. For example, C_{20+} represents a group of components having a carbon number of 20 and higher. Traditionally, compositional data in the petroleum industry were reported to only C_{7+}, using low-temperature fractional distillation techniques. However, this level was found to be inadequate for accurate modeling of phase behavior and physical properties of hydrocarbon mixtures. This inadequacy becomes even more pronounced in the case of gas condensate fluids because the small amounts of the heavier hydrocarbons almost entirely dominate phase behavior and properties of these fluids at various pressure and temperature conditions. Therefore, compositional analysis of gas condensate fluids is conducted generally in more detail than that of oils. However, it is a currently normal practice to extend the compositional description up to a SCN of at least C_{19}. In certain cases, where gas condensate fluids in which the heavy components control the phase behavior, it is advantageous to extend the compositional description beyond C_{20+}. The breakdown of C_{20+} then results in pseudo fractions up to C_{29}, whereas the residue is C_{30+}. The plus fraction or residue is also characterized on the basis of its average boiling point, average molecular weight, and average density.

14.4 GAS CHROMATOGRAPHY

Gas chromatography as an analytical technique is widely used in many industries to resolve and analyze gas and liquid samples into their constituents. In the petroleum industry, gas composition is determined, invariably, by gas chromatography. With recent advances in gas chromatography, it is also possible to extend the technique with comparable accuracy for compositional analysis of hydrocarbon liquids. One of the biggest advantages of gas chromatography is its ability to identify components as heavy as C_{80}[7] in a matter of hours using only a small fluid sample.

Although there are variations in the many chromatograph instruments available in the market, all GCs share similar basic components. A schematic of the typical GC setup is shown in Figure 14.1. The essential elements of a GC setup include the injection valve, a porous packed column, a carrier gas, temperature-programmed oven, and detectors.

The sample is injected through the injection valve into a heated zone, vaporized, and transported by a carrier gas (usually helium) into a column that is packed or internally coated with a stationary liquid or solid phase, resulting in the partitioning of the injected sample components. The partitioning of components occurs mostly according to their boiling points, hence low boiling or volatile components elute first, followed by the relatively heavy components. The temperature of the oven is programmed according to the boiling ranges of various components. The eluted components are carried by the carrier gas into the detectors, where their concentration is related to the area under the detector response–retention time curve shown in

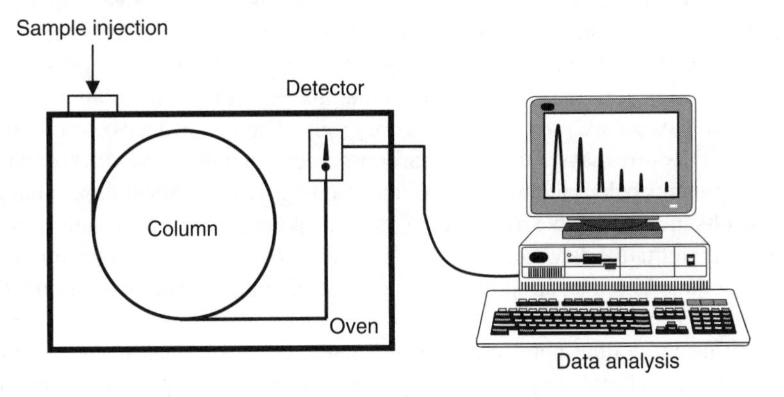

FIGURE 14.1 Schematic representation of a typical gas chromatography setup.

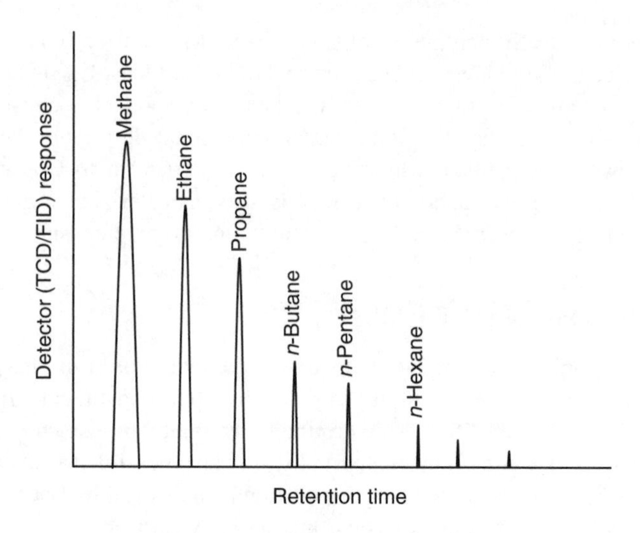

FIGURE 14.2 A gas chromatogram showing peaks of eluted components.

Figure 14.2. The individual peaks may be identified by comparing their retention times inside the column with those of known components previously analyzed at the same GC conditions.[2]

The two most commonly used detectors in a typical GC setup are called the *thermal conductivity detector (TCD)* and *flame ionization detector (FID)*. The TCD is primarily employed for detecting the nonhydrocarbon constituents, nitrogen, carbon dioxide, and hydrogen sulfide, whereas the FID is used for detecting only the hydrocarbon constituents because it cannot detect the nonhydrocarbon components. Also, the FID is a destructive detector; hence for that reason both detectors are placed in series (i.e., the TCD precedes the FID).

The analysis of the low boiling components is relatively straightforward because the majority of the components can be easily identified and analyzed by gas chromatography (methane through normal pentane and nonhydrocarbon components). However, other components are eluted as a continuous stream of overlapping peaks. The components detected by a GC between two neighboring normal alkanes are usually grouped together, measured, and reported as a pseudo fraction or a single carbon number, equal to that of a higher normal alkane. For example, all components that eluted between nC_9 and nC_{10} peaks are grouped and named a pseudo fraction or SCN, C_{10}. However, as soon as the carbon number increases, many more isomers and components belonging to other groups also begin to appear on a chromatogram. This results in considerable overlapping of the peaks at higher carbon numbers, making good quantitative determination of the heavier components very difficult. In such cases, capillary columns are preferred to packed columns, to improve the separation efficiency and peak recognition.[2,6]

Very heavy boiling-point components cannot be eluted; hence, they cannot be detected by GC. These very heavy components possess low volatility and remain on the column. However, this amount of material remaining in the column must be accounted for to determine the overall composition of the fluid. To accurately quantify the nondetectable heavy end, the following mass balance equation[8] can be used:

$$\text{Plus fraction} = \text{Mass of sample injected} - \sum_{i=1} \text{mass of detectable components} \tag{14.1}$$

Gas chromatography analysis, however, suffers from one major drawback: the lack of information such as the molecular weight and the specific gravity of the SCN groups. Molecular weight data are essential in converting the FID response (proportional to mass concentration) to a molar basis to obtain the compositions in terms of mol%. In order to overcome this limitation, four options exist:

1. Normal alkane properties are used.
2. Material balance equations are applied to components identified in each SCN group.[2]
3. Generalized properties based on the data proposed by Katz and Firoozabadi[9] are used.
4. Data from true boiling-point (TBP) distillation (see Section 14.5) are used.

The molecular weight and density of the residue or the plus fraction can be determined by material balance from the mass composition of the fluid obtained from GC analysis, the assumed properties of the SCN groups (from any of the four options), and the respective properties of the total sample. However, significant errors can result in the plus fraction properties depending upon how close the assumed properties of the SCN fractions are to the actual values.

14.5 TRUE BOILING-POINT DISTILLATION

The true boiling-point distillation (TBP) technique offers an effective alternative for analyzing the liquid phase of a petroleum reservoir fluid. This technique has been a standard analytical tool in the industry for many years.

A Fischer distillation apparatus equipped with an HMS 500 Spaltrohr column with 90 theoretical plates is commonly employed to distill the liquid hydrocarbon samples.[10] A schematic of the typical distillation setup is shown in Figure 14.3. A carefully weighed batch of liquid is loaded to the distillation flask and heated up, vaporizing its components according to their boiling points. The liquid bath temperature is set according to the desired temperature ramp of the reboiler (distillation flask) so that the temperature is gradually increased as light components vaporize and concentration of the heavier fraction in the feed increases. The boiled off fractions are collected as distillates in small sample vials (usually 5 mL volume) in an automatic fraction collector.

The boiled off fractions cover the range from 0.5°C above the boiling point of a normal alkane nC_{n-1} to 0.5°C above the next alkane, nC_n, and are named after the latter as, SCN, pseudo fraction, or TBP-cut C_n.[11] For example, C_9 means the fraction collected between 0.5°C above the boiling point of nC_8 and 0.5°C above the boiling point of nC_9. The distillation process begins at atmospheric pressure up to 151.3°C, at which fractions C_6 to C_9 are collected. However, due to the high boiling temperatures of the heavy components, subsequent distillation is carried out under a vacuum of 20 mmHg

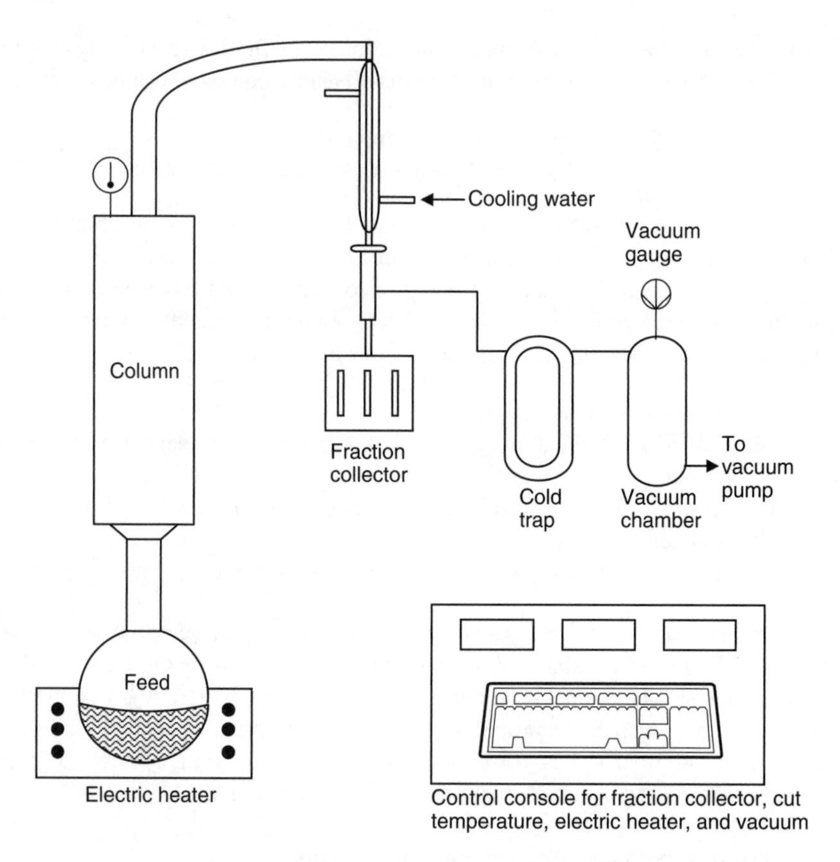

FIGURE 14.3 Schematic representation of a typical true boiling-point distillation setup.

to suppress the boiling points and reduce the heat requirement to avoid thermal cracking of the sample. The distillation process continues in this fashion until the collection of the C_{19} fraction. Because liquid hydrocarbons generally contain very heavy components, a certain amount of the loaded sample does not boil off, and thus is left in the distillation flask and termed the *residue* or *the plus fraction*. When TBP cuts up to C_{19} are collected, the remaining sample in the distillation flask is denoted C_{20+}. If further breakdown of the residue or the C_{20+} is desired, the distillation process can be continued at reduced pressure or in a vacuum of 2 mmHg, so that TBP cuts up to C_{29} are also collected and the remaining residue is C_{30+}. However, this is a very time-consuming and difficult analysis because the process conditions are not easily controlled.[6]

14.5.1 Properties of TBP Cuts and Residue

After completion of the distillation process, sample vials are removed from the fraction collector and weighed to determine the mass of the distilled fractions. The mass of the residue is determined from the difference between the mass of the distillation flask containing the residue and mass of the empty distillation flask.

Each TBP cut and residue is usually characterized by its molecular weight and density. The average molecular weight is often determined by the freezing-point depression apparatus, whereas the density data are measured using the oscillating tube densitometer.[6,10] The freezing-point depression apparatus is calibrated using standard solutions of normal alkanes in solvents such as benzene or *p*-xylene. The oscillating tube densitometer is usually calibrated using air and double distilled water at a standard temperature of 60°F.

Based on the measurement of molecular weights and densities of the TBP cuts, residue, and their measured masses, all data are converted to mass%, mol% and vol% basis. The TBP distillation data of a given hydrocarbon liquid sample are then reported in a logsheet, as shown in Table 14.2.

14.5.2 Internal Consistency of TBP Data

As mentioned previously, most of the analytical data from GC analysis (FID response) as well as the basic TBP data are in terms of mass fractions. Equation 14.2 shows that the molecular weights, therefore, play a major role in converting the mass fractions to mole fractions. The molecular weight of the residue or the plus fraction is especially important.

$$Z_i = \frac{\dfrac{W_i}{\mathrm{MW}_i}}{\sum_{i=1}^{N}\left[\dfrac{W_i}{\mathrm{MW}_i}\right]} \tag{14.2}$$

where Z_i is the mole fraction of component i, W_i the mass fraction of component i, MW_i the molecular weight of component i, and N the number of components.

The conversion from mole fraction to mass fraction is carried out by the following equation:

$$W_i = \frac{Z_i \mathrm{MW}_i}{\sum_{i=1}^{N}(Z_i\ \mathrm{MW}_i)} \tag{14.3}$$

Therefore, when detailed TBP distillation data are available, it is possible to check the measured mass fractions, SCN, and residue molecular weights and densities for internal consistency. The only additional measurements required are the molecular weight and the density of the feed charged to the distillation flask.

As a starting point, the masses of all the TBP cuts and the residue are added and the resulting value is compared with the mass of the sample charged to the distillation flask. A difference of few percent indicates the reliability of collected data.

In order to check the consistency of the molecular weight data, first the molecular weight of the whole sample is calculated from Equation 14.4, on the basis of the data measured for the individual TBP cuts and the residue.

$$\mathrm{MW}_{\mathrm{calculated}} = \frac{\sum_{i=1}^{N} M_i}{\sum_{i=1}^{N}\dfrac{M_i}{\mathrm{MW}_i}} \tag{14.4}$$

TABLE 14.2
Logsheet for True Boiling-Point (TBP) Distillation of a Hydrocarbon Liquid Sample

Total mass of the sample charged = ——— g
Volume of the sample at 15.55°C = ——— cm^3
Density of the sample charged = ——— g/cm^3 at 15.55°C and atmospheric pressure
Molecular weight of the sample charged = ——— g/mol

No.	Fraction	Cut temperature (°C)	Pressure	Mass of Fraction (g)	Volume of Fraction (cm^3)	Density (at 15.55°C and atm. pressure)	Molecular Weight (g/mol)	Composition (%)		
								Mass	Mol	Vol
1	C_6	69.2	atm							
2	C_7	98.9	atm							
3	C_8	126.1	atm							
4	C_9	151.3	atm							
5	C_{10}	174.6	20 mmHg							
6	C_{11}	196.4	20 mmHg							
7	C_{12}	217.2	20 mmHg							
8	C_{13}	235.9	20 mmHg							
9	C_{14}	253.9	20 mmHg							
10	C_{15}	271.1	20 mmHg							
11	C_{16}	287.3	20 mmHg							
12	C_{17}	303.0	20 mmHg							
13	C_{18}	317.0	20 mmHg							
14	C_{19}	331.0	20 mmHg							
15	Residue (C_{20+})	>331.0	-							
Total								100.0	100.0	100.0

Calculated average molecular weight = ——— g/mol
Calculated average density = ——— g/cm^3 at 15.55°C and atmospheric pressure.

where $MW_{calculated}$ is the molecular weight calculated from individual TBP cuts and residue data, M_i the mass of individual cuts and residue, and MW_i the molecular weight of individual cuts and residue.

If the calculated molecular weight from Equation 14.4 agrees with the measured molecular weight of the whole sample within a few percent, then the data are considered to be reliable.

A similar consistency check is carried out for the measured densities of TBP cuts and the residue. Equation 14.5 is used to calculate the overall density on the basis of the individual data.

$$\rho_{calculated} = \frac{\sum_{i=1}^{N} M_i}{\sum_{i=1}^{N} \frac{M_i}{\rho_i}} \tag{14.5}$$

where $\rho_{calculated}$ is the density calculated from individual TBP cuts and residue data, M_i the mass of individual cuts and residue and ρ_i the density of individual cuts and residue.

Again, a difference of few percent between the calculated and measured densities indicates the reliability of the TBP data.

14.5.3 Properties of TBP Cuts and Generalized Data

In the absence of detailed TBP data, generalized values proposed by Katz and Firoozabadi[9] are used by default as an alternative. Although, these generalized values do provide a better alternative to using normal alkane values, they can create major errors in phase behavior calculations. These values are based on too few data points on a limited number of oil samples and hence cannot be considered as generalized because a fairly wide variation exists when properties of TBP cuts belonging to various liquid hydrocarbons are compared with the generalized values. Figure 14.4 shows a comparison between the densities of TBP cuts of condensates

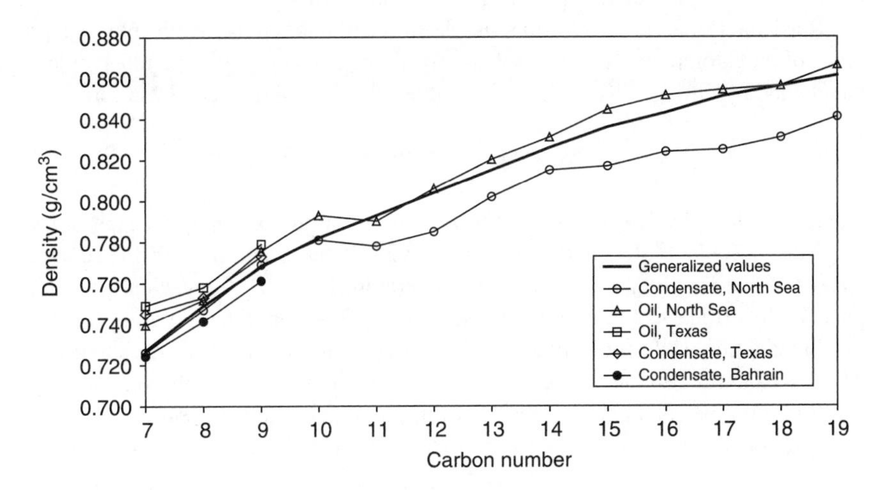

FIGURE 14.4 Comparison of densities of TBP cuts of various condensates and oils with generalized values.

and oils from North Sea, Texas, and Bahrain and the generalized values, clearly indicating the differences between the magnitude and trend.

Dandekar et al.[10] and Pedersen et al.[6] also compared the molecular weight and densities or specific gravities of TBP cuts of various North Sea oils and the generalized values proposed by Katz and Firoozabadi.[9] A wide variation in the measured properties in comparison to the generalized values was clearly seen. Additionally, the generalized molecular weights and densities vs. carbon number do show a gradual increase with the carbon number; both data can be correlated with a straight line and a second-order polynomial with a correlation coefficient close to 1. However, in reality such a well-defined relationship does not exist. This is mainly because every oil is unique in nature, and moreover, the PNA distribution in each TBP cut or the SCN groups is different. Hence, the properties of each SCN group varies according to the PNA distribution.

For the purposes of EOS modeling, it is very important to account for this variation in the SCN properties because if generalized values are employed, it affects not only the calculation of critical constants of SCN groups but also the mole fraction and properties of the plus fraction. These unrealistic values can significantly impact the phase behavior calculations, particularly in the case of gas condensate fluids because the plus fraction almost entirely dominates the phase behavior.[10] Therefore, properties measured on the SCN groups and the residue constitute a very important aspect of compositional analysis of petroleum reservoir fluids.

14.6 CHARACTERIZATION OF PSEUDO FRACTIONS AND RESIDUE

The critical constants, such as critical temperature, critical pressure, critical volume, and acentric factors of the SCN groups and the plus fraction of petroleum reservoir fluids similar to those of the well-defined components, are necessary for EOS modeling. These properties are generally determined from empirical correlations in terms of molecular weight, specific gravity, and boiling points.

Riazi and Daubert[12] developed a simple correlation for predicting the physical properties of pure components and petroleum fractions. The correlation requires molecular weight and specific gravity as the input and has the following generalized form:

$$\theta = a(\mathrm{MW})^b \gamma^c \exp[d(\mathrm{MW}) + e\gamma + f(\mathrm{MW})\gamma] \tag{14.6}$$

where θ is any physical property, a to f are correlation constants for each property as given in Table 14.3, γ is the specific gravity of the fraction, MW the molecular weight of the fraction, T_c the critical temperature in °R, P_c the critical pressure in psia, V_c the critical volume in ft^3/lb, and T_b the boiling point temperature, in °R.

In Equation 14.6, molecular weight and specific gravity data measured on the TBP cuts and the residue can be used to calculate all the required physical properties.

Based on the calculated values of boiling point, critical temperature, and critical pressure, the acentric factor, ω, can be calculated from Edmister's[13] correlation:

$$\omega = \frac{3[\log(P_c/14.70)]}{7[(T_c/T_b - 1)]} - 1 \tag{14.7}$$

TABLE 14.3
Coefficients for Riazi and Daubert[12] Correlation (Equation 14.6)

a	*b*	*c*	*d*	*e*	*f*
			Critical temperature T_c (°R)		
544.4	0.2998	1.0555	-1.3478×10^{-4}	−0.61641	0.0
			Critical pressure, P_c (psia)		
4.5203×10^4	−0.8063	1.6015	-1.8078×10^{-3}	−0.3084	0.0
			Critical volume, V_c (ft³/lb)		
1.206×10^{-2}	0.20378	−1.3036	-2.657×10^{-3}	0.5287	2.6012×10^{-3}
			Boiling point, T_b (°R)		
6.77857	0.401673	−1.58262	3.77409×10^{-3}	2.984036	-4.25288×10^{-3}

TABLE 14.4
Coefficients for Ahmed[14] Correlation (Equation 14.8)

a_1	a_2	a_3	a_4	a_5
		Molecular weight		
−131.11375	24.96156	−0.34079022	2.4941184×10^{-3}	468.32575
		Specific gravity		
0.86714949	3.4143408×10^{-3}	-2.839627×10^{-5}	2.4943308×10^{-8}	−1.1627984
		Critical temperature, T_c (°R)		
915.53747	41.421337	−0.7586859	5.8675351×10^{-3}	-1.3028779×10^3
		Critical pressure, P_c (psia)		
275.56275	−12.522269	0.29926384	$-2.8452129 \times 10^{-3}$	1.7117226×10^3
		Critical volume, V_c (ft³/lb)		
5.223458×10^{-2}	7.8709139×10^{-4}	$-1.9324432 \times 10^{-5}$	1.7547264×10^{-7}	4.4017952×10^{-2}
		Boiling point, T_b (°R)		
434.38878	50.125279	−0.9027283	7.0280657×10^{-3}	−601.85651
		Acentric factor ω		
−0.50862704	8.700211×10^{-2}	$-1.8484814 \times 10^{-3}$	1.4663890×10^{-5}	1.8518106

Ahmed[14] presented a correlation for the determination of Katz and Firoozabadi's generalized SCN fraction properties for carbon numbers ranging from 6 to 45. The proposed correlation has the following form:

$$\theta = a_1 + a_2 n + a_3 n^2 + a_4 n^3 + (a_5/n) \quad (14.8)$$

In Equation 14.8, θ represents any physical property such as molecular weight, specific gravity, critical constants, and so on, while n represents the number of carbon atoms, that is, 6 to 45 of the SCN fraction. The coefficients a_1 to a_5 of Equation 14.8 are given in Table 14.4.

Other characterization procedures have been discussed in detail elsewhere.[2,6]

14.7 OTHER NONCONVENTIONAL METHODS OF COMPOSITIONAL ANALYSIS

In addition to the methods of compositional analysis of petroleum reservoir fluids discussed so far, attempts have also been made to determine the compositional data from various nonconventional methods. Three such approaches reported in the literature are discussed in this section. Although, these approaches have not received much attention as far as practical applications are concerned, the methods may serve as a backup or provide additional opportunity to compare the conventionally measured compositional data.

Fujisawa et al.[15] recently presented a near-infrared spectroscopy-based method to provide *in situ* quantitative characterization of reservoir fluids during wire line sampling. The proposed technique does not provide detailed compositional information, but the data are reported in terms of compositional groups such as C_1, C_2–C_5, C_{6+}, CO_2, and water. The *in situ* analysis by definition provides answers in real time, while compositional analysis, based on sampling and the subsequent laboratory analysis can take several months. Moreover, the fluid sample is most representative in downhole conditions, which is particularly important because changes (sometimes irreversible, if asphaltenes precipitate) that occur in reservoir fluid samples when they are brought to surface can be precluded. Fujisawa et al., however, stated that both *in situ* analysis and the conventional compositional analysis in the laboratory have strengths and weaknesses, that is, the former is robust but coarse and is not detailed whereas the latter is detailed but may not be representative. Therefore, Fujisawa et al. recommended the use of both so that the compositional analysis is both robust and accurate.

Treinen et al.[16] presented an interesting approach for the determination of reservoir fluid composition based on pressurized core samples. This method can also be considered to provide *in situ* compositional data of reservoir fluids because pressure coring is the only method for recovering cores that have not undergone saturation changes caused by depressurization while lifted to the surface. The overall composition of the reservoir fluid is determined by combining the distillation and extraction analysis of fluids removed from the core sample. Treinen et al. presented compositional data based on the analysis of seven different pressure cores from the Prudhoe Bay reservoir on the North Slope of Alaska. The reported compositional analysis provided concentration of components including nitrogen, carbon dioxide, and methane through C_{36+}.

Christensen and Pedersen[17] also presented a very interesting, purely theoretical approach for determining reservoir fluid composition. They showed that the molar composition of a petroleum reservoir fluid is not random but a result of a chemical reaction equilibrium existing in the reservoir, at reservoir temperature. By showing a reasonable comparison between the measured and theoretically estimated compositions of four different reservoir fluids, they demonstrated the possibility of determining compositional data solely on the basis of reservoir temperature.

PROBLEMS

14.1 A wet gas wellhead stream is flashed in a test separator. The molar compositions of the separator gas and separator liquid samples measured by gas chromatography are given in the following table. The separator gas and separator oil ratio is determined to be 0.75 lb/lb. Determine the wet gas reservoir fluid composition (in mol%) that enters the test separator.

Component	Separator Gas (mol%)	Separator Liquid (mol%)
Methane	71.93	6.79
Ethane	11.09	5.40
Propane	8.89	15.19
i-Butane	1.70	6.99
n-Butane	4.10	24.78
i-Pentane	1.20	18.18
n-Pentane	1.10	22.68
	100.00	100.00

14.2 The following table gives the TBP distillation data for a black oil. Check the reliability of the measured distillation data; report the compositions up to C_{10+} in mass%, mol%, and vol.%.

Fraction	Mass (g)	Density at 60°F; 14.696 psia (g/cm³)	Molecular Weight (g/g mol)
C_6	3.1619	0.6823	88.69
C_7	2.2328	0.7201	103.90
C_8	1.5930	0.7474	120.77
C_9	1.7779	0.7683	128.73
C_{10}	1.0167	0.7947	133.94
C_{11}	2.4194	0.8024	144.83
C_{12}	2.3129	0.8146	156.75
C_{13}	2.2291	0.8258	171.59
C_{14}	2.2866	0.8376	179.13
C_{15}	2.3564	0.8449	196.41
C_{16}	2.2447	0.8480	203.51
C_{17}	2.4718	0.8457	214.04
C_{18}	2.2206	0.8501	223.34
C_{19}	2.0932	0.8645	253.77
C_{20+}	42.0743	0.9559	398.90

The density of the sample charged to distillation flask is 0.8763 g/cm³ at 60°F and 14.696 psia; molecular weight of the sample charged to distillation flask = 235.93 g/g mol; and total mass of the sample charged to distillation flask = 74.3370 g.

14.3 Determine the composition of the following fluid in mass%.

Component	Composition (mol%)	Molecular Weight (g/g mol)
Nitrogen	0.83	28.01
Hydrogen sulfide	0.12	34.08
Carbon dioxide	0.93	44.01
Methane	24.62	16.04
Ethane	6.26	30.07
Propane	6.77	44.10
i-Butane	1.86	58.12
n-Butane	4.71	58.12
i-Pentane	2.72	72.15
n-Pentane	1.55	72.15
C_{6+}	49.63	311.5
	100.00	

14.4 If 0.7 mol of the above fluid is mixed with 0.3 mol of the gas with the following composition, what would be the composition of the resultant fluid?

Component	Composition (mol%)
Nitrogen	2.31
Methane	70.27
Ethane	11.75
Propane	7.89
i-Butane	1.52
n-Butane	3.34
i-Pentane	1.18
n-Pentane	0.72
C_{6+}	1.02
	100.00

REFERENCES

1. Danesh, A., Todd, A.C., Somerville, J., and Dandekar, A., Direct measurement of interfacial tension, density, volume and compositions of gas-condensate systems, *Trans. I Chem. E*, 68, Part A, 1990.
2. Danesh, A., *PVT and Phase Behavior of Petroleum Reservoir Fluids*, Elsevier Science, Amsterdam, 1998.
3. Danesh, A. and Todd, A.C., A novel sampling method for compositional analysis of high pressure fluids, *Fluid Phase Equilibria*, 57, 161–171, 1990.
4. Dandekar, A.Y., Interfacial Tension and Viscosity of Reservoir Fluids, Ph.D. thesis, Heriot-Watt University, Edinburgh, 1994.

5. Khan, M.S. and Hatamian, H., Improved method of compositional analysis of liquid at high-pressure condition in PVT study of gas condensate, Society of Petroleum Engineers (SPE) paper number 21429.
6. Pedersen, K.S., Fredenslund, A., and Thomassen, P., *Properties of Oils and Natural Gases*, Gulf Publishing Company, Houston, 1989.
7. Curvers, J. and van den Engel, P., Gas chromatographic method for simulated distillation up to a boiling point of 750°C using temperature-programmed injection and high temperature fused silica wide-bore columns, *J. High Resolution Chromatogr.*, 12, 16–22, 1989.
8. Burke, N.E., Chea, C.K., Hobbs, R.D., and Tran, H.T., Extended analysis of live reservoir oils by gas chromatography, Society of Petroleum Engineers (SPE) paper number 21003.
9. Katz D.L. and Firoozabadi A., Predicting phase behavior of condensate/crude-oil systems using methane interaction coefficients, *J. Pet.Technol.*, 228, 1649–1655, 1978.
10. Dandekar, A., Andersen, S.I.A., and Stenby, E., Compositional analysis of North Sea oils, *Pet. Sci. Technol.*, 18, 975–988, 2000.
11. Osjord, E.H., Rønningsen, H.P., and Tau, L., Distribution of weight, density, and molecular weight in crude oil derived from computerized capillary GC analysis, *J. High Resolution Chromatogr. Chromatogr. Commun.*, 8, 683–690, 1985.
12. Riazi, M.R. and Daubert, T.E., Characterization parameters for petroleum fractions, *Ind. Eng. Chem. Res.*, 26, 755–759, 1987.
13. Edmister, W.C., Applied hydrocarbon thermodynamics, Part 4: compressibility factors and equations of state, *Pet. Refiner*, 37, 173–179, 1958.
14. Ahmed, T., Composition Modeling of Tyler and Mission Canyon Formation Oils with CO_2 and Lean Gases, Report for Montanans, on a New Track for Science (MONTS), 1985.
15. Fujisawa, G, Mullins, O.C., Dong, C., Carnegie, A., Betancourt, S.S., Terabayashi, T., Yoshida, S., Jaramillo, A.R., and Haggag, M., Analyzing reservoir fluid composition in-situ in real time: case study in a carbonate reservoir, Society of Petroleum Engineers (SPE) paper number 84092.
16. Treinen, R.J., Bone, R.L., and Rathmell, J.J., Hydrocarbon composition and saturation from pressure core analysis, Society of Petroleum Engineers (SPE) paper number 27802.
17. Christensen, P.L. and Pedersen, K.S., The molar composition of petroleum reservoir fluids, a result of chemical reaction equilibria, Society of Petroleum Engineers (SPE) paper number 27624.

15 PVT Analysis and Reservoir Fluid Properties

15.1 INTRODUCTION

After studying the basic characteristics of phase behavior, sampling, and compositional analysis, let us turn to Pressure–Volume–Temperature (PVT) analysis and properties of petroleum reservoir fluids.

Almost all petroleum reservoirs are produced by a depletion process in which the reservoir pressure declines as fluids are recovered. The temperature in the reservoir remains practically constant in most recovery methods; however, a reduction in the temperature is generally encountered when fluids from the reservoir arrive at the surface. In addition to the chemistry and the original composition, the two main variables that influence the behavior and the properties of reservoir fluids are pressure and temperature. Therefore, relatively simple laboratory tests that simulate the recovery of hydrocarbon fluids are conducted by varying pressure and temperature, with primary emphasis on the volumetric data at reservoir and surface conditions. However, for certain fluids, instead of just one particular test or few tests, a series of laboratory tests are essential to obtain all necessary data. This type of analysis carried out on most reservoir fluids is called PVT and reservoir fluid properties studies, or simply *PVT Studies* or *Reservoir Fluid Studies*.

Data measured as part of the PVT studies or reservoir fluid studies are key elements of proper management of petroleum reservoirs, which include evaluation of reserves, development of a recovery plan, and also determination of the quantity and quality of the produced fluids. All the data that are obtained for these various purposes are a result of the integration of three different components:

1. High-pressure/high-temperature PVT equipment
2. Different laboratory tests
3. Different properties

These three important components and other elements can be represented by a flowchart such as the one shown in Figure 15.1. Therefore, the primary purpose of this chapter is to describe these three components in detail.

Because this chapter focuses on properties of petroleum reservoir fluids, the various properties are defined first, followed by the generic description of PVT equipment and common PVT tests from which various reservoir fluid properties are obtained. However, when detailed laboratory analysis is not available, fluid properties are estimated from various empirical correlations. The commonly used empirical correlations are also discussed in the concluding section of this chapter.

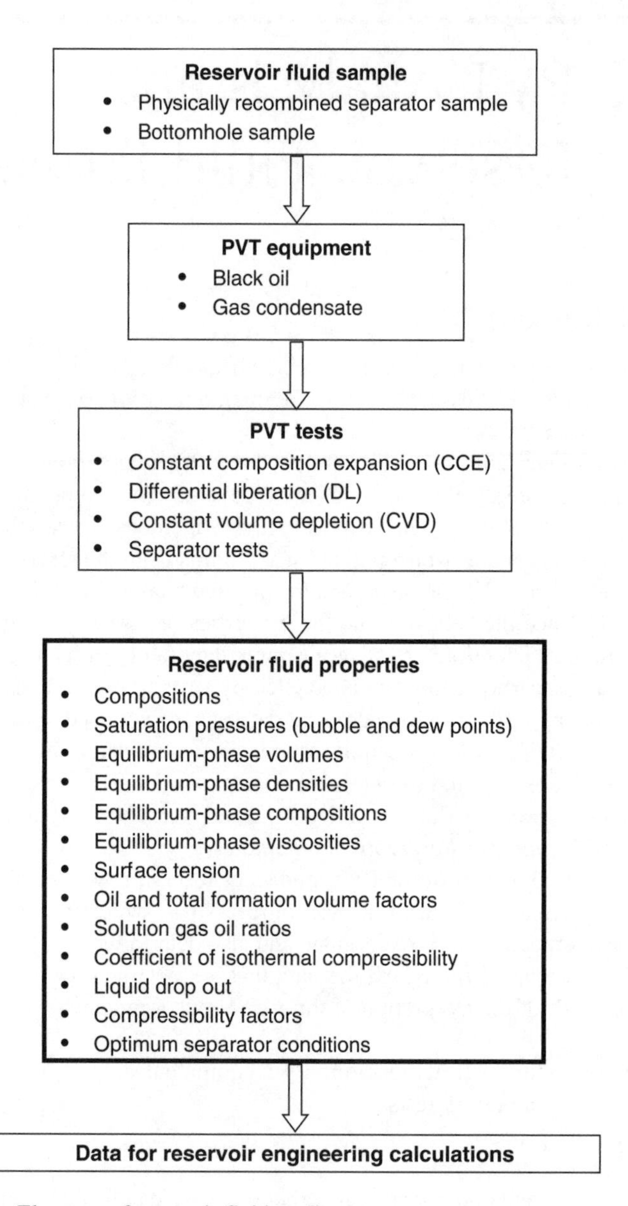

FIGURE 15.1 Elements of reservoir fluid studies.

15.2 PROPERTIES OF PETROLEUM RESERVOIR FLUIDS

This section describes the definitions and characteristics of the important physical properties of petroleum reservoir fluids. The properties described here are commonly used by both reservoir and production engineers and are required for various aspects of overall reservoir management. These properties include pressure–volume (*PV*) relationships for gases and oils, compressibility, expansivity, density, viscosity, surface tension, as well as other reservoir engineering parameters such as formation volume factors and solution gas ratios.

Many of these properties fall naturally into the normal distinctions between gases and liquids. Other properties such as solution gas, formation volume factors, and surface tension are described more appropriately in the context of properties of both gas and liquid (oil or condensate) phases. Before the various properties of petroleum reservoir fluids are defined, the basic distinguishing features between gases and liquids must be studied, which will help us to understand the differences among the various properties.

15.2.1 Gases and Liquids

A gas is generally considered a homogeneous fluid of low density and viscosity with neither independent shape nor volume that expands to assume the shape of the vessel in which it is contained. Gases are important in all aspects of petroleum engineering; therefore, the laws governing their behavior need to be clearly understood. In the case of simple gases, also called *ideal gases*, these laws are straightforward. However, the behavior of hydrocarbon gases at reservoir conditions, also termed real gas behavior, can be much more complicated.

A liquid, by contrast, is considered a homogeneous fluid of moderate to high density and viscosity that also has no independent shape and assumes the shape of the containing vessel, although not necessarily filling it. The key to this distinction between gases and liquids is the relative molecular densities of the phases. In the treatment of gases, it is assumed that the distance between molecules is great enough such that the effect of the attractive forces between them is negligible, while for liquids, the distance between molecules is much less and the forces of attraction are much greater. These differences in the molecular level result in substantial differences in the physical properties between the two fluid states. For instance, a change in pressure has a much greater effect on the density of a gas than of a liquid.

However, it should be noted here that these basic distinctions between gases and liquids become much less applicable in the near-critical region where many reservoir fluids such as gas condensates and volatile oils exist. In fact, at near-critical regions, the gas and liquid phases are sometimes simply referred to as the less dense and more dense phases; the differences between which eventually diminish at the critical point. Nevertheless, it is useful to commence a discussion of fluid properties by reviewing the behavior of the simplest fluid systems, such as ideal gases or low pressure gases, and then developing the approach to deal with increasingly complex systems.

15.2.2 Ideal Gases

An ideal gas does not necessarily mean a particular component, such as methane, nitrogen, or carbon dioxide, but is rather indicative of the behavior exhibited by a given component at certain pressure and temperature conditions. For example, methane can behave as an ideal gas or real gas depending on the prevailing pressure and temperature conditions.

All ideal gases are characterized by the following three properties:

1. The volume occupied by molecules is insignificant with respect to the volume occupied by the gas.
2. No attractive or repulsive forces exist between the molecules or between the molecules and the container.
3. All collisions of molecules are perfectly elastic, that is, no loss of internal energy occurs upon collision.

All gases behave ideally when the pressure approaches zero. The pressure–volume relationship between ideal gases is derived by combining Boyle's, Charle's, and Avogadro's laws, and mathematically expressed by the fundamental equation of state:

$$PV = nRT \tag{15.1}$$

where P is the pressure; V the volume; n the number of moles; R the universal gas constant; and T the temperature.

In Equation 15.1, the value of R depends upon the units employed for other variables. For example, in oil field units, pressure is in psia, volume is in ft^3, quantity of gas is equal to 1 lb-mol, and the temperature is in °R. Therefore, R has a value of 10.732 (ft^3 psia/lb-mol°R). Note that temperature is always in absolute units, that is, either in degrees Rankine (°R = °F + 460), or in Kelvin (K = °C + 273.15) if SI units are used for other variables.

15.2.2.1 Standard Volume

Since the volume of gas varies substantially with pressure and temperature, defining the conditions at which gas volume is reported is necessary. This is especially important in the sales of gas and many other calculations involving gases. Therefore, it is convenient to measure the volume occupied by 1 lb-mol of gas at certain reference conditions. These reference conditions or standard conditions (sometimes referred to as *base conditions*) are usually 14.7 psia and 60°F. The standard volume is then defined as the volume of gas occupied by 1 lb-mol of gas at standard conditions. At standard conditions, the gas is considered to behave ideally, which allows the calculation of volume by applying Equation 15.1:

$$V_{sc} = \frac{RT_{sc}}{P_{sc}} = \frac{\left(10.732\ \text{ft}^3 \dfrac{\text{psia}}{\text{lb-mol}°R}\right)(60+460)°R}{14.7\ \text{psia}} = 379.6\ \frac{\text{scf}}{\text{lb-mol}} \tag{15.2}$$

where V_{sc} is the standard volume, scf /lb-mol; scf the standard cubic feet; T_{sc} the standard temperature, °R; and P_{sc} the standard pressure, psia.

15.2.3 Real Gases

In dealing with gases at low pressures, Equation 15.1 can give satisfactory results. However, the application of Equation 15.1 to gases at elevated pressures may lead to significant errors due to departure from ideality as attractive forces and volume of the molecules become a significant factor.

In order to account for the nonideality, numerous PVT correlations have been developed in the form of equations of state (EOS). However, by far the most simple approach is the use of a correction factor to express a more exact relationship between pressure, volume, and temperature. This correction factor is called *compressibility factor*, *gas deviation factor*, or simply the *Z factor*, and is defined in Equation 15.3 as the ratio of the volume actually occupied by a gas at a given pressure and temperature to the volume the gas would occupy at the same pressure and temperature if it behaved like an ideal gas:

$$Z = \frac{V_{\text{actual}}}{V_{\text{ideal}}} \tag{15.3}$$

such that

$$PV = Zn\text{R}T \tag{15.4}$$

Equation 15.4 is often termed the real gas equation. By its definition it can be applied to any gas that does not undergo a phase change at any temperature and pressure.

However, it should be noted here that the compressibility factor is not a constant, but is a function of pressure, temperature, and gas composition (see Section 15.2.4). The value of Z can be determined experimentally at any given pressure and temperature by measuring the actual volume of some quantity of gas. For example, the Z-factor values for methane, ethane, and propane are shown in Figures 15.2–15.4. All Z factors have been calculated using the highly accurate multiparameter equations of state (not discussed here) specifically developed for these components. The compressibility factor charts presented in Figures 15.2–15.4, are similar to the ones presented by Brown et al.[1]

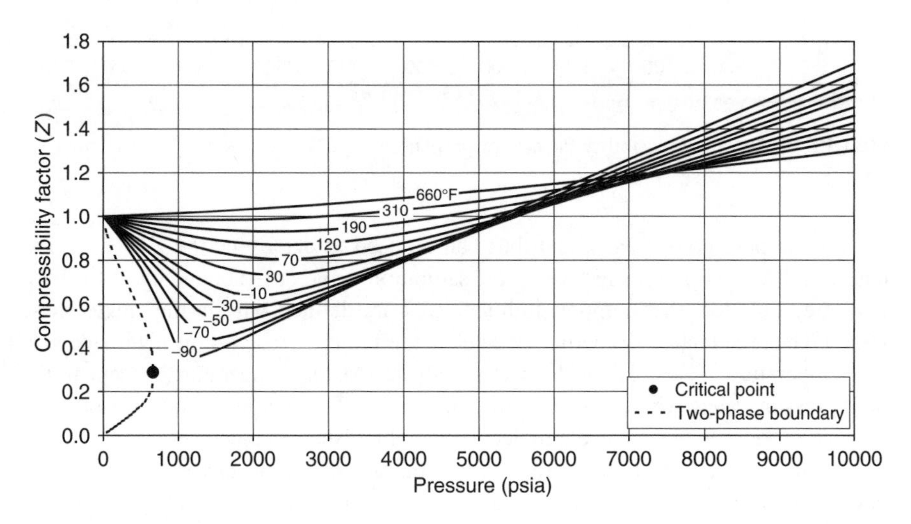

FIGURE 15.2 Compressibility factors for methane.

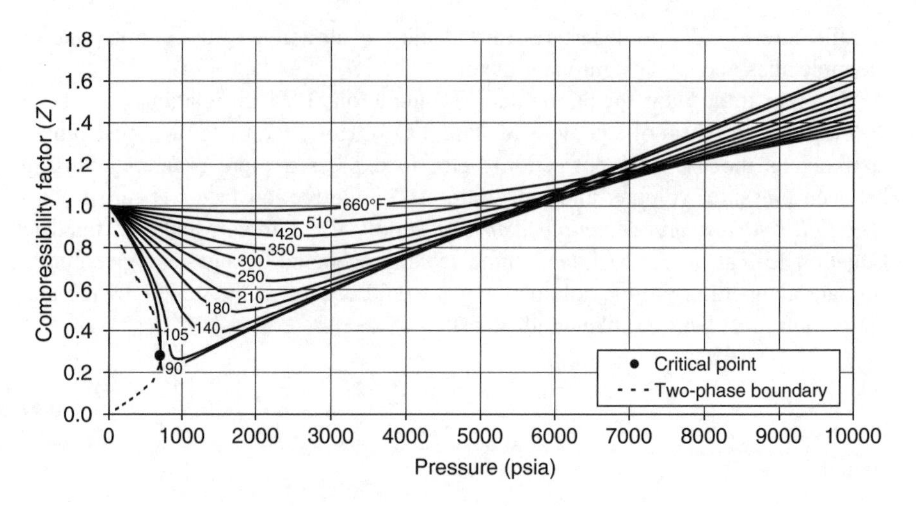

FIGURE 15.3 Compressibility factors for ethane.

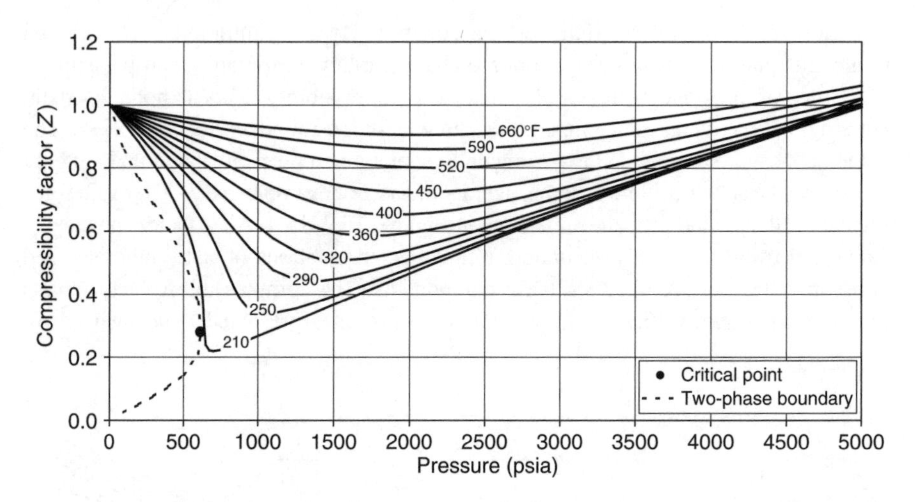

FIGURE 15.4 Compressibility factors for propane.

for various pure components, and thus can be used to determine the Z factors. The dashed curve in all figures represents the saturation curve or the boundary of the two-phase region. Note that compressibility factors are defined only in the single-phase region. The temperature isotherms have distinct minimums, which vanish with increasing temperature. The Z factor decreases with decreasing temperature, except in the high-pressure range where a reversal of trend occurs. The minimums on the isotherms become more pronounced as the molecular weight of the gas increases.

It can also be seen in these figures that the gas deviation factors follow a very definite pattern with similar trend at most pressure and temperature conditions. The similarity in the behavior of Z factor for pure gases led to the development of the principle of corresponding states (also discussed in Chapter 11) and the definitions of reduced or dimensionless pressure, P_r, and temperature, T_r, defined as

$$P_r = \frac{P}{P_c} \text{ and } T_r = \frac{T}{T_c} \tag{15.5}$$

where P is the pressure, psia or in any absolute units; P_c the critical pressure, psia or in any absolute units; T the temperature in any absolute units, such as °R or K; and T_c the critical temperature in any absolute units, such as °R or K.

According to the principle of corresponding states, all pure gases have nearly the same Z factor at the same values of reduced pressure and temperature. Therefore, data presented in Figures 15.2–15.4 can be converted to one generalized compressibility factor chart if pressures and temperatures are expressed in terms of reduced values. Figure 15.5, which is constructed on the basis of the individual charts of methane (Figure 15.2), ethane (Figure 15.3), and propane (Figure 15.4), shows the Z-factor data for these components as a function of reduced pressure and temperature. The application of the corresponding states principle for the determination of Z factors can be readily realized. For example, at a given reduced pressure and temperature, the Z factor from Figure 15.5 and the individual charts of methane, ethane, and propane (after converting P_r and T_r to respective pressure and temperature on the basis of critical constants) are nearly equal.

15.2.3.1 Gas Density

Since density is defined as mass per unit volume, Equation 15.4 can be solved for the density to yield

$$\rho_g = \frac{n}{V}\text{MW} = \frac{P}{ZRT}\text{MW} \tag{15.6}$$

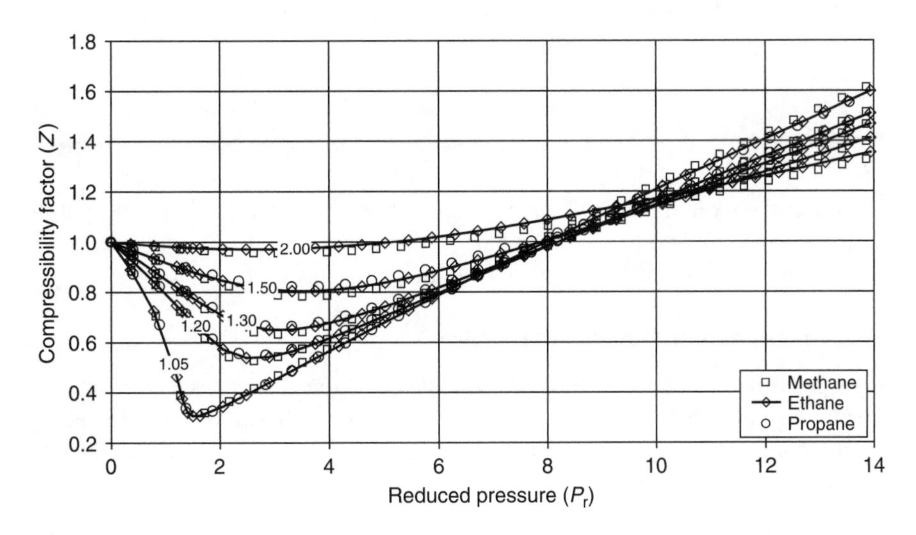

FIGURE 15.5 Generalized compressibility factor chart for pure hydrocarbon components. Note that values on the curves represent reduced temperatures.

where ρ_g is the density of the gas, lb/ft^3; P the pressure, psia; MW the gas molecular weight, lb/lb-mol; Z the compressibility factor of the gas at prevalent pressure and temperature conditions; R the universal gas constant, 10.732 (ft^3 psia/lb-mol°R), T the temperature, °R.

15.2.3.2 Specific Gravity

The specific gravity of a gas is defined as the ratio of the density of the gas to the density of dry air with both measured at the same pressure and temperature, usually the standard conditions. Symbolically,

$$\gamma_g = \frac{\rho_g(T_{sc}, P_{sc})}{\rho_{air}(T_{sc}, P_{sc})} \tag{15.7}$$

where γ_g is the specific gravity of the gas; ρ_g the density of gas at standard conditions, lb/ft^3; and ρ_{air} the density of air at standard conditions, lb/ft^3.

Assuming ideal behavior at standard conditions and using Equation 15.6,

$$\gamma_g = \frac{MW}{MW_{air}} \tag{15.8}$$

where γ_g is the specific gravity of gas; MW the molecular weight of gas, lb/lb-mol; and MW_{air} the molecular weight of air, lb/lb-mol.

15.2.4 Mixtures of Gases

In the previous two sections, the discussion was primarily centered on pure gases. However, petroleum engineers are usually interested in the properties of gas mixtures rather than pure component gases because these are rarely encountered. The composition of a given gas mixture is normally expressed or reported in terms of mol-% or mole fraction and defined as

$$Y_i = \frac{n_i}{n_t = \sum_{i=1}^{n} n_i} \tag{15.9}$$

where Y_i is the mole fraction of component i in the gas mixture (mol-% = $Y_i \times 100$); n_i the moles (lb-mol, gm-mol, or kg-mol) of component i in the gas mixture; and n_t the total moles or summation of moles of all components from 1 to n.

15.2.4.1 Apparent Molecular Weight

Since gas mixtures are composed of molecules of different sizes, saying that a gas mixture has a molecular weight is not strictly correct. Therefore, the molecular weight of a gas mixture is termed *apparent* or *average molecular weight*, defined by Equation 15.10:

$$MW_g = \sum_{i=1}^{n} Y_i MW_i \tag{15.10}$$

where MW_g is the apparent or average molecular weight of a gas mixture; Y_i the mole fraction of component i in the gas mixture; and MW_i = molecular weight of component i in the gas mixture.

For example, dry air is a gas mixture primarily containing 78 mol-% nitrogen, 21 mol-% oxygen, and 1 mol-% argon. Applying Equation 15.10 gives an apparent molecular weight of 28.97 lb/lb-mol for air. Therefore, once the apparent molecular weight of the gas mixture is known from its composition, the specific gravity of the gas mixture can be easily calculated from Equation 15.8.

15.2.4.2 Critical Pressure and Temperature of Gas Mixtures

Similar to the pure components, Z-factor values are necessary for defining the PVT relationships for gas mixtures when they behave as real gases. However, for obtaining the Z-factor values from the corresponding states principle approach requires values of critical pressure and temperature for gas mixtures. For pure component gases, accurate and reliable values of the critical constants such as P_c and T_c are known. Since obtaining the critical point of gas mixtures is somewhat difficult, the concept of pseudocritical pressure and pseudocritical temperature was introduced. These pseudocritical properties do not represent the true or actual critical properties of the gas mixture, but rather serve as correlating parameters in generating gas properties. The different approaches used for the determination of pseudocritical pressure and temperature are discussed in the following subsection.

Kay's Mixing Rules

In 1936 Kay[2] proposed simple mixing rules for the calculation of pseudocritical pressure and pseudocritical temperature for use in place of true critical pressure and temperature of hydrocarbon mixtures. These quantities are defined as

$$P_{pc} = \sum_{i=1}^{n} Y_i P_{ci} \quad \text{and} \quad T_{pc} = \sum_{i=1}^{n} Y_i T_{ci} \tag{15.11}$$

where P_{pc} is the pseudocritical pressure of the mixture; T_{pc} the pseudocritical temperature of the mixture; P_{ci} the critical pressure of component i in the mixture; and T_{ci} the critical temperature of component i in the mixture. In Equation 15.11, P_{pc} and T_{pc} take the units of P_c and T_c.

Pseudocritical Properties from Gas Gravity

In certain cases when the gas composition is unavailable, pseudocritical pressure and pseudocritical temperature can be determined solely on the basis of the specific gravity of the gas mixture. In 1977, Standing[3] presented the following correlations for the determination of P_{pc} and T_{pc} when only the specific gravity of the gas mixture is available.

For natural gas systems,

$$T_{pc} = 168 + 325\,\gamma_g - 12.5\gamma_g^2 \tag{15.12}$$

$$P_{pc} = 677 + 15.0\,\gamma_g - 37.5\,\gamma_g^2 \tag{15.13}$$

For gas condensate systems,

$$T_{pc} = 187 + 330\,\gamma_g - 71.5\,\gamma_g^2 \tag{15.14}$$

$$P_{pc} = 706 - 51.7\,\gamma_g - 11.1\,\gamma_g^2 \tag{15.15}$$

In these correlations, the maximum allowable nonhydrocarbon components are 5% nitrogen, 2% carbon dioxide, and 2% hydrogen sulfide. Other methods that specifically deal with gas mixtures that contain appreciable amounts of nonhydrocarbons are discussed in the following text.

Effect of Nonhydrocarbon Components on Pseudocritical Properties

Natural gas mixtures frequently contain nonhydrocarbon components such as nitrogen, carbon dioxide, and hydrogen sulfide. Presence of these nonhydrocarbon components can affect the accuracy of *Z*-factor determination from the corresponding states principle, mainly because most of the components in gas mixtures are hydrocarbons of the same family, while nonhydrocarbons are not. The approach adopted to remedy this problem is to adjust or correct the pseudocritical properties to account for the presence of nonhydrocarbon components. Two methods were developed to adjust the pseudocritical properties of gas mixtures to account for the presence of nonhydrocarbon components; namely the Wichert–Aziz[4] correction method and the Carr–Kobayashi–Burrows[5] correction method.

Wichert–Aziz Method

The equations used for this method are

$$T'_{pc} = T_{pc} - \varepsilon \tag{15.16}$$

$$P'_{pc} = \frac{P_{pc}\,T'_{pc}}{T_{pc} + B(1 - B)\varepsilon} \tag{15.17}$$

$$\varepsilon = 120\,[A^{0.9} - A^{1.6}] + 15\,[B^{0.5} - B^{4.0}] \tag{15.18}$$

where T'_{pc} is the corrected pseudocritical temperature, °R; P'_{pc} the corrected pseudocritical pressure, psia; T_{pc} the uncorrected pseudocritical temperature, °R; P_{pc} the uncorrected pseudocritical pressure, psia; *B* the mole fraction of hydrogen sulfide in the gas mixture; and *A* the sum of the mole fraction of hydrogen sulfide and carbon dioxide in the gas mixture.

The uncorrected pseudocritical temperature and pressure obtained from Kay's mixing rules or the Standing correlations are corrected in the above manner.

Carr–Kobayashi–Burrows Method

The uncorrected pseudocritical temperature and pressure obtained from Kay's mixing rules or the Standing correlations are corrected by using the following simplified system of equations:

$$T'_{pc} = T_{pc} - 80\,Y_{CO_2} + 130\,Y_{H_2S} - 250\,Y_{N_2} \tag{15.19}$$

$$P'_{pc} = P_{pc} + 440\,Y_{CO_2} + 600\,Y_{H_2S} - 170\,Y_{N_2} \tag{15.20}$$

where T'_{pc} = corrected pseudocritical temperature, °R; P'_{pc} = corrected pseudocritical pressure, psia; T_{pc} = uncorrected pseudocritical temperature, °R; P_{pc} = uncorrected pseudocritical pressure, psia; Y_{CO_2} = mole fraction of carbon dioxide in the gas mixture; Y_{H_2S} = mole fraction of hydrogen sulfide in the gas mixture; and Y_{N_2} = mole fraction of nitrogen in the gas mixture.

Pseudocritical Properties of High-Molecular-Weight Gases

High molecular weight and correspondingly high specific gravity gases basically result when gas mixtures contain appreciable amounts of the C_{7+} fraction. In such cases, where gas specific gravities exceed 0.75, Sutton[6] recommends that Kay's mixing rules should not be used to determine the pseudocritical properties of gas mixtures, since this also affects the determination of the Z factor (discussed later). The approach proposed by Sutton for the determination of pseudocritical properties of high molecular weight gases is based on the mixing rules developed by Stewart et al.,[7] together with empirical adjustment factors related to the presence of the C_{7+} fraction in the gas mixture. The calculation steps for the proposed approach are outlined in the following text.

Calculate the parameters J and K:

$$J = \frac{1}{3}\left[\sum_{i=1}^{n} Y_i \left(\frac{T_c}{P_c}\right)_i\right] + \frac{2}{3}\left[\sum_{i=1}^{n} Y_i \left(\frac{T_c}{P_c}\right)_i^{0.5}\right]^2 \tag{15.21}$$

$$K = \sum_{i=1}^{n} Y_i \left(\frac{T_{ci}}{P_{ci}^{0.5}}\right) \tag{15.22}$$

where J is the Stewart–Burkhardt–Voo correlating parameter, °R/psia; K the Stewart–Burkhardt–Voo correlating parameter, °R/psia$^{0.5}$; and Y_i the mole fraction of component i in the gas mixture.

Calculate the adjustment parameters, F_J, E_J, and E_K:

$$F_J = \frac{1}{3}\left[Y\left(\frac{T_c}{P_c}\right)\right]_{C_{7+}} + \frac{2}{3}\left[Y\left(\frac{T_c}{P_c}\right)^{0.5}\right]^2_{C_{7+}} \tag{15.23}$$

$$E_J = 0.6081\, F_J + 1.1325\, F_J^2 - 14.004\, F_J\, Y_{C_{7+}} + 64.434\, F_J\, Y_{C_{7+}}^2 \tag{15.24}$$

$$E_K = \left[\frac{T_c}{P_c^{0.5}}\right]_{C_{7+}} [0.3129\, Y_{C_{7+}} - 4.8156 Y_{C_{7+}}^2 + 27.3751\, Y_{C_{7+}}^3] \tag{15.25}$$

where $Y_{C_{7+}}$ is the mole fraction of the C_{7+} fraction in the gas mixture; $(T_C)_{C_{7+}}$ the critical temperature of the C_{7+} fraction, °R; $(P_C)_{C_{7+}}$ the critical pressure of the C_{7+} fraction, psia; and $(T_C)_{C_{7+}}$ and $(P_C)_{C_{7+}}$ required in Equations 15.23–15.25 can be calculated from the Riazi–Daubert correlations discussed in Chapter 14.

Adjust the parameters J and K by applying the adjustment factors E_J and E_K according to the following relationships:

$$J' = J - E_J \tag{15.26}$$

$$K' = J - E_K \tag{15.27}$$

Calculate the adjusted pseudocritical temperature and pressure from the following expressions:

$$T'_{pc} = \frac{(K')^2}{J'} \tag{15.28}$$

$$P'_{pc} = \frac{T'_{pc}}{J'} \tag{15.29}$$

15.2.4.3 Determination of Compressibility Factor of Gas Mixtures

The physical properties of gas mixtures are correlated with the reduced pressure and reduced temperature in the same manner as they are in the case of pure component gases. The only difference is the use of pseudocritical pressure and pseudocritical temperature in defining the reduced values

$$P_{pr} = \frac{P}{P_{pc}} \quad \text{and} \quad T_{pr} = \frac{T}{T_{pc}} \tag{15.30}$$

where P_{pr} is the pseudoreduced pressure; T_{pr} the pseudoreduced temperature; P_{pc} (or P'_{pc} if adjusted) T_{pc} (or T'_{pc} if adjusted) values required in Equation 15.30 are determined from the methods described previously. (*Note*: appropriate equations should be used depending on the characteristics of the gas mixture (i.e., relatively sweet gas or gas containing appreciable amounts of nonhydrocarbons or C_{7+} fraction.)

Studies of compressibility factors for gas mixtures of various compositions have shown that compressibility factors can be generalized with sufficient accuracy based on the corresponding states principle by using pseudoreduced properties. In 1942, Standing and Katz[8] presented a generalized compressibility factor chart of sweet natural gas mixtures as a function of P_{pr} and T_{pr}. The Standing and Katz chart is generally reliable for sweet natural gases with minor amounts of nonhydrocarbon components. It is one of the most widely accepted correlations for obtaining Z factors for gas mixtures.

As an alternative to the Z-factor charts, several empirical correlations have been developed over the years for direct determination of the Z factor. Ahmed[9] discussed some of these correlations. However, the most prominent correlation for the determination of compressibility factors is that of Dranchuk and Abu-Kassem,[10] which duplicates the Standing and Katz compressibility factor chart with an average absolute error of 0.585%. The proposed 11-parameter correlation is given by Equation 15.31:

$$Z = \left[A_1 + \frac{A_2}{T_{pr}} + \frac{A_3}{T_{pr}^3} + \frac{A_4}{T_{pr}^4} + \frac{A_5}{T_{pr}^5}\right]\rho_r + \left[A6 + \frac{A_7}{T_{pr}} + \frac{A_8}{T_{pr}^2}\right]\rho_r^2$$

$$- A_9\left[\frac{A_7}{T_{pr}} + \frac{A_8}{T_{pr}^2}\right]\rho_r^5 + A_{10}(1 + A_{11}\rho_r^2)\frac{\rho_r^2}{T_{pr}^3}\exp[-A_{11}\rho_r^2] + 1 \tag{15.31}$$

where ρ_r = reduced gas density; expressed by $0.27\ P_{pr}/ZT_{pr}$.

The constants, A_1 through A_{11} have the following values:

$$A_1 = 0.3265$$
$$A_2 = -1.0700$$
$$A_3 = -0.5339$$
$$A_4 = 0.01569$$
$$A_5 = -0.05165$$
$$A_6 = 0.5475$$
$$A_7 = -0.7361$$
$$A_8 = 0.1844$$
$$A_9 = 0.1056$$
$$A_{10} = 0.6134$$
$$A_{11} = 0.7210$$

Owing to the nonlinear nature of the proposed equation, it is solved by iteration techniques such as the Newton–Raphson method.

Figure 15.6 shows the generalized compressibility factor values as a function of the reduced pressure and temperature, determined by solving Equation 15.31. This chart is similar to the one presented by Standing and Katz. Also shown in Figure 15.6 are the experimental compressibility factors for various other gas mixtures (reported in the work of Brown et al.[1]) that were not part of the data used to construct Standing and Katz's original compressibility factor chart. Figure 15.6 shows that values calculated from Equation 15.31 and the experimental data are in close agreement.

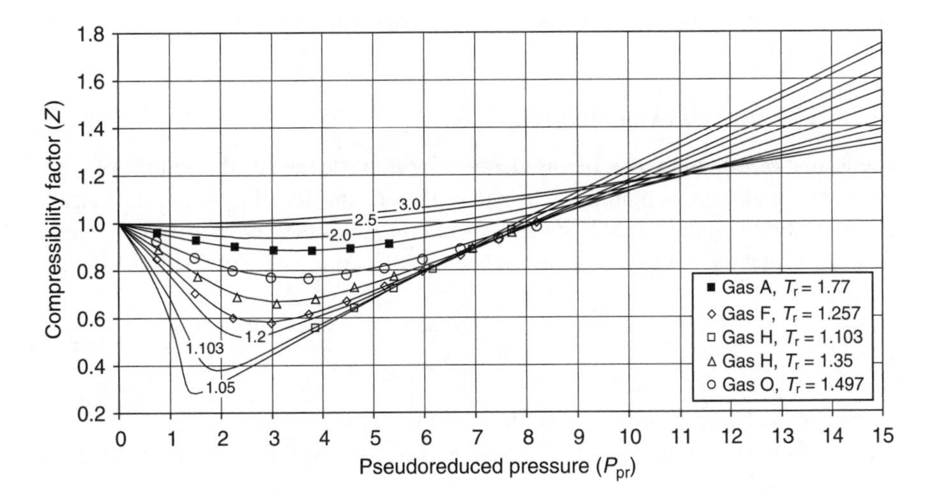

FIGURE 15.6 Generalized compressibility factor chart for hydrocarbon gas mixtures. Note that solid lines are calculated from the Dranchuk and Abu-Kassem[10] equation (Equation 15.31). The data for gases A, F, H, and O at indicated reduced temperatures are taken from Brown et al.[1]

The procedure for obtaining Z factors for gas mixtures is fairly straightforward and similar to the one described earlier for pure component gases. After calculating P_{pr} and T_{pr}, the corresponding value of Z factor is directly read from Figure 15.6 or it can be calculated from Equation 15.31.

15.2.4.4 Determination of Density of Gas Mixtures

For the determination of density of a gas mixture, Equation 15.6 can be expressed as

$$\rho_g = \frac{n}{V} MW_g = \frac{P}{ZRT} MW_g \tag{15.32}$$

where ρ_g is the density of the gas mixture, lb/ft^3; P the pressure, psia; MW_g the molecular weight of gas mixture, lb/lb-mol; Z the compressibility factor of the gas mixture at prevalent pressure and temperature conditions; R the universal gas constant, 10.732 (ft^3 psia/lb-mol°R); and T the temperature, °R.

15.2.5 Dry Gases

After having discussed the basic properties of gas mixtures, we now address the properties of gas mixtures that are of reservoir engineering significance such as the formation volume factor, coefficient of isothermal compressibility, and viscosity. Dry gases are considered in this section, while wet gases are discussed in Section 15.2.6.

Dry gases are the easiest to deal with because no liquid condensation takes place from the gas as it moves from the reservoir to the surface. Therefore the composition of the gas in the reservoir and the surface remains the same, meaning the specific gravity of the surface gas also equals the specific gravity of the reservoir gas.

15.2.5.1 Formation Volume Factor

The formation volume factor of gas, denoted by B_g, is defined as the volume of gas at reservoir conditions required to produce 1 standard ft^3 of gas at the surface. The formation volume factor is sometimes referred to as the *reservoir volume factor*. The gas formation volume factor is mathematically expressed by Equation 15.33:

$$B_g = \frac{V_{P,T}}{V_{SC}} \tag{15.33}$$

where B_g is the gas formation volume factor, ft^3/scf; $V_{P,T}$ the volume of gas at reservoir pressure and temperature, ft^3; and V_{sc} the volume of gas at standard conditions, scf.

Applying the real gas equation to this relationship gives

$$B_g = \frac{ZnRT/P}{Z_{sc}nRT_{sc}/P_{sc}} \tag{15.34}$$

Assuming the ideal behavior of gas at standard conditions ($Z_{sc} = 1$), and substituting the values of standard pressure (14.7 psi) and standard temperature (520°R),

$$B_g = 0.02827 \frac{ZT}{P} \text{ ft}^3/\text{scf} \tag{15.35}$$

In other oil field units, the volume of gas at reservoir conditions can be expressed in terms of barrels (abbreviated as bbl) as

$$B_g = 0.005035 \frac{ZT}{P} \text{ bbl/scf; } 1 \text{ bbl} = 5.615 \text{ ft}^3 \tag{15.36}$$

The reciprocal of the gas formation volume factor is called the *gas expansion factor* E_g and formulated

$$E_g = 35.37 \frac{P}{ZT} \text{ scf/ft}^3 \tag{15.37}$$

or

$$E_g = 198.6 \frac{P}{ZT} \text{ scf/bbl} \tag{15.38}$$

In Equations 15.36 and 15.37, if an experimental value of Z factor is available for the gas of interest, it should be used. Otherwise, various techniques discussed previously can be employed to obtain the value of Z factor. A typical dry gas formation volume factor as a function of pressure at constant temperature is shown in Figure 15.7.

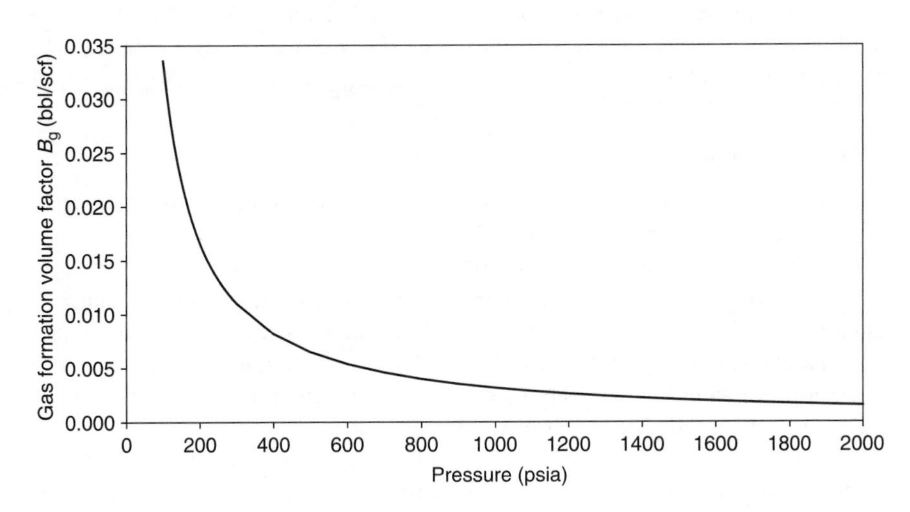

FIGURE 15.7 Variation of dry gas formulation volume factor with pressure at constant reservoir temperature.

15.2.5.2 Coefficient of Isothermal Compressibility

The coefficient of isothermal compressibility, denoted by C_g, for a gas, is defined as the fractional change in volume as pressure is changed at constant temperature and is mathematically expressed as

$$C_g = -\left(\frac{1}{V}\right)\left(\frac{\partial V}{\partial P}\right)_T \tag{15.39}$$

Using Equation 15.4 in the relationship shown in Equation 15.39,

$$\left(\frac{\partial V}{\partial P}\right)_T = \left(\frac{nRT}{P^2}\right)\left[P\left(\frac{\partial Z}{\partial P}\right)_T - Z\right] \tag{15.40}$$

Therefore,

$$C_g = -\left(\frac{P}{ZnRT}\right)\left(\frac{nRT}{P^2}\right)\left[P\left(\frac{\partial Z}{\partial P}\right)_T - Z\right] \tag{15.41}$$

$$C_g = \frac{1}{P} - \left(\frac{1}{Z}\right)\left(\frac{\partial Z}{\partial P}\right)_T \tag{15.42}$$

For ideal gases, $Z = 1$, and $(\partial Z/\partial P)_T = 0$, therefore,

$$C_g = \frac{1}{P} \tag{15.43}$$

where C_g is the coefficient of isothermal compressibility, psi^{-1} and P the pressure, psi. The reciprocal of psi, psi^{-1}, is sometimes called *sip*. An example of the relationship of C_g to reservoir pressure for a typical dry gas at constant temperature is shown in Figure 15.8.

15.2.5.3 Viscosity

The viscosity of a fluid, usually denoted by μ_g (for gas), is a measure of the resistance to fluid flow. The fluid viscosity is commonly given in units of centipoise, or cp, which are equivalent to gm mass/100 s-cm.

The viscosity of gas as a function of pressure and temperature is shown in Figure 15.9, which shows that gas viscosity decreases with decreasing reservoir pressure at all temperatures. However, with regard to the effect of temperature on viscosity, at low pressures gas viscosity increases as temperature increases, while at high pressures, gas viscosity decreases as temperature increases.

Although gas viscosity is a function of pressure, temperature, fluid density, and composition, unlike the other two gas properties discussed earlier (B_g and C_g), gas viscosity cannot be directly expressed as a function of pressure, temperature, or composition, that is, no exact and well-defined or obvious relationships can be used

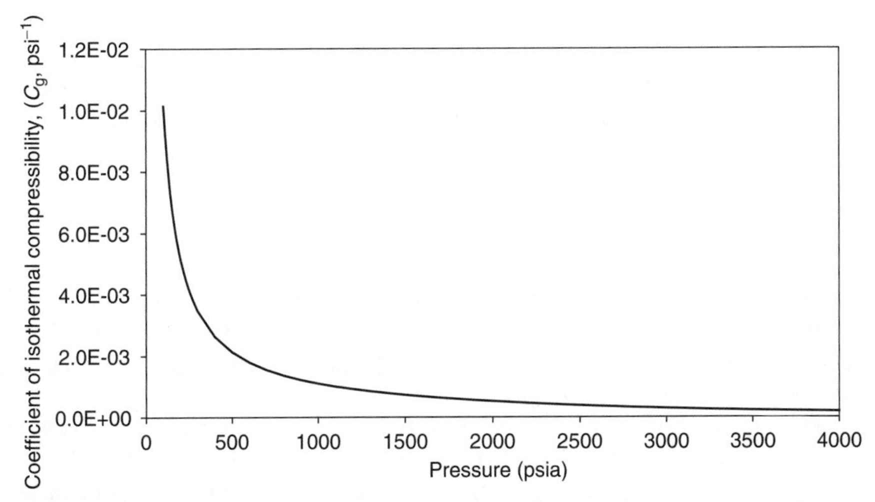

FIGURE 15.8 Variation of the coefficient of isothermal compressibility of a gas with pressure at constant reservoir temperature.

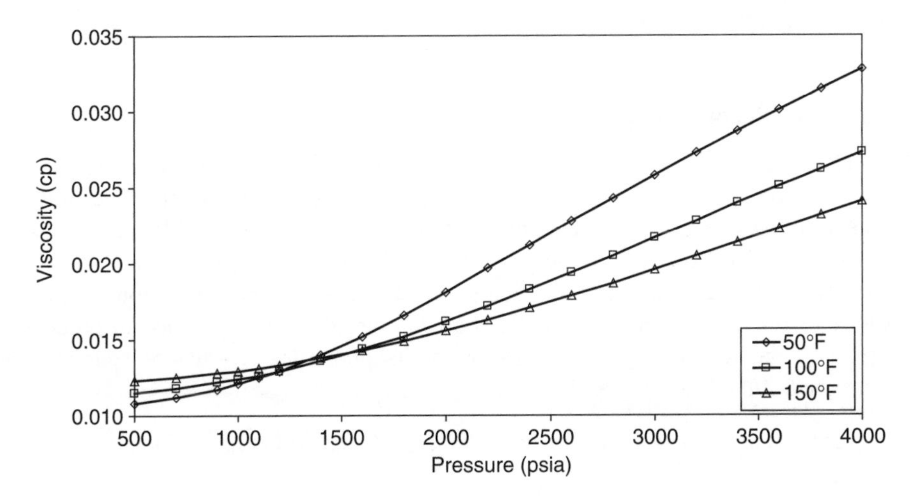

FIGURE 15.9 Variation of gas viscosity with pressure and temperature.

to determine gas viscosity. Additionally, experimental determination of gas viscosity is generally difficult. Therefore, gas viscosity is commonly estimated from empirical correlations developed over the years which relate viscosity to pressure, temperature, and composition in various forms. Some of these correlations are graphical while others are analytical expressions. Many of these correlations are discussed by Ahmed.[9]

However, the most popular and commonly used correlation for predicting viscosity of petroleum reservoir fluids is the Lohrenz–Bray–Clark[11] correlation, also known as *the LBC method*. In the LBC method viscosity is basically correlated as a function of fluid density that can be obtained from Equation 15.32 for gas mixtures. The LBC method is in fact also applicable to liquids, unlike some of the correlations that are specifically developed for either gases or liquids. Therefore, considering the versatility of the LBC method, this is the only correlation that is discussed in this chapter and is described in Section 15.5.2.

15.2.6 Wet Gases

In wet gases, condensation of liquid takes place from the reservoir gas when it moves from the reservoir to the surface. This results in a significantly different surface gas composition compared to the reservoir gas, due to loss of intermediate and heavy components in the condensate. Therefore, properties of the surface gas are not the same as properties of reservoir gas.

Wet gases are usually separated in a two-stage separation system, shown in Figure 13.3, in order to knock down the reservoir gas pressure and recover the condensate. At the surface, the well stream is separated or split into three different separator streams called *separator gas, stock tank gas,* and *stock tank liquid* (condensate). Therefore, in order to determine the properties of the reservoir gas, all these separator streams must be added or recombined in the appropriate ratio in which they are produced.

15.2.6.1 Recombination Cases

McCain[12] basically considered two different cases of recombinations. In the first case, recombination is performed based on the compositional analysis of separator fluids. In the second case, when compositional analysis is unavailable, recombinations are performed based on the properties of the separator fluids, such as specific gravity. The strategy for the recombination process involves the conversion of separator streams to either mass or molar basis that are then simply added in order to arrive at the reservoir gas composition or its properties. These procedures are described in the following text.

Separator Gas, Stock Tank Gas and Stock-Tank Liquid Composition Are Known

This procedure is in fact similar to the laboratory recombination process used for compositional analysis of petroleum reservoir fluids. In this particular case, compositions of all three streams, the gas to condensate ratios of the separator and the stock tank, and the specific gravity of the stock tank condensate are known. Additionally, properties such as the specific gravity and the molecular weight of the C_{7+} fraction are also known.

Let Y_{SPi} is the mole fraction of components $i = 1$ to n in the separator gas, Y_{STi} the mole fraction of components $i = 1$ to n in the stock tank gas; X_{STi} the mole fraction of components $i = 1$ to n in the stock tank condensate; R_{SP} the separator gas to condensate ratio, scf/stock tank bbl (STB); R_{ST} the stock tank gas to condensate ratio, scf/stock tank bbl (STB); and γ_{STC} the specific gravity of the stock tank condensate.

As far as Y_{SPi}, Y_{STi}, and X_{STi} are concerned, usually all components will be present in all three streams in varying quantities. For example, if methane is the lightest component, its concentration in the separator gas is the highest, followed by the stock tank gas, while the least amount of methane is present in the stock tank condensate. The exact opposite is the case for C_{7+}, that is, its highest concentration is in the stock tank condensate, followed by stock tank gas and the least amount present in the separator gas.

Since the separator and stock tank-to-condensate ratios are expressed in terms of a barrel of stock tank condensate, for recombination calculations, the basis starts with one STB of condensate. Next, the specific gravity of the stock tank condensate is converted to density from Equation 15.44:

$$\gamma_{STC} = \frac{\rho_{STC}}{\rho_{water}} \tag{15.44}$$

where ρ_{STC} is the density of stock tank condensate, lb/ft^3 and ρ_{water} the density of water, lb/ft^3.

Since both densities are at standard conditions, a value of 62.43 lb/ft^3 for water density can be used, so that, $\rho_{STC} = 62.43\ \gamma_{STC}$ lb/ft^3. Therefore

$$1 \text{ STB} = (5.615 \text{ ft}^3) \times (\rho_{STC} \text{ lb/ft}^3) = (5.615 \text{ ft}^3) \times 62.43\ \gamma_{STC} \text{ lb/ft}^3 = 350.5\gamma_{STC} \text{ lb.}$$

Next, average molecular weight of the stock tank condensate is calculated

$$\text{MW}_{STC} = \sum_{i=1}^{n} X_{STi}\, \text{MW}_i \tag{15.45}$$

where MW_{STC} is the average molecular weight of the stock tank condensate, lb/lb-mol and MW_i the molecular weight of all the individual components, lb/lb-mol.

Hence,

$$1 \text{ STB} = \frac{350.5\gamma_{STC} \text{ lb}}{\text{MW}_{STC} \text{ lb/lb-mol}} = \frac{350.5\gamma_{STC}}{\text{MW}_{STC}} \text{ lb-mol ST condensate}$$

Since 1 lb-mol of gas occupies 379.6 scf (Equation 15.2), the separator and stock tank-to-condensate ratios can also be converted to a molar basis

$$\left(\frac{R_{SP} \text{ scf/STB}}{379.6 \text{ scf/lb-mole}}\right) = 0.002634\ R_{SP} \frac{\text{lb-mole SP gas}}{\text{STB}}$$

$$= \frac{0.002634\ R_{SP} \text{ lb-mole SP gas}}{\dfrac{350.5\gamma_{STC}}{\text{MW}_{STC}} \text{ lb-mole ST condensate}}$$

$$= \frac{0.000007516\ R_{SP}\, \text{MW}_{STC}}{\gamma_{STC}} \frac{\text{lb-mole SP gas}}{\text{lb-mole ST condensate}}$$

and

$$\frac{0.000007516\, R_{ST}\, MW_{STC}}{\gamma_{STC}}\ \frac{\text{lb-mole ST gas}}{\text{lb-mole ST condensate}}$$

Since the basis is 1 STB of condensate, the molar ratio for stock tank condensate is

$$\frac{\dfrac{350.5\gamma_{STC}}{MW_{STC}}\ \text{lb-mole ST condensate}}{\dfrac{350.5\gamma_{STC}}{MW_{STC}}\ \text{lb-mole ST condensate}} = 1\ \frac{\text{lb-mole ST condensate}}{\text{lb-mole ST condensate}}$$

All three streams can now be recombined. For component i

$$Y_{SPi} \times \frac{0.000007516\, R_{SP}\, MW_{STC}}{\gamma_{STC}}\ \frac{\text{lb-mole component } i \text{ in SP gas}}{\text{lb-mole ST condensate}}$$

$$+ Y_{STi} \times \frac{0.000007516\, R_{ST}\, MW_{STC}}{\gamma_{STC}}\ \frac{\text{lb-mole component } i \text{ in ST gas}}{\text{lb-mole ST condensate}}$$

$$+ (X_{STi}) \times 1\ \frac{\text{lb-mole component } i \text{ in ST condensate}}{\text{lb-mole ST condensate}}$$

$$= \left[\left(Y_{SPi} \times \frac{0.000007516\, R_{SP}\, MW_{STC}}{\gamma_{STC}}\right) + \left(Y_{STi} \times \frac{0.000007516\, R_{ST}\, MW_{STC}}{\gamma_{STC}}\right) + (X_{STi})\right] \frac{\text{lb-mole component } i \text{ in reservoir gas}}{\text{lb-mole ST condensate}}$$

Finally, the composition of the reservoir gas is determined by normalizing the composition to mole fraction of 1 or 100 mol-%. For example, mole fraction of the ith component in the reservoir gas Y_{RGi} is given by

$$Y_{RGi} = \frac{\left[\left(Y_{SPi}\times \dfrac{0.000007516\, R_{SP}\, MW_{STC}}{\gamma_{STC}}\right)+\left(Y_{STi}\times \dfrac{0.000007516\, R_{ST}\, MW_{STC}}{\gamma_{STC}}\right) + (X_{STi})\right]}{\sum_{i=1}^{n}\left[\left(Y_{SPi}\times \dfrac{0.000007516\, R_{SP}\, MW_{STC}}{\gamma_{STC}}\right)+\left(Y_{STi}\times \dfrac{0.000007516\, R_{ST}\, MW_{STC}}{\gamma_{STC}}\right) + (X_{STi})\right]} \quad (15.46)$$

Based on these developed generalized equations, if compositional data and other supporting data are available, the composition of the reservoir gas can be easily calculated. Once the reservoir gas composition is known, Z factors, gas viscosity, and other properties can be determined.

In another case when samples are collected from the primary separator, that is , separator gas and separator liquid, they can be recombined in a somewhat similar manner as described here on the basis of compositions of separator fluids.

Compositions Unknown

If compositional data are unavailable, production data are used to determine the specific gravity of the reservoir gas. The calculations are performed on the basis of separator gas and the stock tank gas-to-condensate ratio and the specific gravities of the separator and the stock tank gas.

First, the surface gas is represented by a weighted average of the specific gravities of the separator and the stock tank gas by Equation 15.47:

$$\gamma_g = \frac{R_{SP}\gamma_{gSP} + R_{ST}\gamma_{gST}}{R_{SP} + R_{ST}} \tag{15.47}$$

where γ_g is the specific gravity of the surface gas; γ_{gSP} the specific gravity of the separator gas; γ_{gST} the specific gravity of the stock tank gas; and R_{SP} and R_{ST} the separator- and stock tank-to-condensate ratio, scf/STB.

The total producing gas-to-condensate ratio, R, is given by

$$R = R_{SP} + R_{ST} \tag{15.48}$$

If a three-stage separation system is used, the gas-to-condensate ratio and specific gravities of all three gas streams are used in Equations 15.47 and 15.48. However, the case of a simple two-stage separator system comprised of a separator and a stock tank is now considered.

Again considering the basis of one stock tank barrel of condensate, mathematical expressions for mass of reservoir gas, m_R, and the moles of reservoir gas, n_R, can be developed as shown in the following steps:

$$m_R = \left(0.002634\, R \frac{\text{lb-mole SP \& ST gas}}{\text{STB}}\right)\left(28.97\gamma_g \frac{\text{lb SP \& ST gas}}{\text{lb-mole SP \& ST gas}}\right) + \left(350.5\gamma_{STC}\frac{\text{lb ST condensate}}{\text{STB}}\right)$$

$$m_R = 0.0763\, R\gamma_g + 350.5\gamma_{STC}\frac{\text{lb reservoir gas}}{\text{STB}} \tag{15.49}$$

$$n_R = \left(0.002634\, R \frac{\text{lb-mole SP \& ST gas}}{\text{STB}}\right) + \left(\frac{350.5\gamma_{STC}}{MW_{STC}} \frac{\dfrac{\text{lb ST condensate}}{\text{STB}}}{\dfrac{\text{lb ST condensate}}{\text{lb-mole ST condensate}}}\right)$$

$$n_R = 0.002634\, R + \frac{350.5\gamma_{STC}}{MW_{STC}}\frac{\text{lb-mole reservoir gas}}{\text{STB}} \tag{15.50}$$

The ratio of m_R/n_R thus gives the molecular weight of the reservoir gas, from which specific gravity of the reservoir gas, γ_{gR}, can be expressed by Equation 15.51:

$$\gamma_{gR} = \frac{m_R/n_R}{28.97} \tag{15.51}$$

Finally, substituting Equations 15.49 and 15.50 in Equation 15.51, the mathematical expression for specific gravity of the reservoir gas is obtained:

$$\gamma_{gR} = \frac{R\gamma_g + 4594\gamma_{STC}}{R + 133{,}068\dfrac{\gamma_{STC}}{MW_{STC}}} \tag{15.52}$$

If molecular weight of the stock tank condensate is unknown, it can be estimated from the following empirical correlation:[13]

$$MW_{STC} = \frac{5954}{°API - 8.8} = \frac{42.43\gamma_{STC}}{1.008 - \gamma_{STC}} \tag{15.53}$$

Once the specific gravity of the reservoir gas is calculated from Equation 15.52, the Z factor, density, viscosity, and so on, can be determined from previously discussed techniques.

The other subset of this type of recombination when stock tank gas properties are unknown is discussed by McCain[12]. However, the method requires extensive use of various graphical correlations.

15.2.6.2 Formation Volume Factor

The formation volume factor of a wet gas, denoted by $B_{wg,}$ is defined as the volume of reservoir gas required to produce one stock tank barrel of condensate at the surface. By definition

$$B_{wg} = \frac{V_{P,T}}{V_{STC}@SC} \tag{15.54}$$

where B_{wg} is the wet gas formation volume factor, usually in barrels of gas at reservoir conditions per barrel of stock tank condensate; $V_{P,T}$ the volume of reservoir gas at reservoir conditions, bbl; V_{STC} the volume of stock tank condensate at standard conditions, bbl or STB.

The procedure for calculating B_{wg}, when compositions are known is:

1. Once the composition of the reservoir gas is determined, its pseudoreduced properties are calculated at given reservoir pressure and temperature conditions for the estimation of the Z factor. After the determination of the Z factor, the molar volume of the reservoir gas is calculated by using the real gas equation (Equation 15.4), i.e., $V_{P,T}$ ft^3/lb-mol reservoir gas or

$0.1781 V_{P,T}$ bbl/lb-mol reservoir gas. After substituting the value of universal gas constant, this gives, 1.911 ZT/P bbl/lb-mol reservoir gas.

2. Recalling Equation 15.46, the denominator gives the lb-mol of reservoir gas per lb-mol of stock tank condensate. However, 1 lb-mol of stock tank condensate equals $MW_{STC}/350.5\ \gamma_{STC}$STB.
3. Finally, using all these conditions and rearrangement, mathematical expression for B_{wg} is obtained:

$$B_{wg} = \left(0.005035\frac{ZT}{P}\right) \times \sum_{i=1}^{n}\left[(Y_{SPi}R_{SP}) + (Y_{STi}R_{ST}) + \left(\frac{133{,}068\ X_{STi}\ \gamma_{STC}}{MW_{STC}}\right)\right], \frac{\text{bbl reservoir gas}}{\text{STB}} \tag{15.55}$$

where P is the reservoir pressure, psia; T the reservoir temperature, °R; and other variables have been defined earlier.

Recall Equation 15.50, in which an expression for the lb-mol of reservoir gas on the basis of one stock tank barrel of condensate was obtained. This equation can be substituted into the real gas equation that directly leads to an expression that can be used to calculate the wet-gas formation volume factor when compositions are unknown:

$$B_{wg} = V = Z\left(0.002634R + \frac{350.5\gamma_{STC}}{MW_{STC}}\right)\frac{10.732\ T}{5.615\ P}\ \frac{\text{bbl reservoir gas}}{\text{STB}}$$
$$= 0.005035\frac{ZT}{P}\left(R + \frac{133{,}068\gamma_{STC}}{MW_{STC}}\right)\frac{\text{bbl reservoir gas}}{\text{STB}} \tag{15.56}$$

In fact Equation 15.56 can also be used when compositions are known.

15.2.7 Gas Condensates

The various calculation procedures described here also apply to gas condensate fluids as long as the reservoir pressure is above the dew-point pressure. At reservoir pressure below the dew point, none of the recombination procedures works because of the change in the overall composition of the reservoir fluid due to condensate precipitation in the reservoir. Therefore, special laboratory studies are required to describe the properties of gas condensate fluids, which are described later in this chapter.

15.2.8 Black Oils and Volatile Oils

In this section oil properties of reservoir engineering interest are considered. The properties discussed here are also called black oil properties, which include the oil formation volume factor, solution gas–oil ratio, coefficient of isothermal compressibility, oil viscosity, and surface tension. The physical processes involved in the way black oil properties change as reservoir pressure is reduced at constant temperature are explained. In defining these properties, the subscript "o" is used to indicate the various oil properties.

15.2.8.1 Formation Volume Factor

The oil formation volume factor, denoted by B_o, is defined as the ratio of the volume of oil at the prevailing reservoir conditions to the volume of oil at standard conditions. The volume of oil at reservoir conditions also includes the gas in solution. Therefore, the formation volume factor is a measure of the volume of reservoir fluid required to produce one unit of saleable product. The oil formation volume factor is mathematically expressed as

$$B_o = \frac{(V_o)_{P,T}}{(V_o)_{sc}} \tag{15.57}$$

where B_o is the oil formation volume factor, res. bbl/STB; $(V_o)_{P,T}$ the volume of oil at reservoir pressure and temperature, bbl; and $(V_o)_{sc}$ the volume of oil at standard conditions, stock tank barrel (STB, always reported at standard conditions).

A typical oil formation volume factor curve for a black oil, as a function of reservoir pressure at constant temperature, is shown in Figure 15.10. The overall curve is a combination of two distinct curves that are joined at the bubble-point pressure of the oil. This peculiar nature of the B_o curve is a result of the following three contributing factors.

1. The most dominant factor is the evolution of gas from the oil as pressure is decreased when the oil moves from the reservoir to the surface that causes a rather large decrease in the volume of the oil when significant amount of gas is in solution.
2. The reduction in pressure causes slight expansion of oil volume.
3. As the oil travels from usually high reservoir temperature to low surface temperature, a contraction in the oil volume takes place due to the temperature reduction.

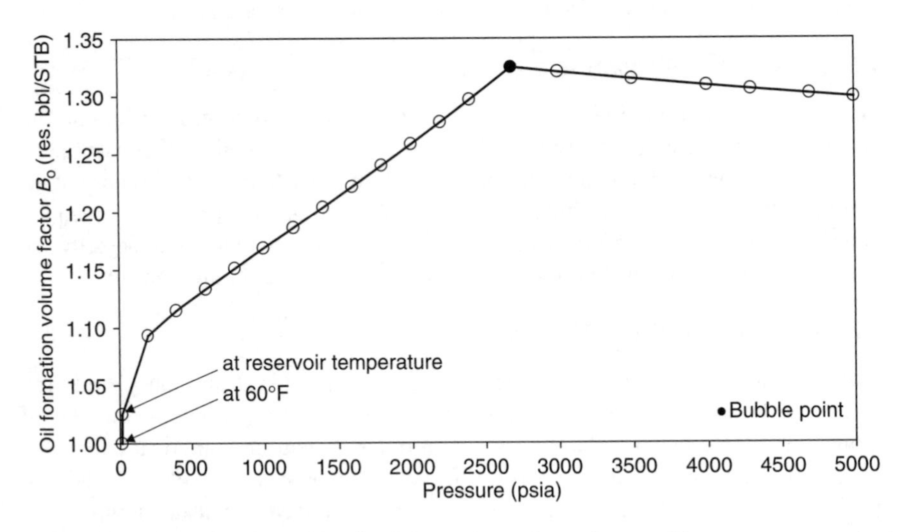

FIGURE 15.10 Variation of black oil formation volume factor with pressure at constant reservoir temperature.

The pressure and temperature effects are, however, somewhat offsetting, leaving the dissolved gas as the most dominant factor controlling the nature of the B_o curve.

For B_o values above the bubble-point pressure, a slight increase is observed as reservoir pressure is reduced. This happens because the oil in the reservoir simply expands and this expansion takes place due to the gas that still remains in solution. At bubble-point pressure, the oil reaches its maximum expansion and consequently attains a maximum value of oil formation volume factor. However, as soon as reservoir pressure falls below the bubble-point pressure, gas begins to come out of solution, thus leaving less gas in the oil, and consequently resulting in decreasing formation volume factors. For instance, if the reservoir pressure could be reduced to atmospheric pressure, the value of formation volume factor would be nearly equal to 1 res. bbl/STB. However, a reduction in temperature to 60°F would be necessary to bring the formation volume factor to exactly 1 res. bbl/STB (see Figure 15.10).

The reciprocal of oil formation volume factor is called the *shrinkage factor*. Black oils generally contain relatively small amounts of gas in solution, resulting in smaller values of B_o, or in other words, relatively less shrinkage is observed. Hence, black oils are sometimes called *low shrinkage oils*.

15.2.8.2 Solution Gas–Oil Ratio or Gas Solubility

The solution gas–oil ratio, denoted by R_s, is defined as the quantity of gas dissolved in an oil at reservoir pressure and temperature, which is also called the *dissolved gas–oil ratio* and occasionally *gas solubility*. The solution gas–oil ratio is usually expressed in terms of scf of gas per stock tank barrel of oil. The amount of gas dissolved in the oil is limited by the reservoir conditions of pressure and temperature and the quantity of light components present in the gas and oil phase.

The significance of R_s is best illustrated by considering a typical solution gas–oil ratio curve for a black oil as a function of pressure at constant reservoir temperature, such as the one shown in Figure 15.11. At all pressures above the bubble-point pressure, obviously the solution gas–oil ratio remains constant because no gas is evolved. However, as reservoir pressure falls below the bubble point; the oil is saturated and cannot contain all the gas in solution, resulting in release of some gas and consequently leaving less gas dissolved in the oil.

The concept of solution gas–oil ratio can be further illustrated by considering a hypothetical example where the reservoir pressure and temperature is reduced to standard conditions. If all the gas that evolved during this reduction in pressure and temperature is determined as Y scf and the volume of oil is X STB then the ratio of (Y/X) scf/STB represents the solution gas–oil ratio at bubble-point pressure and all pressures above the bubble-point pressure. However, at a certain pressure below the bubble-point pressure, if the volume of gas evolved is measured as Y_1 scf, then the solution gas–oil ratio or the gas remaining in solution at that particular pressure is given by $[(Y - Y_1)/X]$ scf/STB.

15.2.8.3 Total Formation Volume Vactor

To describe the pressure–volume relationship of oil below its bubble-point pressure, it is convenient to express this relationship in terms of the total formation volume

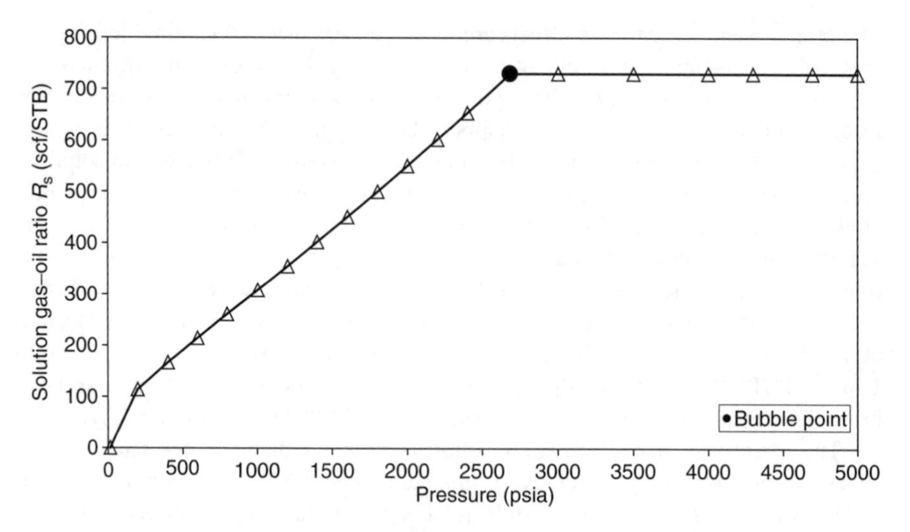

FIGURE 15.11 Variation of solution gas–oil ratio with pressure at constant reservoir temperature for a black oil.

factor as a function of pressure. The total formation volume factor, denoted by B_t, defines the total volume of an oil system at the prevailing pressure and temperature per unit volume of the stock tank oil, regardless of the number of phases present. Since, the oil splits into two phases below the bubble point, the total formation volume factor is sometimes also referred to as *two-phase formation volume factor*.

Mathematically, B_t is defined by Equation 15.58:

$$B_t = B_o + B_g(R_{sb} - R_s) \tag{15.58}$$

where B_t is the total formation volume factor, res. bbl/STB; B_o the oil formation volume factor, res. bbl/STB; B_g the evolved gas formation volume factor, res. bbl/scf (see Equation 15.36); R_{sb} the solution gas oil ratio at bubble point, scf/STB; R_s the solution gas–oil ratio at pressures below the bubble point, scf/STB.

Note that in Equation 15.58, the difference between R_{sb} and R_s is multiplied by B_g in order to express the volume of evolved gas in consistent units of res. bbl/STB.

A typical plot of total formation volume factor for a black oil as a function of pressure at constant reservoir temperature is shown in Figure 15.12, which also shows a plot of B_o. At bubble-point pressure and all pressures above the bubble point, B_o and B_t are equal because only one phase (i.e., liquid (oil)) exists at these pressures. Or in other words, since no gas is evolved, B_g is 0 and R_{sb} and R_s are equal. However, as pressures fall below the bubble point, the single-phase and two-phase formation volume factors are significantly different; B_t rapidly increases, while B_o decreases because B_t is dominated by the large amount of evolved solution gas. Also from Equation 15.58, the difference between R_{sb} and R_s increases rapidly because less amount of gas remains in solution or a large amount of gas is evolved, consequently increasing the value of B_t.

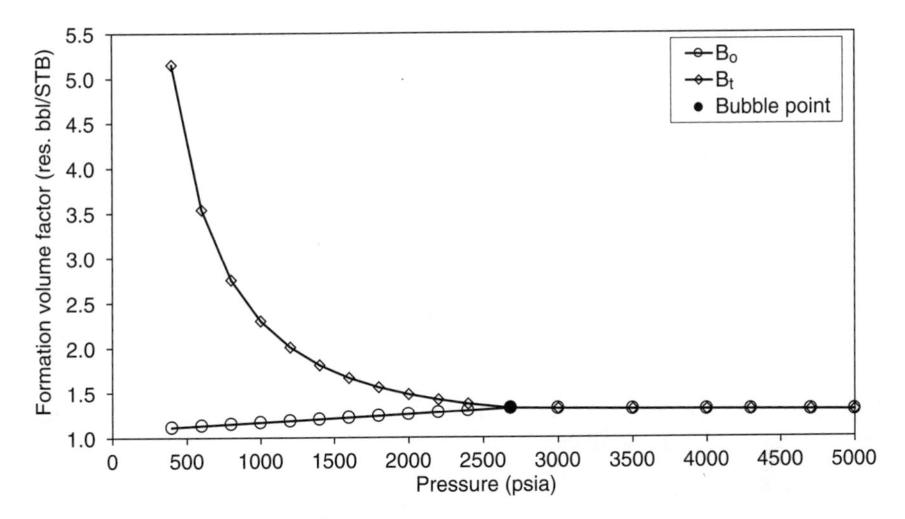

FIGURE 15.12 Comparison of oil formation volume factor and total formation volume factor as a function of pressure at constant reservoir temperature for a black oil.

15.2.8.4 Coefficient of Isothermal Compressibility

The coefficient of isothermal compressibility of oil, denoted by C_o, is defined exactly in the same manner as the coefficient of isothermal compressibility of a gas (Equation 15.39) at pressures above the bubble point. However, at pressures below the bubble point, an additional term must be included in the definition in order to account for the volume of gas that evolves.

Pressures Above the Bubble Point

As mentioned earlier, the mathematical expression for C_o is identical to Equation 15.39, the only difference being the replacement of subscript "g" by "o" to denote the value for oil.

$$C_o = -\left(\frac{1}{V}\right)\left(\frac{\partial V}{\partial P}\right)_T \tag{15.59}$$

where C_o = isothermal compressibility of oil, psi^{-1} and $(\partial V/\partial P)_T$ = slope of the isothermal pressure–volume curve.

If the experimental pressure–volume data for the oil is available, C_o, can be calculated at a given pressure by determining the volume V and slope $(\partial V/\partial P)_T$ at that particular pressure.

Another mathematical expression of C_o can also be obtained by direct substitution of formation volume factor in Equation 15.59 resulting in

$$C_o = -\left(\frac{1}{B_o}\right)\left(\frac{\partial B_o}{\partial P}\right)_T \tag{15.60}$$

If C_o is assumed to remain constant as pressure changes, Equation 15.59 can be integrated to derive an analytical expression relating C_o with pressure and volume at different stages

$$C_o \int_{P_1}^{P_2} \partial P = -\int_{V_1}^{V_2} \frac{\partial V}{V} \tag{15.61}$$

that result in

$$C_o = \frac{-\ln(V_2/V_1)}{(P_2 - P_1)} \tag{15.62}$$

Equation 15.62 can be used to determine the value of C_o if pressure–volume data at two different stages are available.

Pressures Below the Bubble Point

Since both B_o and R_s, decrease with decreasing pressures below the bubble point, they can be represented by $(\partial B_o/\partial P)_T$ and $(\partial R_s/\partial P)_T$ respectively. These changes need to be accounted for in determining the value of C_o at pressures below the bubble point.

Since both B_o and R_s, decrease with decreasing pressures below the bubble point, they can be represented by $R_{sD}R_{sSb}/R_{sDb}$, scf/STB and, respectively. However, the change in volume of free gas is given by, $-(\partial R_s/\partial P)_T$, since it is inversely proportional to R_s.

Thus, at reservoir pressures below the bubble point, the overall change in volume is the summation of the change in oil volume and the change in free gas volume, given by $[(\partial B_o/\partial P)_T - B_g(\partial R_s/\partial P)_T]$. Again note that B_g is inserted here to convert the volume of evolved gas to consistent units, res. bbl/TSB. Consequently, the fractional change in volume as pressure changes can be given by[12]

$$C_o = -\frac{1}{B_o}\left[\left(\frac{\partial B_o}{\partial P}\right)_T - B_g\left(\frac{\partial R_s}{\partial P}\right)_T\right] \tag{15.63}$$

Equation 15.63 obviously reduces to Equation 15.60 at pressures above the bubble point because R_s remains constant, which means that its derivative with respect to pressure is 0.

Considering the fact that the compressibility changes in the evolved gas are also taken into account, the coefficient of isothermal compressibility of oil at pressures below bubble point can also be called *total compressibility*. The coefficient of oil compressibility as a function of pressure (above and below the bubble-point pressure) at constant reservoir temperature for a black oil is shown in Figure 15.13. As seen in this figure, the trend is quite similar to the plot of B_o and B_t shown in Figure 15.12.

15.2.8.5 Viscosity

Oil viscosity, denoted by μ_o usually specified in centipoise, or cp, like other physical properties is affected by pressure as well as temperature. An increase in temperature causes a decrease in viscosity. Ordinarily, a decrease in pressure causes a decrease in viscosity and vice versa, provided that the only effect of pressure is to either decompress

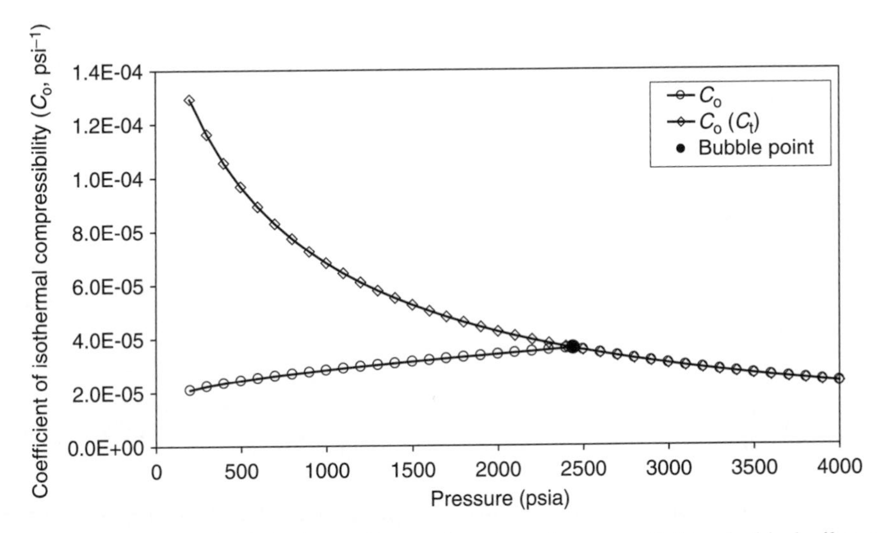

FIGURE 15.13 Variation in the coefficient of isothermal compressibility of a black oil as a function of pressure at constant reservoir temperature.

or compress the oil. In addition to the pressure and temperature effects, in case of oils, a third parameter affects the viscosity. This particular parameter is gas solubility. The amount of gas in solution is a direct function of pressure, that is, lower the pressure, lower is the amount of gas that remains in solution. Therefore, in case of oils, as the solution gas oil ratio decreases with decreasing pressure, the oil becomes heavier and thus its viscosity in fact *increases*, despite the decreasing pressure.

The relationship between the viscosity of a black oil as a function of pressure at constant temperature is shown in Figure 15.14. As seen in this figure, at pressures above the bubble point, the oil remains in single phase, therefore, the only factor that affects the viscosity is pressure. An almost linear decrease in the viscosity with pressure is evident. However, as reservoir pressure falls below the bubble point, the overall original composition of the oil changes. The gas that evolves takes the smaller molecules or the lighter components from the oil, leaving the remaining oil with relatively larger concentration of the heavier components. It is this changing composition that causes a large increase in viscosity of the oil as pressure continues to fall below the bubble-point pressure.

As black oil reservoir is depleted, not only does the production decrease due to the decrease in the driving force of pressure and the competition between free gas, but also because of the increase in the oil viscosity. Therefore, oil viscosity is probably one of the most significant properties that is targeted when considering enhanced oil recovery schemes involving the injection of a miscible gas, which actually acts as a solvent that reduces oil viscosity.

15.2.8.6 Surface Tension

The fundamental concepts of surface and interfacial tension were already discussed in Chapter 7. In Chapter 9 on relative permeability, the significance of surface

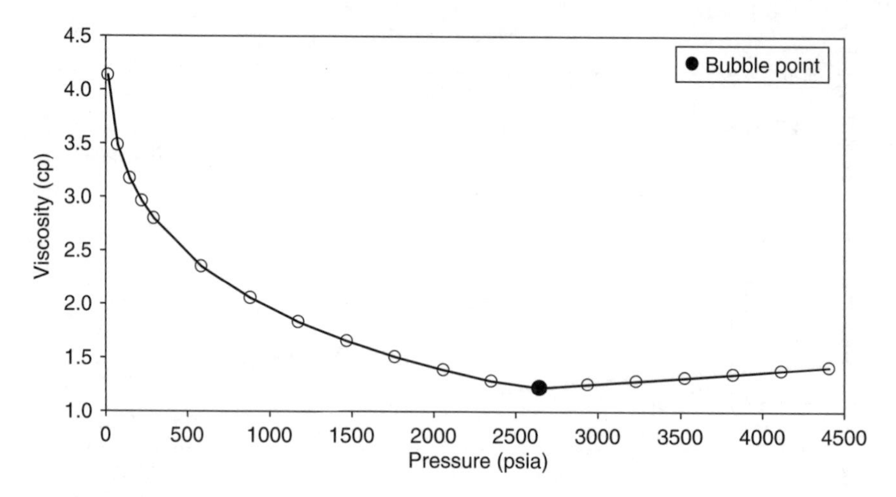

FIGURE 15.14 Variation in the viscosity of a black oil with pressure at constant reservoir temperature.

tension on oil recovery was discussed through the dimensionless capillary number. However, when miscible injection processes are considered for enhanced oil recovery, they rely to a great extent on the interaction between the displacing (gas) and the in-place fluids (oils) producing low surface-tension values. Therefore, surface tensions between hydrocarbon gas and hydrocarbon liquid, such as a gas–oil system, also assumes significant importance for properties of petroleum reservoir fluids.

The surface tension between a gas and oil is normally denoted by σ_{go}, and is specified in dynes/cm or mN/m. Unlike the other properties discussed so far, surface tension is almost equally affected by the properties of the two phases, such as their compositions and densities. The effect of pressure and temperature is normally reflected through the compositions and densities of the gas and liquid phases. Figure 15.15 shows the surface tensions between the equilibrium vapor and oil for a black oil system as a function of pressure below the bubble point at constant reservoir temperature. Needless to say that at pressures above bubble point, since only oil is present in single phase and no gas has evolved, the surface tension is obviously 0. As seen in Figure 15.15, surface tension increases as pressure declines below the bubble point. This behavior is a direct reflection of the density and compositional changes that take place in the equilibrium vapor and liquid phases at pressures below the bubble point. At relatively high pressures, just below the bubble point, the density difference between the equilibrium phases is small compared to relatively low pressures much below the bubble point, where density difference is large. Also, considering the fundamental theory behind the pendant drop technique, since the liquid phase is much lighter at relatively high pressures, the magnitude of surface forces is small resulting in much smaller droplets, while the opposite is true in the case of the heavier liquid at relatively low pressures where the magnitude of surface forces is large.

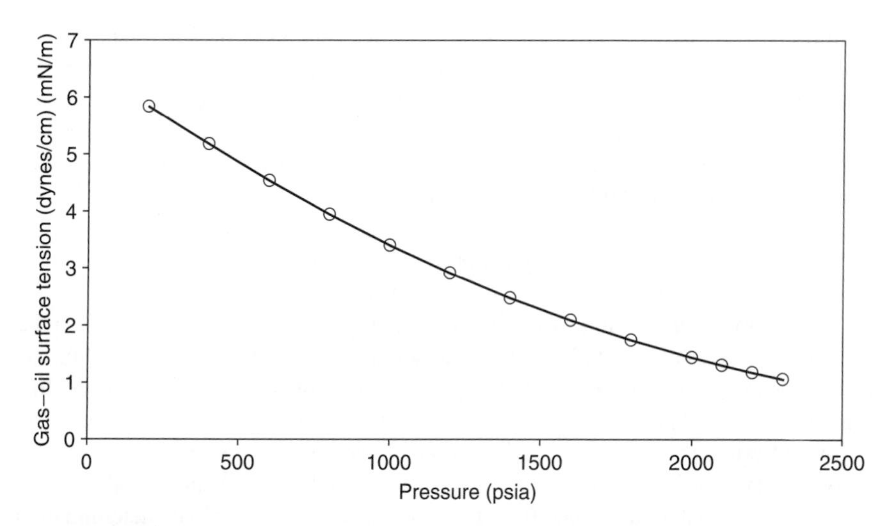

FIGURE 15.15 Variation in surface tension of a black oil system with pressure at constant reservoir temperature.

15.2.8.7 Volatile Oils

All the properties discussed previously for black oils are defined exactly in the same manner for volatile oils. The only difference between black oils and volatile oils is the difference between the magnitudes of these properties. For example, volatile oils contain significant proportion of dissolved gas, which means large quantities of gas evolution take place at pressures below the bubble point indicated by the closely spaced iso-vols (see Figure 12.2). This causes a large decrease in the curves of formation volume factor and solution gas–oil ratio. Also due to the presence of a significant amount of dissolved gas, and its subsequent evolution at pressures below the bubble point; the properties such as coefficient of isothermal compressibility and viscosity are quite important.

15.3 LABORATORY TESTS

The previous sections discussed the various physical properties of petroleum reservoir fluids as well as defined the various parameters of interest to engineers. In this section, the experiments through which these parameters are determined in the laboratory are described because the laboratory is the best source of obtaining the values of these various properties.

It is assumed that a valid fluid sample is available for the tests, either through a single-phase sample in the field (bottomhole sample) or through recombination of separator gas and liquid in the laboratory. The validation of field samples and recombination are also integral parts of the laboratory tests because the representativity of measured data is dependent on the quality of the fluid samples.

Basically, all laboratory tests or reservoir fluid studies are designed to characterize the phase behavior and properties of reservoir fluids at simulated reservoir conditions. Although, water is almost always present in petroleum reservoirs, its effect on phase behavior and properties are ignored in most tests; hence, reservoir fluid studies are conducted in the absence of water. The properties of formation or reservoir waters are discussed separately in Chapter 17. Similarly, almost all reservoir fluid studies are conducted in the absence of a porous medium. Danesh[14] reviewed hydrocarbon fluid-phase behavior studies in porous media and concluded that neglecting porous media effects on fluid phase behavior is a reasonable engineering approach. This has greatly simplified the experimental and theoretical studies of the phase behavior of petroleum reservoir fluids.[14]

The majority of the laboratory tests are depletion experiments, during which the pressure of the reservoir fluid is lowered in successive steps either by expanding the sample or increasing the fluid volume. The reduction of pressure results in the formation of a second phase. The two phases, usually gas and liquid are equilibrated and various physical properties are measured. However, some of these tests are specifically designed for certain types of reservoir fluids, so that their behavior and properties can be adequately described from the reservoir to the surface. The determination of single-phase reservoir fluid composition is also considered as part of the laboratory tests. However, this topic was already covered in Chapter 14 so it is not repeated here.

Specialized PVT equipment capable of handling high-pressure and high-temperature conditions are employed to conduct the reservoir fluid studies. This section first studies the basic features of PVT equipment and follows with a description of the commonly employed laboratory tests for obtaining the reservoir fluid properties.

15.3.1 PVT Equipment

All laboratory tests are generally conducted in conventional PVT equipment, which are commercially available from a variety of manufacturers. Some PVT equipment is used as purchased, while others are modified to enhance the measurement capabilities to include a full suite of measurements, such as *in situ* viscosity and surface tension studies and other exotic PVT studies that involve mixing of reservoir fluids with secondary fluids, such as injection gases.

Regardless of which equipment is used, the most essential component of any PVT equipment is a PVT cell (usually cylindrical and made of special grade stainless steel or titanium) that is equipped with a mechanism to increase or decrease the cell volume thereby altering the pressure. The alteration of the cell volume is usually achieved via a mechanically driven piston or mercury. However, the use of mercury is becoming virtually nonexistent due to the various hazards associated with its use. The PVT cell is housed in a thermostatic enclosure or an air bath, where air can be cooled or heated to maintain a constant temperature. This basic pressure–volume (PV) alteration mechanism allows depletion experiments to be conducted on any given reservoir fluid.

The volume of the PVT cell, ranging from between 50 to about 5000 cc, is usually a function of the type of reservoir fluid to be tested. For example, gas condensates generally require large volumes because the idea is to measure small liquid-phase volumes in equilibrium with a very large volume of vapor phase. Therefore, having a small cell volume would obviously result in a proportionally smaller volume of liquid phase, and the error in measuring that volume would be pronounced. Additionally, due to the compressible nature of gas condensate fluids (single-phase vapor at original conditions), sufficient volume is necessary for carrying out the pressure depletion. Similarly, in the case of black oils that are single-phase liquid at original conditions, a relatively small volume cell would suffice because a small increase in the cell volume causes relatively large pressure reduction due to incompressibility. Therefore, PVT equipment is sometimes characterized as *gas condensate PVT equipment* and *black oil PVT equipment*. However, in some multipurpose PVT cell designs, it is possible to conduct the laboratory studies of both types of reservoir fluids.

A schematic of a multipurpose PVT cell is shown in Figure 15.16. The two PVT cells (generally equal in volume) are arranged in the form of an hourglass shape.

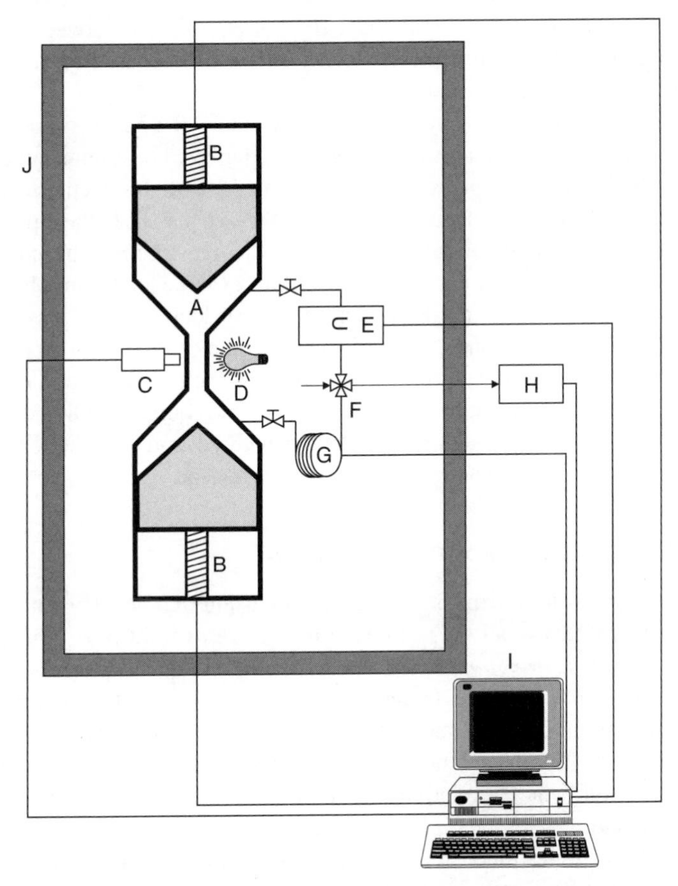

A Space available for hydrocarbon fluids; varied by the mechanically driven pistons
B Mechanically driven pistons
C Video camera
D Light source
E Online densitymeter
F Sampling valve for compositional analysis
G Capillary tube viscometer
H Gas chromatograph
I PC for PVT cell control
J Air bath

FIGURE 15.16 Schematic of a multipurpose PVT cell used in reservoir fluid studies.

The stem of the hourglass connects with the two PVT cells and basically serves as a window through which visual observations can be made (described in the next paragraph). Mechanically driven pistons control the volume available for hydrocarbon fluids within the two cells. The entire assembly and the associated tubing are housed in a forced convection air bath for maintaining constant test temperature.

The hourglass stem consists of two sapphire windows mounted opposite each other. One sapphire window is used for a light source, while the other is used for a video camera to facilitate the achievement of visual information. Although the sapphire window usually has small dimensions, the image obtained can be magnified and projected on a monitor. The cell fluids can be visually observed or video recorded while conducting the pressure depletion studies based on which saturation pressures can be obtained. An agitator or a stirring mechanism is included to speed up the homogenization of the single-phase fluid or equilibrium between the gas and liquid phases. The known dimensions of the stirrer can be used to determine the exact value of magnification used.

The total equilibrium vapor and liquid phase volumes can be determined by monitoring the volume of the mechanically driven piston pumps. For example, the interface between the vapor and liquid phase can be located in the hourglass stem (in view of the sapphire windows) by simultaneous movement of the two pistons in the top and bottom cells at the same rate and in the same direction, so that pressure remains constant. The measurement of properties of the single-phase fluid or the equilibrium phases is normally achieved by passing the fluids through an analysis loop that consists of a densitometer, a gas chromatograph, and a capillary tube viscometer. A pendant drop device can also be incorporated into the hourglass stem, to allow the formation of a droplet, surrounded by the equilibrium vapor phase, by pumping the equilibrium liquid phase. The droplets can be observed through the window, magnified, and recorded on video that can be dimensioned for the determination of surface tension.

15.3.2 Constant Composition Expansion

The constant composition expansion (CCE) or constant mass expansion (CME) test is carried out in virtually all PVT studies irrespective of fluid type. This particular test is also called *flash vaporization*, *flash liberation*, *flash expansion*, or simply *pressure–volume (PV) relation*. As the name suggests, the overall composition of the reservoir fluid or its original mass always remains constant because no reservoir fluid, either gas or liquid, is removed from the PVT cell. The primary objective of CCE tests is to study the PV relationship of a given reservoir fluid and determine its saturation pressure.

Figure 15.17 shows the CCE process. A single-phase sample of the reservoir fluid is placed in a PVT cell. Pressure is adjusted to a value equal to or greater than the initial reservoir pressure. Temperature is set at reservoir temperature. After the pressure and temperature conditions are stabilized, a pressure depletion experiment is carried out by increasing the volume in increments. The PVT cell is agitated regularly to ensure that the cell contents are at equilibrium. The total and phase volumes of the hydrocarbon system are recorded at each pressure step. The depletion process continues in this fashion until a predetermined low pressure or the capacity of the

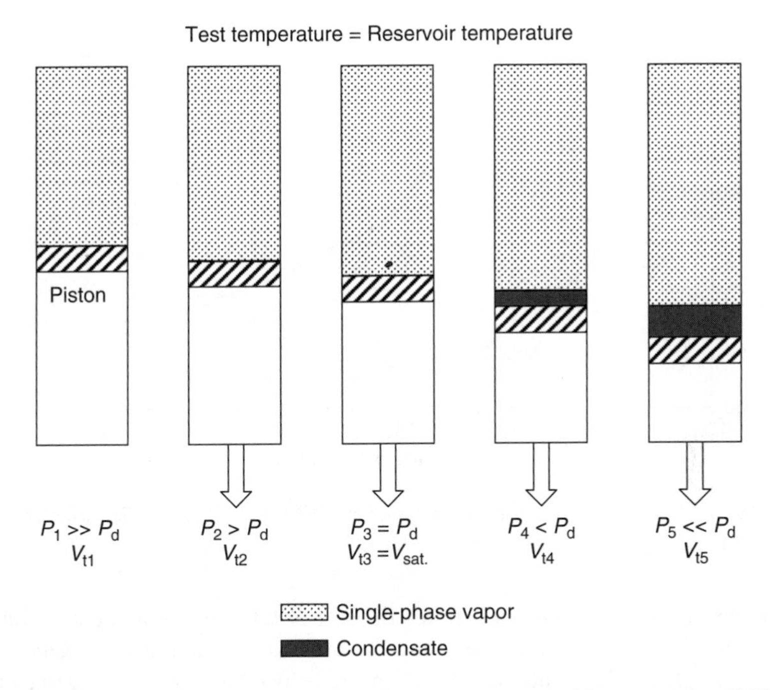

FIGURE 15.17 Schematic representation of a constant composition expansion (CCE) test on a gas condensate.

cell is reached. In case of windowed PVT cells, saturation pressures are also determined by visual observation of cell contents.

After completion of the test, the pressure–volume (PV) data are plotted. A PV relationship plot for a black oil from a CCE experiment is shown in Figure 15.18. As seen in this figure, for a black oil system, the transition from single-phase to a two-phase system is readily apparent in the plot. The pressure at which the slope changes is the bubble-point pressure of the oil. This change in slope occurs for two reasons:

1. In the single-phase region the oil is incompressible, resulting in a large pressure reduction with small volume increase.
2. As soon as bubble point is reached; due to the compressibility of the newly formed vapor phase, the same magnitude of volume increase causes less reduction in pressure.

In case of volatile oils, the change of slope at the bubble point is less pronounced, that is., PV curves are rather flat and a sharp break is not clearly evident (e.g., see Figure 15.18). This happens mainly because volatile oils are relatively compressible compared to black oils. Therefore, an element of uncertainty remains in the determination of the bubble point of volatile oils from PV relationships. In dealing with these types of fluids, it is recommended that more accurate and reliable methods, such as the optical detection techniques[15], should be used.

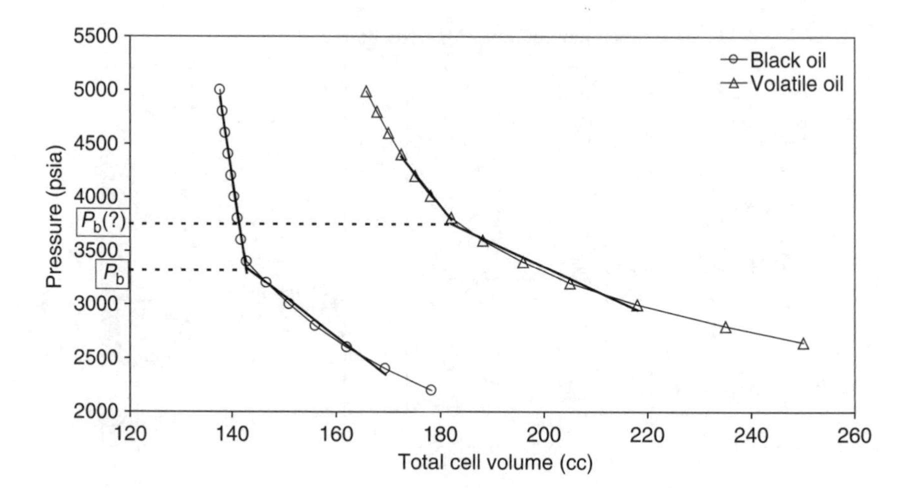

FIGURE 15.18 Comparison of pressure–volume (PV) behavior of Middle Eastern black oil and volatile oil.

In the case of gas condensate fluids the plot of total fluid volume as a function of pressure does not have a sharp change of slope at the saturation pressure unless substantial retrograde condensation takes place below the dew point. Again, similar recommendations to those for volatile oil apply to gas condensate fluids for determining the dew-point pressures.

In addition to the PV relationship and saturation pressure determination, additional data, such as densities, compositions, volumes, viscosities of the equilibrated phases, and surface tensions below the saturation pressure, can also be measured.

For dry gases and wet gases, since no phase change occurs in the reservoir, their compositions remain unchanged during production. Therefore, the PV relationships for these reservoir fluids are identical to the one shown in Figure 15.7. In fact CCE is the only main PVT test carried out for dry and wet gases. However, in case of wet gases, separate additional tests are necessary to determine the amount and properties of produced fluids at the surface conditions. In case of dry gases, the measured PV data are employed to calculate the Z factor and the gas formation volume factor using equations described earlier. The coefficient of isothermal compressibility factor can be calculated using the PV relationship or from Z-factor values. A combination of produced fluid data and the PV relationship can be used to determine the required properties of wet gases.

15.3.3 Differential Liberation

The differential liberation (DL) experiment is the classical depletion experiment carried out on reservoir oils. The experiment is carried out at reservoir temperature to evaluate the volumetric and compositional changes that take place in the oils during the primary production process (pressure depletion). The process is also called *differential vaporization*, *differential depletion*, or *differential expansion.*

As opposed to the CCE test in which the equilibrated phases are always in contact with each other at the reservoir conditions, in DL the solution gas that is liberated from an oil sample during a decline in pressure is continuously removed from contact with the oil. This type of liberation, as presented in Figure 15.19, is characterized by a varying composition of the overall hydrocarbon system.

As shown in Figure 15.19, pressure is reduced by increasing the cell volume at a pressure less than the bubble-point pressure. After achieving stabilization of pressure and temperature conditions and the equilibrated volumes of the gas and the liquid phases, all the evolved gas is expelled at constant pressure by reducing the cell volume. The procedure is repeated in 10–15 pressure stages down to the atmospheric pressure. At each pressure stage the remaining oil volume, the expelled gas volume at the cell conditions and standard conditions, and the gas specific gravity are measured. In the final step, cell temperature is reduced to 60°F, and the volume of remaining liquid is measured. This volume of oil is called the *residual oil volume* by differential liberation. Alternatively, a thermal contraction coefficient of 0.00046(v/v)/°F[14] is applied at atmospheric pressure and the cell temperature to determine the residual oil volume.

On the basis of collected experimental data, the properties determined from DL experiments are gas deviation factor Z, formation volume factor B_{oD}, total formation volume factor B_{tD}, and the solution GOR, R_{sD} as a function of pressure. Note that a

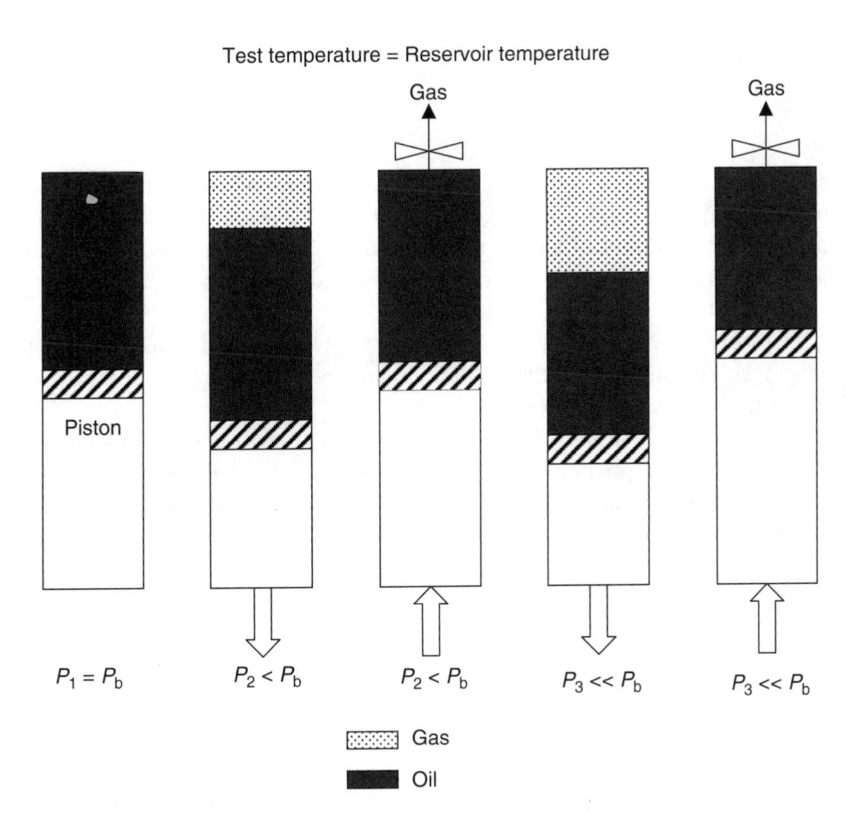

FIGURE 15.19 Schematic representation of a differential liberation (DL) test on an oil sample.

subscript "D" is used to indicate that all properties measured are from a differential liberation process.

The compressibility factor Z of the produced gas at any pressure stage is determined from Equation 15.64:

$$Z = \frac{V_R P_R T_{sc}}{V_{sc} P_{sc} T_R} \quad (15.64)$$

where V_R = expelled gas volume at cell conditions and P_R and T_R = cell pressure and temperature, in absolute units.

Alternatively, if the density of gas and its composition are measured at the cell conditions, the number of moles of gas can be calculated and the real gas equation can be applied to determine the value of Z factor

$$n_1 = \text{lb-mol of gas at pressure stage } P_{R1} = (V_{R1} \times \rho_{g1})/\text{MW}_{g1} \quad (15.65)$$

where V_{R1} is the volume of gas at cell conditions, ft^3; ρ_{g1} the gas density at cell conditions, lb/ft^3; and MW_{g1} the average molecular weight determined from gas composition at cell conditions, lb/lb-mol.
and

$$Z_1 = \frac{P_{R1} V_{R1}}{n_1 R T_R} \quad (15.66)$$

Subsequently, the volume of gas at standard conditions, V_{sc1} can be calculated by using the definition of standard conditions, that is, 1 lb-mol of gas occupies 379.6 scf. Therefore, n_1 lb-mol of gas occupies $379.6 \times n_1$ scf. After calculation of the Z factor, the formation volume factor B_g of the expelled gas is calculated from Equation 15.36.

The formation volume factor B_{oD} at each stage is calculated from the ratio of oil volume at cell conditions and the residual oil volume at standard conditions. This is also referred to as *relative oil volume*.

The total volume of gas removed during the entire process is the amount of gas in solution at the bubble point and all pressures above the bubble point. This total volume is divided by the residual oil volume and the resulting value is converted to standard cubic feet per barrel of residual oil, or scf/STB:

$$R_{sDb} = \frac{\sum_{i=1}^{n} V_{sci}}{(V_o)_{scD}} \quad (15.67)$$

where R_{sDb} is the solution gas–oil ratio at bubble point and all pressures above bubble point, scf/STB; V_{sci} the volume of gas removed at pressure stage i, scf (n represents the last pressure stage); $(V_o)_{scD}$ the residual oil volume at standard conditions, barrels or STB.

The gas remaining in solution at any pressure, P_{R1}, lower than the bubble point is calculated as

$$R_{sD1} = \frac{\sum_{i=1}^{n} V_{sci} - V_{sc1}}{(V_o)_{scD}} \tag{15.68}$$

where R_{sD1} is the solution gas–oil ratio at pressure P_{R1}, scf/STB and V_{sc1} the volume of gas removed at pressure P_{R1}, scf.

Similarly, for calculating the volume of gas remaining in solution at the next pressure stage is given by $(\sum_{i=1}^{n} V_{sci} - V_{sc1} - V_{sc2})/(V_o)_{scD}$.

Finally, the total formation volume factor or the relative total volume B_{tD} is calculated. Equation 15.69 is at pressure stage P_{R1}

$$B_{tD1} = B_{oD1} + B_{g1}(R_{sDb} - R_{sD1}) \tag{15.69}$$

where B_{tD1} is the total formation volume factor at pressure P_{R1}, res. bbl/STB; B_{oD1} the oil formation volume factor at pressure P_{R1}, res. bbl/STB; B_{g1} the evolved gas formation volume factor at pressure P_{R1}, res. bbl/scf; R_{sDb} the solution gas–oil ratio at bubble point and pressures above bubble point, scf/STB; R_{sD1} the solution gas–oil ratio at pressure P_{R1}, scf/STB.

The difference between R_{sDb} and R_{sD1} (Equations 15.67 and 15.68) gives the volume of evolved gas, $V_{sc1}/(V_o)_{scD}$, which after multiplication by B_{g1} results in the volume of liberated gas in res. bbl/STB.

However, it should be noted that the oil formation volume factor and the solution gas–oil ratio from the constant composition expansion and the differential liberation process is not the same. Generally both B_o and R_s for the CCE process is less than the DL process because the residual oil volume obtained by the two processes is not the same. The residual oil volume obtained in the differential liberation process is normally lower than the one obtained in the constant composition expansion process because the oil undergoes greater reduction in volume in the former as gas is continuously removed. However, the exact magnitude of B_o and R_s for the two processes depends primarily on the composition of the reservoir fluid.

15.3.4 Constant Volume Depletion

The constant volume depletion (CVD) test is performed for simulating the production behavior and separation methods of gas condensates fluids. The CVD test is performed on a reservoir fluid in such a manner as to simulate the pressure deletion of the actual reservoir, under the assumption that the retrograde liquid appearing during production remains immobile in the reservoir. The test consists of a series of expansion steps followed by the removal of excess gas at constant pressure in such a way that the cell volume remains constant at the end of each stage. This behavior in a way simulates the production behavior of a gas condensate; the hydrocarbon pore volume remains constant (under the assumption of no water encroachment), which is the constant cell volume, while the excess gas that is removed from the cell represents the gas that is produced from the reservoir and the retrograde liquid that is retained in the cell represents the immobile condensate in the reservoir. Note that the CVD test differs from the DL test in that not all the equilibrium gas is removed at each pressure stage.

A typical CVD process is schematically illustrated in Figure 15.20. The experiment starts with a single-phase reservoir fluid sample of known volume and composition at reservoir temperature and pressure. The pressure is reduced stepwise resulting in an expanded volume for the fluid at each stage. As the reservoir fluid composition remains the same above the dew point during depletion, the test can be simplified by just expanding the cell volume without removing any fluid from it. At some point during the pressure reduction, the fluid passes through the dew point. The volume at the dew point, which is often called the saturation volume, is considered as the reference (constant) volume in this procedure. At later stages, the cell pressure is reduced further below the dew-point pressure and the excess gas is removed from the top of the cell at constant pressure to return to the reference volume. Hence, the name *constant volume depletion*.

The gas phase removed at each stage is flashed to near-standard conditions and analyzed to determine the composition, volume, and compressibility factor. The condensate volume in the cell is also measured at each pressure step below the dew point.

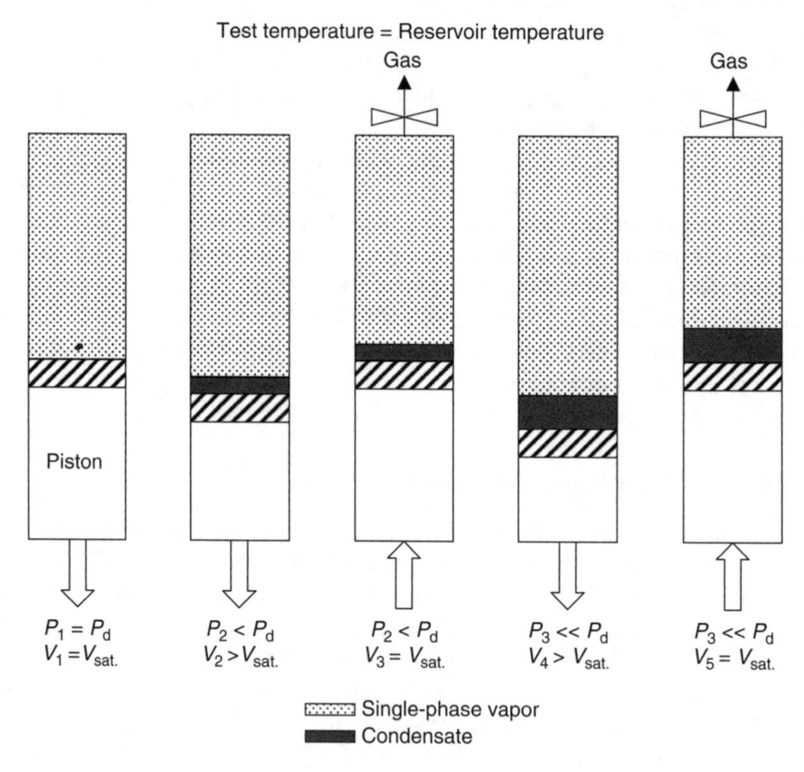

FIGURE 15.20 Schematic representation of a constant volume depletion (CVD) test on a gas condensate.

The entire CVD process is accomplished in 5–10 pressure reduction steps down to atmospheric. On termination of the test, the liquid remaining in the cell is analyzed.

15.3.4.1 Liquid Drop Out

The liquid drop-out volume in a CVD test is determined by the ratio of condensate volume at cell conditions at each pressure stage and the reference volume. Figure 15.21 shows the liquid drop-out behavior for a gas condensate fluid at a constant reservoir temperature for a CVD test. Also shown in Figure 15.21 is the liquid drop-out behavior for the same gas condensate at the same reservoir temperature but for a CCE test. As seen in the figure, the magnitude of liquid drop out for the two processes is different. The liquid drop out for a CCE test is generally higher than that of a CVD test at pressures far below the dew point because at these conditions a significant amount of gas has already been removed, thus reducing the potential for formation of condensate.

15.3.4.2 Material Balance for Condensate Composition

Condensate compositions at various pressure stages of the CVD test can be determined from a material balance approach. Table 15.1 shows data from a laboratory CVD test on a gas condensate[14] that is used to describe the material balance calculation procedure.

Individual Component Balance

Once the moles of condensate in cell are known, their composition can be calculated from individual component balance as follows (shown for methane):

Step 1

$$\text{Moles of methane in 100 lb-mol} = 0.7917 \times 100 = 79.17 \text{ lb-mol.}$$

Since composition of the gas removed and the one in the cell is the same,

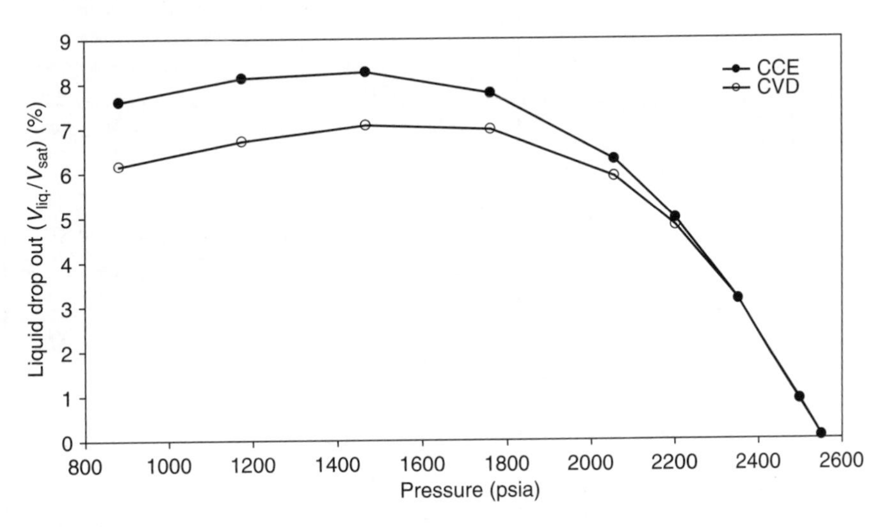

FIGURE 15.21 Comparision of a liquid drop-out behavior of a gas condensate in constant composition expansion (CCE) and constant volume depletion (CVD) tests.

TABLE 15.1
Laboratory CVD Data[a]

	$P > P_d$	$P > P_d$	$P > P_d$	P_d	Step 1	Step 2	Step 3	Step 4	Step 5	Step 6
Pressure (psig)	7183	7105	6960	6822	5800	4930	3915	3045	2030	1160
CGP (mol-% original fluid)	0	0.4	1.51	2.17	9.67	17.66	29.89	42.9	60.29	76.17
Liquid sat (% of V_{sat})	0	0	0	0	4.31	7.53	10.18	11.28	11.32	10.49
Gas Z factor	1.2074	1.1991	1.1876	1.1718	1.0767	1.0056	0.9479	0.9176	0.9171	0.9476
N_2	0.3	0.3	0.3	0.3	0.3	0.31	0.32	0.32	0.33	0.32
CO_2	1.72	1.72	1.72	1.72	1.71	1.71	1.72	1.73	1.75	1.77
C_1	79.17	79.17	79.17	79.17	79.93	80.77	81.61	82.33	82.71	82.58
C_2	7.48	7.48	7.48	7.48	7.44	7.41	7.46	7.54	7.64	7.79
C_3	3.29	3.29	3.29	3.29	3.22	3.21	3.2	3.19	3.22	3.32
iC_4	0.52	0.52	0.52	0.52	0.51	0.5	0.5	0.49	0.49	0.51
nC_4	1.25	1.25	1.25	1.25	1.23	1.21	1.18	1.15	1.15	1.22
iC_5	0.36	0.36	0.36	0.36	0.35	0.34	0.33	0.32	0.32	0.33
nC_5	0.55	0.55	0.55	0.55	0.54	0.52	0.5	0.48	0.48	0.5
C_6	0.62	0.62	0.62	0.62	0.58	0.55	0.52	0.49	0.46	0.46
C_{7+}	**4.74**	**4.74**	**4.74**	**4.74**	**4.19**	**3.47**	**2.66**	**1.96**	**1.45**	**1.20**
Total	100	100	100	100	100	100	100	100	100	100

Overall Balance

Basis of 100 lb-mol of original reservoir fluid										
Pressure (psig)	7183	7105	6960	6822	5800	4930	3915	3045	2030	1160
Gas removed (cum.; lb-mol)	0	0.40	1.51	2.17	9.67	17.66	29.89	42.9	60.29	76.17
Cell volume (ft^3)[b]	127.73									
Gas volume in cell (ft^3)[b]	127.73	127.73	127.70	127.68	122.18	118.07	114.68	113.28	113.23	114.29
Gas lb-mol remaining in cell[c]	100	99.6	98.49	97.83	86.65	76.24	62.44	49.61	33.15	18.61
Cond. lb-mol remaining in cell[d]	0	0	0	0	3.68	6.10	7.67	7.49	6.56	5.22

[a] Reservoir temperature = 249.5°F; CGP is cumulative gas produced.

[b] Up to dew-point pressure, gas volumes are calculated using the real gas equation. For example, at $P = 7197.7$ psia, $Z = 1.2074$, $T = 709.53°R$, and $n = 100$ lb-moles (basis); at dew point, $P = 6836.7$ psia, $Z = 1.1718$, $T = 709.53°R$, and $n = (100 - 2.17)$ lb-mol. For steps 1 to 6, gas volumes are calculated from volume at dew point and liquid drop out. For example, step 1, volume of gas = $127.68 \times (1 - 0.0431)$, and for step 6, volume of gas = $127.68 \times (1 - 0.1049)$.

[c] Up to dew-point pressure, gas lb-mol remaining in cell calculated from [(100 lb-mol) − gas removed (cum.) lb-mol]. Step 1 to 6 calculated using the real gas equation, for example, step 1, $P = 5814.7$ psia, $V = 122.18 ft^3$, $Z = 1.0767$, T = 709.53°R.

[d] Step 1 to 6 calculated from, [100 lb-mol (basis)] − [gas removed (cum.) lb-mol] − [Gas lb-mol remaining in cell]. For example, step 1 is $100 - 9.67 - 86.65 = 3.68$.

$$\text{Moles of methane in cell and removed gas} = 0.7993 \times (9.67 + 86.65) = 76.99 \text{ lb-mol}$$

$$\text{Moles of methane in condensate} = X_{C1} \times 3.68 \text{ lb-mol}$$

$$\text{Composition of methane in condensate} = [(79.17 - 76.99)/3.68] \times 100\% = 59.24 \text{ mole\%}$$

Step 2

Since 86.65 lb-mol of gas and 3.68 lb-mol of condensate progress to step 2, that is, flashed to the next pressure step:

$$\text{Moles of methane in feed (for step 2)} = 0.7993 \times 86.65 + 0.5924 \times 3.68 = 71.44 \text{ lb-mol}$$

$$\text{Moles of methane in gas in cell} = 0.8077 \times 76.24 = 61.58 \text{ lb-mol}$$

$$\text{Moles of methane removed} = 0.8077 \times (17.66\text{–}9.67) = 6.45 \text{ lb-mol}$$

(between steps 1 and 2, not cumulative)

$$\text{Moles of methane in condensate} = X_{C1} \times 6.10 \text{ lb-mol}$$

$$\text{Composition of methane in condensate} = [(71.44 - 61.58 - 6.45)/6.10] \times 100\% = 55.90 \text{ mol-\%}$$

Steps 3–6

Condensate composition for steps 3 to 6 can be calculated in a similar manner.

Finally, the composition of all the components in the condensate phase for steps 1 to 6 are calculated and are shown in Table 15.2.

In order to check the reliability of the laboratory data, the composition determined from the material balance for the last stage is compared with the measured composition of the condensate phase after completion of the test. However, it should be noted here that reliable material balance calculations can only be obtained by an accurate analysis of the produced gases. Sometimes, material balance calculations may also lead to negative composition of the light components, such as nitrogen that

TABLE 15.2
Condensate Compositions Calculated from Material Balance

Component	Step 1 5880 psig	Step 2 4930 psig	Step 3 3915 psig	Step 4 3045 psig	Step 5 2030 psig	Step 6 1160 psig
N_2	0.30	0.16	0.09	0.09	−0.02	−0.04
CO_2	1.98	1.87	1.74	1.66	1.50	1.30
C_1	59.27	55.86	52.80	46.10	37.99	27.44
C_2	8.53	8.51	7.80	7.14	6.31	4.98
C_3	5.12	4.51	4.34	4.45	4.40	4.04
iC_4	0.78	0.81	0.75	0.84	0.89	0.86
nC_4	1.77	1.83	2.00	2.27	2.43	2.29
iC_5	0.62	0.65	0.69	0.78	0.84	0.91
nC_5	0.81	0.98	1.08	1.26	1.37	1.47
C_6	1.67	1.65	1.72	2.00	2.44	2.95
C_{7+}	19.14	23.15	27.00	33.42	41.85	53.81
Total	100.00	100.00	100.00	100.00	100.00	100.00

has low concentrations in the mixture. In order to alleviate this problem, Danesh[14] recommends the use of direct compositional analysis for measurement of the compositions of the equilibrated phases at different stages of a CVD test so that material balance calculations can instead be used to cross check the measured condensate compositions.

15.3.4.3 Two-Phase Compressibility Factor

Another important property that is obtained from a CVD test is the two-phase compressibility factor. The two-phase compressibility factor represents the total compressibility of the remaining fluid (gas and retrograde condensate) in the cell and is computed from the real gas equation as

$$Z_{\text{two phase}} = \frac{PV_i}{(n_i - n_p)RT} \tag{15.70}$$

where $Z_{\text{two phase}}$ is the two-phase compressibility factor; P the pressure; V_i the initial gas volume; n_i the initial moles in the cell; n_p the cumulative moles of gas removed; and $(n_i - n_p)$ the moles of fluid in the cell.

The two-phase compressibility factor is a significant property because it is used when the P/Z vs. cumulative gas produced plot is constructed for evaluating gas condensate production.

15.3.5 Separator Tests

Petroleum reservoir fluids existing at high-pressure and high-temperature conditions in the reservoir experience pressure and temperature reduction when they are produced at the surface. As a result of these changes, gases evolve from the liquids and the well stream changes its character. Separator tests are carried out on reservoir fluids to simulate potential production separator stages and provide volumetric and other data on the stock tank oil and liberated gas streams. These tests also belong to the category of depletion tests described previously; however, in this case the temperature is also reduced at each stage and there are only a few pressure steps (normally just one between the reservoir pressure and atmospheric pressure). Separator tests are primarily carried out on black oils, and are usually the final tests that are conducted in the laboratory.

Figure 15.22 illustrates the laboratory separator test where a sample of reservoir oil at its bubble-point pressure is flashed in two stages. The first stage represents a separator, while the second stage represents the stock tank usually at atmospheric pressure. For oils that contain a high gas in solution, usually the separation is carried out in multiple stages to reduce the pressure on the reservoir fluid a little at a time and also to obtain a stable final liquid product. The separator test is conducted at temperatures that represent the average field conditions. The separator test is usually carried out at a number of separator pressures in order to determine the optimum, field separator conditions.

In laboratory separator tests, three main parameters are usually determined for a pressure reduction path of $P_{\text{res}}\,(P_{\text{b}}) \rightarrow P_{\text{sep}} \rightarrow P_{\text{atm}}$:

1. The formation volume factor of oil B_{oSb};
2. The solution gas–oil ratio R_{sSb}; and
3. The specific gravity of the stock tank oil.

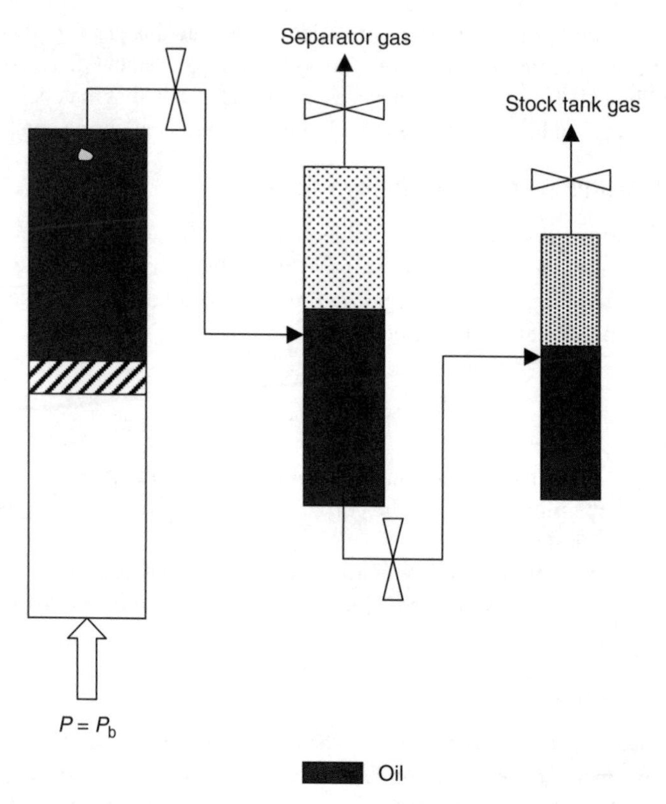

FIGURE 15.22 Schematic representation of a separator test on an oil sample.

The formation volume factor of oil is defined as[12]

$$B_{oSb} = \frac{\text{volume of liquid expelled from the cell}}{\text{volume of liquid arriving in the stock tank}}, \text{ res. bbl/STB} \quad (15.71)$$

In Equation 15.71, the volume of liquid expelled from the cell is measured at bubble-point conditions, while the volume of stock tank liquid is measured at standard conditions. The subscript "S" indicates that the value is a result of the separator test.

The solution gas–oil ratio is defined as[12]

$$R_{sSb} = \frac{\text{volume of separator gas} + \text{volume of stock tank gas}}{\text{volume of liquid in the stock tank}}, \text{ scf/STB} \quad (15.72)$$

The measured density of the stock tank oil is converted into specific gravity, which is subsequently expressed in terms of API gravity.

The other data measured include specific gravities of the separator gas and the stock tank gas. A separator volume factor, defined by the ratio of the volume of separator liquid at separator conditions and the volume of stock tank oil at standard conditions, SP bbl/STB, is also determined.

15.3.5.1 Optimum Separator Conditions

Separator test results are usually reported in a tabular format and basically include separation pressure and temperature conditions and values of various measured parameters. Table 15.3, an example of such a table for a North Sea black oil separator test, shows that the magnitudes of the various parameters vary with separator conditions. This variation is generally a result of the overall composition of the reservoir fluid; and the relative distribution of light, intermediate, and heavy components between the produced gas and the stock tank oil that are affected by the separator conditions. The stock tank oil, however, generally contains only a trace of methane, and an insignificant amount of ethane, regardless of the separator conditions. Similarly, the concentration of the heavy components (C_{7+}) in the gas phase is very small in most cases and obviously very significant in the liquid phase. It is the relative distribution of the intermediate components between the separated phases that determines the optimum separator conditions.

Nevertheless, based on the laboratory separator test data shown Table 15.3, an optimum separator pressure is selected. The identification of optimum separator pressure is based on the following criteria:

- A minimum of the total gas–oil ratio (column 3 of Table 15.3);
- A maximum in the API gravity of stock tank oil (column 4 of Table 15.3); and
- A minimum in formation volume factor of oil at bubble-point conditions (column 5 of Table 15.3).

Finally, for quick identification, these three properties are plotted as a function of separator pressure, as shown in Figure 15.23, to determine the optimum separator pressure. However, it should be noted here that circumstances for a particular field may dictate a specific separator pressure, which may be different than the selected optimum separator pressure. If not, the separator pressure that produces the maximum amount of stock tank liquid is selected.[12]

15.4 ADJUSTMENT OF BLACK OIL LABORATORY DATA

Various laboratory tests, such as the CCE, DL, and separator tests, that are carried out on black oils, from which properties such as formation volume factors, solution gas–oil ratios, total formation volume factors are obtained, were described earlier. However, no single laboratory test alone adequately represents or describes the properties of a black oil from the reservoir to the surface. In other words, the actual reservoir process is neither constant composition expansion, nor differential liberation. Therefore, results obtained from all three tests are usually combined or adjusted with a bubble-point constraint (i.e., for reservoir pressure above and below P_b) in such a manner that the combined data represent the properties of black oils from the reservoir to the surface. It should, however, be noted that the separator test results that are used in the adjustment are optimum or selected separator conditions. These adjusted or corrected properties are then used in reservoir engineering calculations. These adjustment methods that are used for obtaining representative black oil properties are described in this section.

TABLE 15.3
Separator Test Results for a North Sea Black Oil at 80°F[a]

Separator Pressure (psia)	Gas–Oil Ratio[b]	Gas–Oil Ratio[c]	Stock-Tank Oil Gravity (°API)	Formation Volume Factor[d]	Separator Volume Factor[e]	Gas Z Factor	Specific Gravity of Flashed Gas
315	262.78	276.59			1.053	0.963	0.6507
to							
15	51.04	52.33	**43.82**	**1.177**	1.025	0.994	1.1691
		328.92					
215	289.79	301.02			1.039	0.979	0.6796
to							
15	24.87	25.49	**43.85**	**1.175**	1.025	0.994	1.1973
		326.51					
65	314.05	323.26			1.029	0.990	0.7372
to							
15	7.97	8.14	**43.79**	**1.179**	1.025	0.994	1.1479
		331.40					

[a] Separator test results at other pressure conditions and those shown in this table are plotted in Figure 15.23 to determine the optimum separator conditions.
[b] Cubic feet of gas at standard conditions per barrel of oil at indicated pressure and temperature.
[c] Cubic feet of gas at standard conditions per barrel of stock tank oil at standard conditions.
[d] Formation volume factor is barrels of oil at bubble-point pressure of 2522 psia and 212°F per barrel of stock tank oil at standard conditions.
[e] Separator volume factor is barrels of oil at indicated pressure and temperature per barrel of stock tank oil at standard conditions.

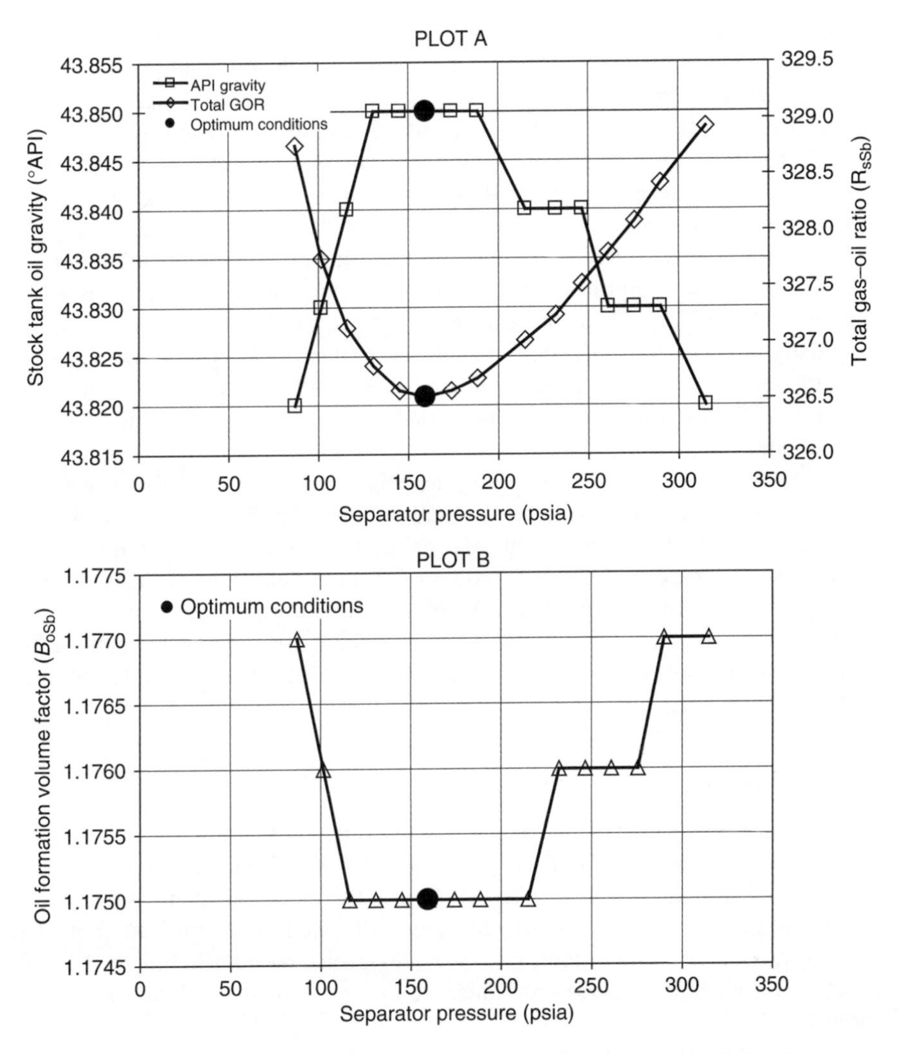

FIGURE 15.23 Determination of optimum separator conditions from black-oil separator tests. Optimum pressure is determined at 159.5 psia at 80°F.

Danesh[14] provides an excellent depiction of phase transition that takes place in an undersaturated oil reservoir and this actually provides an approximate guideline for the manner in which the results from the CCE, DL, and separator tests are combined. The reservoir region is split into three different zones A, B, and C. Zone A is farthest from the well bore and zone C is closest to the well bore, while zone B lies in between zones A and C. When reservoir oil flows from zone A to the separator by passing through zones B and C, the flow is accompanied by a pressure drop, with highest pressure in zone A and the lowest pressure in the separator.

Therefore, far away from the well bore, in zone A, the pressure is still above the bubble-point pressure, due to the highest pressure, and the oil simply expands as a single-phase fluid. In zone B, the pressure is lower than in zone A and the fluid is just below its bubble point and the volume of the liberated gas is too small to allow its mobilization. However, in zone C, where the pressure is even lower than zones A and B, the evolved gas begins to flow toward the producer, but segregates from the oil due to gravity and surface forces. In the well bore, the liberated gas and the liquid phase are considered to flow together due to the dominant mixing[14]. Therefore, the reservoir process in zones A and B is simulated or represented by the constant composition expansion test because the overall composition of the oil essentially remains the same. Since the gas segregates due to gravity and begins to flow, the reservoir process occurring in zone C is simulated by the differential liberation test. The separation that occurs at the surface is represented or simulated by the separator test.

It can be argued that the preceding depiction and simulation of the reservoir process by laboratory tests, such as the CCE, DL and separator test, and their subsequent combination, to some extent are *artificial* or *idealized* representations of the overall reservoir process. However, it does offer a mechanism by which the actual reservoir process may be approached. Moreover, as McCain[12] states, experience has shown that black oil properties calculated under these assumptions are sufficiently accurate for reservoir engineering calculations. These adjustment methods are provided in the next section.

15.4.1 Combination Equations

The adjustment approach and the equations described by McCain[12] is used here for combination equations of the fluid properties from the three laboratory tests. As discussed earlier, the underlying assumption is that at reservoir pressures above the bubble-point pressure, fluid properties are calculated by a combination of CCE or flash vaporization and separator tests. At reservoir pressures below the bubble-point pressure, fluid properties are obtained by a combination of DL and separator tests.

Since properties measured from the three different laboratory tests are combined, it is convenient to first get acquainted with the nomenclature that is used to avoid confusion. The subscripts "D," "F," and "S" represent the differential liberation, flash vaporization or constant composition expansion and separator tests, respectively, while subscript "b" indicates bubble-point conditions. The nomenclature that is used for the combination equations is: B_{oD} is the relative oil volume by differential liberation; B_{oDb} the relative oil volume at bubble point by differential liberation; and B_{oSb} the formation volume factor at bubble point from separator tests (optimum or selected); $(V_t/V_b)_F$ = relative total volume (gas and oil) by constant composition expansion or flash vaporization, where V_t is the total volume, and V_b is the volume at saturation conditions or bubble point; B_{tD} is the relative total volume (gas and oil) by differential liberation; R_{sD} the gas remaining in solution by differential liberation; R_{sDb} the gas in solution at bubble point (and all pressures above) by differential liberation; and R_{sSb} the sum of separator gas and stock tank gas from separator tests (optimum or selected).

15.4.1.1 Formation Volume Factor of Oil

At pressures above the bubble-point pressure, oil formation volume factors are calculated from a combination of CCE and separator tests

$$B_o = \left(\frac{V_t}{V_b}\right)_F B_{oSb} \tag{15.73}$$

The units for B_o are reduced to res. bbl of oil at pressure P/STB, as shown:

$$B_o = \left(\frac{\text{res. bbl of oil at pressure } P}{\text{res. bbl of oil at pressure } P_b}\right)\left(\frac{\text{res. bbl of oil at pressure } P_b}{\text{STB}}\right)$$

$$= \left(\frac{\text{res. bbl of oil at pressure } P}{\text{STB}}\right) \tag{15.74}$$

At pressures below bubble point, a combination of DL and separator test data gives:

$$B_o = B_{oD}\left(\frac{B_{oSb}}{B_{oDb}}\right) \tag{15.75}$$

Again units of B_o are reduced to res. bbl of oil at pressure P/STB as

$$B_o = \left(\frac{\text{res. bbl of oil at pressure P}}{\text{residual bbl by DL}}\right)\frac{\left(\dfrac{\text{res. bbl of oil at pressure P}_b}{\text{STB}}\right)}{\left(\dfrac{\text{res. bbl of oil at pressure P}_b}{\text{residual bbl by DL}}\right)}$$

$$= \left(\frac{\text{res. bbl of oil at pressure P}}{\text{STB}}\right) \tag{15.76}$$

15.4.1.2 Solution Gas–Oil Ratio

Since solution gas–oil ratio at bubble-point pressure and all pressures above bubble-point pressure is constant,

$$R_s = R_{sSb} \tag{15.77}$$

which obviously has the units of scf/STB.

At pressures below the bubble-point pressure, R_s is calculated from a combination of DL and separator test data as:

$$R_s = R_{sSb} - (R_{sDb} - R_{sD})\left(\frac{B_{oSb}}{B_{oDb}}\right) \tag{15.78}$$

The reduction of R_s units to scf/STB is:

$$R_s = \left(\frac{\text{scf}}{\text{STB}}\right) - \left(\frac{\text{scf}}{\text{residual bbl by DL}}\right)\left(\frac{\left(\frac{\text{res.bbl of oil at pressure } P_b}{\text{STB}}\right)}{\left(\frac{\text{res. bbl of oil at pressure } P_b}{\text{residual bbl by DL}}\right)}\right) \tag{15.79}$$

However, in a fairly recent paper, McCain[16] points out that Equation 15.78 is an incorrect formulation for calculating the solution gas–oil ratio at pressures below the bubble point. (Note that in the discussion Equation 15.79 is used instead to show the incorrectness.) Although in Equation 15.79, a distinction is made between the volume of oil (i.e., residual oil from DL and residual oil from a separator test), the equation does not distinguish or specify the process by which the volume of gas is obtained. For example, R_{sSb} denotes the volume of gas (separator and stock tank) from the separator test (which is also the solution gas at bubble point), while the difference ($R_{sDb} - R_{sD}$) denotes the volume of gas evolved in the differential liberation process, thus the gas obtained from the two processes is different. Therefore, if the sources of the data are not taken into account, the units scf/STB in Equation 15.79 appear to be correct.[16] However, McCain[16] points out, that the gas liberated during a separator test is significantly different in quantity and quality from the gas liberated during a differential liberation. Hence, even though the ratio (B_{oSb}/B_{oDb}) in Equation 15.79 takes into account the differences in the oils from the separator test and differential liberation, the differences in the gases are ignored, rendering the material balance expressed in the equation incorrect. In fact, the values of solution gas–oil ratio calculated by Equation 15.79 can even be negative at low pressures.

Considering the discrepancy described previously, McCain[16] proposes a modified equation for calculating the solution gas–oil ratio for reservoir engineering calculations. It is known that R_{sSb} is the gas originally in solution in the reservoir oil at its bubble-point pressure as measured in a separator test, which is *scf of gas from sep.*

test/STB. On the other hand, $(R_{sDb} - R_{sD})$ is the volume of gas liberated in the reservoir during a differential liberation from bubble-point pressure P_b to some pressure P, which is *scf of differentially liberated gas*/residual bbl by DL. The ratio of R_{sSb} and R_{sDb} takes into account the difference in the two oils, as well as the difference in the two gases as shown in Equation 15.80:

$$\frac{R_{sSb}}{R_{sDb}} \frac{\left(\dfrac{\text{scf of gas from separator test}}{\text{STB}}\right)}{\left(\dfrac{\text{scf of gas from DL}}{\text{residual bbl by DL}}\right)} \tag{15.80}$$

Therefore,

$$(R_{sDb} - R_{sD})\frac{R_{sSb}}{R_{sDb}}\left(\frac{\text{scf of gas evolved from DL}}{\text{residual bbl by DL}}\right)$$

$$\left(\frac{\text{scf of gas from separator test, residual bbl by DL}}{\text{scf of gas from DL, STB}}\right) \tag{15.81}$$

is the gas differentially liberated; converted to scf of separator gas/STB. Finally, the difference between the gas originally in solution and the gas liberated during depletion from P_b to some pressure P below the bubble point is the gas remaining is solution at that pressure, and is given by

$$R_s = R_{sSb} - (R_{sDb} - R_{sD})\frac{R_{sSb}}{R_{sDb}} \tag{15.82}$$

which can be rearranged in a simpler form as

$$R_s = R_{sD}\frac{R_{sSb}}{R_{sDb}}, \text{ scf/STB} \tag{15.83}$$

McCain[16] compared the performance of Equations 15.79 and 15.83 and compared the calculated values of R_s with a composite liberation data (described in Section 15.4.2). The values of R_s calculated from Equation 15.83 were found to be in closer agreement with the composite liberation data than Equation 15.79. Furthermore, Equation 15.79 resulted in a negative value of R_s at 0 psig, while Equation 15.83 naturally led to a value of 0. Note, that by definition, R_s should be 0 at 0 psig.

15.4.1.3 Formation Volume Factor of Gas

Gas formation volume factor is calculated from the differential liberation data by using Equation 15.35.

15.4.1.4 Total Formation Volume Factor

Total formation volume factor, B_t, at pressures above the bubble-point pressure equals B_o calculated from Equation 15.73, since no gas is evolved.

At pressures below the bubble-point pressure B_t is calculated from

$$B_t = B_o + B_g(R_{sb} - R_s) \tag{15.84}$$

where B_o is calculated from Equation 15.75, R_{sb} is known from Equation 15.77 (separator test data), and R_s is calculated from Equation 15.78 or, more correctly, from Equation 15.83.

Alternatively, if B_{tD} values are known from the DL data, the total formation volume factors can also be computed from an equation analogous to Equation 15.75:

$$B_t = B_{tD}\left(\frac{B_{oSb}}{B_{oDb}}\right) \tag{15.85}$$

15.4.1.5 Coefficient of Isothermal Compressibility of Oil

The coefficient of isothermal compressibility at pressures above bubble-point pressure is calculated using Equation 15.60 based on B_o values calculated from Equation 15.73. At pressures below bubble point, Equation 15.63 can be applied in which the respective slopes of the adjusted B_o and R_s vs. pressure plots are used.

15.4.2 Composite Liberation

As discussed earlier, the actual liberation process in the reservoir is neither flash nor differential. In certain zones, the process is flash, and in others, the process is differential; hence the data from different laboratory tests are combined by using methods described in the previous section. In 1953, Dodson et al.,[17] proposed a laboratory procedure for directly determining the oil formation volume factors and solution gas–oil ratios that is generally considered to approximate as closely as possible the liberation sequence that occurs in the overall reservoir process (i.e., the producing formation, the well bore, and the surface separators). This particular laboratory test, which is a combination of differential liberation and separator test, proposed by Dodson et al.[17] is called the *composite liberation*. The composite liberation test obviously precludes the need for the adjustment of commonly measured laboratory test data and the data obtained can be used directly in reservoir engineering calculations.

Briefly, the composite liberation process starts with a large volume of reservoir oil sample that is placed in a PVT cell at its bubble-point pressure and reservoir temperature. At some pressure below the bubble-point pressure, a small portion of the oil is removed at constant pressure and flashed at pressures and temperatures equal to those in the surface separator and stock tank. The measured volumes of liberated gas and the stock tank oil are used to directly determine B_o and R_s. This process is repeated at progressively lower pressures so that complete curves of B_o and R_s can be obtained.

Although the composite liberation test is considered a superior method, it requires a rather large sample of reservoir fluid and the test is very time consuming. Thus, it is not used in routine laboratory analysis. Apart from the data of Dodson et al.,[17] no composite liberation data on reservoir oils exist in the current literature.

15.5 OTHER SOURCES OF OBTAINING THE PROPERTIES OF PETROLEUM RESERVOIR FLUIDS

In addition to the detailed laboratory experiments described earlier, various properties of reservoir fluids can also be obtained by employing correlations developed over a number of years. In certain cases, a limited number of properties, such as the bubble point of a black oil and the solution gas–oil ratio, can also be obtained from field or production data. For example, the cumulative oil production data are usually plotted against the reservoir pressure and gas–oil ratio. Based on this data, bubble point and solution gas–oil ratio can be estimated. However, considering the limited use of such data for obtaining reservoir fluid properties, these methods are not discussed here. McCain[12] discusses these procedures in detail. This section presents the various correlations that are commonly used for obtaining reservoir fluid properties.

15.5.1 EMPIRICAL CORRELATIONS

Many empirical fluid property correlations have been developed over the years. Most of these correlations are based on laboratory test results and field data. The properties that are determined from these correlations are the bubble point, gas solubility, formation volume factors, density, compressibility, and viscosity. A number of these correlations are developed specifically for petroleum reservoir fluids originating from a certain geographical region, such as Middle Eastern oils or North Sea oils, and so on. Hence, due to this regional bias, many of these correlations have limited applicability and can produce large errors when applied to oils originating from other localities.

Almost all of these empirical correlations are noncomposition based, that is, reservoir fluid composition is not required in using these correlations. Therefore, these correlations are also commonly referred to as *black oil correlations* because instead of representing the reservoir fluid by their detailed compositions, the correlations treat the oil as essentially a two-component system, that is, —the stock tank oil and the collected gas at standard conditions. Both components are characterized by their respective specific gravities.

Danesh[14] tabulates the range of data used in developing these empirical correlations, which provides a guideline for applicability of these correlations. The various other details of fluid property data, such as geographical origin and number of data points, are also discussed in detail by Danesh.[14] Most of the empirical correlations basically make use of field data, such as gas and oil gravities and the solution gas–oil ratio to correlate a fluid property. The empirical correlations presented in this section are those of Standing; the reader is referred to the text of Ahmed[9] for correlations by other authors.

15.5.1.1 Standing's Empirical Correlations

Most of the empirical correlations that Standing presented initially were in a graphical form[18] and were later expressed by convenient mathematical expressions.[19] Standing's correlations for R_s, P_b, B_o, and oil density are given in the following equations. However, as an alternative to Standing's graphical correlation (no mathematical expression available) for the total formation volume factor B_t, a correlation proposed by Glaso[20] also commonly used is provided.

Solution gas–oil ratio, R_s

$$R_s = \gamma_g\left[\left(\frac{P}{18.2} + 1.4\right)10^{0.0125API-0.0009IT}\right]^{1.2048} \tag{15.86}$$

where R_s is the solution gas–oil ratio, scf/STB; γ_g the gas gravity; T the temperature, °F; and P the pressure, psia.

Note that the R_s (summation of separator(s) and stock tank) and γ_g (calculated from Equation 15.47) values required in all empirical correlations are based on the surface separator data.

Bubble-Point Pressure P_b

$$P_b = 18.2\left[\left(\frac{R_s}{\gamma_g}\right)^{0.83}(10)^a - 1.4\right] \tag{15.87}$$

where P_b bubble-point pressure, psia; a 0.00091T – 0.0125(API); R_s the solution gas–oil ratio, scf/STB; γ_g the solution gas gravity; and T the temperature, °F.

Oil Formation Volume Factor B_o

$$B_{ob} = 0.9759 + 0.000120\left[R_s\left(\frac{\gamma_g}{\gamma_o}\right)^{0.5} + 1.25T\right]^{1.2} \tag{15.88}$$

where B_{ob} is the oil formation volume factor at bubble-point pressure, res. bbl/STB; R_s the solution gas–oil ratio, scf/STB; γ_g the solution gas gravity; and γ_o the stock-tank oil gravity; T the temperature, °F.

Note that Equation 15.88 is also applicable at pressures below the bubble-point pressure. At pressures above the bubble-point pressure, the calculated B_{ob} is adjusted as

$$B_o = B_{ob}\exp[-C_o(P - P_b)] \tag{15.89}$$

where B_o is the oil formation volume factor at pressures P above P_b, res. bbl/STB and C_o the average coefficient of isothermal compressibility, over a pressure range of P to P_b.

Total Formation Volume Factor B_t

$$\log(B_t) = 0.080135 + 0.47257\log(B_t^*) + 0.17351[\log(B_t^*)]^2 \tag{15.90}$$

where (B_t^*) is a correlating parameter and is defined by Equation 15.91:

$$B_t^* = R_s \frac{T^{0.5}}{(\gamma_g)^{0.3}} (\gamma_o)^C P^{-1.1089} \tag{15.91}$$

and

$$C = (2.9)10^{-0.00027R_s} \tag{15.92}$$

Oil Density

Ahmed[9] presented a correlation for calculating the oil density at specified pressure and temperature. The presented correlation is a combination of the Standing correlation and the mathematical definition of the oil formation volume factor presented as

$$\rho_o = \frac{62.4\gamma_o + 0.0136R_s\gamma_g}{0.972 + 0.000147\left[R_s\left(\frac{\gamma_g}{\gamma_o}\right)^{0.5} + 1.25T\right]^{1.175}} \tag{15.93}$$

where ρ_o is the oil density, lb/ft^3 and T the temperature, °F.

Equation 15.93 may be used to calculate the density of the oil at bubble-point pressure and below. At pressures above the bubble-point pressure, the density calculated from Equation 15.93 is adjusted by an equation similar to Equation 15.89, that is , $\rho_o = \rho_o$ (from Equation 15.93) $\times \exp[C_o(P - P_b)]$.

Oil Viscosity

The oil viscosity from empirical correlations is generally determined for the dead oil first, followed by viscosity at bubble-point pressure, and subsequently at other pressures above the bubble point.

The dead oil viscosity can be calculated from Glaso's[20] correlation

$$\mu_{od} = [3.141(10^{10})](T)^{-3.444}[\log(API)]^a \tag{15.94}$$

and

$$a = 10.313[\log(T)] - 36.447 \tag{15.95}$$

where μ_{od} is the dead oil viscosity, cp; T the temperature, °F; and API the oil API gravity.

The oil viscosity at bubble-point pressure (saturated oil viscosity) can be calculated from Standing's correlation

$$\mu_{ob} = (10)^a(\mu_{od})^b \tag{15.96}$$

and

$$a = R_s[2.2(10^{-7})R_s - 7.4(10^{-4})] \tag{15.97}$$

$$b = \frac{0.68}{10^c} + \frac{0.25}{10^d} + \frac{0.062}{10^e} \quad (15.98)$$

$$c = 8.62(10^{-5})R_s \quad (15.99)$$

$$d = 1.1(10^{-3})R_s \quad (15.100)$$

$$e = 3.74(10^{-3})R_s \quad (15.101)$$

where μ_{ob} is the oil viscosity at bubble point, cp.

The viscosity of the undersaturated oil can also be calculated from the Standing correlation

$$\mu = \mu_{ob} + 0.001(P - P_b)(0.024\mu_{ob}^{1.6} + 0.038\mu_{ob}^{0.56}) \quad (15.102)$$

15.5.2 Prediction of Viscosity from Compositional Data

The viscosity prediction model that has earned the most recognition and widespread use is the Lohrenz–Bray–Clark[11] method or the LBC method. The method is an integral part of virtually all reservoir simulators, PVT simulators, and fluid property prediction packages for viscosity prediction. The LBC method is basically that of Jossi et al.[21] for pure components, extended to hydrocarbon mixtures, as described in the following text.

The Jossi et al.[21] method is based on the concept of residual viscosity, defined as the difference between the viscosity at prevailing pressure and temperature conditions and that at low pressure where the viscosity depends only on temperature. The experimental viscosity and density data on a variety of pure components was fitted in a reduced form using the following function:

$$[(\mu - \mu_o)\lambda + 10^{-4}]^{1/4} = a_1 + a_2\rho_r + a_3\rho_r^2 + a_4\rho_r^3 + a_5\rho_r^4 \quad (15.103)$$

where μ is the viscosity to be calculated, cp or mPa.s; ρ_r the reduced density; $a_1 = 0.10230$; $a_2 = 0.023364$; $a_3 = 0.058533$; $a_4 = -0.040758$; and $a_5 = 0.0093324$.

The viscosity at low pressure conditions, μ_o, and the viscosity reducing parameter (or the inverse of critical viscosity), λ, are determined by the following equations.

For the ith component,

$$\mu_{oi} = 34 \times 10^{-5}\, T_{ri}^{0.94}/\lambda_i \quad T_{ri} \le 1.5 \quad (15.104)$$

$$\mu_{oi} = 17.78 \times 10^{-5}(4.58T_{ri} - 1.67)^{5/8}/\lambda_i \quad T_{ri} > 1.5 \quad (15.105)$$

$$\lambda_i \equiv T_{ci}^{1/6}\, MW_i^{-1/2}\, P_{ci}^{-2/3} \quad (15.106)$$

Note that the units of T_c and P_c in Equation 15.106 are Kelvin and atm to obtain the viscosity in cp or mPa.S.

Lohrenz et al.[11] introduced various mixing rules to allow the calculation of μ_o, λ, and ρ_r for hydrocarbon mixtures as

$$\mu_o = \frac{\sum_{i=1}^{n} Z_i \mu_{oi} MW_i^{1/2}}{\sum_{i=1}^{n} Z_i MW_i^{1/2}} \tag{15.107}$$

$$\lambda = \left(\sum_{i=1}^{n} Z_i T_{ci}\right)^{1/6} \left(\sum_{i=1}^{n} Z_i MW_i\right)^{-1/2} \left(\sum_{i=1}^{n} Z_i P_{ci}\right)^{-2/3} \tag{15.108}$$

$$\rho_r = \frac{\rho}{\left(1/\sum_{i=1}^{n} Z_i V_{ci}\right)} \tag{15.109}$$

where Z_i = general representation of the mole fraction of component i in the mixture, which could be single-phase vapor or liquid, or the equilibrated vapor phase or liquid phase; ρ = density of the hydrocarbon mixture, gm/cc; T_{ci}, P_{ci}, and V_{ci} = critical temperature (K), critical pressure (atm), and critical volume (cc/gm); and MW_i is the molecular weight of component i in the mixture.

Lohrenz et al.[11] suggested the following expression for calculating the critical volume of the plus fraction:

$$\begin{aligned}(V_c)_{C7+} &= 1.3468 + 9.4404 \times 10^{-4} MW_{C7+} - 1.72651 SG_{C7+} \\ &\quad + 4.4083 \times 10^{-3} MW_{C7+} SG_{C7+}\end{aligned} \tag{15.110}$$

where $(V_c)_{C_{7+}}$ = critical volume of the plus fraction, m^3/kg-mol, which can be converted to units consistent with the density of the hydrocarbon mixture by using the molecular weight of the plus fraction and appropriate conversion factors; $MW_{C_{7+}}$ and SG_{C7+} = molecular weight and specific gravity of the plus fraction.

As seen by the form of Equation 15.103, viscosities calculated from the LBC method are extremely sensitive to density values. As Dandekar[22] points out; for dense phase fluids, the LBC model can result in under predicting viscosity by as much as 100%. Therefore, the most common practice is to tune or calibrate the method by adjusting the critical volume of the plus fraction to match the measured data. The tuned model with adjusted value of $(V_c)_{C7+}$ can then be used to predict the viscosity at other conditions.

Apart from the LBC method, prediction methods based on the principle of corresponding states have also gained some popularity in recent years and are frequently included as a second option for calculating viscosities. A review of these models is given in the work of Dandekar[22].

15.5.3 Prediction of Surface Tension

The most commonly used method in the petroleum industry for determining the vapor–liquid surface tension is the parachor model. The parachor model relates surface tension, densities, and compositions of the equilibrium phases in the following functional form[23]:

$$\sigma^{1/4} = \sum_{i=1}^{n} P_{\sigma i}(X_i \rho_M^L - Y_i \rho_M^V) \qquad (15.111)$$

where σ is the vapor–liquid surface tension in mN/m or dyne/cm; $P_{\sigma i}$ the parachor of component i in the mixture; X_i the mole fraction of component i in the liquid phase; ρ_M^L the molar density of equilibrium liquid phase, gm-mol/cm^3; Y_i the mole fraction of component i in the vapor phase; and ρ_M^V the molar density of equilibrium vapor phase, gm-mol/cm^3.

The use of Equation 15.111 for determination of surface tension requires parachor values for each component in the mixture, equilibrium phase compositions, and their densities. Molar phase densities are calculated from the ratio of mass phase densities and phase molecular weights. Parachor values required in Equation 15.111 are given in Table 15.4. In order to improve the accuracy of surface tension predictions, it is preferable to use measured phase compositions and densities.

Parachor values of crude oil fractions can be determined from the equation proposed by Firoozabadi,[24] which relates parachor and the molecular weight as

$$\mathrm{P}_{\sigma} = -11.4 + 3.23\,\mathrm{MW}_f - 0.0022\,\mathrm{MW}_f^2 \qquad (15.112)$$

where MW_f is the molecular weight of a crude oil fraction. For example, if TBP distillation data containing molecular weights of SCN fractions and the plus fraction are

TABLE 15.4
Parachor Values of Selected Components

Component	Parachor
N_2	41.0
CO_2	78.0
H_2S	80.1
C_1	77.0
C_2	108.0
C_3	150.3
iC_4	181.5
nC_4	189.9
iC_5	225.0
nC_5	231.5
nC_6	271.0
nC_7	312.5
nC_8	351.5
nC_9	393.0
nC_{10}	433.5

available, Equation 15.112 can be used to estimate the parachor values of various oil fractions. It is, however, preferable to use an experimental surface tension value to back calculate the parachor of the plus fraction or tune Equation 15.111 with a plus fraction parachor value that matches calculated and experimental surface tension. The tuned parachor value of the plus fraction is then used to predict the surface tension values of the same reservoir fluid at other conditions.

The other model commonly referred to as scaling law (not discussed here) that is also used to predict surface tension is based on the corresponding states principle. The major difference between the parachor model and the scaling law is the approach used for the determination of parachors and the value of the exponent used. In scaling law, the parachor value is calculated for the entire vapor phase and liquid phase instead of using the individual component values and an exponent of 0.25568 is used instead of 0.25, used in the parachor model.

Dandekar[22] carried out detailed evaluation of both the parachor and the scaling law methods, revealing that generally, the performance of both methods is similar. However, considering the simplicity of the parachor model, it is in fact the most commonly used method in reservoir and PVT simulators.

PROBLEMS

15.1 A PVT cell contains a natural gas mixture at 1400 psia and 190°F. The cell volume is 10.0 ft^3 and the total lb-moles of gas are determined to be 2.0. Calculate the gas deviation factor.

15.2 A natural gas has the following composition:

Component	Composition (mole fraction)
N_2	0.0062
CO_2	0.0084
H_2S	0.0068
C_1	0.8668
C_2	0.0391
C_3	0.0280
nC_4	0.0224
nC_5	0.0224

Calculate the molecular weight, specific gravity, pseudocritical pressure, and temperature using Kay's mixing rules, and density at 1000 psia and 100°F by assuming ideal gas behavior.

15.3 For the natural gas composition in problem 15.2, calculate the gas density at same pressure and temperature condition but assuming real gas behavior and using corrected and un-corrected pseudocritical properties due to the presence of nonhydrocarbon components.

15.4 A Russian dry gas well is producing gas at a rate of 25,150 ft^3/day. The composition of this gas is given in problem 15.2. The flowing bottomhole pressure is 1100 psia and the reservoir temperature is 112°F. Calculate the gas production rate in scf/day and gas viscosity at flowing bottomhole conditions.

15.5 The PV relationship for a gas at 250°F is shown in the following table. Determine the coefficient of isothermal compressibility at 350 psia.

Pressure (psia)	Volume (ft^3/lb)
480	0.1443
450	0.1705
420	0.1970
390	0.2256
360	0.2574
330	0.2936
305	0.3358
250	0.4476
190	0.6250

15.6 A gas condensate stream from a field in Indonesia is separated in a two-stage separator system. The first stage separator operates at 700 psia and 100°F while the stock tank operates at atmospheric pressure and 90°F. The separator and stock tank gas-condensate ratios are 70,000 scf/STB and 300 scf/STB respectively. The stock tank condensate gravity is 55.0°API. Molar composition of the separated streams is given below. Determine the composition of the reservoir gas. Assume C_6 molecular weight to be the same as *n*-hexane and reservoir pressure above the dew point pressure.

Component	Composition of Separator Gas (mole fraction)	Composition of Stock Tank Gas (mole fraction)	Composition of Stock Tank Condensate (mole fraction)
N_2	0.003	0.004	0.000
CO_2	0.015	0.095	0.013
C_1	0.820	0.381	0.274
C_2	0.065	0.246	0.050
C_3	0.028	0.108	0.040
nC_4	0.015	0.058	0.031
nC_5	0.008	0.030	0.024
C_6	0.005	0.020	0.029
C_{7+}	0.041	0.057	0.538

Properties of C_{7+} are, specific gravity = 0.781, molecular weight 135 lb/lb-mole

15.7 The gas production described in problem 15.6 is commingled from another gas condensate (reservoir pressure higher than dew point pressure) field in the adjoining area after its discovery. Due to high wellhead pressures the commingled streams are separated in three instead of two separator stages. Although compositions of the separated streams are unknown, field data is available and is given in the following table.

Separator	Pressure (psia)	Temperature (°F)	γ	Gas-Condensate Ratio (scf/STB)
Primary	1000	100	0.500	7000
Secondary	250	90	0.900	500
Stock tank	14.7	80	1.300	100

The condensate produced in the stock tank has a API gravity of 60.0. What is the specific gravity of the commingled reservoir gas?

15.8 A wet gas from a field in Iran is processed through two stages of separation; 1st stage separator operates at 220 psia and 71°F, while the stock tank operates at atmospheric pressure and at 69°F. The separator gas-condensate ratio is 41,000 scf/STB and the stock tank gas-condensate ratio is 450 scf/STB. The molecular weight of the separator gas and stock tank gas is 25.0 and 35.0 lb/lb mol respectively, while the stock tank condensate gravity is 0.85. What is the specific gravity of the reservoir gas?

15.9 Calculate the gas formation volume factor at reservoir conditions of 2000 psia and 100°F for the gas in problem 15.8.

15.10 A DL test is conducted on a North Sea black oil. At a certain pressure step in the DL test, 1.90 cm^3 of gas and 88.5 cm^3 of oil are measured in the cell at 910 psig and 110°F. The evolved gas is subsequently displaced at constant pressure and its volume at standard conditions is determined to be 101.9 cm^3. At the termination of the DL test the residual oil volume at standard conditions is measured at 71.9 cm^3. Calculate the Z factor of the gas and the relative oil volume at 910 psig.

15.11 From the data given in the following table, calculate the coefficient of isothermal compressibility factor for this oil at 125°F, at all pressures above the bubble-point pressure of 3000 psig.

Pressure (psig)	Oil Volume (cc)
5000	192
4500	193
3750	195
3500	196
3300	197
3100	198
3000	199

15.12 The following pressure–volume data from a CCE test at 220°F are available for a reservoir oil. Determine the bubble-point pressure of this oil.

Pressure (psia)	Total Cell Volume (cc)
5000	144.6
4100	146.3

(Continued)

PROBLEM 15.12 (*Continued*)

Pressure (psia)	Total Cell Volume (cc)
3500	147.7
2900	149.2
2700	149.8
2605	150.3
2516	152.4
2253	159.7
1897	174.4
1477	204.2
1040	267.6
640	414.0

15.13 A CCE test is carried out on a gas condensate sample at 150°F. The volume of fluid at the saturation pressure of 2330 psia is 2000 cc. The equilibrium liquid volume as a function of pressure is given in the following table. Calculate and plot the liquid drop out as a function of pressure for this gas condensate fluid.

Pressure (psia)	Vol. of Liquid (cc)
2300	0.2
2100	24.2
1900	44.0
1700	56.8
1500	63.8
1300	66.2
1100	64.6
900	59.4
700	50.8
500	38.0
300	21.0

15.14 A differential liberation test is carried out on a crude oil sample taken from an oil field in Alaska. The sample, with a volume of 310 cc, is placed in a PVT cell at its bubble-point pressure of 3015 psia and reservoir temperature of 185°F. At isothermal conditions of 185°F, the cell pressure is reduced to 2500 psia by backward piston movement, resulting in the total volume of the hydrocarbon system of 348.7 cc. The gas is displaced at constant pressure (by forward piston movement) and found to occupy a volume of 0.150 scf. The oil volume shrinks to 291.0 cc. The differential liberation step is repeated at 2000 psia and the remaining oil is flashed through a series of laboratory separators. From the collected experimental data given in the following table, calculate the solution gas–oil ratio at 3015, 2500, and 2000 psia.

Pressure (psia)	Temperature (°F)	Total System Volume (cc)	Volume of Displaced Gas (scf)	Volume of Oil (cc)
2000	185	393.1	0.285	282.1
14.7	60	-	0.441	231.1

15.15 A black oil sample from an Angolan reservoir has a volume of 250 cc in a PVT cell at reservoir temperature and bubble point pressure. The oil is expelled through a laboratory set-up that mimics the separator and stock tank system. The oil volume arriving in the stock tank is 170 cc. The separator and the stock tank produce 0.600 and 0.070 scf of gas. Calculate oil formation volume factor and solution gas oil ratio in the oilfield units.

15.16 The following data are available from a laboratory test carried out on a black oil at 225°F. What is the bubble-point pressure of this oil? Calculate the total formation volume factor and subsequently plot pressure vs. B_o and B_t.

Pressure (psia)	R_s (scf/STB)	B_o (res. bbl/STB)	Gas Z Factor
4500	632	1.3474	
4000	632	1.3575	
3500	632	1.3686	
3000	632	1.3811	
2682	632	1.4040	
2500	584	1.3782	0.8140
2200	509	1.3369	0.8165
2000	460	1.3109	0.8208
1800	414	1.2864	0.8269
1600	369	1.2634	0.8347
1400	326	1.2416	0.8440
1200	285	1.2208	0.8548
1000	245	1.2002	0.8670
800	205	1.1791	0.8808
600	163	1.1566	0.8964
400	119	1.1315	0.9140
200	70	1.1024	0.9339

15.17 The following laboratory data are obtained from a constant composition expansion and differential liberation tests carried out on a black oil at 225°F. The oil formation volume factor and solution gas–oil ratio for the DL test at bubble-point pressure are found to be 1.3298 bbl/STB and 531 scf/STB, respectively. The optimum separator conditions result in the oil formation volume factor and solution gas–oil ratio of 1.3160 bbl/STB and 514 scf/STB, respectively. Determine values of oil formation factor, total formation volume factor and solution gas–oil ratio for use in reservoir engineering calculations.

Pressure (psia)	$(V_t/V_b)_F$	B_{oD} (bbl/STB)	R_{sD} (scf/STB)	B_g (ft^3/scf)
3750	0.9727			
3500	0.9763			
3250	0.9800			
3000	0.9840			
2800	0.9873			
2600	0.9908			
2400	0.9945			
2200	0.9984			
2120	1.0000			
2000	1.0258	1.3119	495	0.0081
1800	1.0790	1.2835	438	0.0091
1600	1.1492	1.2565	383	0.0103
1400	1.2446	1.2309	332	0.0119
1200	1.3787	1.2063	283	0.0140
1000	1.5766	1.1825	236	0.0170
800	1.8898	1.1592	191	0.0216
600	2.4415	1.1357	146	0.0292
400	3.6138	1.1105	100	0.0446
200	7.3979	1.0780	46	0.0910

15.18 Estimate the bubble-point pressure at 225°F using the Standing correlation for the reservoir oil in problem 15.17 and compare the estimated value with the measured value of 2120 psia. The separator and the stock tank gas–oil ratios are found to be 477 and 37 scf/STB. The specific gravities of the separator and the stock tank gas are 0.7964 and 1.2548, respectively. The gravity of the stock tank oil is 37°API.

15.19 The constant volume depletion (CVD) data for a gas condensate are shown in the following table. The measured compositions (mol-%) of the vapor phase removed, the gas deviation factor for the vapor in the cell at cell conditions, the cumulative gas produced, and the liquid drop out for steps 1 to 6 are also shown in the table. The reservoir temperature is 280°F. From the given data determine the compositions of the equilibrium liquid phases at pressure steps 1 to 6.

	Dew Point	Step 1	Step 2	Step 3	Step 4	Step 5	Step 6
Pressure (psia)	6761	5512	4312	3113	2114	1214	714
Cum. gas produced[a]	0	9.637	22.581	39.492	56.196	72.413	81.535
Liquid saturation[b]	0	19.55	26.11	26.65	25.11	23.00	21.58
Gas Z factor	1.238	1.037	0.937	0.890	0.886	0.911	0.936
CO_2	2.37	2.403	2.447	2.497	2.541	2.576	2.583
N_2	0.31	0.323	0.336	0.344	0.343	0.334	0.321
C_1	73.19	75.549	77.644	79.135	79.712	79.242	77.772
C_2	7.8	7.779	7.793	7.878	8.057	8.372	8.711
C_3	3.55	3.474	3.405	3.383	3.444	3.66	3.989

(*Continued*)

PROBLEM 15.19 (*Continued*)

	Dew Point	Step 1	Step 2	Step 3	Step 4	Step 5	Step 6
iC_4	0.71	0.686	0.66	0.644	0.647	0.691	0.778
nC_4	1.45	1.39	1.326	1.281	1.282	1.375	1.567
iC_5	0.64	0.604	0.564	0.53	0.516	0.548	0.638
nC_5	0.68	0.639	0.592	0.55	0.532	0.563	0.659
C_6	1.09	0.996	0.889	0.789	0.727	0.744	0.877
C_{7+}	8.21	6.157	4.343	2.969	2.198	1.895	2.105

[a] mol-% original fluid

[b] % of V_{sat}

15.20 Following two-phase data are available for the equilibrium vapor and liquid phases for a binary system of methane and *n*-butane at pressures below the dew point of 1850 psia at 150°F. Calculate the viscosity of the equilibrium vapor phase and the liquid phases at all pressures using the Lohrenz–Bray–Clark method. Finally, plot the viscosity values (both phases on same plot) as a function of pressure and make appropriate comments.

Pressure (psia)	Vapor Phase Composition (mole fraction)		Liquid Phase Composition (mole fraction)		Vapor Density (gm/cc)	Liquid Density (gm/cc)
	CH_4	nC_4H_{10}	CH_4	nC_4H_{10}		
1800	0.7250	0.2750	0.5940	0.4060	0.1867	0.2773
1700	0.7500	0.2500	0.5420	0.4580	0.1612	0.3044
1600	0.7650	0.2350	0.5000	0.5000	0.1432	0.3241
1500	0.7750	0.2250	0.4620	0.5380	0.1286	0.3404
1400	0.7810	0.2190	0.4270	0.5730	0.1160	0.3546
1300	0.7850	0.2150	0.3920	0.6080	0.1048	0.3675
1200	0.7870	0.2130	0.3590	0.6410	0.0947	0.3793
1100	0.7860	0.2140	0.3260	0.6740	0.0854	0.3902

15.21 For the data given in problem 15.20, calculate the surface tension at all pressures using the parachor method. Parachor values of methane and *n*-butane are 77.0 and 189.9, respectively. Finally, plot the surface tension values as a function of pressure and make appropriate comments on the trend of the plot.

REFERENCES

1. Brown, G.G., Katz, D.L., Oberfell, G.G., and Allen, R.C., *Natural Gasoline and the Volatile Hydrocarbons*, National Gasoline Association of America, Tulsa, OK, 1948.
2. Kay, W.B., Density of hydrocarbon gases and vapors at high temperature and pressure, *Ind. Eng. Chem.*, 28, 1014–1019, 1936.

3. Standing, M.B., Volumetric and Phase Behavior of Oil Field Hydrocarbon Systems, Society of Petroleum Engineers of AIME, 1977.
4. Wichert, E. and Aziz, K., Calculate Z's for sour gases, *Hydrocarbon Process.*, 51, 119–122, 1972.
5. Carr, N., Kobayashi, R., and Burrows, D., Viscosity of hydrocarbon gases under pressure, *Trans. AIME*, 201, 270–275, 1954.
6. Sutton, R.P., Compressibility factors for high molecular weight reservoir gases, Society of Petroleum Engineers (SPE) paper number 14265.
7. Stewart, W.F., Burkhard, S.F., and Voo, D., Prediction of pseudocritical parameters for mixtures, Paper presented at the AIChE meeting, Kansas, MO, 1959.
8. Standing, M.B. and Katz, D.L., Density of natural gases, *Trans. AIME*, 146, 1942.
9. Ahmed, T., *Hydrocarbon Phase Behavior*, Gulf Publishing Company, Houston, 1989.
10. Dranchuk, P.M. and Abu-Kassem, J.H., Calculations of Z-factors for natural gases using equation of state, *J. Canadian Pet. Technol.*, 14, 34–36, 1975.
11. Lohrenz, J., Bray, B.G., and Clark, C.R., Calculating viscosities of reservoir fluids from their compositions, *J. Pet. Technol.*, 1171–1176, 1964.
12. McCain, W.D., Jr., *The Properties of Petroleum Fluids*, PennWell Publishing Co., Tulsa, OK, 1990.
13. Gold, D.K., McCain, W.D., Jr., and Jennings, J.W., An improved method for the determination of the reservoir gas specific gravity for retrograde gases, *J. Pet. Technol.*, 41, 747–753, 1989.
14. Danesh, A., *PVT and Phase Behavior of Petroleum Reservoir Fluids*, Elsevier, Amsterdam, 1998.
15. Dandekar, A. and Stenby, E., Measurement of phase boundaries of hydrocarbon mixtures using fiber optical detection techniques, *Ind. Eng. Chem. Res.*, 39, 2586–2591, 2000.
16. McCain, W.D., Jr., Analysis of black oil PVT reports revisited, Society of Petroleum Engineers (SPE) paper number 77386.
17. Dodson, C.R., Goodwill, D., and Mayer, E.H., Application of laboratory PVT data to reservoir engineering problems, *Trans. AIME*, 198, 287–298, 1953.
18. Standing, M.B., A pressure-volume-temperature correlation for mixtures of California oils and gases, *Drilling and Production Practice*, API, 1957.
19. Standing, M.B., *Volumetric and Phase Behavior of Oil Field Hydrocarbon Systems*, Society of Petroleum Engineers of AIME, 9th ed., 1981.
20. Glaso, O., Generalized pressure-volume-temperature correlations, *J. Pet. Technol.*, 785–795, 1980.
21. Jossi, J.A., Stiel, L.I., and Thodos, G., The viscosity of pure substances in the dense gaseous and liquid phases, *Amer. Inst. Chem. Eng. J.*, 8, 59–63, 1962.
22. Dandekar, A.Y., Interfacial tension and viscosity of reservoir fluids, Ph.D. thesis, Heriot-Watt University, Edinburgh, 1994.
23. Weinaug, C.F. and Katz, D.L., Surface tensions of methane-propane mixtures, *Ind. Eng. Chem.*, 35, 239–246, 1943.
24. Firoozabadi, A., Katz, D.L., Soroosh, H., and Sajjadian, V.A., Surface tension of reservoir crude oil/gas systems recognizing the asphalt in the heavy fraction, *SPE Reservoir Engineering*, 265–272, 1988.

16 Vapor-Liquid Equilibria

16.1 INTRODUCTION

Chapter 12 explains that petroleum reservoir fluids can exist in the reservoir and on the surface either in a single- or in two-phase conditions. The primary variables that dictate the state of the reservoir fluids are system pressure, system temperature, fluid composition, and the chemistry of the components. For example, reservoir fluid exists as a single phase outside the phase envelope or outside the area bounded by the bubble- and dew-point curves. The boundary of the phase envelope or the area bounded by the bubble- and dew-point curves of a reservoir fluid defines the conditions for the vapor or gas phase and the liquid phase to exist in equilibrium. Furthermore, as pressure and temperature conditions change, the quantities, compositions, and properties of the equilibrium vapor and liquid phases vary at different points within the phase envelope.

All petroleum reservoir fluids undergo pressure and temperature changes during the process of production. These pressure and temperature changes frequently result in the formation of equilibrium vapor and liquid phases. These hydrocarbon vapor and liquid phases are the most important phases occurring in petroleum production. The conditions under which these different phases can exist, and moreover, the quantities, compositions and properties of these phases are matters of considerable practical importance in reservoir engineering calculations, compositional reservoir simulation, and in the design of surface separation facilities. Water is also commonly present as an additional liquid phase. However, the effect of water on hydrocarbon phase behavior can be ignored in most cases.

The most accurate and reliable source of obtaining data on equilibrium conditions and the phases is laboratory studies; including compositional analysis and reservoir fluid studies, which were described in Chapters 14 and 15. However, in the absence of such laboratory studies, calculation methods are relied upon, and these methods are called *phase equilibrium* or *vapor-liquid equilibrium (VLE) calculations*. The input data for the VLE calculations consist of the overall composition of the reservoir fluid, pressure and temperature conditions, and the properties of the individual components (defined as well as pseudo-components and plus fractions). Based on such input data, VLE calculations typically involve the determination of saturation pressures (bubble or dew points), equilibrium phase compositions, and various properties of the equilibrium phases.

VLE calculation models range from those based on simple ideal solution principles, to empirical models, to complicated equations-of-state (EOS) models. The selection of a particular model for VLE calculations depends on factors such as the complexity of the reservoir fluid system, pressure and temperature conditions, and the type of information required. For example, an EOS model is generally employed for simulation of a CCE, DL, or CVD experiment. The primary objective of this chapter is to introduce these VLE models, which enable basic prediction of reservoir fluid behavior and determination of conditions for processing reservoir fluids at the surface. This discussion begins with the ideal solution principles in which the important concept of equilibrium ratios are introduced, followed by the concept of PT flash, and the fundamental molar balance equations that govern VLE calculations. Subsequently, empirical models that are commonly used for the determination of equilibrium ratios and the concept of convergence pressure are introduced. The final sections of the chapter deal with the fundamentals of EOS models and their use in VLE calculations for simulating fluid-phase behavior and the determination of properties of reservoir fluids.

16.2 IDEAL SOLUTION PRINCIPLE

An ideal liquid solution is a solution for which:

1. Mutual solubility results when components are mixed,
2. No chemical interaction occurs upon mixing, and
3. The intermolecular forces of attraction and repulsion are the same between unlike and like molecules.

These characteristics of ideal solutions lead to two important practical results: there is no heating effect and no volume change when the components are mixed. Similar to ideal gases, liquid mixtures of components of the same homologous series approach ideal solution behavior at low pressure conditions. Although, these ideal solutions are not commonly encountered in petroleum reservoir fluids, studies of the phase behavior of ideal solutions help us understand the behavior of real solutions.

16.2.1 RAOULT'S LAW

Raoult's law states that the partial pressure P_i of a component in a multicomponent system is the product of its mole fraction in the liquid phase and the vapor pressure of the component P_{vi}, mathematically expressed as

$$P_i = X_i P_{vi} \tag{16.1}$$

where P_i is the partial pressure of component i in the gas phase, X_i the mole fraction of component i in the liquid phase, and P_{vi} the vapor pressure of component i at the temperature of interest.

16.2.2 DALTON'S LAW

Dalton's law states that for an ideal gas mixture, the partial pressure of a component is the product of its mole fraction and the total pressure of the system, mathematically expressed as

$$P_i = Y_i P \tag{16.2}$$

where P is the total system pressure and Y_i the mole fraction of component i in the vapor phase.

16.2.3 EQUILIBRIUM RATIO

At equilibrium conditions, the partial pressure exerted by a component in the gas phase must be equal to the partial pressure exerted by the same component in the liquid phase. Therefore, the combination of the Raoult's and Dalton's laws leads to a very important concept in VLE calculations, called the *equilibrium ratio*, denoted by K_i:

$$X_i P_{vi} = Y_i P \tag{16.3}$$

or

$$\frac{P_{vi}}{P} = \frac{Y_i}{X_i} = K_i \tag{16.4}$$

Equation 16.4 relates the ratio of the mole fraction of component i in the gas phase to its mole fraction in the liquid phase. This particular ratio or the equilibrium ratio, K_i, signifies the partitioning of component i between the equilibrium vapor phase and liquid phase. Equilibrium ratios are sometimes called *equilibrium constants*, *K factors* or *K values*. However, it should be noted that for ideal solutions, regardless of the overall composition of the hydrocarbon mixture, the equilibrium ratio is only a function of the system pressure and temperature (as indicated in Chapter 11, the vapor pressure of a component is only a function of temperature).

Although, on the basis of vapor pressures and the system pressure, equilibrium ratios for various components can be determined, Equation 16.4 still has two unknowns, Y_i and X_i. In order to determine the values of Y_i and X_i, Equation 16.4 must be combined with other equations relating these two quantities. These relationships can be developed through material balance consideration, based on the concept of PT flash, described in the next section.

16.2.4 CONCEPT OF PT FLASH

Consider the simple flow diagram shown in Figure 16.1 where a stream of material containing n moles of feed with overall molar composition Z_i is flashed at pressure P and temperature T. In general the equilibrium at pressure P and temperature T produce n_V moles of vapor having composition Y_i and n_L moles of liquid having composition X_i. This particular process, resulting in the splitting of the feed or a hydrocarbon mixture into equilibrium vapor and liquid phases at a given pressure and temperature is called *PT flash*. Almost all petroleum reservoir fluids undergo this type of PT flash process either in the reservoir and/or the surface.

By definition, the overall material balance on the feed and the equilibrated vapor and liquid phases lead to

$$n = n_L + n_V \tag{16.5}$$

A similar material balance equation can also be written in terms of the ith component of the mixture:

$$Z_i n = X_i n_L + Y_i n_V \tag{16.6}$$

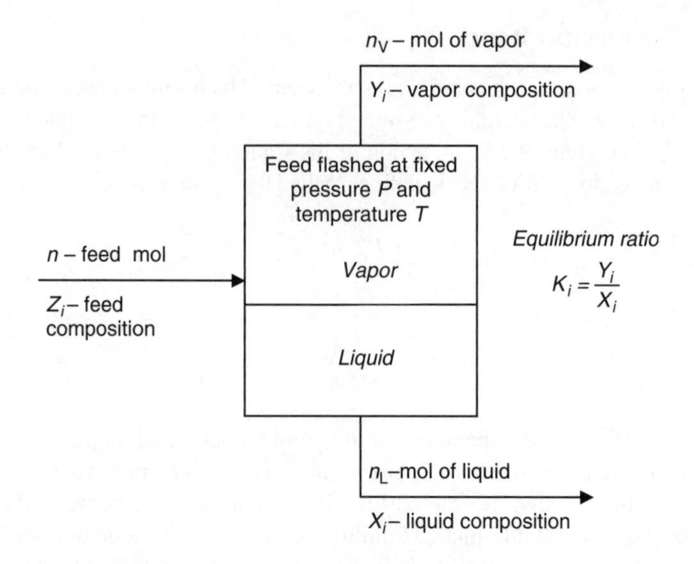

FIGURE 16.1 Schematic illustration of PT flash concept.

where $Z_i n$ represents the moles of component i in the feed, $X_i n_L$ the moles of component i in the equilibrium liquid phase, and $Y_i n_V$ the moles of component i in the equilibrium vapor phase.

Equations 16.5 and 16.6 can be further simplified by considering the basis of 1 mol of feed, that is, $n = 1$:

$$n_L + n_V = 1 \tag{16.7}$$

$$X_i n_L + Y_i n_V = Z_i \tag{16.8}$$

Combining Equations 16.4, 16.7, and 16.8:

$$X_i(1 - n_V) + K_i X_i n_V = Z_i \tag{16.9}$$

$$X_i = \frac{Z_i}{1 + n_V(K_i - 1)} \tag{16.10}$$

Also by definition of mole fraction:

$$\sum_{i=1}^{n} X_i = \sum_{i=1}^{n} \frac{Z_i}{1 + n_V(K_i - 1)} = 1 \tag{16.11}$$

Similarly,

$$\sum_{i=1}^{n} Y_i = \sum_{i=1}^{n} \frac{Z_i K_i}{1 + n_V(K_i - 1)} = 1 \tag{16.12}$$

On the basis of the given mole fraction of the feed Z_i and the calculated equilibrium ratio K_i from Equation 16.4 (from vapor pressure at a given temperature and the system pressure), the only unknown that remains in Equation 16.11 or 16.12 is the moles of the equilibrium phase vapor n_V. However, considering the nature of these equations, a trial-and-error solution is required in either case. Since the calculations are based on 1 mol of feed, a trial value of n_V between 0 and 1 is chosen. If the selected value of n_V results in the summation of 1 in Equation 16.11 or 16.12, the moles of equilibrium phase liquid and the compositions of all the components in the equilibrium vapor phase and the liquid phase are calculated. However, if the summation does not equal 1 with the selected value, a new trial value of n_V is chosen and the computation repeated until the summation equals 1.

16.2.5 Calculation of Bubble-Point Pressure

By definition, bubble point is the point at which the first bubble of gas is formed. Therefore, for all practical purposes, the quantity of gas is negligible. Hence, the number of moles in the vapor phase n_V is 0, and the number of moles in the liquid phase n_L is 1 (assuming the basis to be 1 mol of feed), while pressure P is equal to P_b, the bubble point.

Therefore,

$$\sum_{i=1}^{n} Z_i = \sum_{i=1}^{n} X_i = 1 \tag{16.13}$$

and for the newly formed gas phase:

$$\sum_{i=1}^{n} Y_i = \sum_{i=1}^{n} Z_i K_i = 1 \tag{16.14}$$

By substituting the value of K_i from Equation 16.4:

$$\sum_{i=1}^{n} Z_i \frac{P_{vi}}{P_b} = 1 \tag{16.15}$$

or

$$P_b = \sum_{i=1}^{n} Z_i P_{vi} \tag{16.16}$$

Therefore, the bubble-point pressure of an ideal liquid solution at a given temperature is simply the summation of the product of feed mole fraction and the vapor pressure of each component in the mixture.

16.2.6 Calculation of Dew-Point Pressure

Again, by definition, dew point is the point at which the first drop of liquid is formed. Therefore, for all practical purposes, the quantity of liquid is negligible. Hence, the number of moles in the liquid phase n_L is 0 and the number of moles in the gas phase n_V is 1 (assuming the basis to be 1 mole of feed), while pressure P is equal to P_d, the dew point.

Therefore,

$$\sum_{i=1}^{n} Z_i = \sum_{i=1}^{n} Y_i = 1 \tag{16.17}$$

and for the newly formed liquid phase:

$$\sum_{i=1}^{n} X_i = \sum_{i=1}^{n} \frac{Z_i}{K_i} = 1 \tag{16.18}$$

By substituting the value of K_i from Equation 16.4:

$$\sum_{i=1}^{n} \frac{Z_i}{P_{vi}/P_d} = 1 \tag{16.19}$$

or

$$P_d = \frac{1}{\sum_{i=1}^{n} Z_i/P_{vi}} \tag{16.20}$$

Therefore, the dew-point pressure of an ideal gas mixture at a given temperature is simply the reciprocal of the summation of the ratio of the feed mole fraction and the vapor pressure of each component in the mixture.

16.2.7 Drawbacks of the Ideal Solution Principle

The equations that were developed in the previous sections for VLE calculations are severely restricted for three reasons. First, Raoult's law assumes that the liquid behaves as an ideal solution; which is true only if the components of the liquid mixture are physically and chemically very similar. Second, Dalton's law assumes that the gas behaves as an ideal solution of gases; which is limited to pressures below about 100 psia and moderate temperatures. Third, the most important restriction is that a pure component does not have a vapor pressure above its critical temperature. This means that the VLE equations developed on the basis of ideal solution principle are limited to temperatures less than the critical temperature of the most volatile component of the mixture.

Consider a ternary mixture of overall composition of 60 mol-% propane, 30 mol-% *n*-butane, and 10 mol-% *n*-pentane. The calculated bubble points (bubble-point curve) and dew points (dew-point curve) of this ternary system at various temperatures, using Equations 16.16 and 16.20, are shown in Figure 16.2. The phase envelope constructed from equations-of-state (EOS) predictions is also shown in Figure 16.2. Apart from the obvious difference between the dew-point curve calculated from Equation 16.20 and the one predicted from an EOS model, the calculation of bubble points and dew points from the ideal solution principle terminate at the critical temperature of the most volatile component (propane) of this ternary system. In other words, VLE equations developed on the basis of ideal solution principle cannot be applied at temperatures above 205.97°F (critical temperature of propane).

Clearly, due to these limitations, the ideal solution principle becomes inapplicable for petroleum reservoir fluids. For example, every reservoir fluid contains methane, which is the most volatile hydrocarbon component, having a very low critical temperature of –116°F, limiting the application of ideal solution principle to temperatures lower than –116°F. However, such low temperature and pressure conditions are seldom encountered in either the reservoir or the surface.

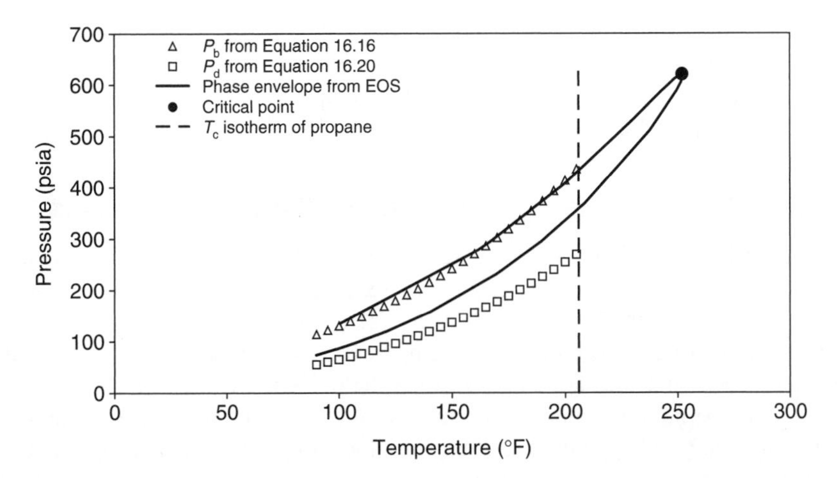

FIGURE 16.2 Comparison of bubble- and dew-point pressures calculated by the ideal solution principle and an EOS model, for ternary mixture of 60 mol-% propane, 30 mol-% *n*-butane, and 10 mol-% *n*-pentane.

16.3 EMPIRICAL CORRELATIONS FOR CALCULATING EQUILIBRIUM RATIOS FOR REAL SOLUTIONS

In order to circumvent the drawbacks associated with the ideal solution principle, the numerous methods proposed have primarily focused on the prediction of the equilibrium ratios of hydrocarbon mixtures. Because, as seen earlier, the major problem lies with the manner in which the equilibrium ratio is calculated from the ideal solution principle. For petroleum reservoir fluids or real solutions, many of these correlations consider the equilibrium ratios to be functions of not only pressure and temperature but also the composition of the hydrocarbon mixture. This observation can be mathematically stated as

$$K_i = f(P, T, Z_i) \tag{16.21}$$

These various correlations eliminate the restrictions posed by the ideal solution principle and enable the calculation of vapor–liquid equilibria of petroleum reservoir fluids at high pressure conditions. Some of these methods are discussed in the following subsections.

16.3.1 Wilson Equation

In 1968, Wilson[1] proposed a simplified thermodynamic expression for estimating the equilibrium ratio values, having the following functional form:

$$K_i = \frac{P_{ci}}{P} \exp\left[5.37(1 + \omega_i)\left(1 - \frac{T_{ci}}{T}\right)\right] \tag{16.22}$$

where K_i is the equilibrium ratio of component i, P_{ci} the critical pressure of component i, P the system pressure, ω_i the acentric factor of component i, T_{ci} the critical temperature of component i, and T the system temperature.

In Equation 16.22, pressures and temperatures in any absolute units can be used.

16.3.2 Methods Based on the Concept of Convergence Pressure

Before we consider the methods for determination of equilibrium ratios, which are based on the concept of convergence pressure, let us first understand what convergence pressure means. The concept of convergence pressure basically resulted from early high-pressure VLE studies. When a hydrocarbon mixture of fixed overall composition is held at a constant temperature as the pressure increases, the equilibrium ratios of all the mixture components converge toward a common value of unity at a certain pressure. This particular pressure at which all K_i values converge to unity is termed *convergence pressure* of the mixture and is commonly denoted by P_k. The convergence pressure is essentially used to account for the effect of the hydrocarbon mixture composition on equilibrium ratios.

The definition of convergence pressure suggests that the composition of both the equilibrium vapor and liquid phases should be the same at the convergence pressure. However, by definition, compositional similarity of both phases can only occur at the critical point. This implies that for any hydrocarbon mixture if the temperature at which the equilibrium ratios are presented is the critical temperature of the mixture, then the convergence pressure is the critical pressure. Thus the critical temperature of any hydrocarbon mixture is the only temperature at which the convergence pressure is the *true convergence* pressure. For all other temperatures other than the critical temperature, the convergence of equilibrium ratios to unity is only an *apparent* convergence.

As mentioned earlier, the bubble- and the dew-point curves on a phase envelope are the outermost boundaries at which equilibrium vapor and liquid phases co-exist, implying that this is the maximum pressure at which equilibrium ratios are actually defined. Therefore, the system either has a bubble or a dew point at some pressure less than the convergence pressure and exists as a single-phase fluid at the conditions expressed by the point of apparent convergence. However, as equilibrium ratios are undefined in the single-phase region, it is merely the extrapolation of the actual values beginning from the saturation pressure to a certain higher pressure at which the equilibrium ratios apparently converge to unity. Therefore, a plot of equilibrium ratios vs. pressure at some constant temperature (other than the critical), for a hydrocarbon mixture has two regions: one is real in which equilibrium ratios are defined, and the other is imaginary because equilibrium ratios are imaginary and do not physically exist. However, at critical temperature the entire plot of equilibrium ratios vs. pressure is real for which the critical pressure simply equals the convergence pressure.

The entire concept of convergence pressure described here can be better appreciated by examining the plots of equilibrium ratios vs. pressure for a five-component synthetic hydrocarbon mixture of fixed overall composition at temperatures of 75 and 126.6°F (critical temperature), as shown in Figures 16.3 and 16.4, respectively. As seen in Figure 16.3, since the temperature does not equal the critical temperature, the plot consists of the two regions, real and imaginary, resulting in a convergence pressure of

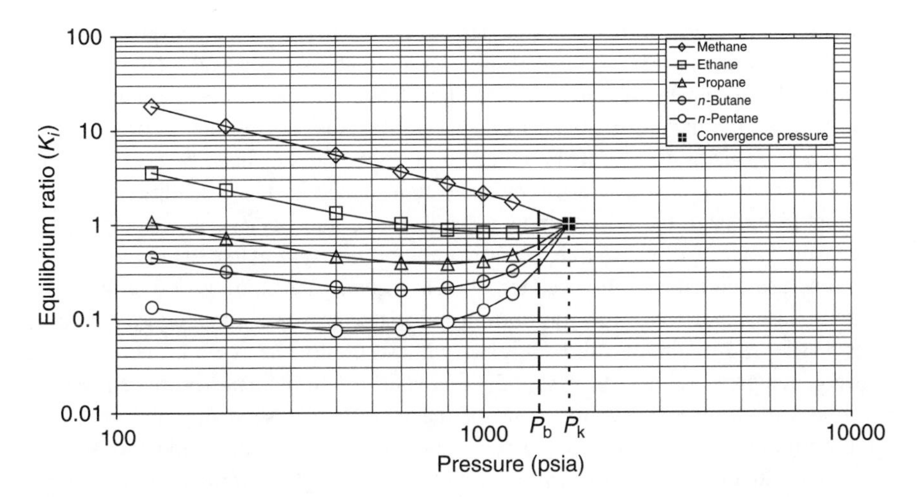

FIGURE 16.3 Equilibrium ratios vs. pressure for a five-component synthetic hydrocarbon mixture of fixed overall composition at a temperature of 75°F. Note that K_i values at pressures greater than the bubble-point pressure are extrapolated values that appear to converge at 1700 psia (apparent convergence).

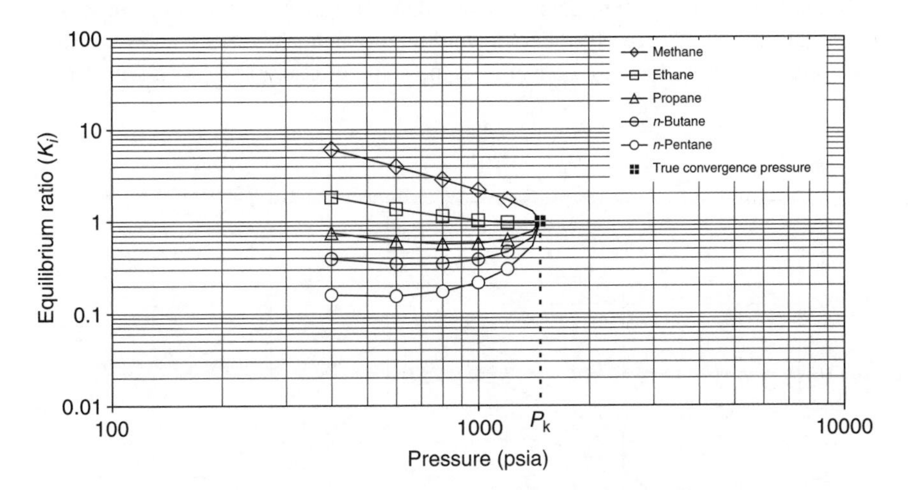

FIGURE 16.4 Equilibrium ratios vs. pressure for a five-component synthetic hydrocarbon mixture of fixed overall composition at a temperature of 126.6°F. Note that this is a true convergence pressure, since the system critical temperature is 126.6°F.

1700 psia. Note that the system has a bubble-point pressure of 1409 psia at 75°F. However, the entire data shown in Figure 16.4 are real and the convergence pressure is the true convergence pressure of 1469 psia, which is equal to the critical pressure of the system.

The concept of convergence pressure is now addressed in more detail. For binary mixtures, the convergence pressure is the critical pressure of a mixture that has a critical temperature equal to the system temperature. Therefore, the convergence pressure for a given binary system is basically represented by its critical locus, shown in Figure 16.5, for binary mixtures of methane–ethane, methane–propane, methane–*n*-butane, and methane–*n*-pentane. For example, the convergence pressure of methane–propane system at 50°F is 1450 psia. It should, however, be noted that the composition of a methane–propane mixture represented by these conditions may be different from that of the system under consideration, unless at its critical temperature. Moreover, by virtue of phase rule, for a two-component system to exist in two phases, only pressure and temperature need to be fixed, meaning the composition of the equilibrated phases, and hence the equilibrium ratio, at a pressure–temperature condition does not depend on the overall or original composition of the system. Therefore, the convergence pressure of 1450 psia at 50°F is valid for all methane–propane mixtures regardless of the original composition of the mixture. Consequently, the generated equilibrium ratios on any binary mixture are valid for all original compositions.

For multicomponent hydrocarbon mixtures, convergence pressures or equilibrium ratios are dependent on the overall composition of the system because the degrees of freedom are more than 2. Therefore, merely specifying pressure and temperature alone does not characterize the system. This means that the convergence pressure at a certain temperature is valid for a fixed overall composition of a given multicomponent system, that is, as soon as the overall system composition is changed, the convergence pressure is also changed. Again, it should be emphasized here that unlike binary systems, a multicomponent mixture having different overall composition does not converge at the same pressure at a given temperature.

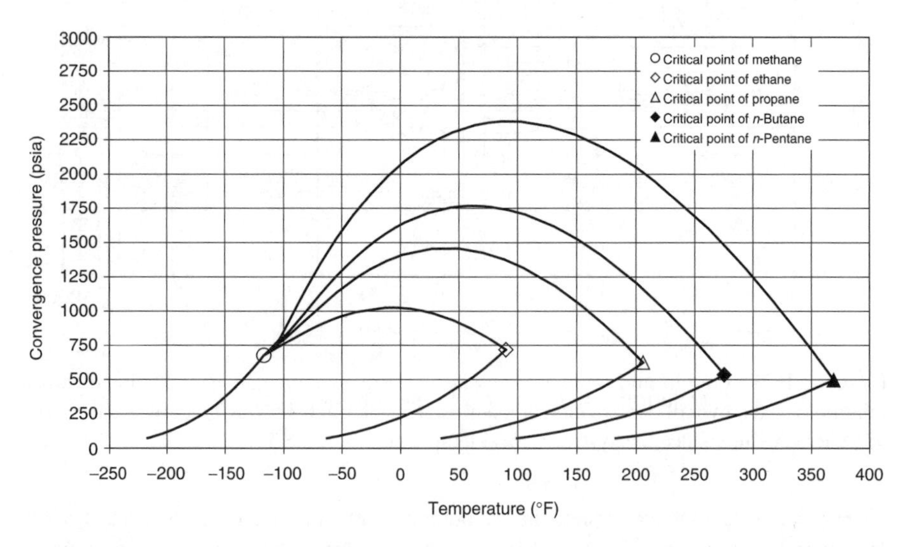

FIGURE 16.5 Critical loci of various methane binary mixtures. The vapor pressure curves of the individual components, ending at their critical points, connect the critical loci of that particular binary mixture. Note that by definition, the various critical loci represent the relationships between the convergence pressure and temperature regardless of the overall compositions of the binary mixtures.

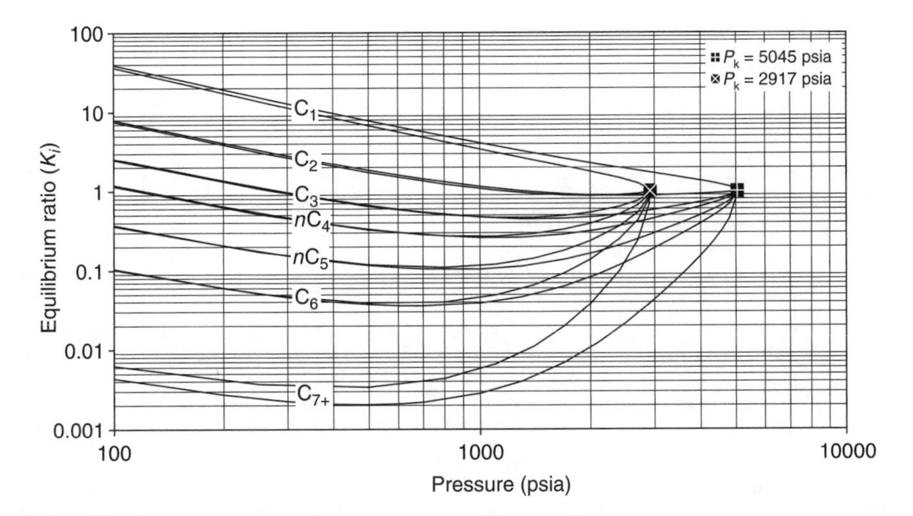

FIGURE 16.6 Comparison of equilibrium ratios at 125°F for a seven-component hydrocarbon system for two different convergence pressures. Unlike a binary system, as the overall composition is changed, the convergence pressure is also changed despite the fact that the temperature is the same.

Figure 16.6 shows the plot of equilibrium ratios as a function of pressure at a temperature of 125°F for a seven-component system having different overall composition and, hence, different convergence pressures. Note that both overall compositions have a critical temperature of 125°F. If this were a binary system, the convergence pressure would be fixed at that particular temperature, regardless of the overall composition. Therefore, in multicomponent hydrocarbon mixtures, equilibrium ratios are often correlated as functions of pressure, temperature, and the overall system composition (convergence pressure).

Based on the convergence pressure (critical) loci of the binary systems, a relationship between the equilibrium ratios as a function of the system pressure and temperature for a fixed convergence pressure can be established. An example of such a relationship for the equilibrium ratios of methane at the convergence pressures of 1000 and 2000 psia at various temperatures is shown in Figures 16.7 and 16.8, respectively.

The literature contains a number of correlations to estimate the equilibrium ratios for multicomponent systems, based on the concept of convergence pressure. Some of these correlations are discussed in the following sections.

16.3.2.1 *K*-Value Charts

The Natural Gas Processors Suppliers Association (NGPSA) presents the most extensive *K*-value graphical correlations for paraffins ranging from methane to decane, ethylene, propylene, and non-hydrocarbons, such as nitrogen and carbon dioxide. Similar to Figures 16.7 and 16.8, the NGPSA *K*-value charts are available for each of these components as a function of system pressure and system temperature at various convergence pressures.

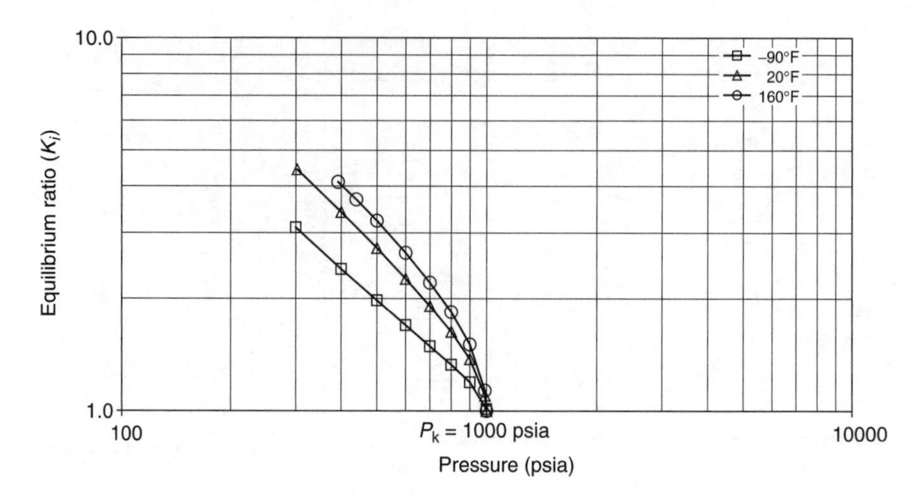

FIGURE 16.7 Equilibrium ratios of methane as a function of temperature for a convergence pressure of 1000 psia.

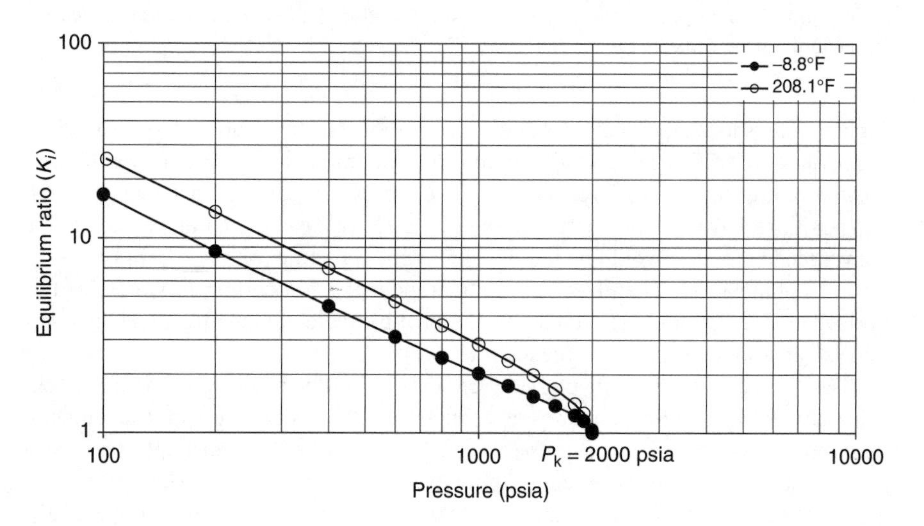

FIGURE 16.8 Equilibrium ratios of methane as a function of temperature for a convergence pressure of 2000 psia.

The NGPSA graphical correlation charts are frequently employed to obtain the equilibrium ratios for performing VLE calculations. However, the problem with using these graphical correlations is that in order to obtain the equilibrium ratios, the convergence pressure must be known before selecting the appropriate charts. Therefore, the use of NGPSA *K*-value charts for performing VLE calculations involves an iterative procedure. This iterative procedure is based on the pseudo or equivalent binary concept proposed by Hadden[2]. The Hadden method basically treats the mixture as a pseudobinary system composed of the lightest component and all other components are grouped as single, heavy pseudocomponents. The procedure for VLE calculations using the Hadden method is described in the following text.

Weighted Average Critical Properties of the Single, Heavy Pseudocomponent

Regardless of the type of VLE calculations desired (i.e., bubble point, dew point or PT flash), the common step in the Hadden method is the determination of weighted average critical pressure and temperature of the heavy components grouped as a single, heavy pseudocomponent. Since most of these heavy components are concentrated in the liquid phase, the weighted average critical pressure and temperature are calculated from the liquid-phase composition. First, the molar composition of the liquid phase is normalized after excluding the lightest component (usually methane) from the liquid phase. The resulting mole fraction of the liquid phase is then converted to weight fraction, from which the weighted average critical pressure and temperature are calculated from the following expressions:

$$P_{\text{cmw}} = \sum_{i=2}^{n} W_i P_{ci} \tag{16.23}$$

$$T_{\text{cmw}} = \sum_{i=2}^{n} W_i T_{ci} \tag{16.24}$$

where P_{cmw} and T_{cmw} are the weighted average critical pressure and temperature of the single, heavy pseudocomponent, W_i is the weight fraction of ith component in the liquid phase, and P_{ci} and T_{ci} the critical pressure and temperature of the ith component in the liquid phase.

Note that the summations in Equations 16.23 and 16.24 begin with $i = 2$ instead of $i = 1$ because the lightest component is excluded.

Once the weighted average critical pressure and temperature is calculated, a plot similar to the one shown in Figure 16.5 is entered to determine the convergence pressure at the temperature of interest. For example, if the weighted average critical values are close to that of n-pentane, then the critical locus of methane–n-pentane binary is used to obtain the convergence pressure at a given temperature. If the pseudocritical values fall between two different critical loci, then interpolation between curves becomes necessary. It should be noted here that since the liquid-phase composition is not known *a priori* in the case of dew-point and PT flash calculations; the determination of convergence pressure becomes iterative. However, in the case of bubble-point calculations, obviously the liquid phase composition is already known; and hence the convergence pressure can be directly determined. The specific procedures for bubble-, dew-point, and PT flash calculations using the weighted average critical properties of the single, heavy pseudo-component are described next.

Calculation of Bubble-Point Pressure Using K-Value Charts

As mentioned earlier, since the liquid phase composition is already known, the convergence pressure can be directly determined from a plot shown in Figure 16.5. For example, if the determined convergence pressure at the given temperature is 2000 psia, then the equilibrium ratio charts for a convergence pressure of 2000 psia for all components present in the mixture are used to obtain the K values at the temperature of interest. Figures 16.7 and 16.8 show examples of methane K values at different temperatures for convergence pressures of 1000 and 2000 psia, respectively. However, the purpose of the calculation is to determine the bubble-point pressure, meaing that the pressure should be known to read the K values at the temperature of interest! Since pressure is implicit in bubble-point calculations, the problem can be solved by reading K values at different pressures and calculating the summation expressed in Equation 16.14. The pressure that satisfies Equation 16.14, that is, $\sum_{i=1}^{n} Z_i K_i = 1$, is the bubble-point pressure.

Calculation of Dew-Point Pressure Using K-Value Charts

For dew-point calculations, the procedure starts with an assumed value of the convergence pressure, since the liquid-phase composition is unknown. Based on assumed convergence pressure, the equilibrium ratios for each component present in the mixture are determined by using the appropriate convergence pressure chart corresponding to the assumed value. Since the dew-point pressure is also unknown, the K values at the temperature of interest at the assumed convergence pressure and the pressure at which Equation 16.18 is satisfied, i.e., $\sum_{i=1}^{n} (Z_i/K_i) = 1$ is considered as the dew-point pressure. However, the dew point that is determined is based on assumed convergence pressure, which needs to be verified against the convergence pressure that is determined from the equivalent binary concept. For this purpose, the procedure described earlier is applied to the composition of the liquid phase ($X_i = Z_i/K_i$) to determine the pseudocritical values, on the basis of which the new convergence pressure is determined. A close agreement between the assumed and the new convergence pressure indicates convergence of the dew-point calculation; otherwise the procedure has to be repeated using the new convergence pressure as a starting value.

PT Flash Calculations Using K-Value Charts

Similar to dew-point calculations, the procedure starts with an assumed value of the convergence pressure, from which equilibrium ratio for each component present in the mixture is obtained at the system pressure and temperature by using the appropriate convergence pressure chart corresponding to the assumed value. Based on these K_i values, PT flash calculations (Equation 16.11) are carried out. Note that the solution of Equation 16.11 itself is also an iterative process. The procedure described earlier is applied to the liquid-phase composition determined from the PT flash calculations, in order to determine the values of P_{cmw} and T_{cmw}. Subsequently, on the basis of calculated pseudocritical values, the new convergence pressure at the temperature of interest is determined (see earlier description). A close agreement between the new convergence pressure and the assumed convergence pressure indicates convergence of the solution. If significant differences exist between the new convergence pressure and the assumed value, then the calculation procedure starts with the new convergence pressure and all steps are repeated until close agreement is obtained.

Drawbacks of the K-Value Charts Procedure for Use in VLE Calculations

Although, it is feasible to perform VLE calculations using the approach of NGPSA K-value charts; the detailed description of the procedures described earlier clearly demonstrate that the method is extremely tedious and cumbersome and cannot be easily automated for computer calculations. Additionally, reading of values from various charts can introduce errors in calculations and interpolation between charts for determining the equilibrium ratios may be necessary, if for example, the new convergence pressure falls between the values for which NGPSA charts are provided. Also, the suggested equivalent binary concept is a highly simplified assumption used in representing a multicomponent mixture in the form of a binary system.

The other significant problem posed by the K-value chart approach is that NGPSA equilibrium ratio charts for single carbon number (SCN) fractions and the plus fraction (C_{20+} or C_{30+}) are not available. Unlike the pure well-defined hydrocarbons, equilibrium ratio curves for different convergence pressures and system pressures and temperatures, cannot be established for SCN fractions and the plus fractions because the properties of these vary for different reservoir fluids. Therefore, SCN fractions and the plus fractions are lumped as a C_{7+} fraction; and its equilibrium ratio is approximated as 15% of heptane, as suggested by Katz et al.,[3] otherwise it can also be estimated as equal to the K values of a hydrocarbon compound, such as nonane or decane.[4] This is again a highly simplified assumption.

16.3.2.2 Whitson–Torp Correlation

Whitson and Torp[5] modified Wilson's equation by incorporating the convergence pressure to yield accurate results for high pressure VLE calculations. The correlation is expressed in the following mathematical form:

$$K_i = \left[\frac{P_{ci}}{P_k}\right]^{A-1}\left[\frac{P_{ci}}{P}\right]\exp\left[5.37A(1+\omega_i)\left(1-\frac{T_{ci}}{T}\right)\right] \tag{16.25}$$

where

$$A = 1 - \left[\frac{P-14.7}{P_k-14.7}\right]^{0.6} \tag{16.26}$$

where P is the system pressure (psia), P_k the convergence pressure (psia), T the system temperature (°R), and ω_i the acentric factor.

The convergence pressure required in the Whitson–Torp correlation can be estimated from Standing's correlation[6]:

$$P_k = 60\,MW_{C_{7+}} - 4200 \tag{16.27}$$

where $MW_{C_{7+}}$ represents molecular weight of the C_{7+} fraction.

In comparison to the K-value charts approach, VLE calculations can be carried out much more efficiently using the Whitson–Torp correlation. Since the method does not involve any iterative procedure for the determination of convergence pressure; VLE calculations are much more simplified. However, it should be noted that iterations are necessary for the determination of bubble-point and dew-point pressures because pressure is implicit in the Whitson–Torp correlation. The equilibrium ratios of PT flash calcuations can be directly calculated from the Whitson–Torp correlation, since pressure is given. The specific procedures for calculation of bubble-point and dew-point pressures and PT flash calculations are described in the following text.

Calculation of Bubble- and Dew-Point Pressures Using the Whitson–Torp Correlation

As mentioned earlier, since pressure is implicitly expressed in Equation 16.25, the calculation of bubble- or dew-point pressures involves an iterative procedure. For these calculations, an initial guess or a good starting value can be obtained from Wilson equation.

For bubble point, using Equation 16.14 and the Wilson correlation:

$$P_b = \sum_{i=1}^{n} Z_i P_{ci} \exp\left[5.37(1+\omega_i)\left(1-\frac{T_{ci}}{T}\right)\right] \tag{16.28}$$

With use of the calculated value of P_b from Equation 16.28, the equilibrium ratios of all the components present in the mixture are calculated from Equation 16.25. The summation expressed by Equation 16.14 is calculated next. A summation equal to 1 indicates convergence of the solution; otherwise the starting value is adjusted until Equation 16.14 is satisfied.

For dew point, using Equation 16.18 and the Wilson correlation:

$$P_d = \sum_{i=1}^{n} \frac{Z_i}{P_{ci} \exp\left[5.37(1+\omega_i)\left(1-\frac{T_{ci}}{T}\right)\right]} \tag{16.29}$$

On the basis of the starting value from Equation 16.29, equilibrium ratios of all components in the mixture are calculated by the Whitson–Torp correlation. The remaining procedure is similar to bubble-point calculations, except that in the case of dew-point calculations, Equation 16.18 has to be satisfied.

PT Flash Calculation Using the Whitson–Torp Correlation

The procedure for PT flash calculations is relatively straightforward and is in fact identical to what has already been described in Section 16.2.4. The only difference is the use of equilibrium ratios in Equation 16.11 or 16.12. Since pressure and temperature conditions for flash calculations are known, equilibrium ratios of all components in the mixture can be directly calculated from Equation 16.25. These equilibrium ratio values and the mixture feed composition are fixed, thus requiring the determination of a correct value of n_V that satisfies Equation 16.11 or 16.12.

16.4 EQUATIONS-OF-STATE (EOS) MODELS

Section 16.3 discussed the application of equilibrium ratio correlations for performing commonly required VLE calculations, such as bubble point, dew point, and PT flash, which are used to predict the phase behavior of petroleum reservoir fluids. However, with the common use of EOS models, the need for these equilibrium ratio correlations has been greatly diminished. Currently, it is a standard practice in reservoir engineering to use EOS models to simulate phase behavior and predict the properties of petroleum reservoir fluids over a wide range of fluid composition, pressure, and temperature conditions, especially in the absence of experimental data. In general, the application of EOS models is preferred because they are more flexible, rigorous, and useful for describing complex hydrocarbon systems. For example, EOS models are easily incorporated into compositional reservoir simulators, where for instance, numerous flash calculations are desired in order to evaluate the mass transfer that takes place on a component basis in every grid block.

The remainder of this chapter focuses on the discussion of EOS models. However, discussion begins with the basic description of an EOS model, starting with the well-known van der Waals equation, followed by the two most commonly used EOS models in the petroleum industry: the Soave–Redlich–Kwong (SRK) equation and the Peng–Robinson (PR) equation. Since all EOS models are basically developed for pure components; their application to single-component systems is studied first. The applicability of EOS models to hydrocarbon mixtures is extended by employing some mixing rules; these are discussed after studying the pure component systems. Finally, the application of EOS models for VLE calculations, and simulation of laboratory PVT and phase behavior experiments is described.

16.4.1 Description of EOS Models

An equation-of-state model is basically an analytical expression that relates pressure, temperature, and volume. As discussed in Chapter 15, the fundamental gas equation is the best known and the simplest example of an equation of state. The fundamental gas equation is, however, limited to describe the volumetric behavior of gases at low pressures for which it was experimentally derived. The limitations imposed by the fundamental gas equation actually prompted numerous attempts to develop an equation of state suitable for describing the behavior of petroleum reservoir fluids at extended ranges of pressures and temperatures.

EOS models basically belong to four different families:

1. The van der Waals (vdW) family
2. Benedict–Webb–Rubin family
3. Reference fluid equations of state
4. Augmented rigid body equations of state

The various EOS models that belong to the vdW family are the most popular EOS models for petroleum reservoir fluids. EOS models that belong to the other three families have not received much attention as far as their application to petroleum reservoir fluids is concerned. Therefore, our discussion of EOS models presented in this chapter is restricted to the vdW family.

16.4.1.1 van der Waals Equation of State

It was the work of van der Waals (vdW)[7] in 1873 that attempted to eliminate the shortcomings of the ideal gas equation and introduced the well-known vdW equation of state:

$$\left(P + \frac{a}{V^2}\right)(V - b) = RT \tag{16.30}$$

where P is the system pressure (psia), T the system temperature (°R), R the gas constant (10.73 psi-ft^3/lb-mol°R), V the molar volume (ft^3/lb-mol), and a and b are constants characterizing the molecular properties of a given component. In Equation 16.30, the terms a/V^2 and b represent the attractive and repulsive forces, respectively.

At low pressures, and correspondingly high volumes, the term a/V^2 is very small and parameter b becomes negligible in comparison with V, which actually reduces the vdW equation to the ideal gas equation ($PV = RT$). However, as pressure approaches infinity, the volume V becomes very small and approaches the value of b, which is considered as an apparent molecular volume called *covolume*.

Equation 16.30 can also be expressed in an alternative form as

$$P = \frac{RT}{(V - b)} - \frac{a}{V^2} \tag{16.31}$$

or

$$P = P_{\text{repulsive}} - P_{\text{attractive}} \tag{16.32}$$

Figure 16.9 shows the pressure–volume (PV) relationship for a pure component at saturation conditions along the vapor pressure curve. Also shown in Figure 16.9 is the PV relationship at critical temperature, which indicates that the critical isotherm has a horizontal inflection point as it passes through the critical point. This observation can be mathematically expressed as

$$\left(\frac{\partial P}{\partial V}\right)_{T_c} = \left(\frac{\partial^2 P}{\partial V^2}\right)_{T_c} = 0 \tag{16.33}$$

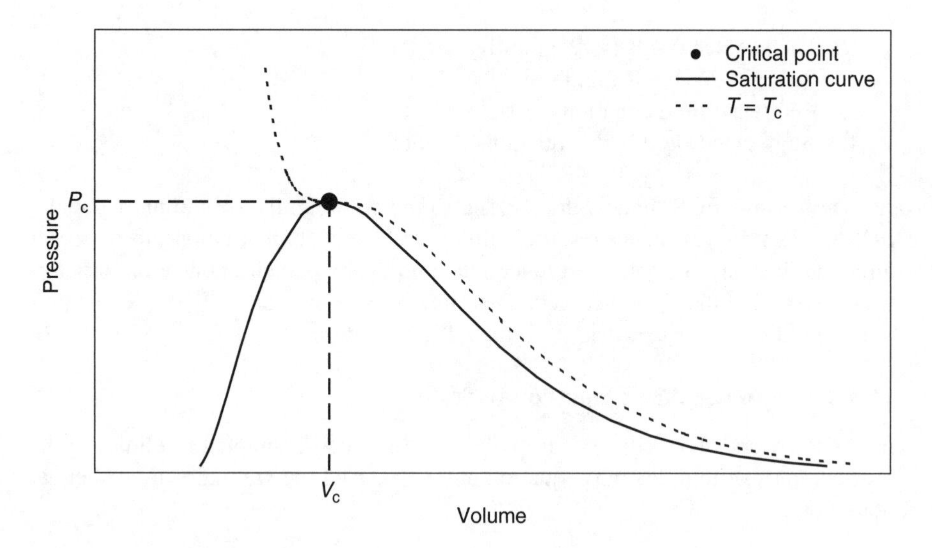

FIGURE 16.9 Pressure–volume (PV) relationship for a pure component at saturation conditions and at the critical isotherm.

By applying these conditions to the vdW EOS, generalized expressions for the two constants a and b can be obtained:

$$\left(\frac{\partial P}{\partial V}\right)_{T_c} = \frac{-RT_c}{(V_c - b)^2} + \frac{2a}{V_c^3} = 0 \tag{16.34}$$

$$\left(\frac{\partial^2 P}{\partial V^2}\right)_{T_c} = \frac{2RT_c}{(V_c - b)^3} - \frac{6a}{V_c^4} = 0 \tag{16.35}$$

Solving Equations 16.34 and 16.35:

$$a = \left(\frac{9}{8}\right)RT_cV_c = \left(\frac{27}{64}\right)\frac{R^2T_c^2}{P_c} \tag{16.36}$$

and

$$b = \left(\frac{1}{3}\right)V_c = \left(\frac{1}{8}\right)\frac{RT_c}{P_c} \tag{16.37}$$

In Equations 16.34, 16.35, and 16.36, subscript c refers to the values at the critical point.

Therefore the vdW EOS results in a critical compressibility factor, Z_c, of 0.375 for all components while Z_c values of pure components range between 0.23 and 0.31.

Equation 16.31 can also be expressed in a cubic form in terms of volume as

$$V^3 - \left(b + \frac{RT}{P}\right)V^2 + \left(\frac{a}{P}\right)V - \left(\frac{ab}{P}\right) = 0 \tag{16.38}$$

which is usually referred to as the van der Waals *two-parameter cubic equation of state*. The term *two parameter* refers to the two pure component constants a and b, while cubic refers to cubic form in terms of volume.

The cubic equation of vdW can also be expressed in terms of the compressibility factor Z as

$$Z^3 - (1 + B)Z^2 + AZ - AB = 0 \tag{16.39}$$

where

$$A = \frac{aP}{(RT)^2} \tag{16.40}$$

$$B = \frac{bP}{RT} \tag{16.41}$$

Let us now study the characteristic features of the pressure–volume (PV) relationship obtained from the vdW EOS. Figure 16.10 shows the PV relationship (hatched line) for a pure component at a temperature less than its critical temperature. The vdW EOS actually results in the peculiar PV curves shown in Figure 16.10 for all temperatures less than the critical temperature. However, these loops disappear at the critical isotherm and at all temperatures above the critical temperature.

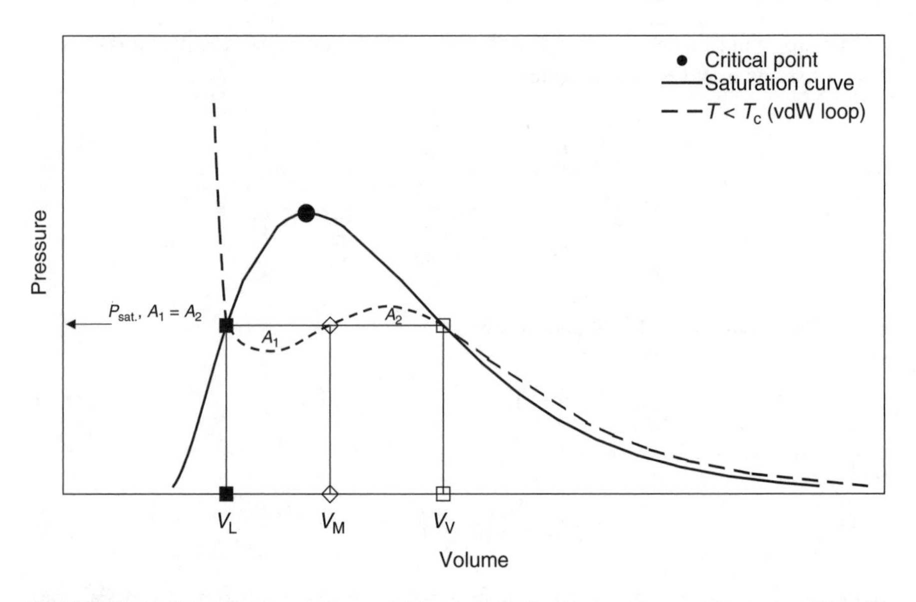

PRESSURE 16.10 Pressure–volume (*PV*) relationship for a pure component predicted by the vdW equation of state. The hatched line is the *PV* relationship at a subcritical temperature, which shows the van der Waals loop.

The characteristic *PV* curves at subcritical temperatures are sometimes referred to as *van der Waals loops*. Clearly the vdW EOS does describe the liquid condensation phenomenon and the passage from vapor or gas to liquid phase as the gas is compressed, characterized by the transition from large phase volumes to small phase volumes. The predicted maximum and minimum volumes, within the two-phase region, however, indicate the pressure limits within which the fluid can be compressed or expanded while it remains a metastable single-phase fluid.[4] However, given the nature of the *PV* curve in Figure 16.10, it is difficult to precisely distinguish the *PV* points that identify the vapor phase and the liquid phase. Maxwell proposed the "equal area rule", which allows the determination of the saturation pressure and *PV* points corresponding to the vapor and liquid phase. As shown in Figure 16.10, the pressure at which the two areas A_1 and A_2 are equal is the saturation pressure and the corresponding vapor and liquid volumes are indicated by V_V and V_L, respectively.

The vdW loops and the Maxwell equal area rule clearly indicate that for a pure component below its critical temperature, the vdW EOS may give three roots for volume at a pressure, for example, corresponding to the horizontal hatched line shown in Figure 16.10. As mentioned earlier, the highest value corresponds to the vapor volume V_V, while the lowest volume V_L corresponds to that of liquid. The middle root indicated by the intersection of the horizontal hatched line and the vdW loop results in a middle root V_M, that is of no physical significance. In the single-phase region (gas phase or liquid phase), the vdW EOS yields 1 real root.

Although, despite its simplicity, the vdW EOS provides a correct description, at least qualitatively, of the phase behavior of a pure component in the gaseous and liquid states, it is not accurate enough for practical applications. These drawbacks associated with the vdW EOS actually prompted the development of new and accurate equations-of-state models. Many of these EOS models are modifications of the vdW equation of state and range in complexity from simple expressions containing two or three parameters to complicated forms containing more than 50 parameters. In the following subsections, two of the most commonly used EOS models in the petroleum industry are presented.

16.4.1.2 Redlich–Kwong Equation of State

In 1948, Redlich and Kwong (RK)[8] proposed a simple modification of the attractive term (a/V^2) of van der Waals EOS model as

$$P = \frac{RT}{(V - b)} - \frac{a}{V(V + b)T^{0.5}} \tag{16.42}$$

The generalized expressions for the constants a and b in RK EOS can be determined by imposing the critical point conditions (Equation 16.33) on Equation 16.42 and solving the resultant equations simultaneously

$$a = \Omega_a \frac{R^2 T_c^{2.5}}{P_c} \tag{16.43}$$

and

$$b = \Omega_b \frac{RT_c}{P_c} \tag{16.44}$$

where Ω_a and Ω_b are constants with values of 0.42727 and 0.08664, respectively. Note that for the vdW EOS, these values are 0.421875 and 0.125, respectively (see Equations 16.36 and 16.37).

Similar to Equation 16.39, the RK EOS can also be expressed in terms of compressibility factor as

$$Z^3 - Z^2 + (A - B - B^2)Z - AB = 0 \tag{16.45}$$

where

$$A = \frac{aP}{(R^2 T^{2.5})} \tag{16.46}$$

$$B = \frac{bP}{RT} \tag{16.47}$$

As seen in the case of the vdW EOS, Equation 16.42 (when expressed in cubic form) or Equation16.45 yields 1real root in the single-phase region, while 3 real roots in the two-phase region are obtained. As discussed earlier; in the two-phase region, the largest and the smallest V or Z root correspond to the vapor phase and the liquid phase, respectively, while the middle root has no physical significance.

Although, Equation 16.42 is quite accurate for predicting molar volumes of pure substances, calculations of the properties of mixtures and predictions of vapor–liquid equilibrium using this model are not particularly accurate. Therefore, several modifications of the RK EOS have been proposed. One such modification is the well-known Soave–Redlich–Kwong equation of state described next.

16.4.1.3 Soave–Redlich–Kwong Equation of State

In 1972, Soave[9] proposed a modification of the Redlich–Kwong equation of state, by replacing the term $a/T^{0.5}$ with a more generalized temperature dependent expression denoted by the product of ($a\alpha$) such that

$$P = \frac{RT}{(V-b)} - \frac{a\alpha}{V(V+b)} \tag{16.48}$$

where α is a dimensionless parameter, which is a function of acentric factor, ω, and reduced temperature, T_r, and is defined by the following expression:

$$\alpha = \left[1 + m(1 - T_r^{0.5})\right]^2 \tag{16.49}$$

The parameter m in Equation 16.49 is in turn correlated with the acentric factor as

$$m = 0.480 + 1.574\omega - 0.176\omega^2 \tag{16.50}$$

The above-modified RK equation is popularly known as the Soave–Redlich–Kwong or SRK equation of state, and is in fact one of the most commonly used EOS models applied to petroleum reservoir fluids.

Similar to the vdW and RK EOS, parameters a and b for any pure component for the SRK EOS can be obtained by imposing the critical point constraints expressed in Equation 16.33:

$$a = \Omega_a \frac{R^2 T_c^2}{P_c} \tag{16.51}$$

and

$$b = \Omega_b \frac{RT_c}{P_c} \tag{16.52}$$

where Ω_a and Ω_b are constants with values of 0.42727 and 0.08664, respectively.

In terms of the compressibility factor Z the SRK EOS is identical to Equation 16.45, with B defined by Equation 16.47, while A is given as

$$A = \frac{a\alpha P}{(RT)^2} \tag{16.53}$$

The characteristics of compressibility factor roots in the single-phase gas or liquid,and the two-phase regions are as described previously, that is, 1 real root in single phase and 2 real roots (largest and the smallest) in the two-phase region.

16.4.1.4 Peng–Robinson Equation of State

In 1975 Peng and Robinson (PR)[10] proposed a slightly different form of the molecular attraction term. Their modification of the attractive term was mainly for the improvement of the liquid density prediction in comparison with the SRK EOS. The functional form of the PR equation is given by the following equation:

$$P = \frac{RT}{(V - b)} - \frac{a\alpha}{V(V + b) + b(V - b)} \tag{16.54}$$

The application of the critical point conditions on Equation 16.54 results in the expressions for a and b that are identical to Equations 16.51 and 16.52; however, with values of Ω_a and Ω_b as 0.45724 and 0.07780, respectively.

Peng and Robinson also adopted the expression proposed by Soave for the determination of α, but with a different expression for calculation of m, given by

$$m = 0.3746 + 1.5423\omega - 0.2699\omega^2 \tag{16.55}$$

Equation 16.55 was later modified to improve the prediction for heavier components[11]

$$m = 0.379642 + 1.48503\omega - 0.1644\omega^2 + 0.016667\omega^2 \tag{16.56}$$

The PR EOS takes the following form in terms of the compressibility factor:

$$Z^3 - (1 - B)Z^2 + (A - 2B - 3B^2)Z - (AB - B^2 - B^3) = 0 \tag{16.57}$$

where A and B are given by Equations 16.53 and 16.47, respectively.

16.4.2 Concept of Fugacity

Phase equilibria relationships using EOS models are expressed most commonly in terms of fugacity or fugacity coefficients. Specifically, the fugacity or fugacity coefficient is introduced as a criterion for thermodynamic equilibrium and is an important aspect of EOS applications to phase equilibria of petroleum reservoir fluids. Therefore, it is appropriate at this time to introduce the concept of fugacity and the fugacity coefficient of a component.

Fugacity can be described as a *fictitious pressure*, which may be considered as a vapor pressure modified to represent correctly the escaping tendency of the molecules from one phase into the other.[12] The fugacity of a component, denoted by "f" is related to the system pressure and compressibility factor according to the following mathematical expression:

$$\ln\left(\frac{f}{P}\right) = \int_0^P \left(\frac{Z - 1}{P}\right) dP \tag{16.58}$$

where f is the fugacity (psia), P the system pressure (psia), and Z the compressibility factor.

The ratio of the fugacity to pressure f/P, is called the *fugacity coefficient* and is denoted by Φ. Thus, Equation 16.58 in terms of the fugacity coefficient is expressed as

$$\ln(\Phi) = \int_{o}^{P}\left(\frac{Z-1}{P}\right)dP \tag{16.59}$$

Defined by the generalized expression in Equation 16.59; equations for fugacity coefficients can be derived for various EOS models.

For SRK EOS,:

$$\ln(\Phi) = Z - 1 - \ln(Z-B) - \frac{A}{B}\ln\left(\frac{Z+B}{Z}\right) \tag{16.60}$$

For PR EOS,

$$\ln(\Phi) = Z - 1 - \ln(Z-B) + \frac{A}{2\sqrt{2}B}\ln\left[\frac{Z+(1-\sqrt{2})B}{Z+(1+\sqrt{2})B}\right] \tag{16.61}$$

The generalized expressions in Equations 16.60 and 16.61 (depending on the EOS model used) are used for both the equilibrium vapor and the liquid phase by using the pertinent phase compressibility factor. For calculation of the vapor-phase fugacity coefficient, the vapor-phase compressibility factor Z_V is used, while for the liquid-phase, compressibility factor Z_L is used. These fugacity coefficient expressions constitute perhaps one of the most important phase equilibria relationships used as a criterion for the evaluation of thermodynamic equilibrium between the vapor and liquid phases. The application of these relationships is discussed in the following section.

16.4.3 Application of Equations of State to Pure Components

As mentioned earlier, the fugacity coefficient is used to evaluate the thermodynamic equilibrium between the vapor and liquid phases. Specifically, the conditions that result in equal fugacity coefficients for the vapor and liquid phases indicate that the system is in thermodynamic equilibrium. For example, in the case of a pure component at its saturation pressure, the fugacity coefficients of the vapor and liquid phases are equal since the system is in equilibrium. The application of the equations-of-state models and the concept of fugacity coefficient to pure components are illustrated by the following example.

For example, if the pure component is *n*-butane and its saturation pressure (vapor pressure) and the densities of the equilibrium vapor phase and the liquid phase are desired at 100°F, using the PR EOS, the calculations would proceed as follows.

For *n*-butane, T_c = 305.69°F (765.69°R), P_c = 551.1 psia, and ω = 0.193; a = 56024.9 and b = 1.16007, from Equations 16.51 and 16.52, with values of Ω_a and Ω_b at 0.45724 and 0.07780, respectively; m = 0.66025, from Equation 16.56, which results in a value of α = 1.200349, from Equation 16.49.

At this stage, since the saturation pressure is unknown; the calculation begins with an assumed value. From the assumed saturation pressure, values of A (Equation 16.53) and B (Equation 16.47) are calculated. With an assumed saturation pressure of 70 psia, A and B are calculated as 0.1303 and 0.0135, respectively.

The calculated A and B values result in the following cubic equation for Z (Equation 16.57):

$$Z^3 - 0.9865Z^2 + 0.1028Z - 0.001576 = 0$$

This equation has three roots, $Z_1 = 0.87052$, $Z_2 = 0.01859$, and $Z_3 = 0.09737$, out of which the intermediate root is rejected since it has no physical significance. The other two roots, Z_1 (Z_V) and Z_2 (Z_L), are assigned to the vapor phase and the liquid phase, respectively.

The substitution of Z_V and Z_L in Equation 16.61 yields the following fugacity coefficient of the vapor and the liquid phases, respectively

$$\Phi_V = 0.88459 \text{ and } \Phi_L = 0.68669$$

As mentioned earlier, the fugacity coefficients for a pure component should be equal at equilibrium conditions (saturation or vapor pressure); the different value of Φ_V and Φ_L using the first guess of 70 psia indicates that the solution did not converge. The solution, however, converges using a value of 52.375 psia with $\Phi_V = \Phi_L = 0.9135$. Therefore, the saturation pressure of n-butane at 100°F is calculated as 52.375 psia, using the PR EOS.

The cubic equation at the saturation pressure of 52.375 psia is

$$Z^3 - 0.98989Z^2 + 0.07699Z - 0.000883 = 0$$

with the compressibility factors of the vapor phase $Z_V = 0.90598$ and the liquid phase $Z_L = 0.01392$.

The compressibility factor values can now be used to determine the molar volume of the vapor phase and the liquid phase, from which the densities of the equilibrium phases are calculated. The calculations are

$$V_V = \frac{Z_V RT}{P} = \frac{0.90598 \times 10.732 \times (100 + 460)}{52.375} = 103.96 \text{ ft}^3/\text{lb-mol}$$

$$V_L = \frac{Z_L RT}{P} = \frac{0.01392 \times 10.732 \times (100 + 460)}{52.375} = 1.597 \text{ ft}^3/\text{lb-mol}$$

$$\rho_v = \left(\frac{1}{V_V}\right) \times MW_{n\text{-butane}} = \frac{58.123}{103.96} = 0.559 \text{ lb/ft}^3$$

$$\rho_L = \left(\frac{1}{V_L}\right) \times MW_{n\text{-butane}} = \frac{58.123}{1.597} = 36.395 \text{ lb/ft}^3$$

16.4.4 Extension of EOS Models to Mixtures

All equations-of-state models are basically developed for pure components and are extended to mixtures by employing mixing rules. These mixing rules are simply means of calculating the mixture parameters equivalent to those of pure components.

In the SRK and PR equations-of-state models, since a, b, and α are component dependent constants, their values are determined for each component that is present in the mixture. After determining these constants for each component, the following mixing rules are used for both the SRK and the PR equations of state, for the calculation of the mixture parameters

$$(a\alpha)_{\mathrm{m}} = \sum_{i=1}^{n}\sum_{j=1}^{n} Z_i Z_j (a_i a_j \alpha_i \alpha_j)^{0.5}(1 - k_{ij}) \tag{16.62}$$

and

$$b_{\mathrm{m}} = \sum_{i=1}^{n} Z_i b_i \tag{16.63}$$

where $(a\alpha)_{\mathrm{m}}$ represents the product of constant a and α for a given mixture (see Equation 16.53), Z_i and Z_j the mole fraction of component i and j in the mixture, a_i and a_j the constant a for component i and j in the mixture (calculated from pertinent equations), α_i and α_j the parameter α for component i and j in the mixture (calculated from pertinent equations), k_{ij} or k_{ji} the binary interaction parameter (described in the following text), b_{m} a constant b for the mixture; and b_i a constant b for component i in the mixture.

The binary interaction parameter (BIP), k_{ij} or k_{ji}, in Equation 16.62 is an empirically determined correction factor that characterizes the binary formed by components i and j in the hydrocarbon mixture. BIPs are generally determined by minimizing the difference between the predicted and experimental data, mainly the saturation pressure, of binary systems. Therefore, due to the empirical nature of the BIP, it should be considered as a fitting parameter rather than a rigorous physical term. BIPs have different values for each binary pair and also take on different values for each equation of state.

Regardless of the EOS model used, BIPs between components with little differences in size are generally considered to be 0, however, BIP values generally increase as differences in the component size increase. For example, BIP between methane and n-hexadecane (or n-hexadecane and methane) is much higher than that between methane and ethane. Also, in the case of nonhydrocarbon and hydrocarbon pairs, BIPs are generally higher due to differences in the molecules. Given their empirical nature, BIPs play a major role in tuning or calibrating EOS models for a particular reservoir fluid against available experimental data.

Once the mixture parameters are calculated from Equations16.62 and 16.63, values of A and B can be calculated to formulate the cubic equation in terms of the compressibility factor. The resulting equation can then be solved to select the appropriate root (vapor or liquid) for the computation of the mixture density. The following example illustrates the application of PR EOS for calculating the density of a hydrocarbon mixture at given pressure and temperature conditions. The application of EOS models to VLE calculations is described in later sections.

For example, consider a five-component mixture that consists of methane, propane, n-pentane, n-decane, and n-hexadecane having a fixed overall composition (see Table 16.1). This particular mixture is placed in a PVT cell that is maintained at 5000 psia and 150°F, the conditions at which the mixture is found to exist as a

TABLE 16.1

Component	Z_i (Mole%)	T_{ci} (°R)	P_{ci} (psia)	ω	m	α	a	b	$a\alpha$
Methane	0.5449	343	667.2	0.008	0.3915	0.7559	9286	0.4292	7019
Propane	0.1394	666	615.8	0.152	0.6016	1.0524	37935	0.9031	39922
n-Pentane	0.1314	845	489.4	0.251	0.7423	1.2357	76838	1.4417	94947
n-Decane	0.0869	1112	305.7	0.490	1.0698	1.6319	213036	3.0374	347650
n-Hexadecane	0.0974	1291	205.7	0.742	1.3978	2.0649	426610	5.2391	880910

TABLE 16.2

Component	Methane	Propane	*n*-Pentane	*n*-Decane	*n*-Hexadecane
Methane	0	0.009	0.021	0.052	0.080
Propane	0.009	0	0.003	0.019	0.039
n-Pentane	0.021	0.003	0	0.008	0.022
n-Decane	0.052	0.019	0.008	0	0.004
n-Hexadecane	0.080	0.039	0.022	0.004	0

single-phase liquid. If the density of this five-component mixture at the given pressure and temperature conditions is desired using the PR EOS, then the calculations proceed as follows.

Step 1: Using the acentric factor data of all the components, calculate m from Equation 16.56.

Step 2: Using the critical temperature data of all the components, and the given temperature of 150°F, calculate the reduced temperature for each component.

Step 3: Based on step 1 and step 2, calculate the α parameter for all components from Equation 16.49.

Step 4: Using Ω_a and Ω_b as 0.45724 and 0.07780, and the critical pressure, critical temperature, and the universal gas constant, calculate a and b for all components from Equations 16.51 and 16.52, respectively.

Step 5: Calculate the product of a and α.

Step 6: Using the given composition and the mixing rules described by Equations 16.62 and 16.63, calculate $(a\alpha)_{\mathrm{m}}$ and b_{m} for the five-component mixture. The binary interaction parameters used in this example are given in Table 16.2. However, if k_{ij} values are not available they can be set to 0.

The summation for Equation 16.62 is expanded as

$$(a\alpha)_{\mathrm{m}}=(0.5449)(0.5449)(7019\times 7019)^{0.5}(1-0)+(0.5449)(0.1394)(7019\times 3992)^{0.5}(1-0.009)+(0.5449)(0.1314)(7019\times 94947)^{0.5}(1-0.021)+(0.5449)(0.0869)(7019\times 347650)^{0.5}(1-0.052)+(0.5449)(0.0974)(7019\times 880910)^{0.5}(1-0.080)+\ldots\ldots$$

in precisely the same manner, the above is repeated for other components, and the value of $(a\alpha)_{\mathrm{m}}$ is obtained as 64,371.

For Equation 16.63,

$b_m = (0.5449)(0.4292) + (0.1394)(0.9031) + (0.1314)(1.4417) + (0.0869)(3.0374) + (0.0974)(5.2391) = 1.3235$

Step 7: Calculate parameters A and B from Equations 16.53 and 16.47, respectively, using the values of $(a\alpha)_m$ and b_m from step 6, given pressure (5000 psia), temperature (150°F), and the universal gas constant. For the given mixture, $A = 7.5100$ and $B = 1.0108$.

Step 8: Using the values of A and B calculated in step 7, formulate the cubic equation in terms of the compressibility factor (Equation 16.57)

$$Z^3 + 0.01081Z^2 + 2.4232Z - 5.5367 = 0$$

the above cubic equation results in 1 real root with a value of $Z = 1.3225$ and 2 imaginary roots that have no meaning.

Step 9: Using the value of Z, the density is calculated as

$$V = \frac{ZRT}{P} = \frac{1.3225 \times 10.732 \times (150 + 460)}{5000} = 1.7315 \text{ ft}^3\text{/lb-mol}$$

$$\rho = \frac{1}{V} \times \sum_{i=1}^{n} Z_i \mathrm{MW}_i = \frac{58.79}{1.7315} = 33.952 \text{ lb/ft}^3$$

16.4.4.1 Determination of Equilibrium Ratios from EOS Models

Let us now turn to the application of equations-of-state models to VLE calculations. As seen earlier, the equilibrium ratio is an integral component of all VLE calculations. The methods discussed earlier focused on the determination of equilibrium ratios from various empirical methods such as the K-value charts, Wilson correlation, and the Whitson–Torp correlation. The EOS models in conjunction with the concept of fugacity also offer a very effective and rigorous means of determining the equilibrium ratios, on the basis of which VLE calculations for petroleum reservoir fluids are commonly carried out in the petroleum industry.

In the example that was presented in Section 16.4.3, the equality of the fugacity coefficient for the vapor phase and the liquid phase was used as a criterion to determine the thermodynamic equilibrium between the two phases. A similar criterion can also be established for hydrocarbon mixtures. When dealing with petroleum reservoir fluids, we are concerned with the equilibrium between the hydrocarbon vapor mixture with the hydrocarbon liquid mixture at a specified pressure and temperature condition. However, since the vapor or the liquid mixture is made up of various components, the fugacity of each component in the vapor and the liquid phases is used as a criterion for determining the thermodynamic equilibrium. The fugacity of a component in the vapor phase and the liquid phase is basically a measure of the potential for transfer of the component between the phases. For example, a higher fugacity of a component in the vapor phase compared to the liquid phase indicates that the liquid phase accepts the component from the vapor phase. The equality of component fugacities in the vapor phase and the liquid phase means zero net transfer of a component between the two phases.

Therefore, a zero net transfer for all components or the equality of component fugacities in the two phases implies thermodynamic equilibrium of a hydrocarbon system, which can be mathematically expressed as

$$f_i^{\mathrm{V}} = f_i^{\mathrm{L}} \tag{16.64}$$

where f_i^{V} is the fugacity of component i in the vapor phase and f_i^{L} the fugacity of component i in the liquid phase.

Since the fugacity coefficient of component i in a hydrocarbon vapor or liquid phase is a function of the fugacity, system pressure and the respective mole fractions, it can be defined by the following expressions:

$$\Phi_i^{\mathrm{V}} = \frac{f_i^{\mathrm{V}}}{Y_i P} \tag{16.65}$$

$$\Phi_i^{\mathrm{L}} = \frac{f_i^{\mathrm{L}}}{X_i P} \tag{16.66}$$

where Φ_i^{V} and Φ_i^{L} are fugacity coefficients of components i in the vapor phase and the liquid phase, respectively.

Using the thermodynamic equilibrium criterion of equal component fugacities in the vapor and the liquid phases:

$$\Phi_i^{\mathrm{V}} Y_i P = \Phi_i^{\mathrm{L}} X_i P \tag{16.67}$$

which allows the determination of equilibrium ratio,

$$K_i = \frac{\Phi_i^{\mathrm{L}}}{\Phi_i^{\mathrm{V}}} \tag{16.68}$$

Since mathematical expressions, such as Equation 16.60 (SRK EOS) or 16.61 (PR EOS), can be developed for fugacity coefficients for any EOS model; Equation 16.68 constitutes one of the most important relationships in EOS-based VLE calculations. Similar to Equation 16.60 or 16.61, expressions for the fugacity coefficient of the ith component in a hydrocarbon phase can be developed for the SRK EOS and the PR EOS. Danesh[4] presents a generalized equation for the SRK and PR EOS models for the fugacity coefficient of ith component in a hydrocarbon phase. The generalized equation for the vapor phase is given by

$$\ln(\Phi_i^{\mathrm{V}}) = \frac{b_i}{b_{\mathrm{m}}}(Z_{\mathrm{V}} - 1) - \ln(Z_{\mathrm{V}} - B) - \frac{A}{B(\delta_2 - \delta_1)}\left[\frac{2\psi_{\mathrm{i}}}{\psi} - \frac{b_i}{b_{\mathrm{m}}}\right]\ln\left[\frac{Z_{\mathrm{V}} + \delta_2 B}{Z_{\mathrm{V}} + \delta_1 B}\right] \tag{16.69}$$

where

$$\psi_i = \sum_{j=1}^{n}\left[Y_j(a_i a_j \alpha_i \alpha_j)^{0.5}(1 - k_{ij})\right] \tag{16.70}$$

$$\psi = \sum_{i=1}^{n}\sum_{j=1}^{n}\left[Y_i Y_j(a_i a_j \alpha_i \alpha_j)^{0.5}(1 - k_{ij})\right] \tag{16.71}$$

The constants δ_1 and δ_2 in Equation 16.69 have values equal to 1 and 0 in SRK EOS, and $1+\sqrt{2}$ and $1-\sqrt{2}$ in PR EOS. An identical equation is used to determine

Φ_i^L by using the composition of the liquid phase X_i in calculating A, B, Z_L, and other composition dependent parameters.

16.4.5 VLE Calculations Using EOS Models

The various mathematical relationships described in Section 16.4.4 form the basis of VLE calculations using EOS models. As it was described in the introduction, typical VLE calculations can be classified into two categories. In the first category the saturation conditions (i.e., the bubble point or dew point) are required for a reservoir fluid of a given overall composition. Most frequently, it is the saturation pressure at the reservoir temperature that is desired since reservoir temperature is assumed to be constant. In the second category, the composition and the properties of the co-existing or equilibrium phases at a given set of pressure and temperature are required, also referred to as flash calculations. In petroleum engineering, these types of calculations are commonly performed using PVT simulators that employ the popular EOS models, such as the Soave–Redlich–Kwong and the Peng–Robinson. In this section, the application of EOS models for performing the VLE calculations is demonstrated.

16.4.5.1 Calculation of Bubble-Point Pressure

For bubble-point calculations, the liquid composition or the original mixture composition remains unchanged. Therefore, the composition of the newly formed vapor phase can be calculated using the equilibrium ratio, $K_i = Y_i/X_i = \Phi_i^L / \Phi_i^V$. However, both Φ_i^L and Φ_i^V are pressure and composition dependent. The pressure (bubble point) and the vapor-phase composition are both unknown. Therefore, the calculation of bubble point using an EOS model involves an iterative procedure.

The calculation of bubble point typically begins with an assumed value of the bubble-point pressure. On the basis of the assumed value of the bubble-point pressure, the Wilson correlation (Equation 16.22) is used to estimate the initial set of equilibrium ratios for all the components present in the mixture. Using the feed composition (X_i) and the calculated vapor composition (Y_i) from the K_i values, f_i^L and f_i^V, the fugacity of each component present in the liquid phase and the vapor phase is calculated. If the calculated f_i^L and f_i^V values satisfy Equation 16.64; then the assumed value of bubble-point pressure is correct. It should, however, be noted that considering the complicated pressure and composition dependency of the various EOS parameters, convergence in the first step on the basis of the assumed value is achieved very rarely. Therefore, the calculation has to continue in an iterative sequence, using the newly calculated equilibrium ratios and the adjusted pressure, until Equation 16.64 is satisfied. However, considering that petroleum reservoir fluids are generally described using a large number of components, instead of checking the equality of each and every component, an error function defined by Equation 16.72 is used as a convergence criterion:

$$\sum_{i=1}^{n}\left[1 - \frac{f_i^L}{f_i^V}\right]^2 \leq 10^{-10} \tag{16.72}$$

TABLE 16.3
Fugacity Coeffcients, Fugacities, and New Equilibrium Ratios with Assumed Pressure = 2600 psia and Temperature = 100°F

Component	Liquid Composition, mole fraction ($Z_i = X_i$)	Vapor Composition, mole fraction (Y_i)	Φ_i^L (Equation 16.69 for Liquid Phase	Φ_i^V (Equation 16.69)	f_i^L (Equation 16.66)	f_i^V (Equation 16.65)	K_i (Equation 16.68)
Methane	0.7000	1.4633	1.0074	0.8439	1833.45	3210.58	1.1937
n-pentane	0.3000	0.0018	0.0004	0.0387	15.2896	0.1859	0.5070

After achieving this defined convergence criterion, the iteration is terminated. The pressure and vapor composition that satisfies Equation 16.72 represents the bubble-point pressure and the composition of the newly formed vapor phase.

The bubble-point calculation procedure using an EOS model is illustrated by the following example. Consider a binary mixture consisting of 70 mol-% methane and 30 mol-% *n*-pentane. The bubble-point pressure of this mixture at 100°F is desired using the PR equation of state. The calculations are performed as described by the following steps.

Step 1: Assume a bubble-point pressure of 2600 psia at 100°F and calculate the equilibrium ratio of methane and *n*-pentane using the Wilson equation (Equation 16.22), $K_{\text{methane}} = 2.0904$ and $K_{n\text{-pentane}} = 0.0062$, which gives $Y_{\text{methane}} = 1.4633$ and $Y_{n\text{-pentane}} = 0.0018$. Obviously, $\sum_{i=1}^{n}, Y_i \neq 1$, which only occurs at the correct bubble-point pressure.

Step 2: Calculate all the component property dependent EOS (PR) parameters (see example in Section 16.4.4).

Step 3: The PR EOS is set for both the liquid phase and the vapor phase for calculation of the fugacity coefficient and the fugacity of methane and *n*-pentane in the two phases. The A and B parameters for the liquid phase and the vapor phase result in values of $A = 1.7071$, $B = 0.3171$ and $A = 1.1473$, $B = 0.2729$, respectively. With use of the calculated A and B values, the cubic compressibility factor equation is set for the individual phases. Note that both cubic compressibility factor equations result in 1 real root each, $Z_L = 0.5761$ and $Z_V = 0.6572$, respectively. Subsequently, the fugacity coefficients, fugacities and the new equilbrium ratios are calculated, as shown in Table 16.3.

The binary interaction parameters required for the calculations in Table 16.3 are taken from the example shown in Section 16.4.4.

The calculated f_i^L and f_i^V values in Table 16.3 result in an error function of 6602, which is far remote from the objective value of 10^{-10}.

Step 4: With use of the previously calculated values of fugacity coefficients, fugacities, and equilibrium ratios, the pressure and composition are updated for use in the next iteration. At equilibrium.

$$Y_i = \left(\frac{1}{P}\right)\left(\frac{f_i^L}{\Phi_i^V}\right)$$

However

$$\sum_{i=1}^{n} Y_i = 1 = \left(\frac{1}{P}\right)\sum_{i=1}^{n}\left(\frac{f_i^L}{\Phi_i^V}\right)$$

Therefore, the updated pressure value is given by $\sum_{i=1}^{n}\left(\frac{f_i^L}{\Phi_i^V}\right)$, which is equal to 2568 psia from Table 16.3. The vapor phase composition is updated next, using $Y_i = K_i \times X_i$, which is $Y_{\text{methane}} = 0.8356$ and $Y_{n\text{-pentane}} = 0.1521$ from Table 16.3.

Step 5: With use of the updated pressure and composition from step 4, the calculations described in step 3 are repeated and the convergence criterion is again evaluated. If the calculations fail to meet the convergence criterion, the iteration continues with the updated values until Equation 16.72 is satisfied. The iteration results for this particular example at the second, intermediate, and the final levels are shown in Table 16.4.

TABLE 16.4
Results for Iteration 2, Pressure = 2568 psia, Z_L = 0.5710, and Z_V = 0.6510

Component	Y_i (mole fraction)	Φ_i^L	Φ_i^V	f_i^L	f_i^V	K_i
Methane	0.8356	1.0122	0.8381	1819.50	1798.48	1.2077
n-Pentane	0.1521	0.0197	0.0391	15.2009	15.2892	0.5041

$$\sum_{i=1}^{n}\left[1-\frac{f_i^L}{f_i^V}\right]^2 = 1.7\text{E}^{-04}$$

Iteration 7, Pressure = 2518 psia, Z_L = 0.5638, and Z_V = 0.6401

Component	Y_i (mole fraction)	Φ_i^L	Φ_i^V	f_i^L	f_i^V	K_i
Methane	0.8409	1.0190	0.8484	1800.14	1800.69	1.2011
n-Pentane	0.1565	0.0199	0.0381	15.0807	15.0429	0.5230

$$\sum_{i=1}^{n}\left[1-\frac{f_i^L}{f_i^V}\right]^2 = 6.4\text{E}^{-06}$$

Iteration 50, Pressure = 2436.739 psia, Z_L = 0.5498, and Z_V = 0.6736

Component	Y_i (mole fraction)	Φ_i^V	Φ_i^V	f_i^L	f_i^V	K_i
Methane	0.8799	1.0330	0.8218	1761.967	1761.960	1.2570
n-Pentane	0.1201	0.0203	0.0507	14.8544	14.8547	0.4004

$$\sum_{i=1}^{n}\left[1-\frac{f_i^L}{f_i^V}\right]^2 = 4.3\text{E}^{-10}$$

Therefore, the bubble-point pressure of this binary system is determined as 2436.739 psia at 100°F.

16.4.5.2 Calculation of Dew-Point Pressure

The calculation of dew-point pressure also begins with an assumed value of the dew-point pressure. Again, on the basis of the assumed value of the dew-point pressure, the Wilson correlation (Equation 16.22) is used to estimate the initial set of equilibrium ratios for all the components present in the mixture. With use of the feed composition (Y_i) and the calculated liquid composition (X_i) from the K_i values, f_i^V and f_i^L, the fugacity of each component present in the vapor phase and the liquid phase is calculated. The updated pressure value is given by $\sum_{i=1}^{n}(f_i^V/\Phi_i^L)$, while the liquid phase composition is updated using $X_i = Y_i/K_i$. The remaining iterative sequence is similar to the bubble-point calculation procedure outlined in the previous section.

16.4.5.3 PT Flash Calculations

The PT flash calculations for the determination of equilibrium phase compositions, based on EOS models, also require an iterative procedure. For this type of calculation, the system pressure, temperature, and the overall feed composition are necessary. The entire process is summarized in a flow-chart shown in Figure 16.11 and is described in the following paragraphs.

Similar to the saturation pressure calculations, the iterative procedure begins with an assumed value of the equilibrium ratios at the given pressure and temperature conditions. The Wilson correlation can provide reasonable starting values of the equilibrium ratios K_i^A, for the mixture components. Based on these starting values of equilibrium ratios, flash calculations outlined in Section 16.2.4 (Equations 16.11 and 16.12) are performed.

With use of the calculated vapor- and liquid-phase compositions, the fugacity coefficients and fugacities of each component in the vapor and liquid phases are determined. The existence of thermodynamic equilibrium is evaluated by comparing the fugacities of each component in the vapor phase and the liquid phase, respectively. If Equation 16.72 is satisfied, then the solution has converged. Otherwise the newly calculated values of the equilibrium ratios, K_i^N, from the fugacity coefficients are used in the next iteration and the calculations continue in this manner until Equation 16.72 is satisfied.

The PT flash calculation procedure using an EOS model is illustrated by the following example. Consider a five-component mixture consisting of 82.32 mol-% methane, 8.71 mol-% propane, 5.05 mol-% *n*-pentane, 1.98 mol-% *n*-decane, and 1.94 mol-% *n*-hexadecane. This particular mixture is flashed at 1500 psia and 100°F and the resulting equilibrium vapor- and liquid-phase compositions and densities are desired using the PR equation of state. The various calculation steps are outlined in the following steps.

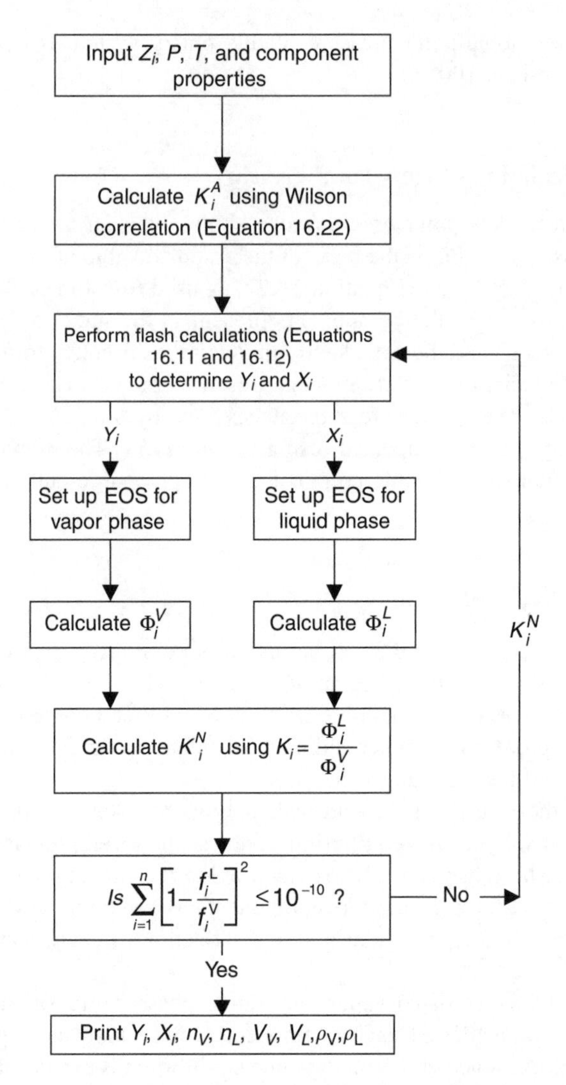

FIGURE 16.11 Flow chart of iterative sequence for flash calculations, using an equation of state.

Step 1: Using the given pressure and temperature conditions of 1500 psia and 100°F, the component critical properties and acentric factors, calculate the equilibrium ratios of all components from the Wilson equation (Equation 16.22).

Step 2: On the basis of the calculated equilibrium ratios, perform flash calculations (Equations 16.11 and 16.12) to determine the vapor and liquid phase compositions and set up the PR EOS for both phases. Subsequently, calculate the fugacity coefficients and the fugacities of all the components and evaluate Equation 16.72. The calculations up to this point are shown in Table 16.5.

TABLE 16.5
$n_V = 0.8028$ and $n_L = 0.1972$

Z_i (mole fraction)	K_i^A (Equation 16.22)	X_i (mole fraction)	Y_i (mole fraction)	Φ_i^L (Equation 16.69 for Liquid Phase)	Φ_i^V (Equation 16.69)	f_i^L (Equation 16.66)	f_i^V (Equation 16.65)	K_i^N
0.8232	3.623314	0.2650	0.960	2.1042	0.8401	836.5548	1210.2162	2.5046
0.0871	0.127285	0.2909	0.037	0.1758	0.3987	76.6966	22.1432	0.4409
0.0505	0.010684	0.2454	0.003	0.0168	0.2119	6.1857	0.8331	0.0793
0.0198	0.000077	0.1004	0.000	1.19E-04	4.89E-02	1.79E-02	5.63E-04	0.0024
0.0194	0.000001	0.0984	0.000	6.66E-07	9.75E-03	9.83E-05	9.82E-07	0.0001
		1.0000	1.000					

$$\sum_{i=1}^{n}\left[1-\frac{f_i^L}{f_i^V}\right]^2 = 1.1E^{+04}$$

TABLE 16.6
Iteration 2; $n_V = 0.8425$ and $n_L = 0.1575$

X_i (mole fraction)	Y_i (mole fraction)	Φ_i^L	Φ_i^V	f_i^L	f_i^V	K_i^{N+1}
0.3630	0.909	2.1041	0.8470	1145.7927	1155.1322	2.4843
0.1647	0.073	0.1828	0.3616	45.1516	39.3772	0.5055
0.2251	0.018	0.0172	0.1745	5.8046	4.6737	0.0985
0.1241	0.000	1.18E-04	3.20E-02	2.19E-02	1.45E-02	0.0037
0.1231	0.000	6.14E-07	4.91E-03	1.13E-04	6.20E-05	0.0001
1.0000	1.000					

$$\sum_{i=1}^{n}\left[1-\frac{f_i^L}{f_i^V}\right]^2 = 1.0E^{00}$$

Step 3: Clearly as seen in Table 16.5; $K_i^A \neq K_i^N$, and obviously $f_i^L \neq f_i^V$. The second iteration then proceeds with the K_i^N values used to perform flash calculations to determine the new vapor- and liquid-phase compositions. On the basis of these vapor- and liquid-phase compositions, the PR EOS is set up for both phases for calculation of the fugacity coefficients and the fugacities. Subsequently, Equation 16.72 is evaluated; this results in a value of 1 in the case of second iteration, which is still far away from the target value of 10^{-10}. See Table 16.6.

Step 4: The iterative sequence thus continues on the basis of the newly calculated equilibrium ratios until Equation 16.72 is satisfied. In this particular example, the solution converges after 10 iterations, as seen in Table 16.7. The converged values of the equilibrium vapor- and liquid-phase compositions, fugacities, and equilibrium ratios are shown in the following table. The compressibility factors of the equilibrium phases are used to calculate the phase densities.

TABLE 16.7
Iteration 10 (Converged Values); $n_V = 0.8602$ and $n_L = 0.1398$, $Z_V = 0.7684$, and $Z_L = 0.5408$, $\rho_V = 6.3574$ lb/ft³ , and $\rho_L = 36.5161$ lb/ft³

X_i (mole fraction)	Y_i (mole fraction)	Φ_i^L	Φ_i^V	f_i^L	f_i^V	K_i^{N+9}
0.35630	0.89911	2.1440	0.8496	1145.8620	1145.8556	2.52344
0.14651	0.07744	0.1867	0.3532	41.0315	41.0311	0.52858
0.22041	0.02288	0.0173	0.1663	5.7067	5.7066	0.10379
0.13818	0.00055	1.15E-04	0.0288	2.39E-02	2.39E-02	0.00401
0.13860	0.00002	5.83E-07	0.0041	1.21E-04	1.21E-04	0.00014
1.00000	1.00000					

$$\sum_{i=1}^{n}\left[1-\frac{f_i^L}{f_i^V}\right]^2 = 5.4\text{E}^{-10}$$

16.4.5.4 Separator Calculations

As discussed in Chapter 15, for efficient surface processing of petroleum reservoir fluids, the knowledge of optimum separator conditions is essential. Usually, these conditions (maximum API gravity, minimum GOR, and minimum formation volume factor) are determined on the basis of laboratory studies that simulate a two- or three-stage separator system. However, in the absence of such laboratory studies, EOS-based VLE calculations can be carried out to determine the various properties from which optimum separator conditions can be selected. These applications are described in this subsection.

The separator calculations performed to determine the optimum separator conditions based on EOS models predominantly involve PT flash calculations. In a two-stage separation system typically used for black oils, the PT flash calculations are carried out twice, for the first stage and the second stage (stock tank), respectively. For example, a liquid mixture (feed) of given overall composition either at or above its bubble-point pressure is flashed in the first stage at a fixed pressure and temperature, resulting in certain number of moles of an equilibrium vapor phase and a liquid phase of certain composition. The liquid phase in the first stage in turn becomes the feed for the second stage, which is flashed generally at atmospheric pressure and a given temperature resulting in yet another set of equilibrium vapor and liquid phases. These equilibrium phases formed in the first stage and the second stage are the separator gas, separator liquid and stock tank gas, and stock tank liquid, respectively. The equations employed for the determination of gas–oil ratio and the formation volume factor that use PT flash calculations are developed as follows.

Let n_F be the liquid mixture feed into first-stage separator, lb-mol feed; n_{V1} the moles of equilibrium vapor in the first-stage separator, lb-mol SP gas; n_{L1} the moles of equilibrium liquid in the first-stage separator, lb-mol SP liquid; n_{V2} the moles of equilibrium vapor in the second-stage separator (stock tank), lb-mol ST gas; and n_{L2} the moles of equilibrium liquid in the second-stage separator (stock tank), lb-mol ST oil.

Since all PT flash calculations are performed on the basis of 1 lb-mol of feed, feed to the first-stage separator or the separator liquid feed to the second-stage separator, the individual moles can be expressed as

$$\bar{n}_{V1}\frac{\text{lb-mol SP gas}}{\text{lb-mol feed}},\quad \bar{n}_{L1}\frac{\text{lb-mol SP liquid}}{\text{lb-mol feed}},$$

$$\bar{n}_{V2}\frac{\text{lb-mol ST gas}}{\text{lb-mol SP liquid}},\quad \bar{n}_{L2}\frac{\text{lb-mol ST oil}}{\text{lb-mol SP liquid}}$$

The following ratio gives the lb-mol of separator gas per lb-mol of stock tank oil[13]

$$\frac{\left(\bar{n}_{V1}\dfrac{\text{lb-mol SP gas}}{\text{lb-mol feed}}\right)}{\left(\bar{n}_{L1}\dfrac{\text{lb-mol SP liquid}}{\text{lb-mol feed}}\right)\left(\bar{n}_{L2}\dfrac{\text{lb-mol ST oil}}{\text{lb-mol SP liquid}}\right)}=\left(\frac{\bar{n}_{V1}}{\bar{n}_{L1}\bar{n}_{L2}}\right)\frac{\text{lb-mol SP gas}}{\text{lb-mol ST oil}} \quad (16.73)$$

Equation 16.73 can be expanded further by considering the fact that 1 lb-mol of gas occupies 379.6 scf, and using the molecular weight and density of the stock tank oil so that the gas–oil ratio is expressed in terms of standard cubic feet of gas per stock tank barrel of stock tank oil:

$$R_{SP}=\left(\frac{\bar{n}_{V1}}{\bar{n}_{L1}\bar{n}_{L2}}\frac{\text{lb-mol SP gas}}{\text{lb-mol ST oil}}\right)\left(\frac{379.6\text{ scf SP gas}}{\text{lb-mole SP gas}}\right)\left(\frac{5.615\rho_{STO}}{MW_{STO}}\frac{\text{lb-mol ST oil}}{\text{STB}}\right) \quad (16.74)$$

In Equation 16.74, 5.615 is the volume conversion factor, 1 barrel = 5.615 ft^3.

$$R_{SP}=\frac{2131.45\,\bar{n}_{V1}\rho_{STO}\text{ scf SP gas}}{\bar{n}_{L1}\bar{n}_{L2}\,MW_{STO}\text{ STB}} \quad (16.75)$$

where R_{SP} = separator gas–oil ratio in scf/STB; ρ_{STO} = density of stock tank oil, lb/ft^3; and MW_{STO} = molecular weight of stock tank oil, lb/lb-mol.

An expression for the gas–oil ratio, R_{ST}, of the stock tank can be developed on the basis of the following ratio:[13]

$$\frac{\left(\bar{n}_{V2}\dfrac{\text{lb-mol ST gas}}{\text{lb-mol SP liquid}}\right)}{\left(\bar{n}_{L2}\dfrac{\text{lb-mol ST oil}}{\text{lb-mol SP liquid}}\right)} \quad (16.76)$$

Again using an approach similar to the one in Equation 16.74

$$R_{ST}=\left(\frac{\bar{n}_{V2}}{\bar{n}_{L2}}\frac{\text{lb-mol SP gas}}{\text{lb-mol ST oil}}\right)\left(\frac{379.6\text{ scf SP gas}}{\text{lb-mole SP gas}}\right)\left(\frac{5.615\rho_{STO}}{MW_{STO}}\frac{\text{lb-mol ST oil}}{\text{STB}}\right) \quad (16.77)$$

$$R_{ST}=\frac{2131.45\,\bar{n}_{V2}\rho_{STO}}{\bar{n}_{L2}MW_{STO}}\frac{\text{scf ST gas}}{\text{STB}} \quad (16.78)$$

The total gas–oil ratio R is the sum of R_{SP} and R_{ST}.

The application of Equations 16.75 and 16.78 involves two sets of PT flash calculations, one for the feed (well stream) for the first-stage separator which gives $\bar{n}_{V1}$, $\bar{n}_{L1}$ and the compositions of the separator gas and the separator liquid, and the second for the separator liquid which gives $\bar{n}_{V2}$, $\bar{n}_{L2}$ and the compositions of the stock tank gas and the stock tank liquid. The calculated composition of the stock tank liquid can then be used to determine its density at standard conditions using an EOS model and also its molecular weight. A substitution of these pertinent values in Equations 16.75 and 16.78 yields the separator and the stock tank gas–oil ratios. The API gravity of the stock tank oil can be calculated using Equations 10.1 and 10.2.

The mathematical expression for oil formation volume factor B_{oSb} on the basis of these calculated quantities and the properties of the reservoir oil is developed as[13]

$$B_{oSb} = \frac{\text{reservoir barrel/lb mole of reservoir oil (feed)}}{\text{STB/lb mole of reservoir oil (feed)}} \tag{16.79}$$

With use of the density ρ_{RO} and the molecular weight MW_{RO} of reservoir oil, the numerator can be expressed as

than

$$\frac{\text{reservoir barrel}}{\text{lb mole of reservoir oil}} = \frac{\left(MW_{RO}\,\dfrac{\text{lb of reservoir oil}}{\text{lb mole of reservoir oil}}\right)}{\left(\rho_{RO}\,\dfrac{\text{lb of reservoir oil}}{\text{ft}^3\text{ reservoir oil}}\right)\left(5.615\,\dfrac{\text{ft}^3\text{reservoir oil}}{\text{reservoir barrel}}\right)} \tag{16.80}$$

and the denominator can be expressed using $\bar{n}_{L1}$, $\bar{n}_{L2}$ and the stock tank oil properties:

$$\frac{\text{STB}}{\text{lb mole of reservoir oil}}$$

$$= \frac{\left(MW_{STO}\,\dfrac{\text{lb of ST oil}}{\text{lb mole of ST oil}}\right)\left(\bar{n}_{L1}\,\dfrac{\text{lb-mole SP liquid}}{\text{lb-mole reservoir oil}}\right)\left(\bar{n}_{L2}\,\dfrac{\text{lb-mole ST oil}}{\text{lb-mole SP liquid}}\right)}{\left(\rho_{STO}\,\dfrac{\text{lb of ST oil}}{\text{ft}^3\text{ ST oil}}\right)\left(5.615\,\dfrac{\text{ft}^3\text{ ST oil}}{\text{STB}}\right)} \tag{16.81}$$

Finally, the substitution of Equations 16.80 and 16.81 in Equation 16.79 results in

$$B_{oSb} = \frac{MW_{RO}\rho_{STO}}{MW_{STO}\rho_{RO}\bar{n}_{L1}\bar{n}_{L2}}\,\frac{\text{res.bbl}}{\text{STB}} \tag{16.82}$$

The molecular weight and density of the reservoir oil in Equation 16.82 can be calculated from the well stream (feed) composition, component molecular weights, and an EOS model, respectively.

The application of an EOS model for separator calculations is illustrated by the following example. Consider a five-component hydrocarbon mixture consisting of 25.59 mol-% methane, 9.31 mol-% propane, 9.60 mol-% *n*-pentane, 23.12 mol-% *n*-decane, and 32.37 mol-% *n*-hexadecane. This particular mixture exists at 1100 psia (slightly above its bubble-point pressure) at 100°F. A two-stage separation system is used to separate this mixture. The first-stage separator operates at 200 psia and 75°F, while the stock tank operates at 14.7 psia and 60°F. The separator and stock tank gas–oil ratios, API gravity of the stock tank oil, and the formation volume factor are desired using the PR equation of state. The calculations are carried out as described by the following steps.

Step 1: From the given well stream (feed) composition, first calculate the molecular weight and the density of the mixture at 1100 psia and 100°F. The molecular weight is calculated using the simple molar mixing rules, which results in MW_{RO} = 121.35 lb/lb-mol. The density at reservoir conditions is calculated by setting the PR EOS for the given well stream composition, pressure, and temperature (see Section 16.4.4), which results in ρ_{RO} = 38.7716 lb/ft^3.

Step 2: Using the given feed composition, perform flash calculation at 200 psia and 75°F. These PT flash calculations are carried out in precisely the same manner as shown in subsection 16.4.5.3. Therefore, they are not repeated here; only the final results are shown in Table 16.8.

Step 3: Next, using the composition calculated in step 2 (last column of Table 16.8), flash the separator liquid at 14.7 psia and 60°F and again perform PT flash calculations. The results are shown in Table 16.9.

TABLE 16.8
First-Stage Separator Flash of Well Stream at 200 psia and 75°F

Component	Feed Composition (mole fraction)	Separator Gas Composition (mole fraction)	Separator Liquid Composition (mole fraction)
Methane	0.2559	0.88509	0.05812
Propane	0.0931	0.10505	0.08928
n-Pentane	0.0960	0.00978	0.12313
n-Decane	0.2312	0.00008	0.30393
n-Hexadecane	0.3237	0.00000	0.42554
	1.0000	1.00000	1.00000
		$\bar{n}_{V1}$ = 0.2392	$\bar{n}_{L1}$ = 0.7608
		γ_{gSP} = 0.675	

TABLE 16.9
Second-Stage Stock Tank Flash of Separator Liquid at 14.7 psia and 60°F

Component	Feed Composition (mole fraction)	Stock Tank Gas Composition (mole fraction)	Stock Tank Liquid Composition (mole fraction)
Methane	0.05812	0.46694	0.00245
Propane	0.08928	0.45511	0.03947
n-Pentane	0.12313	0.07755	0.12933
n-Decane	0.30393	0.00040	0.34526
n-Hexadecane	0.42554	0.00000	0.48349
	1.00000	1.00000	1.00000
		$\bar{n}_{V2}$ = 0.1198	$\bar{n}_{L2}$ = 0.8802
		γ_{gST} = 1.147	

Step 4: Using the composition of the stock tank liquid (last column of Table 16.9) set up PR EOS and calculate the density of the stock tank oil at 14.7 psia and 60°F. The calculated density is $\rho_{STO} = 40.8389$ lb/ft^3; consequently the API gravity = 84.8°API. The molecular weight of the stock tank oil is also calculated from its composition, resulting in a value of $MW_{STO} = 169.72$ lb/lb-mol.

Step 5: Finally, using Equations 16.75, 16.78, and 16.82 calculate the gas–oil ratios and the formation volume factor

$$R_{SP} = \frac{2131.45 \times 0.2392 \times 40.8389}{0.7608 \times 0.8802 \times 169.72} = 183.2\ \frac{\text{scf SP gas}}{\text{STB}}$$

$$R_{ST} = \frac{2131.45 \times 0.1198 \times 40.8389}{0.8802 \times 169.72} = 69.8\ \frac{\text{scf ST gas}}{\text{STB}}$$

$$R = 183.2 + 69.8 = 253 \text{ scf/STB}$$

$$B_{oSb} = \frac{121.35 \times 40.8389}{169.72 \times 38.7716 \times 0.7608 \times 0.8802} = 1.125\ \frac{\text{res.bbl}}{\text{STB}}$$

All the previous calculations could be repeated for different separator pressures in order to determine the optimum separator conditions.

16.4.5.5 A Note About the Application of EOS Models to Real Reservoir Fluids

As seen in the previous sections, EOS models can be effectively applied to model or synthetic systems that are comprised of well-defined components. However, the real challenge is the application of EOS models to naturally occurring hydrocarbon fluids or real reservoir fluids. Almost all petroleum reservoir fluids contain some quantity of the heavy fractions that are usually lumped together as a C_{7+} fraction or are sometimes defined in terms of the single carbon number (SCN) or pseudofractions and a much heavier plus fraction, such as C_{20+} or C_{30+}.

Unlike the well-defined components in model systems; for real reservoir fluids, obviously SCN fractions and the plus fraction are not very well defined in terms of their critical properties and acentric factors. Although, various characterization procedures are employed to determine their critical properties and acentric factors, an element of uncertainty still remains for the values of these properties. Therefore, the performance of various EOS models generally depends on the characterization (to determine T_c, P_c and ω) of the SCN fractions and the plus fraction because these properties directly enter in the calculation of various EOS parameters. Hence, changing the characterization of the SCN fractions and most importantly the plus fraction can have a profound effect on phase equilibria predicted by an EOS model.

Given the uncertainty in the critical properties and the acentric factor of the SCN fractions and especially the plus fraction, the most commonly adopted approach is to "tune" or "calibrate" the EOS model for a given reservoir fluid in an attempt to improve the overall prediction capability. The tuning or calibration of an EOS model is carried out by adjusting the critical properties of the plus fraction to obtain a match with the experimental data available for the mixture. The adjusted critical properties that provide the best match are referred to as regressed values. For example, the critical properties of the plus fraction can be adjusted to match the saturation pressures of the mixture at various temperatures and subsequently the tuned EOS model can be applied to predict other properties. Danesh[4] and Pedersen et al.[14] discuss the tuning of EOS models in detail.

16.5 USE OF EOS MODELS IN PVT PACKAGES

By far the biggest application of EOS models is in a variety of PVT packages that are used to simulate the PVT and phase behavior of petroleum reservoir fluids. PVT packages are basically computer programs that use and offer the choice of various equations-of-state models to carry out a number of tasks associated with the modeling of the PVT properties and phase behavior of petroleum reservoir fluids. The most common applications are:

- Simulation of laboratory PVT experiments, such as generation of a phase envelope, constant composition expansion (CCE), constant volume depletion (CVD), differential liberation (DL), and separator tests using an EOS model. As seen in Chapter 15, the primary data obtained in these various PVT experiments are the properties of the original hydrocarbon fluid, the equilibrated phases at various pressure and temperature conditions. These properties are mainly compositions, volumes, densities, moles, viscosities, and surface tension, which in fact take input from the EOS-generated values of compositions and densities. Therefore, in a broad sense, these properties for a given PVT experiment can be easily determined by a series of EOS-based VLE calculations.
- Adjustment of selected EOS parameters to match results from laboratory PVT measurements by use of nonlinear regression analysis, also called tuning or calibration of the EOS model.
- Reduction of a multicomponent representation of a reservoir fluid to a model containing fewer components for use in compositional simulation. This particular process is called lumping. As the name suggests, in order to reduce the computation time, the detailed compositional representation is sometimes lumped into a group of components. After termination of the calculations, these lumped components are then delumped to determine the compositional characteristics of the various components that were lumped initially. For example, SCN fractions C_7 to C_{12} can be lumped as 1 fraction C_7-C_{12}.
- Generation of interface files containing input tables for black oil simulation or PVT data for compositional simulation. The generated interface files are then linked to the black oil reservoir simulators or compositional simulators.
- As part of advanced PVT modeling, calculations of the variation of fluid properties with depth are also carried out by inclusion of the gravitational potential.

- In addition to the conventional applications, PVT packages containing EOS models are also used for modeling of surface process plant and pipelines.

Almost all modern PVT packages include the Soave–Redlich–Kwong (SRK) and the Peng–Robinson (PR) equations of state. In addition to these commonly used EOS models, many PVT simulators also offer the option of other EOS models that are basically various modifications of the RK and PR EOS models.

PROBLEMS

16.1 Prepare a plot of pressure vs. volume for propane at 122°F, using the SRK equation of state and subsequently determine the saturation pressure based on the Maxwell equal area rule.

16.2 Calculate the density of *n*-decane using SRK and PR equations of state at 700°F and 1500 psia.

16.3 Calculate the saturation pressure of *n*-pentane at 250°F, using both the SRK and PR equations of state.

16.4 At the saturation pressure calculated in Problem 16.3 for *n*-pentane, determine the density of the saturated vapor and liquid phases, using the SRK and PR EOS.

16.5 Calculate the dew-point pressure of a 85 mol-% methane and 15 mol-% propane, binary system at 0°F, using the PR equation of state. Use binary interaction parameters of $k_{C_1C_3} = k_{C_3C_1} = 0.009$ and $k_{C_3C_3} = 0.000$.

16.6 For the mixture given in problem 16.5 also calculate the bubble point at –50°F, using the PR equation of state.

16.7 A six-component mixture of the following composition is flashed at 25°F and 1000 psia. Calculate the compositions and the densities of the equilibrium vapor phase and the liquid phase, using PR equation of state.

Component	Overall Composition Z_i (mole fraction)
CO_2	0.0728
C_1	0.5812
C_2	0.1821
C_3	0.0728
nC_4	0.0546
nC_5	0.0364

The binary interaction parameters required in the calculations are given in the following table.

Component	CO_2	C_1	C_2	C_3	nC_4	nC_5
CO_2	0.000	0.103	0.130	0.135	0.130	0.125
C_1	0.103	0.000	0.003	0.009	0.015	0.021
C_2	0.130	0.003	0.000	0.002	0.005	0.009
C_3	0.135	0.009	0.002	0.000	0.001	0.003
nC_4	0.130	0.015	0.005	0.001	0.000	0.001
nC_5	0.125	0.021	0.009	0.003	0.001	0.000

16.8 Using the Whitson–Torp correlation, calculate the bubble-point pressure at the reservoir temperature of 250°F for a black oil that has the following composition. The convergence pressure at the reservoir temperature is determined to be 8000 psia.

Component	Overall Composition Z_i (mole fraction)	T_c (°R)	P_c (psia)	ω
C_1	0.5000	343.3	666.4	0.010
C_2	0.0583	549.9	706.5	0.098
C_3	0.0417	666.1	616.0	0.152
iC_4	0.0250	734.5	527.9	0.185
nC_4	0.0167	765.6	550.6	0.200
iC_5	0.0092	829.1	490.4	0.228
nC_5	0.0083	845.8	488.6	0.251
C_6	0.0075	913.6	436.9	0.299
C_{7+}	0.3333	1300.0	235.0	0.660

16.9 A PVT cell contains a homogeneous single-phase mixture of 5 lb-mol propane, 3 lb-mol of *n*-butane and 2 lb-mol of *n*-pentane at a temperature of 200°F. Calculate the bubble-point pressure and dew-point pressure using the ideal solution principle. Subsequently, compare these calculated saturation pressures with those predicted by using the PR equation of state. The binary interaction parameters given in problem 16.7 can be used in the PR equation-of-state calculations.

REFERENCES

1. Wilson, G., A modified Redlich-Kwong EOS, application to general physical data calculations, Paper 15C, presented at the Annual AIChE National Meeting, Cleveland, 1968.
2. Hadden, J.T., Convergence pressure in hydrocarbon vapor-liquid equilibria, *Chemical Eng. Progr. Symposium Series*, 49(7), 1953.
3. Katz, D.L., Cornell, D., Kobayashi, R., Poettman, F.H., Vary, J.A., Elenbans, J.R., and Weinaug, C.F., *Handbook of Natural Gas Engineer*ing, McGraw-Hill, New York, 1959.
4. Danesh, A., *PVT and Phase Behavior of Petroleum Reservoir Fluids,* Elsevier Science, Amsterdam, 1998.
5. Whitson, C.H. and Torp, S.B., Evaluating constant volume depletion data, Society of Petroleum Engineers (SPE) paper number 10067.
6. Standing, M.B., Volumetric and phase behavior of oil field hydrocarbon systems, Society of Petroleum Engineers of AIME, 1977.
7. van der Waals, J.D., On the continuity of the liquid and gaseous state, Ph.D. thesis, University of Leiden, Leiden,Netherlands, 1873.
8. Redlich, O. and Kwong, J.N.S., On the thermodynamics of solutions. V. An equation of state. Fugacities of gaseous solutions, *Chemical Reviews*, 44, 1949.
9. Soave, G., Equilibrium constants from a modified Redlich-Kwong equation of state, *Chemical Engineering Science*, 27 (6), 1972.

10. Peng, D.Y. and Robinson, D.B., A new two constant equation of state, *Industrial and Engineering Chemistry Fundamentals*, 15 (1), 1976.
11. Robinson, D.B. and Peng, D.Y., The characterization of the heptanes and heavier fractions for the GPA Peng-Robinson programs, GPA Research Report 28, Tulsa, 1978.
12. Ahmed, T., *Hydrocarbon Phase Behavior*, Gulf Publishing Company, Houston, 1989.
13. McCain, W.D., Jr., *The Properties of Petroleum Fluids*, PennWell Publishing Co., Tulsa, OK, 1990.
14. Pedersen, K.S., Fredenslund, Aa. and Thomassen, P., *Properties of Oils and Natural Gases*, Gulf Publishing Company, Houston, 1989.

17 Properties of Formation Waters

17.1 INTRODUCTION

Water invariably occurs in petroleum reservoirs where petroleum reservoir fluids are found associated with water. The water present in petroleum reservoirs is commonly referred to as *connate water*, *interstitial water*, *formation water*, *oil field water*, *reservoir water*, and sometimes simply *brine* due to the presence of salts. These terms are used interchangeably in petroleum literature.

The production of hydrocarbon fluids from petroleum reservoirs is frequently accompanied with formation water production. In fact, in certain cases, such as maturing oil fields, the total volume of water produced far exceeds the production of petroleum. The produced formation water generally has no use, and hence, no commercial value other than reinjecting it into the reservoir for pressure maintenance purposes. The handling of produced formation water thus becomes an issue, especially when considering stricter environmental regulations.

In a pore level scenario, the presence of water of particular characteristics or properties can influence system wettability and irreducible water saturation, which in turn can affect relative permeability functions. In reservoir fluid flow equations, formation water is treated as a separate phase along with the hydrocarbon gas phase and the oil phase. Similar to the properties of the hydrocarbon phases, properties of the water phase are necessary to solve the flow equations. In water injection practices, formation water characteristics, such as chemical composition, are also used to ascertain the potential of formation plugging due to incompatibility between the external injection water (e.g., sea water) and the reservoir water.

The associated presence of water with the hydrocarbon phases also leads to mutual solubility, that is., the solubility of hydrocarbons in the water phase and the solubility of water in hydrocarbon phases. The presence of water and some gases, typically in surface facilities and pipelines, under certain conditions of pressure and temperature may also form solid crystalline compounds known as clathrate gas hydrates, or simply hydrates. The hydrate forming pressure and temperature conditions are dependent on the characteristics of the produced water, such as salinity.

Therefore, it is clearly important to have knowledge of the chemical and physical properties of formation waters. In order to make a complete and comprehensive petroleum-reservoir engineering study, it is necessary to have a complete water analysis, including both the chemical and physical property data. The frequently used chemical property data are chemical composition or salinity of water, while physical properties include density, viscosity, compressibility, and formation volume factor. Quite often, it is also desirable to include the mutual solubility data. The primary purpose of this chapter is to examine these chemical and physical properties.

Similar to the properties of petroleum reservoir fluids, the most ideal source of obtaining the properties of formation waters is laboratory analysis of representative samples of reservoir waters. However, this does not seem to be the case; because oil field waters are not as routinely tested in the laboratory, as petroleum reservoir fluids are. Moreover, the properties of oil field waters have not been studied as carefully and systematically as the properties of petroleum reservoir fluids because the latter are obviously of much greater significance. Therefore, it is a common practice to employ various empirical correlations to estimate the properties of formation waters. Some of these correlations presented in this chapter are either based on limited experimental data on formation waters or data on pure water.

17.2 COMPOSITIONAL CHARACTERISTICS OF FORMATION WATER

All formation waters are chemically characterized by their composition in terms of dissolved solids. The most commonly found dissolved solid in formation waters is sodium chloride. Therefore, formation waters are also sometimes referred to as brine or salt water. Other dissolved solids found in formation waters include potassium chloride, calcium chloride, magnesium chloride, strontium chloride, sodium sulfate, and sodium bicarbonate. However, in terms of composition, the most dominant solid is sodium chloride. The solids present in formation waters are also defined in terms of cations and anions, such as sodium, potassium, magnesium, and calcium and chloride, sulfate, carbonate, and bicarbonate, respectively. Cations and anions present in formation waters can be identified using a technique called *ion exchange chromatography* (not discussed in this chapter).

Although, various units[1] are employed to describe the dissolved solid content, the concentration is often expressed in terms of milligrams of each solid per liter or simply in parts per million (PPM) of total dissolved solids. Parts per million usually refer to grams of solids per 1 million grams of brine, which is converted to weight-percent of total dissolved solids by multiplying ppm with 10^{-4} [(g solids/10^6 g brine) $\times$ 100%)].

For oil field waters from different reservoirs, relative distribution of these dissolved solids or cations and anions distinguish the various formation waters. Similar to various petroleum reservoir fluids, characterized by their unique overall compositions; different formation waters are also characterized by uniquely distributed solids concentrations. As an example, Figure 17.1 shows a comparison of formation water composition from various reservoirs. As seen in this figure, the composition varies significantly (in orders of magnitude) from formation to formation. In fact, McCain[1] states that water contained in a producing formation has composition different from any other water, even those in the immediate vicinity of that formation.

The composition of formation waters is in fact one of the most important characteristics because relative distribution of the dissolved solids has a significant impact (in addition to pressure and temperature) on almost all physical properties of formation waters. Some of the physical property correlations developed for formation waters are based on pure water, to which a certain correction factor, depending on the

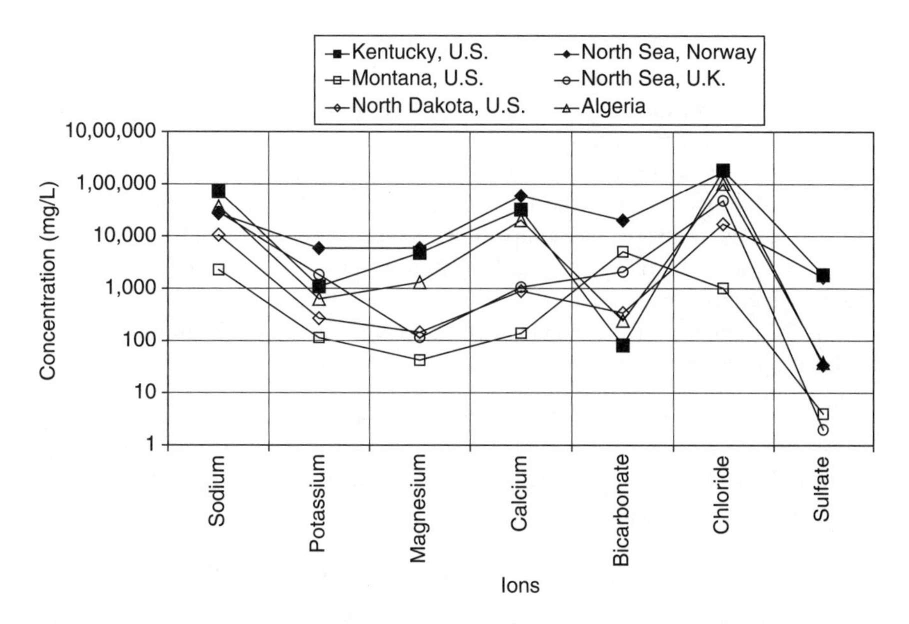

FIGURE 17.1 Ionic concentration of various formation waters of oil fields from various parts of the world.

total dissolved solids present, is applied to determine a given property. Therefore, complete chemical analysis is an important aspect of formation water characterization and should be available on the water from every petroleum reservoir.

17.3 BUBBLE-POINT PRESSURE OF FORMATION WATER

During the process of hydrocarbon migration, the oil is not in equilibrium with the water. However, from that point in time until the maturation of a given hydrocarbon accumulation, the dissolution of light components from the oil into the water can be assumed to be completed. At this particular stage, the oil and the water in a petroleum reservoir can be considered to be in thermodynamic equilibrium. Therefore, the bubble-point pressure of formation water equals the bubble-point pressure of the co-existing oil[1] because water in a petroleum reservoir can be assumed to be in equilibrium with the hydrocarbon phases.[2]

17.4 FORMATION VOLUME FACTOR OF FORMATION WATER

The formation volume factor of water, denoted by B_w, represents the change in volume of the water as it is brought from the reservoir to the surface. Similar to the oil formation volume factor, the following three effects are involved:

(1) The evolution of dissolved gas from the water due to pressure reduction
(2) The slight expansion due to pressure reduction
(3) The slight contraction owing to temperature reduction from reservoir to surface.

The units used for formation volume factor of formation waters are reservoir barrels per surface barrel at standard conditions, res. bbl/STB.

Although these three contributing factors for formation volume factor are common for oils and formation waters, unlike oils, the contribution of dissolved gas is significantly less in the case of water due to considerably less hydrocarbon gas solubility. Additionally, due to the low compressibility of water, expansion and contraction due to pressure and temperature, respectively, are small and somewhat offsetting. Therefore, oil field water formation volume factors are numerically small and are generally close to one.

McCain[1] proposes the following correlation for estimating the formation volume factor of oil field water:

$$B_w = (1+\Delta V_{wP})(1+\Delta V_{wT}) \tag{17.1}$$

where ΔV_{wP} and ΔV_{wT} are the volume changes due to pressure and temperature, respectively, which are correlated[2,3] as

$$\begin{aligned}\Delta V_{wP} = &-(3.58922 \times 10^{-7} + 1.95301\times10^{-9}\,\mathrm{T})\,P \\ &-(2.25341\times10^{-10} + 1.72834 \times 10^{-13}\,T)\,P^2\end{aligned} \tag{17.2}$$

$$\Delta V_{wT} = -1.001 \times 10^{-2} + 1.33391 \times 10^{-4}\,T + 5.50654\times10^{-7}\,T^2 \tag{17.3}$$

The preceding correlation is valid for formation waters with widely varying solids concentrations and is applicable to pressures and temperatures up to 5,000 psia and 260°F. The B_w values from this correlation agree with limited experimental data to within 2%.[3]

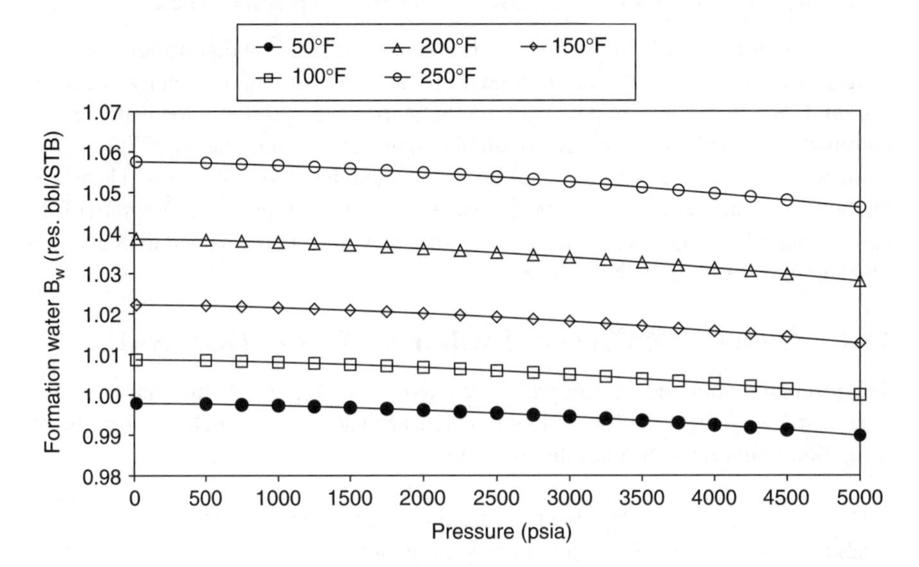

FIGURE 17.2 Formation volume factor of formation water, calculated from Equations 17.1 to 17.3 as a function of pressure and temperature.

The equations presented previously produce a formation volume factor and pressure, and temperature relationships shown in Figure 17.2. As seen in Figure 17.2, in a wide range of pressure and temperature conditions, the value of B_w remains close to 1. In accordance with compressibility and thermal expansion characteristics of water, an increase in pressure produces a decrease in B_w, while at constant pressure, an increase in temperature results in an increase in the value of B_w. It can also be noted that for a given pressure, the change in B_w with temperature is much more significant compared to the change in B_w with pressure at constant temperature.

17.5 DENSITY OF FORMATION WATER

The density of formation waters can be calculated on the basis of the water formation volume factor if the mass of dissolved gas in water at reservoir conditions is neglected. The water formation volume factor relates the volume of water at given reservoir conditions to the volume of water at standard conditions (14.7 psia and 60°F) such that

$$B_w = \frac{V_{wR}}{V_{wsc}} = \frac{\rho_{wsc}}{\rho_{wR}} \quad (17.4)$$

or

$$\rho_{wR} = \frac{\rho_{wsc}}{B_w} \quad (17.5)$$

where ρ_{wR} is the density of formation water at reservoir conditions, lb/ft^3; ρ_{wsc} the density of formation water at standard conditions, lb/std. ft^3; and B_w the formation volume factor, res. bbl/STB or res. ft^3/std. ft^3.

The density of formation water at reservoir conditions, calculated by Equation 17.5, is thus in lb/res. ft^3 or simply lb/ft^3.

The density of formation water at standard conditions can be estimated from the following correlation:[3]

$$\rho_{wsc} = 62.368 + 0.438603S + 0.00160074S^2 \quad (17.6)$$

where S is the weight-percent of total dissolved solids.

17.6 VISCOSITY OF FORMATION WATER

The viscosity of formation waters can be estimated from the following correlations:[3]

$$\frac{\mu_{wR}}{\mu_{w1}} = 0.9994 + 4.0295 \times 10^{-5}P + 3.1062 \times 10^{-9}P^2 \quad (17.7)$$

where μ_{wR} is the viscosity of formation water at reservoir conditions, cp; μ_{w1} the viscosity of formation water at atmospheric pressure and reservoir temperature, cp; and P the pressure, psia.

The viscosity of formation water at atmospheric pressure and reservoir temperature T, estimated from Equation 17.8, is[3]

$$\mu_{w1} = AT^{-B} \tag{17.8}$$

where A and B are defined by Equations 17.9 and 17.10:

$$A = 109.574 - 8.40564S + 0.313314S^2 + 0.00872213S^3 \tag{17.9}$$

$$B = 1.12166 - 0.0263951S + 6.79461 \times 10^{-4}S^2 + 5.47119 \times 10^{-5}S^3 - 1.55586 \times 10^{-6}S^4 \tag{17.10}$$

The coefficients of Equation 17.8, defined by Equations 17.9 and 17.10, include the salinity effects where S is the weight-percent of total dissolved solids.

Equation 17.7 is applicable in the range of 86 to 167°F, and pressures up to 15,000 psia, while Equation 17.8 applies in the range of 100 to 400°F, and weight percent of total dissolved solids up to 26%.[3]

17.7 SOLUBILITY OF HYDROCARBONS IN FORMATION WATER

Figure 17.3 shows the solubility of methane in pure water at various pressures and temperatures. The data presented in Figure 17.3 are predicted by the modified Patel–Teja[4] equation of state as proposed by Zuo et al.,[5] which was shown to accurately predict the solubility of gases in pure water and brines. As seen in Figure 17.3, the solubility is dependent on both pressure as well as temperature. However, at a given temperature, pressure seems to have more of a significant effect on solubility than temperature at a given pressure.

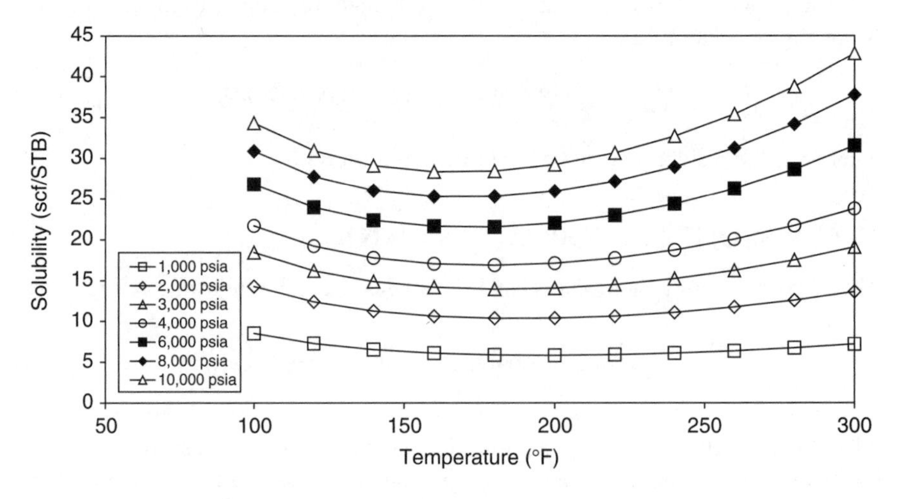

FIGURE 17.3 Solubility of methane in pure water as a function of pressure and temperature.

The predicted[5] solubility data of light hydrocarbon components, methane, ethane, propane, and *n*-butane in pure water at various pressures and a temperature of 150°F are shown in Figure 17.4. As seen in Figure 17.4, the solubility of all components increases as pressure increases; and as expected, methane has the highest solubility, while *n*-butane has the lowest solubility. However, with increasing carbon number, the effect of pressure on solubility appears to be minimal.

The presence of dissolved solids affects the solubility of gases in water. Figure 17.5 shows a comparison of the solubility of methane in pure water and formation waters, containing different weight-percent of total dissolved solids, at 250°F and

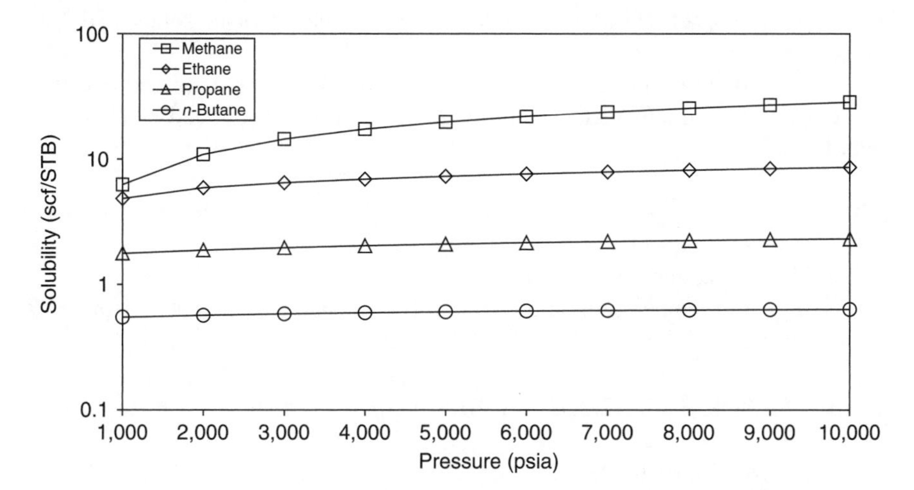

FIGURE 17.4 Solubility of pure hydrocarbon components in pure water as a function of pressure at 150°F.

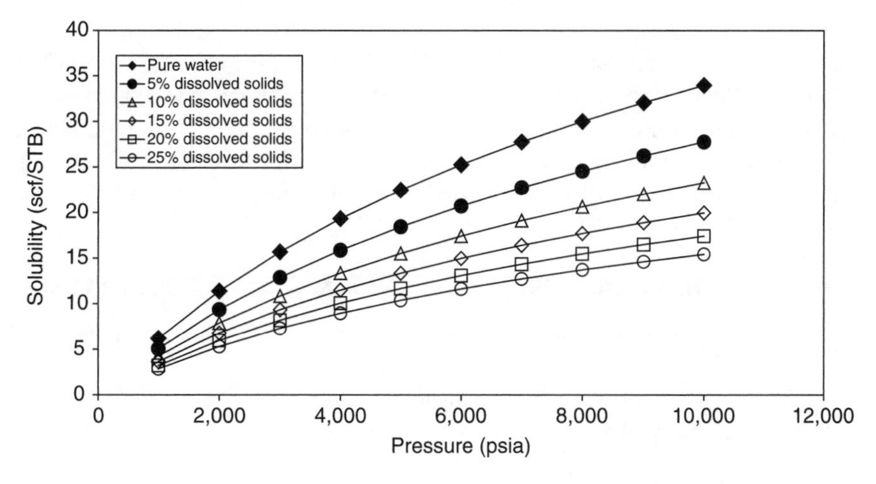

FIGURE 17.5 Solubility of methane in pure water and waters containing various percentages of total dissolved solids, as a function of pressure at 250°F.

various pressures. As the amount of total dissolved solids increases, the solubility decreases.

The empirical correlation proposed[3] for calculation of gas solubility in formation water is based on the solubility of gas in pure water. The two solubilities related by Equation 17.11 are[3]

$$\log\left[\frac{R_{sfw}}{R_{spw}}\right] = -0.0840655ST^{-0.285854} \tag{17.11}$$

where R_{sfw} is the solubility of gas in formation water, scf/STB; R_{spw} the solubility of gas in pure water, scf/STB; S the weight-percent of total dissolved solids; and T the temperature, °F.

The gas solubility in pure water is given by the following empirical equation, which is based on the methane solubility data in pure water reported by Culberson and McKetta:[6]

$$R_{spw} = A + BP + CP^2 \tag{17.12}$$

where P is the pressure, psia.

The effect of temperature on gas solubility is included in the coefficients A, B, and C defined by

$$A = 8.15839 - 0.0612265T + 1.91663\times10^{-4}T^2 - 2.1654\times10^{-7}T^3 \tag{17.13}$$

$$B = 0.0101021 - 7.44241\times10^{-5}T + 3.05553\times10^{-7}T^2 - 2.94883\times10^{-10}T^3 \tag{17.14}$$

$$C = -(10^{-7})[9.02505 - 0.130237T + 8.53425\times10^{-4}T^2 - 2.34122\times10^{-6}T^3 + 2.37049\times10^{-9}T^4] \tag{17.15}$$

Equation 17.11 is valid in the temperature range of 70 to 250°F, for weight-percent of total dissolved solids up to 30%, while Equations 17.12 to 17.15 are valid in the pressure range of 1,000 to 10,000 psia and temperature range of 100 to 340°F.

17.8 SOLUBILITY OF FORMATION WATER IN HYDROCARBONS

For the solubility of water in hydrocarbons, two types of solubilities are considered:

(1) The solubility of water in gaseous hydrocarbons and
(2) The solubility of water in liquid hydrocarbons.

Both solubilities are important since they influence the treating, processing, and transporting of gaseous and liquid hydrocarbons.

17.8.1 Water Content of Gaseous Hydrocarbons

In 1958, McKetta and Wehe[7] published a chart for estimating the water content of sweet natural gas. This chart has been modified[8] to also include the correction for molecular weight and dissolved solids content of the water. This chart is commonly used to estimate the water content of natural gases.

In 1959, Bukacek[9] suggested a relatively simple, yet reasonably accurate, correlation for estimating the water content of a sweet gas. The water content is calculated using the following correlation:

$$W_{pw} = 47484\frac{P_v^{pw}}{P} + B \tag{17.16}$$

where W_{pw} is the water (pure) content of gas, lb/million standard ft^3 (MMSCF); P the system pressure, psia; and P_v^{pw} the vapor pressure of pure water at the given temperature, psia.

The parameter B in Equation 17.16 is made a function of temperature (in °F) by

$$\log(B) = \frac{-3083.87}{460 + T} + 6.69449 \tag{17.17}$$

This correlation is reported to be accurate for temperatures up to 460°F and for pressures up to 10,000 psia. A comparative study reported by Carroll[10] in fact indicates that the method of Bukacek is as accurate as the charts of McKetta and Wehe. Also, McCain[3] states that results obtained from Equations 17.16 and 17.17 are as accurate as moisture content can be measured (about 5%).

The vapor pressure of pure water, between the freezing point and the critical point, required in Equation 17.16, can be estimated from Equation 17.18:[11]

$$P_v^{pw} = \text{EXP}\left[A + \frac{B}{(T + 460)} + \text{C}\ln(T + 460) + \text{D}(T + 460)^{\text{E}}\right] \tag{17.18}$$

where $A = 69.103501$; $B = -13064.76$; $C = -7.3037$; $D = 1.2856 \times 10^{-6}$; $E = 2$; and $T =$ temperature, °F.

Since dissolved solids in the water reduce the moisture content of the gas, the water content calculated in Equation 17.16 should be corrected to account for the dissolved solids. The following equation[3] is used as a correction:

$$W_{fw} = W_{pw}[1 - 0.004920\text{S} - 0.00017672\text{S}^2] \tag{17.19}$$

where W_{fw} is the formation water content of gas, lb/MMSCF and S the weight-percent of total dissolved solids.

17.8.2 Water Content of Liquid Hydrocarbons

The solubility of water in liquid hydrocarbons is generally quite low due to extremely small mutual attraction. No empirical correlations are available to

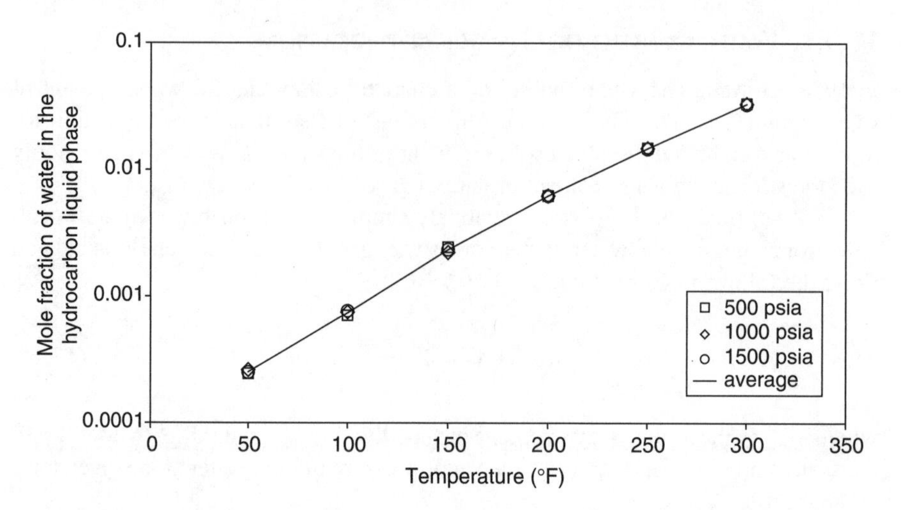

FIGURE 17.6 Solubility of water in the hydrocarbon liquid phase as a function of temperature.

estimate the water solubility in liquid hydrocarbons. Danesh[2] presents a chart for the solubility of water in liquid hydrocarbons at their vapor pressures. The chart contains solubility data as a function of temperature for several normal alkanes. The solubility of water increases with increasing temperature. In the temperature range of 40 to 170°F, on average the solubility of water in various alkanes increases from 0.004 to 0.15% by weight.

Models such as the modified Patel–Teja[4] equation of state as proposed by Zuo et al.[5] can be used to predict the mole fraction of water in the hydrocarbon liquid phase by performing three-phase flash or three-phase VLE calculations. Figure 17.6 shows the predicted values[5] of the mole fractions of water in the hydrocarbon liquid phase of a ten-component (consisting of methane through *n*-decane) synthetic oil mixture at various pressures and temperatures. As seen in Figure 17.6, pressure does not seem to affect the solubility of water in the hydrocarbon liquid phase. The temperature seems to significantly affect the solubility; however, the mole fraction values are quite small.

17.9 COMPRESSIBILITY OF FORMATION WATER

The correlation proposed by Meehan[12] is used for estimating the coefficient of compressibility of water. For gas-free water,

$$C_{\mathrm{wgf}} = 10^{-6}(A_1 + A_2T + A_3T^2) \tag{17.20}$$

where C_{wgf} is the coefficient of compressibility of gas-free water, psi^{-1} and T the temperature, °F.

The coefficients in Equation 17.20 are related to pressure by Equations 17.21, 17.22, and 17.23:

$$A_1 = 3.8546 - 0.000134P \tag{17.21}$$

$$A_2 = -0.01052 + 4.77 \times 10^{-7}P \tag{17.22}$$

$$A_3 = 3.9267 \times 10^{-5} - 8.8 \times 10^{-10}P \tag{17.23}$$

where P is the pressure, psia.

The compressibility of gas saturated water is related to the compressibility of gas-free water as

$$C_w = C_{wgf}(1 + 0.0089R_{spw}) \tag{17.24}$$

where C_w is the coefficient of compressibility of gas saturated water, psi^{-1} and R_{spw} the gas solubility, scf/STB.

Numbere et al.[13] propose the following correction to C_w for presence of dissolved solids:

$$SC = 1 + S[-0.052 + 2.7 \times 10^{-4}T - 1.14 \times 10^{-6}T^2 + 1.121 \times 10^{-9}T^3] \tag{17.25}$$

such that

$$C_{wfw} = C_w SC \tag{17.26}$$

where C_{wfw} is the coefficient of compressibility of formation water, psi^{-1}; SC the correction for dissolved solids; and S the weight-percent of total dissolved solids.

Equation 17.26 can also be applied[14] to gas-free water, that is, $C_{wgf}SC$.

PROBLEMS

17.1 The following compositional data are available for a formation water sample from a Middle Eastern oil field. Calculate the weight-percent of total dissolved solids.

Constituents	Concentration (ppm)
$NaCl$	25,755
$NaHCO_3$	3795
KCl	345
$MgCl_2.6H_2O$	921
$CaCl_2$	1011
$SrCl_2.2H_2O$	131

17.2 Calculate the density of the formation water in problem 17.1 at standard conditions.

17.3 Calculate the formation volume factor of the formation water in problem 17.1 at the reservoir conditions of 5,000 psia and 160°F.

17.4 Calculate the density and the viscosity of formation water in problem 17.1 at the reservoir conditions of 5,000 psia and 160°F.

17.5 Calculate the solubility of a natural gas mixture that is in contact with a formation water, containing 50,000 ppm of total dissolved solids, at 5,000 psia and 200°F.

17.6 Calculate the coefficient of compressibility of formation water in problem 17.5, at 5,000 psia and 200°F.

17.7 A natural gas mixture is in equilibrium with a formation water at 7,500 psia and 175°F. The formation water contains 15% of total dissolved solids by weight. Estimate the water content in the natural gas at equilibrium conditions.

17.8 Repeat problem 17.7 for equilibrium pressure of 5,000 psia and equilibrium temperatures in the range of 100 to 300°F (with a spacing of 25°F). Subsequently plot the water content as a function of studied temperatures at 5,000 psia, and comment on the observed trend.

REFERENCES

1. McCain, W.D., Jr., *The Properties of Petroleum Fluids*, PennWell Publishing Co., Tulsa, OK, 1990.
2. Danesh, A., *PVT and Phase Behavior of Petroleum Reservoir Fluids,* Elsevier Science, Amsterdam, 1998.
3. McCain, W.D., Jr., Reservoir fluid property correlations – state of the art, *SPE Res. Eng.*, 266–272, 1991.
4. Patel, N.C. and Teja, A.S., A new cubic equation of state for fluids and fluid mixtures, *Chem. Eng. Sci.*, 37, 1982.
5. Zuo, Y-X., Stenby E.H., and Guo, T.M., Simulation of the high-pressure phase equilibria of hydrocarbon-water/brine systems, *J. Pet. Sci. Eng.*, 15, 201–220, 1996.
6. Culberson, O.L. and McKetta, J.J., Phase equilibria in hydrocarbon-water systems, Part 3-the solubility of methane in water at pressures to 10,000 psia, *Trans. AIME*, 192, 223–226, 1951.
7. McKetta, J.J. and Wehe, A.H., Use this chart for water content of natural gases, *Petroleum Refiner*, 37, 1958.
8. GPSA *Engineering Data Book*, 11th ed., Gas Processors Suppliers Association, Tulsa, OK, 1998.
9. Bukacek R.F., Equilibrium moisture content of natural gases, Research Bulletin Series, Institute of Gas Technology, 1959.
10. Carroll, J.J., The water content of acid gas and sour gas from 100 to 220°F and pressures to 10,000 psia, Presented at the 81st Annual GPA Convention, Dallas, 2002.
11. Daubert, T.E. and Danner, R.P., *DIPPR Data Compilation Tables of Properties of Pure Compounds*, AIChE, NY, 1985.
12. Meehan, D.N., A correlation for water compressibility, *Pet. Eng.*, 125–126, 1980.
13. Numbere, D., Brigham, W.E., and Standing, M.B., Correlations for physical properties of petroleum reservoir brines, Petroleum Research Institute Report, Stanford University, 1977.
14. Ahmed, T., *Hydrocarbon Phase Behavior*, Gulf Publishing Company, Houston, 1989.

Author Index

Subject Index

J

K

L

M

N

S

T